ROUTLEDGE HANDBOOK OF REWILDING

This handbook provides a comprehensive overview of the history, theory, and current practices of rewilding.

Rewilding offers a transformational paradigm shift in conservation thinking, and as such is increasingly of interest to academics, policymakers, and practitioners. However, as a rapidly emerging area of conservation, the term has often been defined and used in a variety of different ways (both temporally and spatially). There is, therefore, the need for a comprehensive assessment of this field, and the *Routledge Handbook of Rewilding* fills this lacuna. The handbook is organised into four sections to reflect key areas of rewilding theory, practice, and debate: the evolution of rewilding, theoretical and practical underpinnings, applications and impacts, and the ethics and philosophy of rewilding. Drawing on a range of international case studies the handbook addresses many of the key issues, including land acquisition and longer-term planning, transitioning from restoration (human-led, nature enabled) to rewilding (nature-led, human enabled), and the role of political and social transformational change.

Led by an editorial team who have extensive experience researching and practising rewilding, this handbook is essential reading for students, academics, and practitioners interested in rewilding, ecological restoration, natural resource management, and conservation.

Sally Hawkins is an environmental social scientist at the University of Cumbria, UK. She is a core member of the IUCN CEM Rewilding Thematic Group and a founding trustee of the Lifescape Project.

Ian Convery is Professor of Environment & Society at the University of Cumbria, co-chairs the IUCN CEM Rewilding Thematic Group, and is Chair of IUCN CEM western Europe.

Steve Carver is Director of the Wildland Research Institute at the University of Leeds, UK, and Co-Chair of the IUCN CEM Rewilding Thematic Group.

Rene L. Beyers is a Research Associate in the Beaty Biodiversity Research Centre at the University of British Columbia, Canada.

ROUTLEDGE HANDBOOK OF REWILDING

*Edited by Sally Hawkins, Ian Convery,
Steve Carver, and Rene Beyers*

LONDON AND NEW YORK

Cover image: Getty

First published 2023
by Routledge
4 Park Square, Milton Park, Abingdon, Oxon OX14 4RN

and by Routledge
605 Third Avenue, New York, NY 10158

Routledge is an imprint of the Taylor & Francis Group, an informa business

British Library Cataloguing-in-Publication Data
A catalogue record for this book is available from the British Library

Library of Congress Cataloging- in- Publication Data
Names: Hawkins, Sally (Environmental pyschologist), editor. |
Convery, Ian, 1965– editor. | Carver, Steve, editor. |
Beyers, Rene, 1961– editor.
Title: Routledge handbook of rewilding / Sally Hawkins,
Ian Convery, Steve Carver, Rene Beyers.
Description: New York, NY : Routledge, 2023. |
Includes bibliographical references and index.
Identifiers: LCCN 2022025284 (print) | LCCN 2022025285 (ebook) |
ISBN 9780367564483 (hardback) | ISBN 9780367564490 (paperback) |
ISBN 9781003097822 (ebook)
Subjects: LCSH: Wildlife reintroduction. | Nature conservation.
Classification: LCC QL83.4 .R68 2022 (print) | LCC QL83.4 (ebook) |
DDC 333.95/4–dc23/eng/20220721
LC record available at https://lccn.loc.gov/2022025284
LC ebook record available at https://lccn.loc.gov/2022025285

ISBN: 978-0-367-56448-3 (hbk)
ISBN: 978-0-367-56449-0 (pbk)
ISBN: 978-1-003-09782-2 (ebk)

DOI: 10.4324/9781003097822

Typeset in Bembo
by Newgen Publishing UK

This book is dedicated to Dave Foreman, Alison Parfitt, and Michael Soulé.

CONTENTS

EDITORS

Sally Hawkins is an environmental social scientist at the University of Cumbria. She is a core member of the IUCN CEM Rewilding Thematic Group, a founding trustee of the Lifescape Project and has a background in STEM publishing and environmental management. In research and practice she focuses on social-ecological approaches to promote landscape or systemic change.

Ian Convery is Professor of Environment & Society at the University of Cumbria. He leads the 'Back on our Map' UK multi-species translocation project, co-chairs the IUCN CEM Rewilding Thematic Group, and is Chair of IUCN CEM western Europe.

Steve Carver is a geographer with interests in wilderness, rewilding, and the application of spatial modelling to landscape assessment. He is Director of the Wildland Research Institute at the University of Leeds and Co-Chair of the IUCN CEM Rewilding Thematic Group.

Rene L. Beyers is Research Associate in the Beaty Biodiversity Research Centre at the University of British Columbia, Canada. He has a PhD in ecology and is interested in wildlife monitoring, ecosystem functioning and recovery. He has worked as a medical doctor, wildlife manager, and researcher in Africa, Asia, and Canada.

Section editor

Kate Rawles is a former lecturer in environmental philosophy. Kate now works freelance as a writer and environmentalist, using adventurous journeys to help raise awareness and inspire action on our most urgent environmental challenges. More info at www.outdoorphilosophy. co.uk.

CONTRIBUTORS

Stuart Adair is a fully qualified biological surveyor and habitat ecologist with over 20 years' experience, with particular interest in semi-natural and near-natural vegetation with reference to both plant and vegetation succession and the effects of certain treatments. He has been involved in several ecological restoration schemes in Scotland and northern England over the past two decades.

Julia Aglionby is Executive Director of the Foundation for Common Land, Chair of the Uplands Alliance and a practising Agricultural Valuer. Julia is a Professor in Practice at the University of Cumbria's Centre for National Parks and Protected Areas. Julia was from 2014 to 2019 a Board Member of Natural England.

Angela Andrade is an anthropologist with a specialisation in geography. She obtained an MSc in Rural and Land Ecology and is a Senior Climate Change and Biodiversity Policy Director for Conservation International, Colombia. Angela serves as Chair for the IUCN Commission on Ecosystem Management.

Koen Arts is a lecturer at the Forest and Nature Conservation Policy group at Wageningen University, the Netherlands. He is also an instructor in elementary nature skills and writes popular nature books on experiencing and valuing the natural world.

Philip Ashmole has a doctorate in seabird ecology and taught ornithology, ecology and evolution at Yale and Edinburgh universities for 30 years. With his wife Myrtle he has undertaken ecological and conservation projects on various Atlantic islands and published books on natural history. He is co-ordinator of the Carrifran Wildwood project and a founding Trustee of Borders Forest Trust.

Simon Ayres is a forester and founder of Wales Wild Land Foundation, Wales; and director of Cambrian Wildwood.

Marc Bekoff is Professor Emeritus of Ecology and Evolutionary Biology at the University of Colorado, Boulder. His homepage is marcbekoff.com.

Joel Berger is University Chair in Wildlife Conservation at Colorado State University, and a senior scientist for the Wildlife Conservation Society. He has written about the science and protection of migration corridors, helped to establish a federally protected one in the USA, and has concentrated on carnivore reintroduction and biodiversity in extreme environments.

Ananta Ram Bhandari has worked for over 15 years in integrated landscape management, biodiversity conservation, forestry, and natural resource management in Nepal. He currently leads WWF-Nepal's Forest & Landscape Program.

Elsie Blackshaw-Crosby is Managing Lawyer at The Lifescape Project. She is responsible for all of the Lifescape Project's legal projects, including the Litigation for Nature project and its work as part of the Forest Litigation Collaborative, as well as work to protect wild landscapes through novel legal mechanisms such as conservation covenants and easements. The 'Rewilding Law' programme also falls under Elsie's remit – this work focusses on identifying, explaining, and resolving legal barriers to rewilding across eight countries.

Shiv Raj Bhatta has worked for 22 years in the Government of Nepal as a manager of various Protected Areas. For over a decade he has worked for WWF Nepal to oversee the management of conservation landscapes and is currently their Conservation Program Director.

Neil Burrows BSc (For), PhD, AFSM is a fire scientist and was Director of Science with the Western Australian government land management agencies from 1977 to 2019. He now works as a bushfire consultant.

Yue Cao is Assistant Professor at the Institute for National Parks and Department of Landscape Architecture, School of Architecture, Tsinghua University.

Jonathan Carruthers-Jones has worked on transdisciplinary approaches to the research and conservation of wilderness for over a decade. He is currently Post-Doctoral Research Fellow at the University of Leeds as part of 'Corridor Talk: Conservation Humanities and the Future of Europe's National Parks'.

Charles C. Chester teaches global environmental politics at Brandeis University and at the Fletcher School of Tufts University. He serves as the Board Chair of Bat Conservation International and as Chair of the Yellowstone to Yukon Conservation Council. He is currently building the website, GEP-guide.net, an online guide to global environmental politics.

T.J. Clark-Wolf is UW Data Science Postdoctoral Fellow at the University of Washington, and graduated with his PhD in Wildlife Biology from the University of Montana in 2021. His research focuses on developing quantitative methods to conserve wildlife and biodiversity in an increasingly changing world.

Alex Cooper is a lawyer at the Commonwealth Climate and Law Initiative, focusing on the implications of nature crises under corporate and securities laws. Alex carries out legal analysis for the Lifescape Project on rewilding issues, including comparative developments in UK and international law.

Paul Cryer is a British and South African citizen, born in Zimbabwe. He is a career conservationist who works on expanding protected areas, especially for species and landscapes under anthropogenic threat. He likes reading, listening to music, and living outside as much as possible.

Sebastián Di Martino holds a bachelor in Zoology and a Masters in Management of Protected Areas. He worked for over 15 years in the Agency of Protected Areas of Neuquén province until 2015 when he became the Director of Conservation of Fundación Rewilding Argentina.

Emiliano Donadio holds a bachelor and a Masters in Zoology and a PhD in ecology. He was a researcher at the Argentine Research Council until 2019, when he joined Fundación Rewilding Argentina as its Science Director.

Martin Drenthen is Associate Professor of Environmental Philosophy at the Institute for Science in Society (ISiS) at Radboud University in Nijmegen (Netherlands). His most recent research focuses on ethical issues related to coexistence with unruly wildlife, notably wolf resurgence in Western Europe. He published extensively on environmental philosophy in both Dutch and English.

Adam Eagle is CEO of the Lifescape Project, which uses the law and other tools to protect and restore wild landscapes. Adam has advised on legal issues relating to rewilding, ranging from bringing litigation to protect keystone species to regulations regarding the reintroduction of carnivores. He is also a member of the IUCN Rewilding Thematic Group.

Rob Espin is a rewilding consultant for the Lifescape Project, a UK based rewilding charity. Rob is also the co-chair of the UK Centre for Animal Law's wildlife working group, advising on a variety of legal issues impacting wildlife.

Lisa Fenton is an ethnobiologist and environmental anthropologist. An internationally respected Bushcraft practitioner, Lisa is a lecturer at the Institute of Science & Environment, University of Cumbria, where she is the pathway leader for MA Outdoor and Experiential Learning (in Bushcraft).

Hannah Field is PhD Researcher at the University of Cumbria, focusing on commoning and common land. She is also a consultant in land management decision making and supports place-based action, working collaboratively cross-sector. This includes working with farmers, land managers, NGOs, education institutions, and many more.

Mark Fisher circulated a personal manifesto for rewilding the UK after walking the open spaces of North America for ten weeks in 2003. He has since then written for his advocacy website—Self-willed Land—as well as joined with others to promote rewilding.

Dominique Gonçalves is PhD candidate in the Durrell Institute of Conservation & Ecology at the University of Kent and manages the Elephant Ecology Project at Gorongosa National Park.

Jack Gould is a lawyer at Clifford Chance LLP, specialising in financial regulation and private funds. Jack has advised the Lifescape Project on issues relating to the legal protection and restoration of wild landscapes, including through the use of conservation leases/covenants.

Andrew Gregory is Assistant Professor of Wildlife Conservation and Spatial Ecology at the University of North Texas in Denton. His primary research focus is wildlife population viability and dispersal across anthropic landscapes and is particularly interested in how land cover and land use heterogeneity interact with species traits to influence conservation corridor functionality.

Adam Griffin co-founded Moor Trees, on Dartmoor, and is a director of Red Earth Landscapesand a dance teacher at Soul Motion.

Adrien Guetté is a geographer, specialising in Conservation. His work focuses on the mapping and modelling of human landscape pressures and landscape naturalness. He is currently lecturer at ISTOM–France and is responsible for the Master's degree 'Territorial Risks and Planning'.

Mark Hebblewhite is Professor in Large Mammal Ecology in the Wildlife Biology Program in the University of Montana where he has served since 2006. Mark and his students in the Ungulate Ecology Lab have conducted research on large carnivores and their large herbivore prey since 1994 across Canada, Europe, and Asia.

Anja Heister is an independent researcher and writer. Originally from Germany, she lives in Montana/USA and is co-founder of Footloose Montana and recently authored a book that calls for an end of hunting and trapping of wild animals and for more empathy and compassion in our relations with other animals.

Jodi Hilty is president and chief scientist of the Yellowstone to Yukon Conservation Initiative, a joint US-Canada non-profit organisation dedicated to connecting and protecting the region so that people and nature can thrive.

Shuyu Hou is PhD candidate at the Department of Landscape Architecture, School of Architecture, Tsinghua University.

Ian Kealley, OAM BSc (forestry), was Regional Manager for Western Australia's conservation, forestry, environmental, wildlife, and land management agencies for the inland rangeland and desert Goldfields Region between 1984 and 2017.

Helen Kopnina (PhD, Cambridge) is currently employed at both Northumbria University in the UK, and The Hague University of Applied Sciences in The Netherlands, co-ordinating their Sustainable Business programmes and conducting research within the inter-related areas of environmental sustainability, environmental education, biological conservation, and animal ethics.

Joanna E. Lambert is Professor of Wildlife Biology at the University of Colorado–Boulder. Her research centres on the evolutionary ecology and conservation biology of endangered primates and Carnivora in equatorial Africa and in the American West. Among her proudest recent accomplishments was contributing to the effort to restore grey wolves to Colorado.

Simon Leadbeater is a woodland owner and sheep keeper, and trustee of the British Association of Nature Conservationists (BANC), which publishes *ECOS*. He is writing for Dixi Books (dixibooks.com), recounting the trials and tribulations of an off-grid life shared with his wife Toni, and how they set about rewilding (mostly) their wood.

Emilia A. Leese, owns Birchfield, the site of the Natural Capital Laboratory. She is co-author of *Think Like a Vegan: What Everyone Can Learn from Vegan Ethics* (Unbound, 2021), host of the eponymous podcast and speaker on vegan ethics at a variety of events. Emilia developed life skills and ethics workshops for underserved youth and has been a corporate finance lawyer for over 20 years.

Alexandra Locquet is a doctor in geography and a researcher at the CNRS. She defended a PhD thesis on strategies for the protection of wilderness in Europe. She works on the issues of wilderness, rewilding, free evolution, and nature protection.

Chris Loynes has been an outdoor educator throughout his professional life. He is currently Professor of Human Nature Relations at The Centre for National Parks and Protected Areas at the University of Cumbria.

Eric Maddern is a storyteller, musician, and founder of the Cae Mabon educational centre in Snowdonia.

Georgina Maffey is the chair of Stichting Wildeor a charity focused on Bringing Nature Back to Life. She works across disciplines to embed the value of the natural world in educational and outreach settings.

Cara Nelson is Professor of Restoration Ecology at University of Montana, Lead of IUCN CEM's Ecosystem Restoration Thematic Group, and past Chair of the Society for Ecological Restoration. Her research focuses on the effects of large-scale disturbance on vegetation, the efficacy and ecological impacts of ecological restoration, and the selection of native plant materials for restoration.

Zoë Playdon is Emeritus Professor of Medical Humanities at the University of London. Her research interests include the relationship between Indigenous medicine and Western biomedicine.

Heather Prince is Professor of Outdoor and Environmental Education at the University of Cumbria, UK. She is interested in creative and innovative pedagogic practice of outdoor learning in schools and higher education, adventure, and sustainability.

Robert Pringle is Professor of Ecology and Evolutionary Biology at Princeton University and the author of over 100 articles on ecology and conservation, focusing mainly on African ecosystems.

Philip Rooney is Environmental AI/ ML for the UK and Europe at AECOM. He has a strong interest in building and exploiting high value cross disciplinary data sets, particularly in the field of natural capital.

Meredith Root-Bernstein, CNRS, National Museum of Natural History, Paris, France, co-founded the rewilding NGO Kintu in Chile; she researches ecology and ethnobiology.

David Satori holds MSc in Plant and Fungal Taxonomy, Diversity and Conservation from Queen Mary University of London and the Royal Botanic Gardens, Kew. He is the founder of Rewilding Mycology, where he advocates for greater recognition of fungi in rewilding through biodiversity surveys, consultation, and public outreach.

Anthony R.E. Sinclair, now Professor Emeritus, was Director of the Beaty Biodiversity Research Center, University of British Columbia. He has conducted ecological research on how ecosystems work and the role of biodiversity in many ecosystems around the world including Africa, Australia, New Zealand, and Canada.

Mark Stanley-Price is a senior research associate at the Department of Zoology at the University of Oxford. He has had a career-long interest in translocation, starting with the Arabian oryx in Oman, and he now follows and contributes to the evolution of species translocation to rewilding practice.

Peter Taylor is an ecologist and author of *Beyond Conservation, Rewilding,* and *The Spirit of Rewilding;* and is a founding member and trustee of the Wales Wild Land Foundation.

Kristine Tompkins is the former CEO of Patagonia, is the co-founder and president of Tompkins Conservation and the UN Patron of Protected Areas.

Heather VanVolkenburg received her MSc in Ecology and Evolution from Brock University where she is now a manager and contributor with the Vasseur Ecosystem Ecology Lab. With a research focus on agroecology, she has contributed to projects related to sustainable agriculture, food insecurity in Africa, Species at Risk in Ontario, and mature women in science.

Liette Vasseur is Professor in the Department of Biological Sciences at Brock University, a member of the Environmental Sustainability Research Centre and since 2014, she holds the UNESCO Chair on Community Sustainability: from Local to Global at Brock. She is Deputy Chair of the IUCN CEM and co-lead of the thematic group on Climate Change and Biodiversity Policy and Practice.

Matt Wainhouse has a research focus on the ecology and conservation of wood-decay fungi at Cardiff University. He is the fungi and lichen specialist at Natural England and was awarded a Churchill Fellowship for integrating fungal conservation in spatial planning.

Alan Watson Featherstone is an ecologist, photographer, and founder of *Trees for Life*, based at Findhorn, Scotland.

Chris White is a multi-award winning Associate Director of Environmental Economics at AECOM, London. His primary area of work focuses on natural capital and ecosystem services. Chris is also a Member of the IUCN Commission on Ecosystem Management and a Trustee of the Lifescape Project.

Pamela Wright is a faculty member in the Ecosystem Science and Management programme at the University of Northern British Columbia. A conservation scientist, her research focus on parks and protected areas design and management.

Rui Yang is Dean and co-founder of the Department of Landscape Architecture, School of Architecture, Tsinghua University. He is also director and founder of the Institute for National Parks, Tsinghua University.

Talía Zamboni holds a bachelor in Biology and a Masters in Wildlife Management and Conservation. Since 2015, she has participated in different aspects of the species reintroduction programme in Iberá, where she currently coordinates the rewilding team of Fundación Rewilding Argentina.

Zhicong Zhao is Assistant Professor at the Institute for National Parks and the Department of Landscape Architecture, School of Architecture, Tsinghua University.

FOREWORD
PIONEERING A REWILDING PARADIGM

For me the road to rewilding has been a long one. In 1990, my husband Doug left Esprit, the company he co-founded, moving from global capitalist to dedicated ecologist. His next mission in life was what he called, 'paying rent for his life on Earth', which included saving the world, species by species and place by place. Soon enough, I retired from my job as CEO of Patagonia, and joined him in the southern extreme of South America.

At first, we worked to create large new national parks in Chile and Argentina, with the mission to restore wild nature where human destruction had taken its toll and protect still-pristine areas. Over time, we realised that protecting the land was not enough; we came to learn that ecosystems needed all their members present to be healthy and fully functioning. In some places, like the Ibera wetlands, so many native species had already disappeared. When we arrived there in 1997, no one was keeping track of these losses. But we knew that bringing back these species could make these environments self-sustaining again.

In modern times, the loss of keystone species was evident at least 150 years ago. With some exceptions, the work to bring back extirpated species only began over the last 45 years. In our case, the commitment towards rewilding work had to begin in the 1990s with the recognition that we humans are inextricably connected with the natural world, all of it. When nature thrives, so do we. We were first inspired by an example out of North America. In 1991, the she-wolf Pluie was fitted with a radio collar in Alberta, Canada. Her subsequent journeys revealed the little-known lives of wolves. Traveling over 10,000 miles, she showed us the role wolves play as architects of ecosystems, revealing the mystery and beauty these predators bring to our forests and grasslands. Unfortunately, she also showed us how these great predators meet abrupt ends at human hands. Pluie herself was killed in a hunting party after travelling for four and a half years between the United States and Canada. Many species, especially large predators, and others key to the proper functioning of our natural ecosystems, have suffered a dizzying decline, causing environments to lose their integrity, making environmental crises soar to intolerable levels. We witnessed the great beauty of nature fading fast.

These messages were not ignored. Scientists of a high academic level were recognising the destruction but, most importantly, people of great social commitment and respect for all forms of life such as Dave Foreman, Michael Soule, and Reed Noss began to meet to form what would become known as the Wildlands Project. The Wildlands Project was a point of pride and joy for Doug who funded the Project from the beginning. He saw early on that if territories were to become whole again, they needed to be 'Big, Wild, and Connected', a concept that was new and allowed conservationists to realise goals much more strategically than was understood before the Wildlands Project birthed these ideas.

Very quickly those meetings grew to include a large group of enthusiastic thinkers, scientists, and activists. It was here that the conceptual basis of rewilding was defined; we must conserve or restore key species, especially large carnivores, in extensive core areas connected by a system of corridors. Although the definition was later mutated and new variants of the term appeared, all of them refer to the conservation of large spaces where key species are present and in sufficient numbers to carry out their ecological roles that structure and regulate the ecosystems they inhabit. Within the Wildlands Project, we not only set out to discuss and define this new way of doing conservation, but we also committed to disseminating it widely. For this we printed thousands of copies of a newsletter called Wild Earth and distributed it to different key actors, especially NGOs and government agencies that had to embrace this conservation strategy for it to have an impact.

The reintroduction of the wolf to Yellowstone was a global conservation milestone and a great accolade for highlighting the role of keystone species and the need for rewilding. Of course, it was not the only one: the recovery of the sea otter on the Pacific coast or the bison on the vast American prairies are other examples of species that returned and along with them, verifying the rebirth of ecosystems. In other parts of the world and especially in Africa, rewilding had begun to be implemented several decades before, although without calling it that way and perhaps with less diffusion. The spectacular return of the wildebeest in the Serengeti is a very good example. One can also cite other incredible recoveries of numerous species in many countries of this continent, among which we find elephants, lions, rhinos, leopards, cheetahs, and wild dogs.

In the Southern Cone of South America, our foundation acquired vast tracts of land to eliminate the threats that had degraded these incredible environments so that nature would recover. It was important to donate the land in good condition as national parks, which belong to all citizens. So far, we have helped protect nearly 15 million acres of land, an area larger than the country of Costa Rica, through the creation or expansion of 15 national parks in Chile and Argentina.

However, from the first moment Doug had realised that some of the species would not return simply by eliminating the threats that had caused them to disappear. This was especially notable for some species of large carnivores, herbivores, and frugivores that play key roles in the ecosystems they inhabit and that without their presence in sufficient numbers, these ecosystems degrade and even collapse. This was particularly evident in northeastern Argentina, where we began acquiring land to build the huge Iberá park, which today encompasses 1.8 million acres of wetlands, grasslands, and dry and humid forests. There we began to execute an ambitious rewilding programme in 2006 aimed at bringing back key species lost in this territory. This programme is continued today by Rewilding Argentina and its evolution occupies one of the chapters of this book.

Rewilding is revolutionising conservation culture worldwide because it implies a shift towards active management practices, among which the translocation of key species for reintroduction or supplementation purposes plays a very important role. This implies a change in the way of perceiving conservation, which has traditionally been focused on conserving natural environments and remaining species, above all through the elimination of threats. We must continue to expand protection in all corners of our planet, but we must go one step further and regain what we have lost. That is what rewilding is all about, to recover extinct or rare species, especially those that are found in the upper trophic levels of the food chain, in order to recover their ecological roles and the interactions of these species.

In this sense, rewilding challenges and improves on already established tools such as the red list of endangered species, since recovering ecological roles often implies recovering species that are not necessarily threatened at a global or national level. Rewilding has also shown us that many of the areas that we consider pristine or well conserved (including many national parks) are actually deprived of many of their keystone species and therefore suffering varying levels of degradation. In this sense, it challenges the traditional ways of measuring the effectiveness of the management of these areas, more focused on the type of actions that we carry out to manage them than on the integrity of their ecosystems. Even the much-needed 30x30 movement is questioned: do we want 30% of the planet legally protected but without most of its key species present in these ecosystems?

The proactive and active conservation agenda of rewilding generates hope and inspiration for the widespread conservation movement. By necessity, rewilding implies changing how we relate to nature, and learning to coexist with species that demand large territories and that even compete with us. It forces us to become more dedicated and make more concessions, accentuating the need to understand that all species have the right to live on this planet.

Rewilding not only delegates conservation to public institutions but also incorporates private owners or community lands in this conservation strategy, especially through wildlife observation and ecotourism activities on private and community-held lands, helping communities build durable futures through nature-based economies.

Finally, rewilding constitutes a nature-based solution to face the grave environmental crises on our planet, such as the loss of biodiversity, climate change, and the appearance of pandemics, all related to the degradation of ecosystems and the loss of key species.

After 30 years of working in conservation, my own motivation remains stronger than ever. I've been fortunate enough to experience my connection with the wild in tangible ways, such as bottle-feeding baby giant anteaters, hoping that these orphans will become strong enough to join those already released into the wilds of the Ibera wetlands. They deserve a wild future. There is no greater feeling than being part of these rewilding teams whose daily effort is making jaguars and red-and-green macaws roam free again in Ibera, and bringing other species back to the diverse array of ecosystems where we work in throughout southern Chile and Argentina.

Rewilding is undoubtedly part of this change for ecological and cultural health, for our health. I see the *Routledge Handbook of Rewilding* as a valuable contribution arriving just in time, addressing our most urgent crises at a time when the practice of this conservation strategy becomes more mainstream. Its editors are all founding members of the Rewilding Thematic Group, created by the IUCN (International Union for the Conservation of Nature), which already embraces this strategy. The *Rewilding* Handbook offers a comprehensive overview of the history, theory, and current practice of rewilding, organising it in four parts: the evolution of the concept, its theoretical and practical underpinnings, the application of rewilding principles, and rewilding ethics.

The content was developed by an outstanding group of academics in social and natural sciences researching the practice of rewilding. As experienced practitioners and wildlife managers, they are on the frontlines of the battle against the biodiversity crisis, making key decisions about rewilding projects on a daily basis. We hope that their experience in bringing species and ecosystems back to life can help a growing number of emerging enthusiasts entering the fascinating world of restoring nature through rewilding. They represent the hope of creating large scale rewilding actions in a world more in need of it every day.

In Doug's words, 'Are you ready to do your part? Everyone is capable of taking on their role and using their energy, political influence, talent, and financial or other resources to be part of a global movement for ecological and cultural health. Everything will be useful. There is important and significant work to be done. To change everything, everyone is needed. Everyone is welcome.'

Kristine Tompkins
President of Tompkins Conservation
UN Patron of Protected Areas

ACKNOWLEDGEMENTS

This book originates from various workshops and discussions held by the IUCN Commission for Ecosystem Management Rewilding Thematic Group (which we will now thankfully abbreviate to IUCN CEM RTG) as we worked on developing a set of rewilding guiding principles between 2017 and 2019. In that process, we saw the need for a handbook on rewilding that took a global perspective, started at the beginning, didn't shy away from the difficult questions, and offered practical guidance to those considering rewilding projects that is true to its foundational principles. To achieve these wide-ranging aims we have been fortunate to bring together a distinguished field of authors, all of whom have addressed rewilding from different experiences and perspectives.

Little did we know, of course, back in our early book meetings in Autumn 2018, that a global pandemic was just around the corner. COVID-19 has thrown a lot at the contributors to this book, and we are forever grateful that you have stuck with us and delivered some fabulous chapters despite experiencing setbacks that would derail most book projects. Your commitment and enthusiasm for rewilding is inspiring.

Special thanks are due to Katie Stokes, John Baddeley, and Hannah Ferguson at Routledge, who have been extremely patient with the Editors, keeping us on track and always responding supportively to our various questions and concerns. We would also like to thank our colleagues in the IUCN CEM RTG, especially Zoltan Kun, Mark Fisher, Adam Eagle, Jessica Rothwell, Simon Whitehead, Rob Morley, and Cao Yue, all of whom have provided encouragement and guidance, and our IUCN CEM chair Angela Andrade, who has consistently championed our rewilding group and provided wisdom when we need it the most. A huge thanks is also due to Steve Edwards, our IUCN guru and counsellor, for taking us under his wing.

Finally, we dedicate this book to the memory of Dave Foreman, Michael Soulé, and Alison Parfitt. They both contributed greatly to rewilding science, and their presence can be felt throughout the pages of this book. We cherish their friendship and mourn their passing.

PART I

The evolution of rewilding

1

INTRODUCTION

What is rewilding?

Sally Hawkins, Rene Beyers, Steve Carver, and Ian Convery

Researchers and practitioners active within the field of rewilding have been grappling with this question for many years. Simultaneously, there has been a continual increase in the number of rewilding-related projects, organizations, and research, all bringing different characteristics, cultures, and intentions to the field. During this time society has become increasingly aware of the severe and complex nature of global threats, including climate change and biodiversity loss. At the same time, the foundations of our societies have been thrown into question and confidence in our political systems and science has eroded. This is a challenging context for rewilding to evolve in as a discipline, but it provides opportunities to bring forth innovative solutions for the testing, wicked, problems of our time.

The IUCN CEM Rewilding Thematic Group, of which the editors are founding members, spent over two years reviewing the literature and talking with rewilding and restoration experts before publishing, together with many of these experts, the following definition of rewilding (from Carver et al., 2021):

> Rewilding is the process of rebuilding, following major human disturbance, a natural ecosystem by restoring natural processes and the complete or near complete food web at all trophic levels as a self-sustaining and resilient ecosystem with biota that would have been present had the disturbance not occurred.
>
> This will involve a paradigm shift in the relationship between humans and nature. The ultimate goal of rewilding is the restoration of functioning native ecosystems containing the full range of species at all trophic levels while reducing human control and pressures. Rewilded ecosystems should—where possible—be self-sustaining. That is, they require no or minimal management (i.e., *natura naturans* [nature doing what nature does]), and it is recognized that ecosystems are dynamic.

While these words on paper seem straightforward and capture the history and trajectory of rewilding, there remain many questions about the task at hand: Which biota? From when? At what point does resource use become unsustainable? How does one start a paradigm shift within entrenched cultures? Who is best placed to answer these questions? Rewilding, it would seem, is simultaneously unwieldy and visionary.

DOI: 10.4324/9781003097822-2

At its most basic definition, rewilding is about affecting change. John C. Maxwell (2007) famously noted that 'most people don't like change. They revolt against it unless they can clearly see the advantage it brings. For that reason, when good leaders prepare to take action or make changes, they take people through a process to get them ready for it.' Given the scale of challenges presented by the climate and biodiversity crises, it is vital that we make a compelling, positive, and inclusive case for change to broaden support and increase the potential to achieve rewilding's transformative goals.

The chapters in this book all explore the kinds of change that rewilding is looking to promote—ecological, socio-cultural, or systemic—at various levels from local to global, looking inwards at how we undertake our research and practice, and outwards at ecosystems, landscapes, and society at large. The book is organised into four parts to reflect key areas of rewilding theory, practice, and debate: the evolution of rewilding, theoretical and practical underpinnings of rewilding, applications and impacts of rewilding, and the ethics and philosophy of rewilding. Within these sections we explore who or what has influenced the rewilding concept, why it is necessary, what it is trying to achieve, how it is practised and then look more deeply into the fundamental (and sometimes difficult or conflicting) values and ethics that rewilding embodies.

The concept of rewilding emerged in North America in the 1980s, where it was originally concerned with safeguarding and restoring native biodiversity through large-scale, interconnected networks of reserves established primarily to protect and restore interacting keystone species and their trophic relationships (Power et al., 1996). The concept was largely developed by a group of conservationists, academics, and activists involved with the Wildlands Network, driven by increasing knowledge of ecological processes and the failures of traditional conservation practices (especially confined protected areas and single-species conservation) to stem the tide of biodiversity loss. Soule' and Noss (1998) published a landmark paper describing the scientific basis for rewilding. They describe three key features: large core protected areas, ecological connectivity between those areas, and keystone species, especially carnivores, for their influential roles in the ecosystem. This is the '3Cs' model of cores, corridors, and carnivores that continues to inform rewilding practice in the Americas. Fisher and Carver discuss the roots and history of rewilding in North America in more detail in Chapter 2, while Hilty et al. (Chapter 14) and Donadio et al. (Chapter 16) present examples of this in North and South America.

During this period conservationists in Europe began to question conservation policies which relied on static reserves and intensive management to 'freeze living systems in time' (Monbiot, 2013: 152) and on maintaining prescribed ecological conditions based on agricultural practices (Vera, 2000; Taylor, 2005). Increasing rural land abandonment created areas where nature flourished without management, spurring on many to advocate for ecological restoration based on passive rewilding or letting nature take care of itself. Locquet and Carver give an overview of European rewilding history and perspectives in Chapter 3. Rewilding in Europe diverged somewhat from North America to adapt to a landscape dominated for millennia by intensive exploitation, sparking debate about what constitutes natural ecosystems. However, while the roots of these rewilding approaches differ, what they share is a desire to change both the practice and culture of conservation biology, towards a more proactive, optimistic, and ecologically sound foundation to tackle the seemingly unstoppable ecological degradation of the Anthropocene era.

Rewilding has continued to grow in popularity and is now practised or considered as a conservation option around the world. It has also continued to adapt and evolve to suit different conditions. Despite these differences, there are similarities which can be used to unify the field of rewilding. Hawkins in Chapter 5 proposes a framework for rewilding based on its social-ecological aims. There are certain values and principles that have become commonplace within the concept of rewilding through experience and improved ecological understanding. One example, as demonstrated in the case studies presented in this book (e.g. Pringle and Goncalves, Chapter 17; Donadio et al., Chapter 16), is that ecosystems change over time and that change is often unpredictable. Rewilding practice, therefore, must be adaptable. Species in an ecosystem have co-evolved, they interact and are inter-dependent. Rewilding therefore relies on restoring these interactions based on reference ecosystems from which we can know which species work together in a system. The complexities of this logic are explored by Clarke and Hebblewhite, Chapter 6 and Stanley-Price, Chapter 7. Another key concern is that the causes of ecological degradation are mainly socio-cultural, and therefore in order to affect transformational change, rewilding must address the drivers of that degradation. Rewilding must therefore include people. These and other principles of rewilding can be found in Box 1.1 and these were used to inform the structure of the book and many of the discussions herein. This book therefore forms part of our continued work towards mainstreaming these principles in human society and engaging the rewilding concept with other evolving principles and standards for ecological restoration (Gann et al., 2019), Nature-Based Solutions (Cohen-Schacham et al., 2019), and the Convention on Biological Diversity (CBD) Ecosystem Approach (Smith & Maltby, 2001).

Box 1.1 Guiding principles for rewilding (from Carver et al., 2021)

Principle 1: Rewilding utilises wildlife to restore trophic interactions.

Successful rewilding results in, or leads to, a self-sustaining ecosystem in which native species' populations are regulated through predation, competition, and other biotic and abiotic interactions. It is crucial that consideration be given to the role large herbivores and apex predators play in maintaining and enhancing the biodiversity within landscapes. Keystone species (organisms that influence the functioning of an ecosystem disproportionate to their abundance) and ecosystem engineers (organisms that directly or indirectly modulate the availability of resources to other species by causing physical state changes in biotic or abiotic materials) are also important in securing the integrity of the ecosystem and thus enhancing ecosystem resilience. Where appropriate, strongly interacting keystone species that have roles in maintaining the ecosystem should be reintroduced or depleted populations reinforced to an ecologically effective level.

Principle 2: Rewilding employs landscape-scale planning that considers core areas, connectivity, and co-existence.

At the landscape scale, it is crucial that core areas provide a secure space that accommodates the full array of species that comprise a self-sustaining natural ecosystem. These areas may be either legally designated or under private management. Restoring connectivity between core areas promotes movement and migration across the wider landscape and improves resilience to the impacts of climate change. Rewilding can build on existing core areas, such as designated wilderness areas, national parks, or privately managed natural areas. Plans for rewilding at the landscape scale should accommodate the need for coexistence between wild species and humans (and livestock) through careful integration of cores and connectivity in functioning ecological networks and zoned systems of compatible low-intensity human land use (e.g., buffers and extensive multiple-use landscapes).

Principle 3: Rewilding focuses on the recovery of ecological processes, interactions, and conditions based on reference ecosystems.

Rewilding should aim to restore self-sustaining and resilient ecosystems and specifically the natural patterns and dynamics of abundance, distribution, and interactions between native species. To do this, rewilding should make use of an appropriate ecological reference. Any reference point is ultimately arbitrary, but it is expected to be self-sustaining and resilient. A reference can be based on carefully selected contemporary near-natural reference areas with relatively complete biota where these still exist or appropriate scientific or historical evidence supported by expert indigenous and local knowledge. Rewilding should allow for natural disturbance within an evolutionary relevant range of variability and take environmental change into account. Key native species that have become globally extinct can be replaced by suitable carefully selected wild surrogates, where legislation permits and their ecological role is deemed important. The surrogate should, where possible, be phylogenetically close to and have similar ecological and trophic functionality as the extinct species and appropriate management and monitoring should be put in place.

Principle 4: Rewilding recognises that ecosystems are dynamic and constantly changing.

Temporal change, both allogenic (external) and autogenic (internal), is a fundamental attribute of ecosystems and the evolutionary processes critical to ecosystem function. Allogenic factors include storms, floods, wildfire, and large-scale changes in climate. Equally important are changes from autogenic processes, such as nutrient cycles, energy and genes flows, decomposition, herbivory, pollination, seed dispersal, and predation. Conservation planning for rewilding should consider the dynamic nature of ecosystems and be responsive to individual species range shifts and the disaggregation and assembly of genes, species, and biotic communities. Rewilding should facilitate the space and connectivity needed for these processes to have free reign, allowing the wider processes of succession, disturbance, and biotic interactions to determine ecological trajectories without impediment or constraint. Rewilding programmes must take both genetic and ecologically effective population sizes into account and employ strategies (e.g., connectivity) that ensure ecologically sustainable and genetically healthy populations of animals, plants, and other organisms. Where species of concern are globally rare and in danger of extinction, intervention may be required to prevent this from happening, including more traditional conservation measures, such as reserves and captive breeding.

Principle 5: Rewilding should anticipate the effects of climate change and where possible act as a tool to mitigate impacts.

Anthropogenic impacts of climate change are rapid and pervasive, creating the need to anticipate the likely impacts on rewilding. Rewilding projects have medium- to long-term time scales that inevitably span the predicted scales and magnitudes of global climate change as regards warming trends, ice sheet collapse, sea-level rise, storm events, and so forth; thus, climate change needs to be considered when planning such projects. Rewilding can also be considered an example of an NbS with the potential to absorb, ameliorate, and tackle the effects of climate change. This includes mitigating the impacts of climate change on ecosystems and increasing the capture of atmospheric carbon (e.g., through natural regeneration following land abandonment and replacing livestock with wild herbivores) as well as providing ample space and connectivity along environmental and climatic gradients to enhance opportunities for species movements.

Principle 6: Rewilding requires local engagement and support.

Rewilding should be inclusive of all stakeholders and embrace participatory approaches and transparent local consultation in the planning process for any project. Rewilding should encourage public understanding and appreciation of wild nature and should address existing concerns about coexisting with wildlife and natural processes of disturbance. Stakeholder engagement and support can reinforce the use of rewilding as an opportunity to promote education and knowledge exchange about the functioning of ecosystems. Although everyone is a potential stakeholder, no one strategy will satisfy everyone all the time and rewilding projects will need to address barriers to acceptance.

Principle 7: Rewilding is informed by science, traditional ecological knowledge (TEK), and other local knowledge.

Traditional ecological knowledge provides a complementary body of knowledge to science and collaborations between researchers. Holders of TEK and other local experts can generate benefits that maximise innovation and best management guidance through knowledge exchange, transparency, and mutual learning. This can include, for example, the role of customary institutions that rely on cultural values, such as sharing and eco-reciprocity in relation to transmission of ecological knowledge. All these forms of knowledge are important for the success of rewilding projects and can help inform adaptive management frameworks and gather evidence. Local experts can provide detailed knowledge of sites, their histories, and processes, all of which can inform rewilding outcomes. It is important to acknowledge knowledge gaps and be aware of shifting baselines and the implications of these for rewilding projects while ensuring that traditional practices are sustainable and supported by appropriate evidence. Projects themselves can form the basis for knowledge generation, data, and information of use to future projects.

Principle 8: Rewilding is adaptive and dependent on monitoring and feedback.

Monitoring is essential to provide evidence of short- and medium-term results with long-term rewilding goals in mind. This is required to determine whether rewilding trajectories, such as a particular treatment, are working as planned. Participatory monitoring based on (SSG, using) simple crowd-sourced methods with local volunteers coupled with more detailed scientific monitoring can be used to provide the necessary data and information. Rewilding projects should use these data to identify problems and possible solutions as part of an appropriate adaptive management framework. These need to be adequately resourced such that further interventions can be implemented without loss to project budgets and resources.

Principle 9: Rewilding recognises the intrinsic value of all species and ecosystems.

Although there is increasing recognition that natural ecosystems, and the species within them, provide valued goods and services to humans, wild nature has its own intrinsic value that humanity has an ethical responsibility to both respect and protect. This principle emphasises the values of compassion and coexistence. Rewilding should primarily be an ecocentric, rather than an anthropocentric, activity. Where management interventions are required, these should focus on removal of human control and restoring native species with minimal intervention and nonlethal means wherever possible.

Principle 10: Rewilding requires a paradigm shift in the coexistence of humans and nature.

In alliance with the global conservation and restoration communities, rewilding means transformative change and provides optimism, purpose, and motivation for engagement alongside a greater awareness of global ecosystems that are essential for life on the planet. This should lead to a paradigm shift in advocacy and activism for change in political will and help shift ecological baselines towards recovering fully functioning trophic ecosystems, such that society no longer accepts degraded ecosystems and overexploitation of nature as the baseline for each successive future generation. This paradigm shift will also help create new sustainable economic opportunities, delivering the best outcomes for nature and people.

One fundamental principle of rewilding is the required paradigm shift in the relationships between humans and nature which is elaborated on in practical terms of co-existence by Lambert in Chapter 23 and from a philosophical and practical view in Part IV edited by Kate Rawles. Many of the human aspects of rewilding extend beyond rewilding and the culture of conservation and are being grappled with in other fields promoting change—whether ecological, systemic, or cultural—and there is knowledge and experience to be shared across disciplines. Several chapters in the book are devoted to these connections and how different sectors, such as agriculture (Aglionby and Field, Chapter 23), health (VanVolkenburg et al., Chapter 25), resource exploitation (Beyers and Hawkins, Chapter 26), education (Prince, Chapter 27), and recreation and adventure travel (Loynes, Chapter 28) can engage with the rewilding concept.

This book does not seek to present rewilding as a neat, complete concept. We have purposefully avoided the 'echo-chamber' approach that bedevils so many edited collections. Many chapters highlight uncertainties or question assumptions about the concept or culture of practice, for example epistemological paradigms (Fenton and Playdon, Chapter 12), ethical stances (Kopnina et al., Chapter 32), and measures of success (Root-Bernstein, Chapter 11). Even those chapters that deal with ecological theories underpinning rewilding, such as trophic cascades (Clarke and Hebblewhite, Chapter 6), demonstrate that there is still much need for research and careful experimentation to deal with uncertain ecological outcomes. Rewilding itself assumes limitations to human knowledge of how ecosystems work, inherent in its aims to let nature be itself and look after itself, and therefore some humility on our part is required. Rewilding and conservation research reveal biases towards the most obvious or appealing biota and we must broaden our views to include all the important ecosystem actors. Satori and Wainhouse help us with this in their treatment of the role of fungi in rewilding in Chapter 24.

There remain contradictions within the concept of rewilding. For example, while Bekoff in Chapter 30 explores how biocentrism might inform a rewilding ethic that values life above all, Kealley and Burrows in Chapter 18 demonstrate that lethal control of invasive species is essential for rewilding in parts of Australia. While we are aware that change is most likely to work when it is instigated by communities or stakeholders, rewilding or conservation is often instigated or influenced by external organisations, government policy, or funding streams. In reality, we must therefore rely on both top-down and bottom-up approaches to change. The case studies presented here demonstrate different degrees of community or stakeholder involvement.

Rewilding is an evolving term and we expect (and hope) that this continues to be the case; uncertainty, adaptability, and experimentation are part of the DNA of rewilding. We acknowledge, therefore, the impossibility of trying to present a complete picture, and admit that there may be themes, such as broader economic and political aspects of rewilding, which were missed or given insufficient coverage. To address this and to acknowledge the value of knowledge gained through experience, we have particularly called on rewilding practitioners to share their experiences in the rewilding case studies.

Rewilding will continue to adapt to global, national and local forces, such as with the recent Covid-19 pandemic, climate change, political instability, food insecurity, or with emerging or changing understanding, such as with rewilding's continued engagement with problems related to colonialism and imperialism (e.g. Taylor, 2005; Ward, 2019). The original '3Cs' model has itself evolved to include climate resilience (Carroll & Noss, 2021), compassion (Bekoff, 2014; Kopnina, Leadbeater, & Cryer, 2019), and coexistence (Johns, 2019). We hope that the field of rewilding continues to evolve and lead to new insights and profound solutions to avert the collapse of ecosystems.

We see the need for rewilding guidelines that provide steer but that are likewise adaptable across a range of landscapes, ecosystems, and social contexts. Hawkins (Chapter 5) offers a framework based on the aims of rewilding, but other chapters look at elements which can inform guidelines and frameworks from an ecological (e.g. Beyers and Sinclair, Chapter 10) or social (e.g. Root-Bernstein, Chapter 11) point of view. Because the definition and guiding principles require broad support to provide a basis for the advancement of rewilding as adapted to differing ecological and socioeconomic systems around the world, our overarching objective was to provide unifying focus for the field of rewilding to enable identification of gaps in knowledge and tailoring of research so that concepts can be further refined and contribute to robust rewilding guidelines. The purpose of this book is to continue this work.

We are also concerned that rewilding risks becoming mainstreamed to such an extent that it effectively just becomes conservation by another name. This should not be allowed to happen, and whilst we recognise the need for rewilding to work within a range of human contexts (Principles 6 & 7), we will not get the paradigm shift we so urgently require (Principle 10) if we persist with an approach posited in some corners that restrict our thinking and ambitions in a way that ultimately frames rewilding as yet another anthropocentric approach. This is wrong, and risks rendering rewilding meaningless and empty. Rewilding is not a one-size-fits-all model, it cannot be everywhere and everything. Rewilding challenges the shifting baselines towards accepting degraded nature as the norm and confronts cultural norms of narrow targets and controls across both conservation and society. It instead reintegrates nature into our lives and across landscapes, allowing it the time and space to heal itself from human-driven degradation, to thrive. This work needs to be done urgently. Time is running out because the more degraded natural ecosystems there are, the more difficult it becomes to rewild ecosystems that support the biospheric functions vital to all life on Earth.

The time is now, and the place is here.

References

Bekoff, M. (2014) *Rewilding Our Hearts: Building Pathways of Compassion and Coexistence.* New World Library.

Carroll, C. and Noss, R.F. (2021) 'Rewilding in the face of climate change', *Conservation Biology*, 35(1), pp. 155–67. doi:10.1111/cobi.13531

Carver, S. et al. (2021) 'Guiding principles for rewilding', *Conservation Biology* [Preprint].doi:10.1111/cobi.13730

Cohen-Shacham, Emmanuelle, Angela Andrade, James Dalton, Nigel Dudley, Mike Jones, Chetan Kumar, Stewart Maginnis, Simone Maynard, Cara R. Nelson, Fabrice G. Renaud, Rebecca Welling, Gretchen Walters. (2019) Core principles for successfully implementing and upscaling Nature-based Solutions, *Environmental Science & Policy*, 98, pp. 20–9. https://doi.org/10.1016/j.envsci.2019.04.014

Gann, G.D., McDonald, T., Walder, B., Aronson, J., et al. (2019) International principles and standards for the practice of ecological restoration. Second edition. *Restoration Ecology*, 27(S1), pp. S1–S46, doi: 10.1111/rec.13035

Johns, D. (2019) 'History of rewilding: Ideas and practice', in Pettorelli, N., Durant, S.M., du Toit, J.T. (eds), *Rewilding*. Cambridge University Press, pp. 12–33. Available at: www.scopus.com/inward/record.uri?eid=2-s2.0-85087049760&partnerID=40&md5=caa6ac23672cf4a9a02d039d5d3cab65.

Kopnina, H., Leadbeater, S. and Cryer, P. (2019) 'The golden rules of rewilding—examining the case of Oostvaardersplassen—ECOS—Challenging Conservation'. Available at: www.ecos.org.uk/ecos-406-the-golden-rules-of-rewilding-examining-the-case-of-oostvaardersplassen/ (accessed: 10 May 2022).

Maxwell, J.C. (2007) *21 Irrefutable Laws of Leadership: Follow Them and People Will Follow You.* Nashville: Thomas Nelson.

Monbiot, G. (2013) *Feral*. Penguin.

Power, M.E. et al. (1996) 'Challenges in the Quest for Keystones: Identifying keystone species is difficult—but essential to understanding how loss of species will affect ecosystems', *BioScience*, 46(8), pp. 609–20. doi:10.2307/1312990

Smith, R.D. and Maltby, E. (2003) *Using the Ecosystem Approach to Implement the Convention on Biological Diversity: Key Issues and Case Studies*. Cambridge, UK: IUCN

Soule, M.E. and Noss, R.F. (1998) 'Rewilding and biodiversity: complementary goals for continental conservation', *Wild Earth*, 8(3), pp. 18–28.

Taylor, P. (2005) *Beyond Conservation*. London: Earthscan.

Vera, F. (2000) *Grazing Ecology and Forest History*. CABI Publishing.

Ward, K. (2019) 'For wilderness or wildness? Decolonising rewilding', In Pettorelli, N., Durant, S.M., du Toit, J.T. (eds), *Rewilding*. Cambridge University Press, pp. 34–54. Available at: www.scopus.com/inward/record.uri?eid=2-s2.0-85070630066&partnerID=40&md5=3814873620da51edd8d5c84caafeecfa.

2

THE EMERGENCE OF REWILDING IN NORTH AMERICA

Mark Fisher and Steve Carver

Introduction

The early evolution of rewilding in the literature spans the years from 1990 when the word 'rewild' first appeared in print (Foote, 1990), to 2004 when Dave Foreman—a central character in that evolution—drew its conceptual foundations and actions together in his book on 'Rewilding North America' (Foreman, 2004). Driving this was a core group of highly motivated activists working for nature conservation, amongst which was a group of influential scientists working in the developing field of landscape ecology and conservation biology. The fact that this original evolution of the idea of rewilding eludes many contemporary exponents of rewilding (e.g. Svenning et al., 2016; Pettorelli et al., 2018; Perino et al., 2019; Pedersen et al., 2019) as well as its critics (Jørgensen, 2015; Nogués-Bravo, 2016; Corlett, 2016; Hayward et al., 2019) is evidence of the paucity of back referencing to its early origins, such that frequently only just the one article is cited: that by Michael Soulé and Reed Noss on 'Rewilding and Biodiversity: Complementary Goals for Continental Conservation' (Soulé & Noss, 1998). This was published in *Wild Earth* in 1998, a magazine that during its lifetime was the primary publication for articles on rewilding. Many contemporary works citing this article seem unaware of its context as regards its origins and what came afterwards in the further evolution of the rewilding concept by The Wildlands Project. While complete sets of paper copies of *Wild Earth* exist in repository libraries (WorldCat, ND), the recent availability of digital archives of *Wild Earth* published between 1991 and 2004 (ESP, ND) and of the Wildlands Network Designs developed with the assistance of The Wildlands Project (WN, ND) has allowed for a full contextual analysis of this timeline. Together these identify a number of objectives in the conservation of wild nature that became axiomatic with rewilding (Fisher, 2020). A synthesis derived from these axioms is given in the conclusions to this chapter. Taken together, the axioms and synthesis can be set against the plethora of contemporary definitions and criticisms that are silent on this evolved meaning of rewilding, providing an underpinning that brings clarity and focus.

DOI: 10.4324/9781003097822-3

Early years and the evolution of a new discipline

The terms 'rewild' and 'rewilding' must have been, by association, common parlance amongst activists involved in the radical environmentalism of the late 1980s in America, along with their sympathisers in the Sierra Club and Wilderness Society. These groups can be characterised by a commitment to bioregionalism and Deep Ecology thus giving recognition to the intrinsic value of wild nature (Foote, 1990; Cramer, 1998; Taylor, 2008). During that decade, a group of scientists were shaping a new field in ecology that became known as conservation biology (Meine, 2005) the founder of which is generally considered to be Biologist Michael Soulé. It was Soulé who organised the first international conference in 1978, and, two years later, edited the first book on the topic (Soulé & Wilcox, 1980; Noss, 1999).

Conservation biology was described in the book as a 'new rallying point for biologists wishing to pool their knowledge and techniques to solve [ecological] problems' and that it was 'a mission-oriented discipline comprising both pure and applied science' (Soulé & Wilcox, 1980). A shared concern for recovery and conservation of wild nature led several of conservation biology's foremost thinkers, including Soulé, to develop ties and become engaged with the radical environmental movement (Taylor, 2008). Thus, Soulé described conservation biology as crisis oriented, in that practitioners were often having to react quickly to advancing endangerment of habits and species (Soulé, 1985).

Soulé further showed that conservation biology was a holistic, multidisciplinary approach to ecosystem recovery that encompassed island biogeography, ecophilosophy, population biology, genetics, environmental monitoring, and hazard evaluation, as well as the social sciences, all of which were radical ideas for the time. This was a new interdisciplinary field in biological sciences that was mission-oriented, crisis-driven, and problem solving (Meine et al., 2006). The following years saw a second international conference on conservation biology and another book (Soulé, 1986) as well as formation of the Society for Conservation Biology with Soulé being its first President. *Conservation Biology*—the Society's own journal—followed in 1987 (Noss, 1999).

The evolution in meaning of rewilding would, however, start with the establishment of a new, non-profit conservation periodical called *Wild Earth*. Dissent had been growing within Earth First!—a radical environmental activist group in America—at a dilution of its biocentric vision and agenda by a more humanistic view based on social priorities, and which was leading some of the more conservation-minded Earth First! leaders—dubbed the 'Wilders' by Bron Taylor—to consider breaking away to establish a new conservation magazine (Foreman, 1987; Foreman & Morton, 1990; Taylor, 1991; Sessions, 1992; Balser, 1997). Thus after leaving Earth First!, Foreman and John Davis met in December 1990, to plan a new conservation magazine that they would call Wild Earth (Foreman, 2003). The first issue was published in the spring of 1991, and within its Statement of Purpose were commitments to 'provide a voice for the many effective but little-known regional and ad hoc wilderness groups and coalitions in North America' and to 'render accessible the teachings of conservation biology, that activists may employ them in defense of biodiversity' (WE, 1991).

The aim of the magazine was to blend traditional wilderness and wildlife conservation with the science of conservation biology, to act as a platform for dissemination of the latest conservation science, and to provide a voice for those working on restoration and protection of all the natural elements of wild nature (WE, 1991; Foreman, 2003). As the first issue was being published in 1991, a small meeting of environmental activists and conservation biologists came together to talk about an ecological vision for North America, out of which was created The Wildlands Project (Davis, 1991; Johns, 1991; Foreman, 2004; Taylor, 2008). They saw this non-profit conservation organisation as a way of merging sound science with practical action in their recognition of the need for conservation planning, restoration and protection at a large scale. This represented a significant move beyond a concern with the preservation of islands of wilderness to articulating a broader vision for restoring wildness on a large scale (Foreman, 2004; Smalley, 2017). It was thus a cross-fertilisation between the outputs of those scientists in their professional life with the strategy and actions of The Wildlands Project and its followers, and the articles that they and others wrote for Wild Earth, that paved the way for rewilding to be a part of the conservation strategies of today.

John Davis was the first to use the word rewilding in Wild Earth in what was its fourth issue (Davis, 1991). He was writing in the editorial of the winter issue of 1991 about the meeting that led to the creation of The Wildlands Project (see above). It was thus through the pages of *Wild Earth* that the meaning of rewilding was shaped by its users. An etymology of rewilding is revealed through a contextual analysis of the writings in *Wild Earth* that cite rewilding, showing how its meaning became clearer from the contexts within which it was used. Those contexts themselves occupied more and more space in *Wild Earth*, and saw translation into action in a series of regional Wildlands Network Designs. No definition was given in the earlier citations of rewilding in *Wild Earth*. Instead, The Wildlands Project was developing its ideas on how to design connected wildlands and putting them out for consideration and discussion in both *Wild Earth*, and in journal articles, as well as presenting to an annual meeting of the Society for Conservation Biology in 1993 (Johns, 1993; Mann & Plummer, 1993; Noss, 1995; Turner, 2012).

The ecological foundations of rewilding

The ecological foundations of conservation biology emerging over the 1980s were enlisted over the following decade of the 1990s into the approach of rewilding through *Wild Earth* and by The Wildlands Project. A first stage could be seen in the first Special issue of *Wild Earth* in 1992, which had a focus on The Wildlands Project and its adoption of conservation biology in its restoration strategy. Introducing the Project, Soulé observed that an instantaneous ecological metamorphosis in North America was impossible because the continent was now too disrupted and fragmented, noting that the mountain ranges of the Southwest were isolated, that the national parks and wilderness were islands that were too small and too poached to sustain viable populations of predators (Soulé, 1992). However, he sought to give hope for restoring wild nature in the face of the impact of a burgeoning human population through land-use planning on spatial and temporal scales never attempted before. He noted that this land-use planning had to occur at the regional level, and that it must be participatory—'The restoration of the wildlands network will depend on the knowledge of people intimate with the mountains, canyons, forests, coves, rivers and creeks. Such planning will not work without grass-roots education and empowerment. Over time, each regional planning group will develop a map-based program for their bioregion. Later, representatives of the bioregional groups will meet and integrate their plans into a national, then continental strategy.'

In the same issue, Noss drew on his own published research, as well as that from fellow conservation biologists, to enunciate The Wildlands Project Land Conservation Strategy in which he presented a conceptual overview, the scientific background, and general guidelines for applying conservation biology to wilderness recovery (Noss, 1992). He asserted that goal-setting must be the first step in the conservation process, and gave four fundamental objectives that he considered were consistent with the overarching goal of maintaining the native biodiversity of a region in perpetuity: capture all native ecosystem types and seral stages across their natural range of variation; maintain viable populations of all native species in natural patterns of abundance and distribution; maintain ecological and evolutionary processes; and designing the system to be responsive to short-term and long-term environmental change. He went on to give a detailed description of the components of a wilderness recovery network and their purpose and function. These would include strictly protected core areas; buffer or multiple use zones; and connectivity through linkages or corridors that were habitat specific, or for dispersal or seasonal movements, the sum total of the network enclosing and linking biologically critical areas in a continuous system of natural habitat.

Noss noted, in relation to survey and selection for core reserves and primary corridors, that there were critical steps in selecting the most strictly protected areas and primary linkages in a wilderness recovery network. This would be field reconnaissance and interpretation of maps, aerial photographs, or satellite images to identify areas that appeared to be roadless, undeveloped, or otherwise in an essentially natural condition, and recognising which of these areas were public lands. He emphasised the need for large scale in regard to species-area relationships, and to island biogeographic theory, but felt the most compelling argument for large reserves was to do with population viability and habitat diversity in the face of environmental change.

A next stage confirmed the enlisting of conservation biology in the approach of rewilding. Soulé and Noss presented the scientific basis for rewilding in the landmark article—'Rewilding and Biodiversity: Complementary Goals for Continental Conservation'—published in the autumn edition of *Wild Earth* in 1998 (Soulé & Noss, 1998). They began by placing rewilding in the context of the history of conservation 'currents' in the stream of nature protection in America. They described the first current as monumentalism, the wish to preserve places of extraordinary natural beauty—the grand spectacles of nature that were the foundation of the National Park System. They noted that, over time, monumentalism had evolved into the wilderness movement. They traced the next important current as biological conservation, the protection of representative samples of all features, landforms, or vegetation types and successional stages in a reserve network that captured and protected most of a region's species in separate reserves—it was a compositional approach to reserve identification. However, they noted that a representational approach might not be adequate because it did not justify the protection of sufficient space for a viable, regional network of natural areas. Thus in locations where vegetation diversity was low, a system of ecological reserves based only on vegetational diversity could end up being small, fragmented, and vulnerable. The third current had arisen with the elucidation of island biogeography and its species area relationship, but more importantly the implications it had for quantitative prediction of extinctions in isolated habitat remnants and nature reserves. They noted that the principles of island biogeography were soon incorporated into the emerging new science of conservation biology, its adherents having identified weaknesses with the existing conservation approaches, based on an understanding of the scale on which ecological processes operated. Thus small, isolated populations of animals were vulnerable to accidents of demography and genetics, and to environmental fluctuations and catastrophe, underlining the need for large scale and connectivity.

Soulé and Noss explained that rewilding was the fourth current. While it was science-based like biological conservation, the second current, it was a more inclusive strategy that incorporated special elements and phenomena such as hotspots of endemism, important migratory stopovers or breeding areas, old-growth patches, or roadless areas, elements that had such restricted distributions that they would not be captured by a representational approach alone. It was a non-representational methodology that emphasised the restoration and protection of large areas and wide-ranging, large animals—particularly carnivores. They noted that although all species interacted, the interactions of some species were more profound and far-reaching than others, such that their elimination from an ecosystem often triggered cascades of direct and indirect changes on more than a single trophic level, leading eventually to losses of habitats and extirpation of other species in the foodweb. These were the keystone species that enriched ecosystem function in unique and significant ways, and were central to the rewilding argument.

They gave an example of unpopulated or sparsely settled '*frontier*' areas, such as most of Canada, where map-based reserve planning proceeded from a basis of securing entire unlogged or undeveloped watersheds, in part because such large, topographically diverse watersheds would contain virtually all of the vegetational diversity within the region. In finer scale, they noted the different approach that conservationists designing a nature reserve network for the Sky Island–Greater Gila region of southwestern America were taking that emphasised rewilding and ecological restoration rather than representation, or other biodiversity-focused goals. The reserve design was based on the needs of focal species, some of which were large carnivores and ungulates, and some of which were indicators of the ecological resilience and restoration of particular systems or processes that had suffered from mismanagement, such as the extirpation of some ungulates and large carnivores, the suppression of fire, and extensive overgrazing, particularly in riparian zones.

Soulé and Noss then asserted that three major scientific arguments established the case for rewilding and justified the emphasis on large predators: the 'structure, resilience, and diversity of ecosystems is often maintained by "top-down" ecological (trophic) interactions that are initiated by top predators'; large areas are justified by wide-ranging predators because they require large cores of strictly protected landscape for secure foraging and seasonal movements; because core reserves were typically not large enough in most regions, they must be connected to ensure long-term viability of wide-ranging species. They understood that the ecological argument for rewilding was bolstered by research on the roles of large animals, particularly top carnivores and other keystone species in many continental and marine systems, where studies were demonstrating that their disappearance often caused these ecosystems to undergo dramatic changes, many of which led to biotic simplification and species loss. Their view was that extensive networks of cores and habitat linkages also sustained a vast range of natural processes, and thus rewilding was a 'critical step in restoring self-regulating land communities' that minimised the need for human management.

Soulé and Noss also claimed two non-scientific justifications for rewilding. First, there was the ethical issue of human responsibility in relation to the history of persecution and local extirpation of large carnivores, noting that their capacity to recover from over-hunting or extirpation campaigns was relatively limited. Because of that, there was a need for benign human intervention in the form of translocation or augmentation of carnivores. The second was the aesthetic and cultural value of large predators, as it would restore the subjective, emotional essence of 'the wild'or wilderness. They argued that wilderness could hardly be wild in the absence of these large carnivores, that nature would seem 'somehow incomplete, truncated, overly tame. Human opportunities to attain humility are reduced'. There was, however, further goal set alongside rewilding in most regional reserve design efforts, of redressing the major wounds or ecological insults caused by abusive land uses of the past, a notion they said was easily traced to Aldo Leopold and other early ecologists. Amongst the list of these wounds to wildlands were the extirpation of large predators; overgrazing and destruction of riparian habitats; introduction of exotic species; draining or pollution of wetlands; and habitat changes stemming from decades of fire suppression

A synthesis of the meaning of rewilding

The objectives identified in the conservation of wild nature that became axiomatic with rewilding during the lifetime of *Wild Earth* from 1991 to 2004 were (Fisher, 2020):

- Habitat fragmentation and island biogeography as a justification for core wild areas, wildlife movement linkages, and compatible-use lands: mapping, public lands, roadless areas, connectivity
- Ecological and evolutionary processes
- Natural disturbance regimes
- Healing ecological wounds
- Scientific justification for rewilding
- Highly, strongly, or ecologically interactive species
- Ecologically effective populations
- Carnivores
- Focal species planning
- Food chain, food web, trophic level and trophic cascades including the role of carnivores

The explanations of rewilding and its application in developing Wildland Network Designs in the four Wildlands Network Visions between 2000 and 2003—Sky Islands Wildlands Network (Foreman et al., 2000abc), Maine Wildlands Network Vision (Long et al., 2002), Southern Rockies Wildlands Network Vision (Miller et al., 2003), and New Mexico Highlands Wildlands Network Vision (Foreman et al., 2003)—were indicative of the way that the term had evolved and taken on meaning over the years, since they exhibited an updated synthesis of the foregoing axioms. Rewilding was commonly framed as the scientific approach to nature restoration and conservation that sought to heal ecological wounds. The emphasis was on a map-based spatial approach grounded in focal species planning based on species distribution to design a connected wildland reserve system. It was a Wildlands Network Design that was comprised of large, strictly protected, core wild areas on predominantly public lands that encompassed intact food webs; compatible-use lands around the cores; and functional connectivity across the wider landscape by way of wildlife movement linkages. The aim of the design was to maintain native species distributions, their natural range of variation and natural patterns of abundance. The scale and connections of the wildland reserve system would ensure the operation of natural processes, including the vital role of keystone species, especially highly or strongly interactive species, such as large carnivores at ecologically effective populations, in the maintenance of ecological and evolutionary processes. Amongst these processes would be gene flow and exchange, disturbance regimes, hydrological processes, nutrient cycles, and biotic interactions that included predation.

Foreman wrote that this ecological renaissance in conservation had come about largely because of new research and theory in several branches of biology (Foreman, 1998, 2004). In a retrospective synthesis of rewilding in his book, he saw that six interrelated lines of scientific inquiry had led to the sort of Wildlands Networks that were proposed by The Wildlands Project and its partners: namely extinction dynamics, island biogeography, metapopulation theory, natural disturbance ecology, large carnivore ecology, and landscape-scale ecological restoration. Detailed explanations were given on each inquiry, and which mirror the objectives associated with rewilding that were identified in *Wild Earth* (see above).

Foreman also wrote about the culture of enablement within which rewilding operated when he explained that The Wildlands Project and *Wild Earth* magazine had worked to bring together citizen conservationists and conservation biologists to craft an evolved idea of conservation, and to apply science to the design and stewardship of protected areas (Foreman, 2004). As well as working directly with many regional conservation groupings, The Wildlands Project had acted as a source of information for how to carry out reserve network designs, such as an early article by Foreman in the first Special Issue of *Wild Earth* that was a primer on how to start designing a regional recovery plan for wilderness (Foreman, 1992). A later issue provided a guide for the Wildlands Reserve Design Process by Johns and Soulé based on an assessment of work underway in some regions, as well as on extensive discussion with regional groups throughout the American continent (Foreman, 1995; Johns & Soulé, 1995; Mondt, 1995). A Reserve Design Framework Package would accompany the guide [see p. 35 in (Johns & Soulé, 1995)]. Soulé would later have an article in *Wild Earth* about the emerging theme of The Wildlands Project vision of reaching a healthier balance between Nature and human society in which he asserted that it was necessary to cultivate a sense of participation and ownership in Nature protection through personal involvement in the development of regional wildlands networks—it would be a nurturing of networks of people to nurture networks of wildlands (Soulé, 1999). Following on from this, a substantial part of the third Special Issue of *Wild Earth* was given over to articles describing the process of design and implementation of the Sky Islands Wildlands Network Conservation Plan (Foreman et al., 2000abc). Noss would later publish a checklist for Wildlands Network Designs (Noss, 2003).

It is interesting to note that Mary Ellen Hannibal, writing about connectivity between the wild spaces along the Rocky Mountains in her book *The Spine of the Continent*, likened the Wildlands Network Design workshop approach of The Wildlands Project to 'collective impact' (Hannibal, 2013). This has been defined elsewhere as a collaborative approach using a structured process that leads to a common agenda, shared measurement, continuous communication, and mutually reinforcing activities among all participants (Kania & Kramer, 2011). This collective impact approach infused the evolution of rewilding within the Wildlands Project and the Wildlands Network Designs. It was a process that built respect for wild nature, a citizen engagement activity that is sorely needed across the whole spectrum of nature conservation. This uniqueness comes down to modus operandi, to the wildlands network design planning meetings where everyone worked together, each contributing their talent and knowledge. This participatory dimension to rewilding is as important as its scientific underpinnings.

Rewilding as originally conceived is of its time and place, a factor that could have been delimiting for its reach into the 21st century. Its subsequent transferability elsewhere needed identification of comparable opportunities on public lands, or land tenures where exploitation is removed, rather than a compromising of its critical mission. These opportunities exist in Europe where there are established national systems of strictly protected areas in public ownership, and with areas that are without exploitation (Fisher et al., 2010; Carver, 2019). However, the boundaries of rewilding have undergone constant stretching and redrawing in Europe as practitioners and stakeholders attempt to dictate what rewilding means in a series of compromises that has resulted in its domestication (Thomas, 2021). Foreman was aware of the dangers of uncritically redrawing such boundaries. He was sensitive to the censure that the emerging approach of ecological restoration and protection through rewilding should not disregard traditional conservation, arguing that good lessons from the past should not be ignored. In writing about the history of protected areas in North America, he pointed out that 'what some may think is new is, in fact, based on ideas and strategies from long ago' (Foreman, 2004: 146). He cautioned that knowing the history of protected areas was important for today's conservation biologists and citizen conservationists alike—'If you don't know whose shoulders you're standing on, you are standing in a void.' It could be said that this also applies to contemporary discourse on rewilding since participants blindly stand on the shoulders of giants through their unawareness of its origins, its scientific basis in conservation biology, its ecocentric vision, and its delivery within the public domain, so that by comparison their words and actions lack any ecological coherence, rooted as they are in a continuing human dominance of wild nature.

References

Balser, D.B. (1997). The impact of environmental factors on factionalism and schism in social movement organizations. *Social Forces*, 76(1), 199–228.

Carver, S. (2019) Rewilding through land abandonment. In Pettorelli, N., Durant, S.M., du Toit, J.T. (eds), *Rewilding*. Cambridge University Press. 99–122.

Corlett, R.T. (2016). Restoration, reintroduction, and rewilding in a changing world. *Trends in Ecology & Evolution*, 31(6), 453–62.

Cramer, P.F. (1998). *Deep Environmental Politics: The Role of Radical Environmentalism in Crafting American Environmental Policy*. Greenwood Publishing Group.

Davis, J. (1991) A minority view. *Wild Earth* 1(4) (Winter, 1991/92) 5–6.

ESP (ND) *Wild Earth, Earth First! Movement Writings*, Multi-Media Library Collections, Environment & Society Portal, Rachel Carson Center for Environment and Society. www.environmentandsoci ety.org/mml/collection/11571?tid=16422&items_per_page=15&order=field_date_partial_publicat ion&sort=asc

Fisher, M., Carver, S., Kun, Z., McMorran, R., Arrell, K., & Mitchell, G. (2010) *Review of status and conservation of wild land in Europe*. Report: The Wildland Research Institute, University of Leeds, UK. Project commissioned by the Scottish Government.

Fisher, M. (2020) Natural Science and Spatial Approach of Rewilding Evolution in meaning of rewilding in *Wild Earth* and The Wildlands Project. *Self-willed Land* March 2020 www.self-willed-land.org.uk/ rep_res/REWILDING_WILDEARTH_WILDLANDS_PROJECT.pdf

Foote, J. (1990) Trying to take back the planet. *Newsweek* 115(6): 24.

Foreman, D. (1987) Whither Earth First!? *Earth First! Journal* 8(8): 21–2.

Foreman, D. (1992) Developing a regional wilderness program. *Wild Earth* Special Issue, 26–9.

Foreman, D. (1995) Around the campfire. *Wild Earth* 5(4) (Winter1995/1996): 3–4.

Foreman, D. (1998) The Wildlands Project and the rewilding of North America, *Denver University Law Review* 76(2): 535–53.

Foreman, D. (2003) The Rewilding Institute, around the campfire with Dave Foreman. *Wild Earth* 13(4) (Winter 2003/04): 2–3.

Foreman, D. (2004) Rewilding North America: a vision for conservation in the 21st century. Island Press

Foreman, D. and Morton, N. (1990) Good luck, darlin'. It's been great, *Earth First!* 10(8) (September): 5.

Foreman, D., Seidman, M., Howard, B., Humphrey, J., Dugelby, B., and Holdsworth, A. (2000a) The Sky Islands Wildlands Network: diverse, beautiful, wild- and globally important. *Wild Earth* 10(1) (Special issue, Spring 2000): 11–16.

Foreman, D., Dugelby, B., Humphrey, J., Howard, B., and Holdsworth, A. (2000b) The elements of a Wildlands Network Conservation Plan: an example from the Sky Islands. *Wild Earth* 10(1) (Special issue, Spring 2000): 17–30.

Foreman, D., List, R., Dugelby, B., Humphrey, J., Howard, B., and Holdsworth, A. (2000c) Healing the wounds: an example from the Sky Islands. *Wild Earth* 10(1) (Special issue, Spring 2000): 31–42.

Foreman, D., Daly, K., Noss, R., Clark, M., Menke, K., Parsons, D.R., and Howard, R. (2003) New Mexico Highlands Wildlands Network Vision: connecting the Sky Islands to the Southern Rockies. Wildlands Project and New Mexico Wilderness Alliance. May 2003. https://wildlandsnetwork.sha repoint.com//_layouts/15/download.aspx?SourceUrl=%2FShared%20Documents%2FCommuni cations%2FFor%20Communications%20Team%2FWebsite%2FResources%2FNM%2DHighla nds%2DWND%2Epdf

Hannibal, M.E. (2013) *Spine of the Continent: The Race to Save America's Last, Best Wilderness*. Rowman & Littlefield.

Hayward, M.W., Scanlon, R.J., Callen, A., Howell, L.G., Klop-Toker, K.L., Di Blanco, Y., Balkenhol, N., Bugir, C.K., Campbell, L., Caravaggi, A., and Chalmers, A.C. (2019) Reintroducing rewilding to restoration—rejecting the search for novelty. *Biological Conservation*, 233: 255–9.

Johns, D. (1991) North American Wilderness Recovery Strategy. *Wild Earth*, 1(4) (Winter, 1991/92): 7–8.

Johns, D. (1993) Wildlands Project Update, *Wild Earth*, 3(3) (Fall, 1993): 4.

Johns, D. and Soulé, M. (1995) Getting from here to there: an outline of the Wildlands Reserve Design Process. *Wild Earth* 5(4) (Winter, 1995/1996): 32–6.

Jørgensen, D. (2015). Rethinking rewilding. *Geoforum*, 65: 482–8.

Kania, J. and Kramer, M. (2011) Collective impact. *Stanford Social Innovation Review* (Winter, 2011), 9(1): 36–41.

Long, R., MacKay, P., Reining, C., Dugelby, B., and Daly, K. (2002) Maine Wildlands Network Vision: A Scientific Approach to Conservation Planning in Maine. Wildlands Project June, 2002. https://static1.squarespace.com/static/60b7e4e41506593f7f926fe7/t/60db3ee6185db8375b41b10c/1624981256618/Maine-WND.pdf

Meine, C. (2010) Conservation biology: past and present. In Sodhi, N.S. and Ehrlich, P.R. (eds), *Conservation Biology for All*. Oxford University Press.

Mann, C. and Plummer, M.L. (1993). The high cost of biodiversity. *Science*, 260(5116), 1868–72.

Meine, C., Soulé, M., and Noss, R.F. (2006) 'A mission-driven discipline': the growth of conservation biology. *Conservation Biology*, 20(3): 631–51.

Miller, B., Foreman, D., Fink, M., Shinneman, D., Smith, J., DeMarco, M., Soulé, M., and Howard, R. (2003). Southern Rockies wildlands network vision: A science-based approach to rewilding the southern Rockies. Southern Rockies Ecosystem Project and Wildlands Project, September 2003. https://wildlandsnetwork.sharepoint.com//_layouts/15/download.aspx?SourceUrl=%2FShared%20Documents%2FCommunications%2FFor%20Communications%20Team%2FWebsite%2FResources%2FS%2E%2DRockies%2DWND%2Epdf

Mondt, R. (1995) Real work and wild vision—highlights of Wildlands Network Design. *Wild Earth*, 5(4): 68–70.

Nogués-Bravo, D., Simberloff, D., Rahbek, C., and Sanders, N.J. (2016) Rewilding is the new Pandora's box in conservation. *Current Biology*, 26(3): R87–R91.

Noss, R.F. (1992) The Wildlands Project Land Conservation Strategy. *Wild Earth* Special Issue, 10–25.

Noss, R.F. (1995) Science Grounding Strategy—conservation biology in the wildlands work. *Wild Earth*, 5(4) (Winter, 1995/1996): 17–19.

Noss, R.F. (1999). Is there a special conservation biology? *Ecography*, 22(2), 113–22.

Noss, R.F. (2003) A checklist for wildlands network designs. *Conservation Biology*, 17(5), 1270–5.

Pedersen, P.B.M., Ejrnæs, R., Sandel, B., and Svenning, J.C. (2019) Trophic Rewilding Advancement in Anthropogenically Impacted Landscapes (TRAAIL): A framework to link conventional conservation management and rewilding. *Ambio*,1–14.

Perino, A., Pereira, H.M., Navarro, L.M., Fernández, N., Bullock, J.M., Ceauşu, S., Cortés-Avizanda, A., van Klink, R., Kuemmerle, T., Lomba, A., and Pe'er, G. (2019) Rewilding complex ecosystems. *Science*, 364(6438). doi: 10.1126/science.aav5570

Pettorelli, N., Barlow, J., Stephens, P.A., Durant, S.M., Connor, B., Schulte to Bühne, H., Sandom, C.J., Wentworth, J., and du Toit, J.T (2018) Making rewilding fit for policy. *Journal of Applied Ecology* 55(3), pp. 1114–25.

Sessions, G. (1992) Radical environmentalism in the 90s. *Wild Earth*, 2(3) (Fall, 1992): 64–7.

Smalley, A.L. (2017). *Wild by Nature: North American Animals Confront Colonization*. John Hopkins University Press.

Soulé, M. E. (1985). What is conservation biology?BioScience, 35(11): 727–34.

Soulé, M.E. (ed.) (1986) *Conservation Biology: The Science of Scarcity and Diversity*. Sinauer.

Soulé, M.E. (1992) A vision for the meantime. *Wild Earth* Special Issue, 7–8, 25.

Soulé, M.E. (1999) An unflinching vision: Networks of people for networks of wildlands. *Wild Earth*, 9(4) (Winter, 1999/2000): 38–46.

Soulé, M. and Noss, R. (1998) Rewilding and biodiversity: complementary goals for continental conservation. *Wild Earth*, 8(3) (Fall, 1998): 18–28.

Soulé, M.E., and Wilcox, B.A. (eds) (1980) *Conservation Biology: An Evolutionary-Ecological Perspective*. Sunderland, MA: Sinauer Associates.

Svenning, J.C., Pedersen, P.B., Donlan, C.J., Ejrnæs, R., Faurby, S., Galetti, M., Hansen, D.M., Sandel, B., Sandom, C.J., Terborgh, J.W., and Vera, F.W. (2016) Science for a wilder Anthropocene: Synthesis and future directions for trophic rewilding research. *Proceedings of the National Academy of Sciences*, 113(4), pp. 898–906.

Taylor, B. (1991) The religion and politics of Earth First! *The Ecologist*, 21(6): 258–66.

Taylor, B. (2008) The Tributaries of Radical Environmentalism. *Journal for the Study of Radicalism*, 2(1): 27–61

Thomas, V. (2021). Domesticating rewilding: interpreting rewilding in England's green and pleasant land. *Environmental Values*, 31(5): 515–532.

Turner, J.M. (2012) *The Promise of Wilderness: American Environmental Politics since, 1964*. University of Washington Press.

Vera, F. (2000) *Grazing Ecology and Forest History*. Oxford: CABI Publishing.

WE (1991) Statement of purpose and strategy. *Wild Earth*, 1(1) (Spring, 1991).

WN (ND) Wildlands Network Designs Resources, Wildlands Network. https://wildlandsnetwork.org/resources/category/Wildlands+Network+Design

WorldCat (ND) Wild Earth Association, Cenozoic Society., & Wildlands Project. (1991). Wild Earth. Canton, NY: Wild Earth Association. WorldCat, Online Computer Library Center. www.worldcat.org/title/wild-earth/oclc/23077499

3

THE EMERGENCE OF REWILDING IN EUROPE

Alexandra Locquet and Steve Carver

Introduction

Despite the origins of the rewilding movement being in North America (see Chapter 3), it may be argued that the process of rewilding has a much longer history without being called as much. Wherever humans abandon land after modifying it for their own ends, nature is released and allowed to take its course. If we consider rewilding to include the process of nature reclaiming those lands abandoned by humans—for whatever reason—then rewilding has been going on for thousands of years. The history of humanity is littered with examples of civilisations that flourished and then died out, leaving the land for nature to reclaim.

There is increasing evidence to suggest that even the Amazon rainforest contains more human modification than previously realised. Archaeological investigations have revealed evidence of modified soils, regular settlements, and routeways through the jungle which lends some credence to early reports from European explorers of heavily populated and farmed areas that have since returned to rainforest (Stenborg et al., 2018; Coomes et al., 2021). The theory here is that the first European visitors brought with them diseases that virtually wiped out the indigenous people who had no natural resistance, such that when European explorers returned several decades later, the land had returned to forest giving the impression of a jungle wilderness (Pearce, 2015). Other similar examples can be found elsewhere including Easter Island and the Mayans as described by Jared Diamond in his book *Collapse* (2005).

In Europe, land abandonment and subsequent return of nature has long been a recurring and ongoing phenomenon driven by economics, land degradation, and other changes (Carver, 2019). Examples include the Portuguese island of Madeira and Norway. In the case of Madeira, agriculture was focused initially on grain and later sugar cane and wine for export together with fruit and vegetables. Production on the island is limited by extremes of topography and need for irrigation water, and a complex landscape of terraces and irrigation canals (known locally as 'levadas') was constructed to make agriculture possible. Many of these areas have since been abandoned in favour of tourism, leading to the recovery of native vegetation on former terraced fields and gardens. This is an example of circumstantial, medium scale passive rewilding following farmland abandonment. In southwest Norway out migration of rural population to the Americas during the mid to late 19th century and early 20th century has led to many sheep farms being abandoned. These have since reforested because of reductions in grazing pressure

DOI: 10.4324/9781003097822-4

where there are no native herbivores to supress tree regrowth. This is a good example of human out migration leading to land abandonment and subsequent 'passive rewilding' of native forest.

More recently, rewilding has gained popularity in Europe, influenced by the emergence of the rewilding movement in North America. Here we examine the history of the development of rewilding in Europe and show how, in crossing the Atlantic, rewilding has transformed into a somewhat different concept in its application to European landscapes. This is being driven by differences in geography, politics, and culture, leading to a more hands-on approach to rewilding in Europe. This 'active rewilding' retains human decision-making in a more central role deter- mining project areas, species and habitats as well as desired outcomes in terms of landscape, biodiversity, ecosystem trajectories, and benefits to humans. The latter may include delivery of ecosystem goods and services but has also been extended to include fiscal and economic benefits from government support and inward investment from ecotourism and conservation.

Early influences

As described in Chapter 3, the word 'rewilding' was first coined in North America and championed by, among others, Dave Forman, Michael Soulé, and Reed Noss. A key driver of rewilding strategies developed in the United States and Canada appears to be the recovery of wilderness (Soulé & Noss, 1998), with an emphasis on the ecological feasibility of predators as keystone species. As Locquet and Héritier (2020) have shown, issues linked to wilderness and the associated concepts have been disseminated throughout the world via the 'ecologisation' and internationalisation of the idea of wild.[1]

In the European context, rewilding as a concept emerged first in some Northern European countries (Bastmeijer, 2016). In the 2000s, rewilding was promoted in debates and reflections in Europe through the work of organisations such as the NGO Wild Europe Initiative (created in 2005), which promotes and encourages strategies for the protection and restoration of wil- derness areas in Europe (Wild Europe, 2010). The mobilisation of this NGO, and of around 100 organisations with various interests, including wildlife protection, environment, tourism, and government, contributed to the promotion of the concept to the European Parliament in 2008 (Locquet & Héritier, 2020), until the ultimate signing of the Wilderness Resolution (Kun, 2013). Following this phenomenon, the idea of rewilding spread throughout Europe in response to the realisation that current conservation practice and protection arrangements are insufficient to achieve conservation objectives (Lorimer et al., 2015).

It is important to acknowledge the influence of the Dutch concept of 'Natuurontwikkeling' (Natural Development) and other pre-rewilding influences in the move towards rewilding in Europe. While not called 'rewilding' until later, conservationists principally in the UK and Netherlands began in the 1990s to question more traditional conservation practices which sought to maintain certain ecological conditions in statis through human interference and pro- motion and/or control of selected natural processes. This led to changes in conservation prac- tice and rise in popularity of new theories [e.g. Frans Vera's (2000) wood-pasture hypothesis] and the development of organisations which later adopted the term rewilding. This strand of rewilding history has been very influential in the UK and The Netherlands, led to the foun- dation of the organisation Rewilding Europe and an approach to rewilding that focuses more on the role of large herbivores in enclosed (fenced) areas and realising benefits to humans such as ecotourism initiatives supporting local economies. This is very different to the 'original' North American concept and often stops short of end goals of recreating wilderness ecosystems in favour of a more anthropocentric model based around the idea of 'kept wild' (Jepson et al., 2018). We return to this and these ideas later in the chapter.

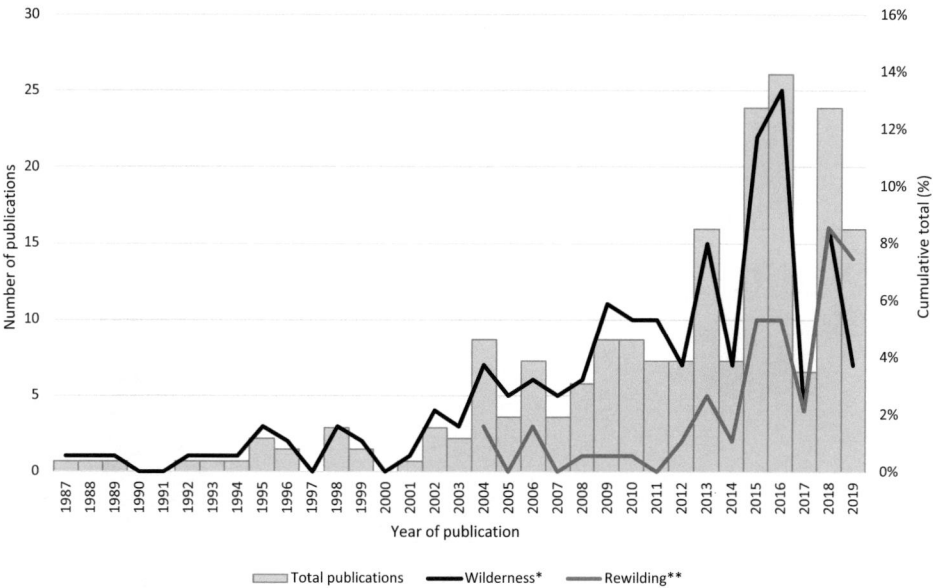

Figure 3.1 Evolution of article publications according to the terms used on a European scale.
* all publications on wilderness, on a European scale (n =187)
** all publications on rewilding, on a European scale (n = 72)
Source: Scopus 2019 (adapted from Locquet and Héritier (2020), Locquet (2021)).

The growth of interest in the rewilding concept at a global scale can be observed through the analysis of the evolution of the number of scientific publications on rewilding. Lorimer et al. (2015), have shown that articles and papers on rewilding around the world have not stopped growing since the 2000s. The analysis of the scientific literature carried out by Locquet and Héritier (2020)[2] also shows that most publications relating specifically to rewilding in Europe emerged from 2005 onwards (Figure 3.1). The authors observe a net increase in scientific production relating to rewilding over the period 2010–2019 (Figure 3.2), which corresponds to the development of the handling of rewilding issues by the European organisation Rewilding Europe (Locquet, 2021).

Emerging European rewilding groups

On a global scale, there are several variations of rewilding (Jørgensen, 2015), ranging from passive management (Norgués-Bravo et al., 2016), which favours the return of natural processes without fixed objectives, through to active interventions including species reintroduction practices, sometimes utilising non-native species (Fernandez et al., 2017).

Rewilding raises questions about the reference states used in current conservation practices (Lorimer et al., 2015). On the one hand, some rewilding projects consider ancient (Pleistocene) reference states by arguing the fundamental role of megafauna on ecosystems (Du Toit, 2019). On the other hand, some stakeholders consider that rewilding is a way of 'looking to the future[3] rather than relying on the past'[4] (Locquet, 2021). Taking past ecosystems as a reference point has its limitations, as climatic and ecological conditions have changed (both because of human activities and from natural developments). This is why some stakeholders prefer to talk about

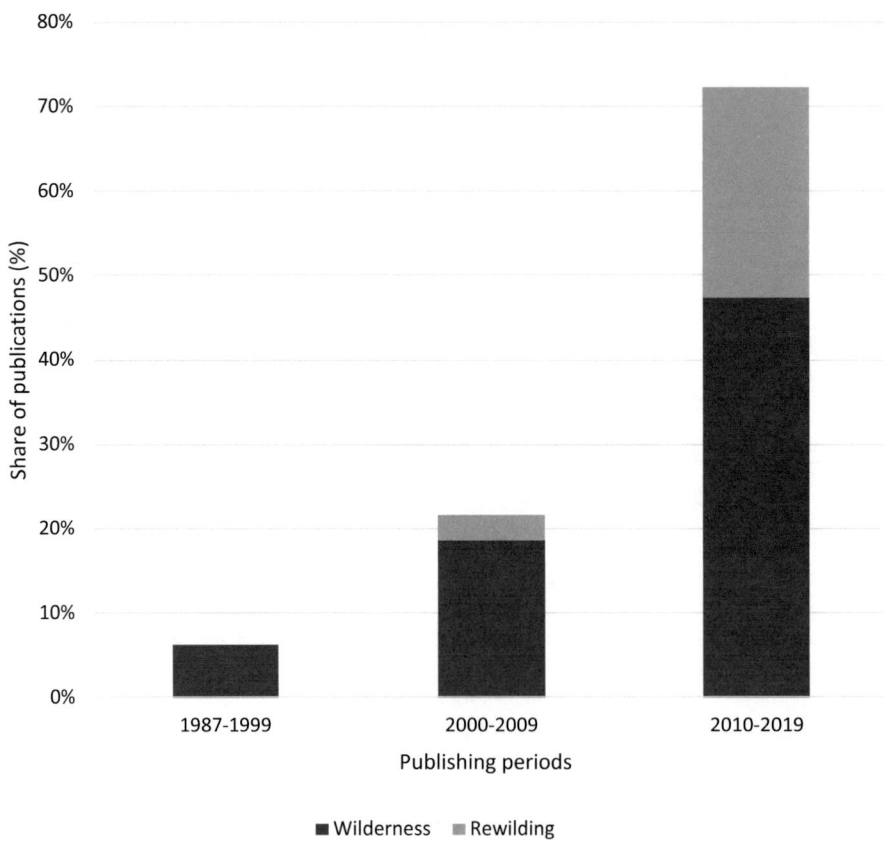

Figure 3.2   Evolution of the distribution of publications between 1981 and 2019.

Number of publications per period:
- 1981–1999: n = 16
- 2000–2009: n = 56
- 2010–2019: n = 187

Source: Scopus, 2019 (adapted from Locquet and Héritier (2020), Locquet (2021)).

novel ecosystems, which refer to environments that have been impacted by human activities, but which are no longer subject to active management (Marris, 2011). As explained by Evers et al. (2018)), this nevertheless implies identifying what it is about these ecosystems that are new in relation to existing ecosystems (Locquet and Héritier, 2020). Therefore, as Locquet (2021), has shown, some of the stakeholders working for wilderness in Europe prefer not to use precise reference states.

Nevertheless, it is possible to observe the influence of certain models on the conceptions and representations of rewilding in Europe. The pan-European organisation Rewilding Europe principally promotes the return of large herbivores to ensure the return of natural dynamics and is recognised as a pioneer and model for rewilding in Europe (Ward, 2019). The dissemination of this vision is ensured by a strategy based on professional marketing and the constitution of a network of pilot sites throughout Europe promoting the idea of linked economic development of rural territories through rewilding projects (Locquet, 2021). This phenomenon is notably reinforced by the development of a 'bank' (Europe Wildlife Bank, ND) for the exchange of animals to constitute the herds in the various sites of the organisation throughout Europe (Rewilding Europe, 2020).

As Locquet (2021), has shown, in the European context the majority of trophic rewilding projects developed are based on the return of large herbivores with the aim of recovering natural grazing and disturbance processes that have largely disappeared since the Pleistocene. These models are based on the principles of ecological management of environments through grazing of large herbivores (Lecomte and Le Neveu, 1986). The underlying idea here is that the European continent would naturally be composed of a mosaic of closed and open environments, the latter being maintained by natural disturbance of grazing pressures exerted by large herbivores (Locquet, 2021).

There are also other initiatives which, although not directly claiming to be rewilding strategies (notably for reasons of linguistic and conceptual transposition of the term), are based on the reintroduction of large herbivores. In France, several projects have been developed around the return of grazing pressures like those induced by megafauna. These actions are based either on hardy breeds such as Highland cattle or on the return of so-called wild species: European bison, moose, horses (Prjevalski horses) or ponies (Konik) (Locquet, 2021). Other initiatives focus on recovering the phenotypic and behavioural characteristics of primitive horses such as the Tarpan (Arthen) (Locquet, 2021).

A different path … how rewilding in Europe is different to its North American roots

Nature conservation in Europe has traditionally focused on protecting remnant natural ecosystems as reserves for wildlife and their habitats. Development elsewhere for human land use (i.e. settlement, agriculture, forestry, etc.) has led to a network of core nature areas embedded in a matrix of human-dominated ecosystems. While Europe's Natura 2000 network is well-developed and among the most extensive in the world with over 27,000 sites and over three-quarters of a million square kilometres of land (and a third of a million square kilometres of marine areas) protected, many of the sites it contains are under some form of human land use. Connectivity remains a problem and many nature sites are fenced and isolated (Ceauşu et al., 2015).

Despite the return of the wolf (*Canis lupus*), Eurasian lynx (*Lynx lynx*), Iberian lynx (*Lynx Pardinus*) and, in some instances, the European brown bear (*Ursus arctos*) in small numbers to their former ranges across mainland Europe, they have not been universally welcomed and where numbers are increasing, political pressure for population control (often by lethal means) is growing. In the Scandinavian countries pressure from a strong farming and hunting lobby within the country have led to governments mandating a cull of their wolves to keep breeding pairs to low numbers to protect livestock and game (Trouwborst et al., 2017). Meanwhile, in the Pyrenees increasing brown bear populations, while welcomed by conservationists and tourism groups, have met with resistance from farmers concerned about attacks on livestock (Herrero et al., 2021). Thus, while the return of large carnivores to European countries has been aided, at least in part, by improvements in wider nature protection through policy mechanisms such as Natura 2000, the Bern Convention, and Emerald Network, it has not been without conflict with public opinion mixed and reactions from traditional land uses largely hostile (Chapron et al., 2014; Kutal et al., 2018; Franchini et al., 2021).

Much of the work on rewilding in Europe has been influenced by the Dutch school of thought focusing on the role of large herbivores in so-called 'nature development' (in Dutch *natuurontwikkeling*). Here the emphasis is on the use of large herbivores as vectors of ecological disturbance through trampling, grazing, browsing, etc. to create an open mosaic of vegetation in rewilded landscapes. Much of this school of thought stems from the work of Frans Vera and the theories he set out in his PhD thesis and subsequent book *Grazing Ecology and Forest History* (Vera, 2000). European landscapes are, in the main, devoid of any significant numbers of large carnivores, such that the role of predation and disturbance and behavioural traits of herbivores created by their presence (i.e. 'landscapes of fear') is largely absent. The theories of Frans Vera on wood-pasture landscapes (known as 'The Vera Hypothesis') wherein semi-open pastures dominated the pre-human European landscape driven by large herbivore grazing and disturbance regimes, has therefore gained popularity among conservation practitioners in Europe where the regulatory role of large carnivores cannot be realised and resulting landscapes are in any case not too dissimilar to those created by traditional agricultural practices.

Even though carnivores are often not involved, rewilding projects can be a source of conflict as the case of the Oostvardersplassen experiment shows (Kopnina et al., 2019). Here, tensions relating to ethical issues have emerged following the development of the project. The project planned to intervene very little on the reserve, which implied not feeding the animals. But the closed nature of the site (it is enclosed by stock-proof fencing) and the absence of predators has meant that the regulation of the herbivore population was wholly dependent on the amount of food resources available and on disease (Carver and Convery, 2021). In 2005 and 2018, particularly harsh winters led to the death of many animals from starvation resulting in widespread public concern and an acrimonious debate in the conservation community (Theunissen, 2019; Barkham, 2018). As a result, preventive culling of individuals that might not survive the winters, as well as winter feeding of the animals, has been implemented after government intervention (Theunissen, 2019; ICMO2, 2010).

The concept of rewilding can thus be seen to have several limitations relevant to its European setting when compared to North America. Firstly, in its very understanding, just as the concept of 'wilderness' is neither translatable nor transposable across many languages and cultures (Locquet, 2021), the term 'rewilding' is also a source of some tension and debate. It is associated with a complex and constraining concept (Tree, 2018; Lorimer, 2015), and as such, some stakeholders refuse to use this term, since it is considered too controversial (Locquet, 2021). The term rewilding is often associated, rightly or wrongly, with the reintroduction of predators or with a change in local uses induced by exclusion or abandonment of agricultural land.

The notion of rewilding, and more broadly the reintroduction projects of large herbivores, are confronted with legal limitations regarding the status of the animals. Indeed, most of the animals reintroduced in rewilding projects are domestic or semi-domestic animals[5] (Locquet, 2021) except for the so-called 'non-domestic' European Bison and are usually constrained by fences. As these animals are 'captive' they are not necessarily perceived as being truly wild, but as 'kept wild'. In Europe, only the herbivores in the Oostvaardersplassen reserve have the status of wild animals, although their movements are still limited by the surrounding fence (Locquet, 2021; Carver & Convery, 2021). While limiting the mobility of these groups of animals may have consequences for population dynamics (reproduction, genetic mixing, etc.), this serves to further highlight the lack of ecological connectivity between the various projects involving such animals and thus questions the relevance and possibility of large-scale programmes in heavily modified landscapes such as that found in Europe due to the severe constraints created by human land use pressures. Thus, the captivity of the 'kept wild' animals used in many rewilding projects raises significant ecological and ethical issues (Locquet, 2021).

Funding of rewilding project in Europe can also be an issue. Here the rewilding projects that have been developed are mainly led by the stakeholders involved or are the result of private or

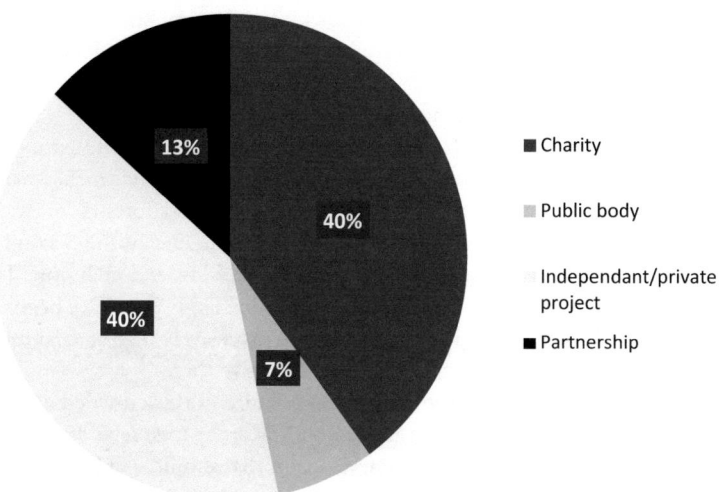

Figure 3.3 Distribution of the stakeholders interviewed according to the type of structure to which they belong (Locquet, 2021).

(This figure is based on the analysis of the nature of the organisations (studied in the framework of Locquet's thesis, 2021) working in favour of the return of large herbivores or rewilding. This information was obtained by means of site and bibliographic research as well as interviews. The data obtained was divided into organisational categories and then counted in order to analyse the percentage distribution in relation to all the types of organisations studied.)

independent initiatives (see Figure 3.3). Indeed, as Locquet (2021) has shown, issues relating to the protection of the wilderness or rewilding in Europe are rarely dealt with exclusively by public bodies but are rather the preserve of NGOs. These organisations mainly rely on donations or private funding from the economic activities developed by the stakeholders as part of the implementation of their projects (Locquet, 2021) leading to possible conflicts of interest.

Most of the rewilding initiatives developed in Europe are promoted to ensure local economic development. These projects would thus generate income mainly from the tourism that emerges around the rewilding areas, either directly through job creation and wildlife observation opportunities, or indirectly through development of accommodation and catering facilities, etc. (Locquet, 2021). Organisations such as the Knepp Estate (UK) and the Mont d'Azur reserve (France) offer accommodation and catering services as well as 'wildlife safari' activities (Locquet, 2021). In this way, *'the wilderness and the labels promoting the wilderness character of an area are presented as guarantees of quality and attraction for a territory'*[6] (Locquet, 2021), particularly in the context of agricultural decline. This leads us to question whether these initiatives are really rewilding or whether the term has simply been appropriated in Europe and used as a marketing tool and so may be seen as lacking in the ecological/philosophical integrity of its North American roots.

Conclusions: The mainstreaming of rewilding in Europe

While rewilding in Europe has geographical, ecological, philosophical, socio-political, and cultural limitations, it has perhaps gained more support and traction than in North America where it is seen by many as more of a sub-discipline of ecological restoration (Hayward et al., 2019). The fact that much of Europe, with perhaps the exception of northern Scandinavia, has lost much of its wilderness over thousands of years of human history and landscape modification (principally deforestation for agriculture), means that rewilding has found fertile ground within a conservation movement that hitherto has been playing a rear-guard action against rapid development and agricultural intensification.

So-called 'fortress conservation' has worked to prevent the complete extinction of many wilderness-dependent European species, while others have found niches within human dominated landscapes, most notably traditional agricultural and forestry practices. As agriculture and forestry has intensified over the decades since the end of the Second World War, even these are coming under increased pressure from habitat loss and pollution. The result has been that many open and semi-open traditional agricultural landscapes have been incorporated into the Natura 2000 network as a means of protecting extensively managed landscapes and the species and habitats assemblages that have managed to thrive there.

Rewilding provides an alternative view of future nature in this crowded continent. As the world faces the joint crises of climate changes and biodiversity loss, rewilding provides a more positive and proactive approach to nature conservation that should—at least in theory—allow nature both the space and time to slow and even reverse biodiversity declines. This relies heavily on the Three C's model of Cores, Corridors and Carnivores, just as developed in North America, with perhaps the exception that in Europe, the keystone species are largely restricted to large herbivores despite the natural recolonisation of some carnivores back into their former ranges. This concentration on the cores and to some extent the corridors can give the impression of rewilding in Europe being something of a three-legged stool but with, in this case, a shorter third leg (the carnivores) which inevitably gives rise to certain imbalances that must be propped up by human intervention. The classic example here is the need for fencing in rewilding projects, both to keep large 'wild' herbivores in (to prevent damage to surrounding croplands) and to keep livestock out (to prevent overgrazing of recovering natural habitats). As such, this can be viewed as a somewhat watered-down version of the original idea as envisaged by the rewilding pioneers in North America where the focus is on continental scale recovery and connections between extant wilderness that allows freedom of movement of keystone species (including large carnivores) in response to climate change and seasons. In Europe, cores are smaller, less connected, often fenced, and lacking keystone predator species.

Some of this nature recovery vision is being 'sold' on the back of economic models of Natural Capital accounting and ecotourism via some very glossy marketing and carbon credit trading. This involves new forms of finance from the private commercial sector wishing to offset their standard operations by buying in to carbon and biodiversity offsetting schemes. Whilst this can be a new and much needed source of revenue for conservation NGOs it risks becoming a form of 'greenwashing' that shifts the impacts from 'Business as Usual' modes of operation in one place by offsetting in another. This is not helped in some instances through a reliance on digital accounting methods underpinned by energy-thirsty computational models and distributed encrypted digital ledgers such as the Blockchain. Nevertheless, policy focus on Nature-based Solution (NbS) and ecosystem service models that incorporate elements of Natural Capital accounting and human agency into the rewilding portfolio are perhaps essential elements of rewilding's appeal and success in a European setting. Anything that improves the lot of wild nature and provides benefits for both wildlife and humans must be a good thing, provided it doesn't simply displace resource demands elsewhere in the already significant environmental and economic footprint of Europeans.

Notes

1 This process has resulted in the progressive inclusion of wilderness in international debates and documents, and more particularly in the creation of the IUCN category Ib wilderness areas in 1984 (Bastmeijer, 2016).
2 The methodology used to establish this analysis and produce these graphs is explained in the article published by Locquet and Héritier (2020) and in the PhD manuscript of (Locquet 2021).
3 Of all the 63 stakeholders working in favour of wilderness or rewilding in Europe within the framework of Locquet's PhD thesis, only 13% indicated that they relied on data relating to disappeared environments and landscapes. While 33% of the stakeholders interviewed said they did not use a reference state (Locquet, 2021).
4 Translated from French: « *de regarder vers le futur plutôt que de s'appuyer sur le passé* » (Locquet, 2021).
5 Because of their domestic status, these animals must be kept in captivity and under sanitary care (each individual must be identified, vaccinated and treated with an antiparasitic). Sanitary care and treatment involve the manipulation of individuals and leads to the disappearance of certain natural dynamics—such as the regulation of populations by disease, or the disappearance of the microfauna attached to these herbivores (Locquet, 2021).
6 Translated from French: « *Le sauvage et les labels promouvant le caractère sauvage d'un espace sont présentés comme des gages de qualité et d'attraction d'un territoire* » (Locquet, 2021).

References

Barkham, P. 2018. Dutch rewilding experiment sparks backlash as thousands of animals starve. *The Guardian*, 27 April 2018, sect. Environment. www.theguardian.com/environment/2018/apr/27/dutch-rewilding-experiment-backfires-as-thousands-of-animals-starve.

Bastmeijer, K. 2016. *Wilderness Protection in Europe The Role of International European and National Law.* Cambridge University Press.

Carver, S. 2019. Rewilding through land abandonment. In Pettorelli, N., Durant, S.M. and du Toit, J.T. (eds), *Rewilding*, 99–122, Cambridge: Cambridge University Press.

Carver, S. and Convery, I. 2021. Rewilding: time to get down off the fence. *British Wildlife*, 32(4): 246–255.

Ceauşu, S., Hofmann, M., Navarro, L.M., Carver, S., Verburg, P.H., and Pereira, H.M. 2015. Mapping opportunities and challenges for rewilding in Europe. *Conservation Biology*, 29(4): 1017–27.

Chapron, G., Kaczensky, P., Linnell, J.D., Von Arx, M., Huber, D., Andrén, H., López-Bao, J.V., Adamec, M., Álvares, F., Anders, O., and Balčiauskas, L. 2014. Recovery of large carnivores in Europe's modern human-dominated landscapes. *Science*, 346(6216): 1517–19.

Coomes, O.T., Rivas Panduro, S., Abizaid, C., and Takasaki, Y. 2021. Geolocation of unpublished archaeological sites in the Peruvian Amazon. *Scientific Data*, 8(1): 1–8.

Diamond, J. 2005. *Collapse: How Societies Choose to Fail or Succeed.* Viking Press.

Du Toit, J.T. 2019. Pleistocene rewilding: an enlightening thought experiment. In N.Pettorelli, S.M. Durant, and J.T. du Toit (eds), *Rewilding* (pp. 55–72). Cambridge University Press.

Evers, C.R., C.B. Wardropper, B. Branoff, E.F. Granek, S.L. Hirsch, T.E. Link, S. Olivero-Lora, and C. Wilson. 2018. The ecosystem services and biodiversity of novel ecosystems: A literature review. *Global Ecology and Conservation,* 13: e00362.

Fernandez, F.A.S., M.L. Rheingantz, L. Genes, C.F. Kenup, M. Galliez, T. Cezimbra, B. Cid, et al., 2017. Rewilding the Atlantic forest: restoring the fauna and ecological interactions of a protected area. *Perspectives in Ecology and* Conservation, 15(4): 308–14. https://doi.org/10.1016/j.pecon.2017.09.004.

Franchini, M., Corazzin, M., Bovolenta, S., and Filacorda, S. 2021. The return of large carnivores and extensive farming systems: a review of stakeholders' perception at an EU level. *Animals*, 11(6): 1735.

Hayward, M.W., Scanlon, R.J., Callen, A., Howell, L.G., Klop-Toker, K.L., Di Blanco, Y., Balkenhol, N., Bugir, C.K., Campbell, L., Caravaggi, A., and Chalmers, A.C. 2019. Reintroducing rewilding to restoration–Rejecting the search for novelty. *Biological Conservation*, 233: 255–9.

Herrero, J., García-Serrano, A., Reiné, R., Ferrer, V., Azón, R., López-Bao, J.V., and Palomero, G. 2021. Challenges for recovery of large carnivores in humanized countries: attitudes and knowledge of sheep farmers towards brown bear in Western Pyrenees, Spain. *European Journal of Wildlife Research*, 67(6): 1–13.

ICMO2. 2010. Natural processes, animal welfare, moral aspects and management of the Oostvaardersplassen. Report of the second International Commission on Management of the Ootsvaardersplassen.

Jepson, P., Schepers, F., and Helmer, W., 2018. Governing with nature: a European perspective on putting rewilding principles into practice. *Philosophical Transactions of the Royal Society B: Biological Sciences*, 373(1761): 20170434.

Jørgensen, D. 2015. Rethinking rewilding. *Geoforum* 65: 482–8. https://doi.org/10.1016/j.geoforum.2014.11.016.

Kopnina, H., Leadbeater, S., and Cryer, P. 2019. Learning to rewild: Examining the failed case of the Dutch 'New Wilderness' Oostvaardersplassen. *International Journal of Wilderness*, 25(3): 72–89.

Kun, Z. 2013. Preservation of wilderness areas in Europe. *European Journal of Environmental Sciences*, 3(1): 54–6.

Kutal, M., Kovařík, P., Kutalová, L., Bojda, M., and Dušková, M., 2018. Attitudes towards large carnivore species in the West Carpathians: Shifts in public perception and media content after the return of the wolf and the bear. In *Large Carnivore Conservation and Management* (pp. 168–89). Routledge.

Lecomte, T., and C. Le Neveu. 1986. Le marais Vernier: contribution à l'étude et à la gestion d'une zone humide. PhD thesis: Biology of organisms and populations, University of Rouen Normandy.

Locquet, A. 2021. Born to be wild? Représentations du sauvage et stratégies de protection de la wilderness en Europe. Geography PhD thesis, University of Paris 1 Panthéon Sorbonne.

Locquet, A., and S. Héritier. 2020. "Questioning nature and wildness in relation to the establishment of *wilderness areas* in Europe!". *Cybergeo: European Journal of Geography*. Online. Environnement, Nature, Paysage, document 946, mis en ligne le 11 juin 2020, consulté le 18 août 2022. URL : http://journals.openedition.org/cybergeo/34986 ; DOI : https://doi.org/10.4000/cybergeo.34986

Lorimer, J. 2015. *Wildlife in the Anthropocene: Conservation after Nature*. University of Minnesota Press.

Lorimer, J., C. Sandom, P. Jepson, C. Doughty, M. Barua, and K.J. Kirby. 2015. Rewilding: science, practice, and politics. *Annual Review of Environment and Resources*, 40: 39–62.

Marris, E. 2011. *Rambunctious Garden: Saving Nature in a Post-Wild World*. Bloomsbury Press.

Norgués-Bravo, D., D. Simberloff, C. Rahbek, and N. Sanders. 2016. Rewilding is the new Pandora box in conservation. *Current Biology*, 26(3): 7–91.

Pearce, F. 2015. Myth of pristine Amazon rainforest busted as old cities reappear. *New Scientist*. July 2015.

Rewilding Europe. 2020. European Wildlife Bank | Rewilding Europe. 2020. https://rewildingeurope.com/european-wildlife-bank/.

Soulé, M., and R.F. Noss. 1998. Rewilding and biodiversity: complementary goals for continental conservation. *Wild Earth*, 8: 19–28.

Stenborg, P., Schaan, D.P., and Figueiredo, C.G., 2018. Contours of the past: LiDAR data expands the limits of late pre-Columbian human settlement in the Santarém region, lower Amazon. *Journal of Field Archaeology*, 43(1): 44–57.

Theunissen, B. 2019. The oostvaardersplassen fiasco. *ISIS*, 110(2): 341–5. https://doi.org/10.1086/703338.

Tree, I. 2018. *Wilding: The Return of Nature out of British Farm*. Picador.

Trouwborst, A., Fleurke, F.M., and Linnell, J.D. 2017. Norway's wolf policy and the Bern Convention on European Wildlife: avoiding the 'manifestly absurd'. *Journal of International Wildlife Law & Policy*, 20(2): 155–67.

Ward, K. 2019. For wilderness or wildness? Decolonising rewilding. In N. Pettorelli, S.M. Durant, and J.T. du Toit (eds), *Rewilding*. Cambridge University Press.

Wild Europe Initiative. 2010. "Towards a Wilder Europe Developing an Action Agenda for Wilderness and Large Natural Habitat Areas." p. 20.

4
ECOLOGICAL RESTORATION AND REWILDING

Integrating communities of practice to achieve common goals

Cara R. Nelson

An important part of understanding the practice of rewilding is understanding its relationship to ecological restoration. Why is this relationship important? These allied practices have similar aims to reverse ecological degradation and recover biodiversity, but emerged from different land management traditions and scientific subdisciplines, and have historically emphasised different taxonomic groups (wildlife versus plants) and levels of the biological hierarchy (populations versus ecosystems) (Young, 2000). As such, the most effective and efficient path to ecological recovery may occur when concepts are integrating across fields. Despite potential advantages of alignment, rewilding and restoration largely have operated in silos, with little exchange among communities of practice. In part, this is due to lack of clear understanding of the definitions and principles of each field (Loth and Newton, 2018; Anderson et al., 2019). The recent publication of globally endorsed principles for both ecological restoration (Gann et al., 2019) and rewilding (Carver et al., 2021) provides an opportunity for an informed comparison between the two fields. Towards that end, the goal of this chapter is to review the relationship between ecological restoration and rewilding, based on their historical roots and scientific bases, as well as their underlying principles.

Historical roots and scientific underpinnings

Ecological Restoration—Although people have been assisting in the recovery of degraded ecosystems for centuries, the practice of ecological restoration picked up steam in the 1970s as a promising solution for reversing land degradation (Clewell & Aronson 2007). The dominant idea behind ecological restoration was applying vegetation and soil management techniques to repair habitat. For instance, Bradshaw and Chadwick in their 1980 landmark book, *The Restoration of Land: The Ecology and Reclamation of Derelict and Degraded Land*, state:'the task of understanding how to restore derelict land is not an easy one, because it involves understanding soils and plants and how they interact with each other'. This initial emphasis on managing plants and their abiotic environment is understandable, given that ecological restoration efforts were focused at the ecosystem level—and vegetation and soils are generally the defining features of an ecosystem (e.g., 'ponderosa pine forest' or 'short-grass prairie') (Young, 2000). During the

DOI: 10.4324/9781003097822-5

1980s, the practice of ecological restoration and the science of restoration ecology self-identified as fields, with the launch of the journal, *Ecological Restoration* (University of Wisconsin Press), in 1981 and the formation of the Society for Ecological Restoration in 1988. The following year, the Society launched its first conference and in 1993 it established the journal *Restoration Ecology* (Wiley-Blackwell).

To date, the most highly cited definition of ecological restoration is 'the process of assisting in the recovery of an ecosystem that has been degraded, damaged, or destroyed' (SER, 2004; Gann et al., 2019). This definition, however, fails to clearly distinguish ecological restoration from other restorative management activities such as rehabilitation or remediation, in that it does not specify what is meant by recovery. Ecological restoration does, in fact, have specific recovery goals, although ideas about these goals have evolved over time, keeping pace with advances in understanding about ecosystem dynamics. In the early years of the field, the recovery goal often was described as returning ecosystems to an 'historic' state (Clewell and Aronson, 2007). However, the relevance and feasibility of this goal has been called into question given that maintaining an ecosystem in an historic state does not align with the fact that change is an *inherent characteristic* of all ecosystems, including ones that have not been disturbed or degraded (Kimmins, 2004). In fact, ecosystems continually change in response to external factors, such as climate, and internal successional processes. Large-scale changes in vegetation over time can be seen in the pollen record (Millar & Brubaker, 2006), and demonstrate that there is not a stable 'natural' condition associated with any location on the planet. Given this increasing understanding that ecosystem are dynamic over time, it is now well recognised that the goal of ecological restoration should not be to constrain ecosystems within past conditions, but rather to remove degradation and assist the ecosystem in recovering to the state it would be in had degradation not occurred (Gann et al., 2019). Restoration practitioners ideally estimate this state through the development of an appropriate reference model, based on a suitable number of reference sites—sites that are similar to the target site, but that have not experienced (or have experienced minimal) degradation. These concepts are described in detail in *International Principles and Standards for the Practice of Ecological Restoration Second Edition* (Gann et al., 2019), which were developed through formal consultations with hundreds of experts in the science, practice, and policy of ecological restoration from all over the globe (Gann et al., 2019).

At the same time as the creation of principles and standards that detail the tenants of ecological restoration, the word 'restoration' is increasingly being used to refer to a wide range of ecosystem management activities and initiatives, such as Forest Landscape Restoration (Laestadius et al., 2015), that aim to improve ecological integrity and human wellbeing. Although these initiatives use the term 'restoration' in their titles, the meaning of restoration is not synonymous with ecological restoration, leading to confusion about the overall breadth of the field and the types of activities that fall within it. To assist with developing a common lexicon, the Society for Ecological Restoration advanced the concept of the 'Restorative Continuum' (Figure 4.1; Gann et al., 2019): at one end of this continuum, ecosystem restoration includes management activities aimed at reducing societal impacts, such as runoff into urban streams, and mitigating threats such as contaminated soils; at the other end of the continuum is *ecological restoration*, which aims to both remove degradation and recover ecosystems to the condition that they would be in had degradation not occurred, while allowing for environmental change (Gann et al., 2019). Creating a 'large tent' for restorative activities in general, but also clearly defining the practice of ecological restoration, is strategic for maximising options for ecosystem repair. It acknowledges the need for multiple types of restorative activities, while at the same time encouraging the use of best practice ecological restoration where possible to achieve the highest level of recovery.

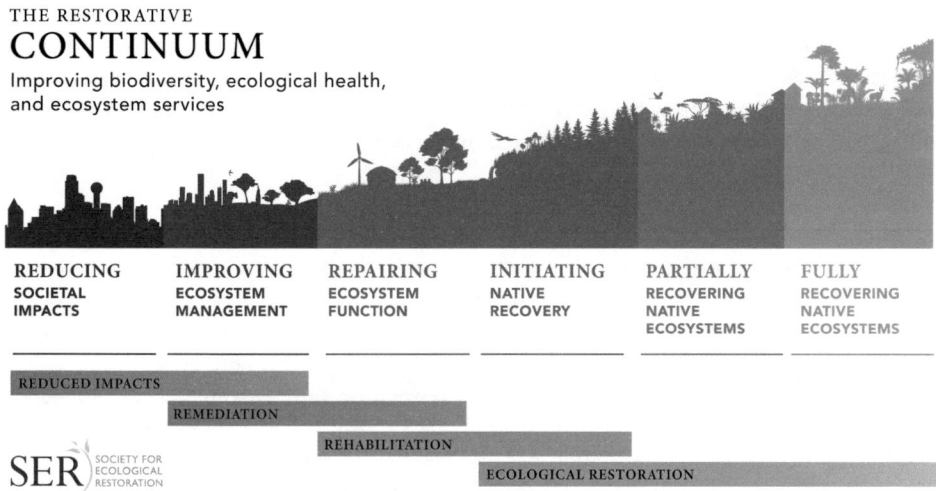

THE RESTORATIVE
CONTINUUM
Improving biodiversity, ecological health,
and ecosystem services

REDUCING	IMPROVING	REPAIRING	INITIATING	PARTIALLY	FULLY
SOCIETAL IMPACTS	ECOSYSTEM MANAGEMENT	ECOSYSTEM FUNCTION	NATIVE RECOVERY	RECOVERING NATIVE ECOSYSTEMS	RECOVERING NATIVE ECOSYSTEMS

REDUCED IMPACTS

REMEDIATION

REHABILITATION

SER SOCIETY FOR ECOLOGICAL RESTORATION

ECOLOGICAL RESTORATION

Figure 4.1 The continuum of restorative ecosystem management. At one end of the continuum are management activities aimed at reducing societal impacts, such as runoff into urban streams, and mitigating threats such as contaminated soils; at the other end is *ecological restoration*, which aims to both remove degradation and recover ecosystems to the condition that they would be in had degradation not occurred, while allowing for environmental change. Reproduced from Gann et al., 2019.

Rewilding—The concept of rewilding developed in the 1980s, during roughly the same time period as ecological restoration. While restoration focused on recovery of degraded and contaminated land, rewilding has its roots in the wilderness recovery movement of North America (Carver et al., 2021) coupled with the science of conservation biology. Conservation biology emerged as a self-identified field in the mid-1980s, with the founding of the Society for Conservation Biology in 1985 and the launch of their journal *Conservation Biology* (Wiley-Blackwell) two years later. What set conservation biology apart from other biological sciences was its integration of concepts from systematics, population biology, ecology, and landscape ecology to specifically develop the knowledge base required for avoiding the extinction crisis (Meine et al., 2006). Although ideas in conservation biology apply broadly to biodiversity, the field developed with a strong taxonomic bias for wildlife (Young, 2000). During the 1980s and 90s, conservation biologists developed methods to quantify species viability and explored relationships between viability and landscape patterns with respect to size and location of core areas and corridors, especially with respect to large carnivores. Rewilders worked to apply these concepts on the ground, for example through the 'Wildlands Project' a large-scale initiative to map cores, corridors, and connectivity (Noss, 1992).

In addition to large landscape conservation strategies, rewilding advanced the idea of reinitiating trophic cascades through the reintroduction of top carnivores (Soulé & Noss, 1998), a concept now referred to as 'trophic rewilding' (Svenning et al., 2016; Garcia-Ruiz et al., 2020). At the same time, other camps in the rewilding movement were advocating for reinitiating trophic cascades by reintroducing modern relatives of megafauna extinct since the Pleistocene (Donlan et al., 2005). This focus on re-establishment of historic processes within the field of rewilding in some senses mirrors the debate about historic fidelity in ecological restoration (Miller & Hobbs, 2016). However, recognition that recreating historic baselines is generally not a feasible, relevant, or desirable goal in modern socio-ecological systems led to movement away from the concept of Pleistocene rewilding. More recently, there has been increasing interest in 'ecological' approaches to rewilding that do not involve species introductions, but rather focus on passive approaches intended to reinstate successional processes, such as abandoning agricultural land or creating areas where hunting is restricted (Pereira & Navarro, 2015). The conceptual basis of ecological approaches to rewilding continue to develop (Perino et al., 2019), while at the same time trophic rewilding remains an important part of rewilding practice.

Relationship between restoration and rewilding

Although rewilding and ecological restoration are clearly allied activities, the relationship between the two has sparked much debate. Authors have variously stated that the foundation of rewilding is ecological restoration (Miller & Hobbs, 2019), that rewilding is an ecological restoration strategy that relies on species reintroductions to promote ecological recovery (Svenning, 2016), and that because ecological restoration fully encompasses rewilding the term rewilding is not needed (Hayward et al., 2019). Yet other authors have argued that rewilding is different from restoration because it relies on passive approaches, while restoration requires active manipulation (Butler, 2021) or that while some forms of ecological restoration may be considered rewilding, restoration that requires ongoing human intervention would not (Svenning, 2020). Authors also have attempted to distinguish the two in terms of their goals, mentioning that restoration looks backwards and aims to recreate past conditions (but see description of ecological restoration above) while rewilding 'has lower fidelity to taxonomic precedent and promotes taxonomic substitutions for extinct native species' to restore lost interactions (Du Toit & Pettorelli, 2019). Clearly some of the confusion rests with whether the type of restoration being considered in the comparison is ecological restoration versus other allied restorative activities (Figure 4.1), as well as whether the type of rewilding being considered is trophic (involving introductions to restore top-down interactions) versus ecological (allowing natural processes to regain dominance) rewilding (Corlett, 2016). But more than that, there is a real lack of clarity about differences and similarities between the two practices, and published opinions often do not reflect ideas and activities within either field.

Until recently, the ability to characterise the relationship between restoration and rewilding has been hampered by lack of broad agreement within each field on their definitions (Anderson et al., 2019), principles and key concepts. With the recent publication of overarching principles for ecological restoration (Gann et al., 2019) and rewilding (Carver et al., 2021)—both of which were based on extensive consultation processes with a large number of experts across the globe—it is now possible to systematically review similarities and differences. Towards that end, I cross-walked the eight principles of ecological restoration with the ten principles of rewilding (Table 4.1). Overall, I found a high degree of commonality among these fields. For instance, the goals for both practices include recovery of ecological processes and development of self-sustaining ecosystems that reflect current conditions and allow for environmental change rather than re-establishment of historical conditions. Both practices also rely on developing a reference model to establish recovery goals, engaging stakeholders, using all types of knowledge, emphasising natural (passive) recovery processes where feasible, and considering landscape-scale processes. Although there are some key differences, like the more in-depth focus within ecological restoration on ecological recovery attributes as well as social goals, this may be due to the fact that the two sets of published principles vary greatly in length: the average length of the description for each rewilding principles is 142 words compared to an average of 544 words for principles of ecological restoration (not including words in boxes, figures, and tables—elements that were not included in the rewilding principles).

Table 4.1 The primary concepts from the 10 rewilding principles (Carver et al., 2021), along with related concepts from the ecological restoration principles (Gann et al., 2019). Text is direct quotations unless in square brackets # refers to specific principle numbers from Carver et al., 2021 (rewilding) or Gann et al., 2019 (restoration); * indicates quotations from the introduction of Gann et al. (2019) rather than from a specific principle

Rewilding		Ecological Restoration	
#	*Primary concepts*	#	*Related concepts*
1	Rewilding utilises wildlife to restore trophic interactions; strongly interacting keystone species that have roles in maintaining the ecosystem should be reintroduced or depleted populations reinforced.	6	Ecological restoration seeks the highest level of recovery attainable... for physical conditions, species composition, structural diversity, ecosystem function, external exchanges, and reduction of threats. [Trophic levels are included within structural diversity]
	Successful rewilding results in, or leads to, a self-sustaining ecosystem in which native species' populations are regulated through predation, competition, and other biotic and abiotic interactions.	4	Ecological restoration supports ecosystem recovery processes. Ecological restoration actions are designed to assist natural processes of recovery that ultimately are carried out by the effects of time on physical processes and the responses and interactions of the biota throughout their life cycles. Restoration activities focus on reinstating components and conditions suitable for these processes to recommence and support recovery of ecosystem attributes, including capacity for self-organisation and for ecosystem resilience to future stresses.
		*	When full recovery is the goal, an important benchmark is when the ecosystem demonstrates self-organisation
2	Rewilding employs landscape-scale planning that considers core areas, connectivity, and co-existence.	7	Ecological restoration gains cumulative value when applied at the landscape scale; planning and prioritising site-level activities are necessary as part of integrated landscape planning efforts.
3	Rewilding focuses on the recovery of ecological processes, interactions, and conditions based on reference ecosystems; reference models should allow for natural disturbance within an evolutionary relevant range of variability and take environmental change into account.	3	Ecological restoration practice is informed by native reference ecosystems, while considering environmental change … [and recovery of] physical conditions, species composition, structural diversity, ecosystem function, external exchanges, and reduction of threats.
	A reference can be based on carefully selected contemporary near-natural reference areas with relatively complete biota where these still exist or appropriate scientific or historical evidence supported by expert indigenous and local knowledge.		Optimally the reference model describes the approximate condition the site would be in had degradation not occurred. Reference models are developed using multiple sources of information. Best practice is to build empirical models based on information on specific ecosystem attributes obtained from multiple modern analogs or reference sites.

(continued)

Table 4.1 (Cont.)

Rewilding		Ecological Restoration	
#	Primary concepts	#	Related concepts
			These sites are environmentally and ecologically similar to the project site, but optimally have experienced little or minimal degradation. Information on past and current conditions at the site as well as consultation with stakeholders can assist in developing reference models, especially where nondegraded local reference sites are unavailable
4	Rewilding recognises that ecosystems are dynamic and constantly changing.	3	[The reference] condition is not necessarily the same as the historic state, as it accounts for the inherent capacity of ecosystems to change in response to changing conditions. Reference models should ... account for temporal change.
	Rewilding programmes must take both genetic and ecologically effective population sizes into account and employ strategies (e.g., connectivity) that ensure ecologically sustainable and genetically healthy populations of animals, plants, and other organisms.		[Ecological restoration should] optimise genetic diversity and potential for populations to adapt, to prevent extirpations from current habitat areas, and that promote migration to new areas. Options include retaining and augmenting genetically diverse populations of existing native floral and faunal species, and ensuring that these populations exist in configurations that increase linkages and improve gene flow where appropriate to boost adaptability to changed conditions.
5	Rewilding should anticipate the effects of climate change and where possible act as a tool to mitigate impacts.	3	....along- side working to improve potential for restoration and other actions to slow climate change, climate change needs to be recognised as part of the current environmental background condition to which many species will either adapt or go extinct.
		*	Ecological restoration, when implemented effectively and sustainably, contributes to protecting biodiversity; improving human health and wellbeing; increasing food and water security; delivering goods, services, and economic prosperity; and supporting climate change mitigation, resilience, and adaptation.
	Rewilding can also be considered an example of an NbS with the potential to absorb, ameliorate, and tackle the effects of climate change.	10	Ecological restoration is part of a continuum of restorative activities.

Table 4.1 (Cont.)

Rewilding		Ecological Restoration	
#	Primary concepts	#	Related concepts
6	Rewilding requires local engagement and support.	1	Ecological restoration engages stakeholders. Recognising the expectations and interests of stakeholders and directly involving them is key to ensuring that both nature and society mutually benefit. Social and human wellbeing goals, including those that reinstate or reinforce ecosystems services, must be identified alongside ecological goals during the planning stage of a restoration project.
7	Rewilding is informed by science, traditional ecological knowledge (TEK), and other local knowledge.	2	Ecological restoration draws on many types of knowledge… ..practitioner experience, Traditional Ecological Knowledge, Local Ecological Knowledge, and scientific discovery
8	Rewilding is adaptive and dependent on monitoring and feedback.	5	Ecosystem recovery is assessed against clear goals and objectives, using measurable indicators.
9	Rewilding recognises the intrinsic value of all species and ecosystems.	*	In addition to their intrinsic value… ..healthy native ecosystems assure the flow of ecosystem services.
	Where management interventions are required, these should focus on removal of human control and restoring native species with minimal intervention and nonlethal means wherever possible.	4	The most reliable and cost-effective way to kick-start restoration is to harness the potential of remnant species (e.g., plants, animals, microorganisms) to regenerate (i.e., to colonise or expand from in situ components), but degraded ecosystems often require substantial intervention to compensate for lost natural recovery potential.
10	Rewilding requires a paradigm shift in the coexistence of humans and nature.	1	Humans benefit from a closer and reciprocal engagement with nature.
	Rewilding means transformative change and provides optimism, purpose, and motivation.		Participating in restoration projects can be transformative.
		8	[Restoration can] transform the way societies interact with nature.

The most visible difference between the two that cannot be explained by text length is the emphasis on wildlife. The very first rewilding principle requires the use of 'wildlife to restore trophic interactions', while the restoration principles do not explicitly state a taxonomic focus and nest trophic interactions within broader recovery goals for structural diversity (there are also goals for species composition, ecosystem function, abiotic properties, and external exchanges, as well as removal of threats). Thus, the principles for ecological restoration include wildlife, but do not give them priority, while the principles of rewilding prioritise wildlife. Although the principles for rewilding place wildlife front and centre, they also include the need to 'ensure ecological sustainable and genetically healthy populations' of not just wildlife, but also 'plants and other organisms' (Principle 4). So the difference is in focus but not in breadth of inclusion.

This difference in the principles is not surprising given the historical emphasis of top-down approaches (e.g., reintroducing carnivores) in rewilding versus bottom up approaches (managing vegetation and soils) in ecological restoration and makes sense given the (still evolving) scientific underpinnings (see chapters 6 and 7) and development of the fields. But which of these approaches is most efficient and effective in achieving the common end goal of full recovery?

Imagine this: Three friends are together in a crowded city and need to get to the airport as quickly as possible to catch the same flight. All three have the exact same goal, arriving at the airport. However, each has a different idea about the most efficient and reliable way to achieve that goal. One advocates taking the subway, because she has years of experience commuting by train to work, knows the station locations and how to transfer between trains, and heard on the radio that the main highway to the airport was closed for construction. Another advocates driving because of her past experience owning a taxi cab and ability to navigate to the airport through back roads that avoid both congestion and construction. In addition, she has a heavy suitcase and doesn't think it will fit through the subway turn style. The third friend has a bicycle but no luggage, and thinks that the only way to make the flight on time is to take the bike path, which is actually the shortest distance and will require no waiting for traffic or trains. Each individual is relying on their knowledge and experience to decide the best plan of action, and based on differences in their knowledge and experience, they have come up with different approaches to reach the same end goal.

There is good reason to think a similar mechanism as the one described above for travel to the airport may be driving differences in approaches to rewilding and ecological restoration. Rewilders tend to be trained in conservation biology, which has historically had a strong bias towards wildlife over plants (at a rate of three to one based on a bibliometric analysis by Young, 2000). In contrast, the ecological restoration practitioners are more likely to be grounded in the science of restoration ecology, which has historically had a greater focus on vegetation and soil science than on wildlife and trophic dynamics (at a rate of four to one based on Young, 2000). There is some evidence that these trends are relaxing over time. For instance, rewilding studies by Morel et al. (2020) and Thompson et al. (2018), respectively, emphasise plant community richness and manipulation of large woody debris, suggesting a high degree of alignment between ecological rewilding and ecological restoration. Increasing alignment between these fields is also evident in the increasing amount of scientific literature that references both terms (Figure 4.2).

Strategic integration

Recent global initiatives have led to an unparalleled opportunity to reverse degradation, recover biodiversity and ecological integrity and improve human wellbeing. For instance, last year the United Nations General Assembly declared 2021–2030 as the United Nations Decade on Ecosystem Restoration (hereafter the 'UN Decade') (United Nations Environment Programme & FAO 2020). Also last year, the United Nations Convention on Biological Diversity (2021) called for 'at least 30% of land and sea areas global (especially areas of particular importance for biodiversity and its contributions to people) conserved through effective, equitably managed, ecologically representative and well-connected systems of protected areas'. While these initiatives are highly motivating (Young Schwartz, 2019), ecological repair is a complex

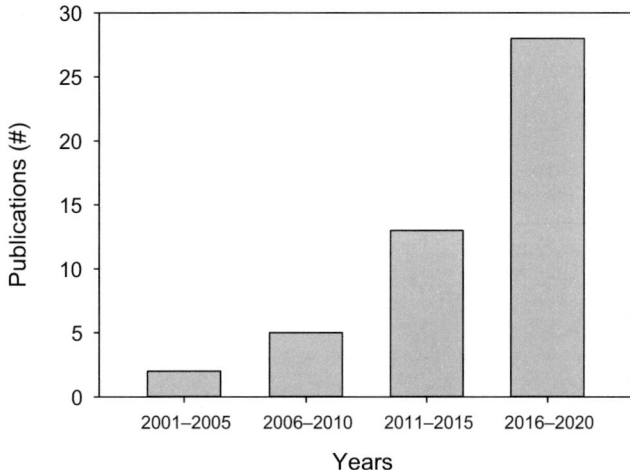

Figure 4.2 Number of articles published with both restoration and rewilding in their titles between 2000 and 2020. Data are based on a 16 February 2022 search of *Web of Science* using the 'all databases' function and the search terms 'rewild*' and 'restor*'. There were no publications found in *Web of Science* with both terms in their titles before 2000; since 2020 there have been 6.

challenge and there is concern that without a paradigm shift (Abhilash, 2021), the UN Decade will result in restoration projects, programmes, and initiatives that waste resources (Cooke et al., 2019) or result in further degradation of high-value ecosystems (Dudley et al., 2020). The extent to which we are successful at leveraging current global initiatives for long-term biodiversity conservation will depend on our ability to leverage the knowledge, experience, and energies of both the rewilding and ecological restoration communities to build strategic alliances (McAlpine, 2016; Anderson et al., 2019).

For example, a central challenge in ecological restoration is the prioritisation of sites and landscapes to include in restoration programmes and initiatives. Although Principle 7 of ecological restoration calls for integrated landscape planning, restoration assessments often do not consider the landscape-level factors that are front and centre in rewilding and large landscape conservation initiatives, namely cores, corridors, and connectivity required to support wildlife. Tools from conservation biology, such as programmes for analysing gene flow across landscapes (e.g., Circuitscape; Duda et al., 2016; McRae et al., 2016), allow increasingly sophisticated analyses of landscape connectivity and could optimise the benefits of restoration for wildlife and trophic cascades, as well as ecological complexity, integrity, and human wellbeing.

While the ecological restoration community would clearly benefit from engagement by rewilders, there are also compelling reasons for rewilders to engage in restoration initiatives. For example, initiatives like the Bonn Challenge aim to restore over 350 million hectares of land. Restoration at this magnitude represents a substantial opportunity to make progress on the UN Convention on Biological Diversity's 30% goal. But unless the rewilding community engages, they will not have an opportunity to promote activities in areas that they have identified as rewilding priorities. Furthermore, integrating bottom-up plant and soil strategies with top-down wildlife ones may increase the chance that restoration will lead to broader recovery of ecological complexity, although there sometimes may be trade-offs and conflicts between approaches that utilise plants versus animals, such as when a rare species relies on an invasive one (McAlpine, 2016).

The fields of ecological restoration and rewilding will no doubt continue to evolve and improve over time. With increasing interaction between them, there may be greater advances in the tools and tactics for ecological repair, ultimately improving biodiversity outcomes.

References

Abhilash, P.C. 2021. Restoring the unrestored: strategies for restoring global land during the UN Decade on Ecosystem Restoration (UN-DER). *Land, 10*(2): 201.

Anderson, R.M., Buitenwerf, R., Driessen, C., Genes, L., Lorimer, J., and Svenning, J.C. 2019. Introducing rewilding to restoration to expand the conservation effort: a response to Hayward et al. *Biodiversity and Conservation, 28*(13): 3691–3.

Bradshaw, A.D. and Chadwick, M.J. 1980. *The Restoration of Land: The Ecology and Reclamation of Derelict and Degraded Land.* Univ. of California Press.

Butler, J.R., Marzano, M., Pettorelli, N., Durant, S.M., du Toit, J.T., and Young, J.C. 2021. Decision-making for rewilding: an adaptive governance framework for social-ecological complexity. *Frontiers in Conservation Science, 2.*

Carver, S., Convery, I., Hawkins, S., Beyers, R., Eagle, A., Kun, Z., Van Maanen, E., Cao, Y., Fisher, M., Edwards, S.R., and Nelson, C. 2021. Guiding principles for rewilding. *Conservation Biology, 35*(6): 1882–93.

Clewell, A.F. and Aronson, J. 2007. *Ecological Restoration: Principles, Values, and Structure of an Emerging Profession.* Washington, DC: Island Press.

Cooke, S.J., Bennett, J.R., and Jones, H.P. 2019. We have a long way to go if we want to realize the promise of the 'Decade on Ecosystem Restoration'. *Conservation Science and Practice, 1*(12): e129.

Corlett, R.T. 2016. Restoration, reintroduction, and rewilding in a changing world. *Trends in Ecology and Evolution,* 31(6): 453–62.

Donlan, J. 2005. Re-wilding North America. *Nature, 436*(7053): 913–14.

Dudley, N., Eufemia, L., Fleckenstein, M., Periago, M.E., Petersen, I., and Timmers, J.F, 2020. Grasslands and savannahs in the UN Decade on Ecosystem Restoration. *Restoration Ecology, 28*(6): 1313–17.

Dutta, T., Sharma, S., McRae, B.H., Roy, P.S., and DeFries, R. 2016. Connecting the dots: mapping habitat connectivity for tigers in central India. *Regional Environmental Change, 16*(1): 53–67.

du Toit, J.T. and Pettorelli, N. 2019. The differences between rewilding and restoring an ecologically degraded landscape. *Journal of Applied Ecology, 56*(11): 2467–71.

Gann, G.D., McDonald, T., Walder, B., Aronson, J., Nelson, C.R., Jonson, J., Hallett, J.G., Eisenberg, C., Guariguata, M.R., Liu, J., and Hua, F. 2019. International principles and standards for the practice of ecological restoration, second edition. *Restoration Ecology.* 27 (S1): S1–S46, 27(S1): S1–S46.

García-Ruiz, J.M., Lasanta, T., Nadal-Romero, E., Lana-Renault, N., and Álvarez-Farizo, B. 2020. Rewilding and restoring cultural landscapes in Mediterranean mountains: Opportunities and challenges. *Land use policy, 99,* p.104850.

Hayward, M.W., Scanlon, R.J., Callen, A., Howell, L.G., Klop-Toker, K.L., Di Blanco, Y., Balkenhol, N., Bugir, C.K., Campbell, L., Caravaggi, A., and Chalmers, A.C. 2019. Reintroducing rewilding to restoration – Rejecting the search for novelty. *Biological Conservation, 233*: 255–9.

Kimmins, J.P. (2004). *Forest Ecology: A Foundation for Sustainable Forest Management and Environmental Ethics in Forestry.* Upper Saddle River, New Jersey: Prentice Hall.

Laestadius, L., Buckingham, K., Maginnis, S., and Saint-Laurent, C. 2015. Before Bonn and beyond: The history and future of forest landscape restoration. *Unasylva, 66*(245): 11

Loth, A.F. and Newton, A.C. 2018. Rewilding as a restoration strategy for lowland agricultural landscapes: Stakeholder-assisted multi-criteria analysis in Dorset, UK. *Journal for Nature Conservation, 46*: 110–20.

McAlpine, C., Catterall, C.P., Nally, R.M., Lindenmayer, D., Reid, J.L., Holl, K.D., Bennett, A.F., Runting, R.K., Wilson, K., Hobbs, R.J., and Seabrook, L. 2016. Integrating plant- and animal-based perspectives for more effective restoration of biodiversity. *Frontiers in Ecology and the Environment, 14*(1): 37–45.

McRae, B.H., Shah, V., and Edelman, A. 2016. Circuitscape: modeling landscape connectivity to promote conservation and human health. *The Nature Conservancy,* Fort Collins, CO. 14 pp 1–14.

Meine, C., Soulé, M., and Noss, R.F. 2006. 'A mission-driven discipline': the growth of conservation biology. *Conservation Biology, 20*(3): 631–51.

Millar, C.I. and Brubaker, L.B. 2006. Climate change and paleoecology: new contexts for restoration ecology. In Falk, D.A. et al., *Foundations of Restoration Ecology*, 315–40, Island Press.

Miller, J.R. and Hobbs, R.J. 2019. Rewilding and restoration. In Pettorelli, N., Durant, S.M. and du Toit, J.T. (eds), *Rewilding*, p.123, Cambridge: Cambridge University Press.

Morel, L., Barbe, L., Jung, V., Clément, B., Schnitzler, A., and Ysnel, F. 2020. Passive rewilding may (also) restore phylogenetically rich and functionally resilient forest plant communities. *Ecological Applications*, *30*(1): e02007.

Noss, R.F. 1992. The Wildlands Project: land conservation strategy. *Wild Earth*, *1*(9): e25.

Pereira, H.M. and Navarro, L.M. 2015. *Rewilding European Landscapes* (p. 227). Springer Nature.

Perino, A., Pereira, H.M., Navarro, L.M., Fernández, N., Bullock, J.M., Ceauşu, S., Cortés-Avizanda, A., van Klink, R., Kuemmerle, T., Lomba, A., and Pe'er, G., 2019. Rewilding complex ecosystems. *Science*, *364*(6438): eaav5570.

SER (Society for Ecological Restoration International Science & Policy Working Group). 2004. *The SER International Primer on Ecological Restoration*. Society for Ecological Restoration International, Tucson, Arizona.

Soulé, M. and Noss, R. 1998. Rewilding and biodiversity: complementary goals for continental conservation. *Wild Earth*, *8*: 18–28.

Svenning, J.C. 2020. Rewilding should be central to global restoration efforts. *One Earth*, *3*(6): 657–60.

Svenning, J.C., Pedersen, P.B., Donlan, C.J., Ejrnæs, R., Faurby, S., Galetti, M., Hansen, D.M., Sandel, B., Sandom, C.J., Terborgh, J.W., and Vera, F.W. 2016. Science for a wilder Anthropocene: Synthesis and future directions for trophic rewilding research. *Proceedings of the National Academy of Sciences*, *113*(4): 898–906.

Thompson, M.S., Brooks, S.J., Sayer, C.D., Woodward, G., Axmacher, J.C., Perkins, D.M., and Gray, C. 2018. Large woody debris 'rewilding' rapidly restores biodiversity in riverine food webs. *Journal of Applied Ecology*, *55*(2): 895–904.

United Nations Convention on Biological Diversity. 2021. First Draft of the Post-2020 Global Biodiversity Framework. [Cited 11 February 2022.] www.cbd.int/doc/c/abb5/591f/2e46096d3f0330b08ce87a45/wg2020-03-03-en.pdf

United Nations Environment Programme (UNEP) & FAO. 2020. Strategy for the UN Decade on Ecosystem Restoration [online]. [Cited 11 February 2022.] https://wedocs.unep.org/bitstream/handle/20.500.11822/31813/ERDStrat.pdf?sequence=1&isAllowed=y

Young, T.P. 2000. Restoration ecology and conservation biology. *Biological conservation*, *92*(1): 73–83.

Young, T.P. and Schwartz, M.W. 2019. The decade on ecosystem restoration is an impetus to get it right. *Conservation Science and Practice*, *1*(12): e145.

5

DEVELOPING A FRAMEWORK FOR REWILDING BASED ON ITS SOCIAL-ECOLOGICAL AIMS

Sally Hawkins

Introduction

Since the emergence of rewilding concepts, as detailed in previous chapters in this book, rewilding practice has become increasingly popular around the world. It has been described as a practical, science-based method to restore functioning ecosystems and reduce conservation management at a landscape scale, and also as a movement for transformational change with the potential to restore human–nature relationships and kerb the effects of the Anthropocene (Carver et al., 2021). Theories on how best to achieve rewilding are still evolving (see for example Martin et al. (2021) and other chapters in this book for the positioning of people in rewilding) and interventions used in rewilding can vary depending on what is suitable in current ecological or cultural contexts, which means that rewilding projects can seem very different, with their long-term, transformational potential often intangible. This has led to some confusion over the term rewilding, leaving it open to misinterpretation and the risk of diluting its longer-term potential to deliver transformational change. There is, therefore, consensus that the concept of rewilding needs some unification to cement its reputation and harness its momentum while maintaining its multivalence and adaptability to different social or ecological conditions.

In 2017 the IUCN Commission for Ecosystem Management commissioned a Rewilding Task Force (now the Rewilding Thematic Group (RTG)) to synthesise and streamline the increasingly global concept of rewilding. Between 2017 and 2020 the RTG undertook a global consultation of rewilding and restoration experts, collecting data through various means, which culminated in the publication of a rewilding definition and ten guiding principles for rewilding (Carver et al., 2021). The RTG continues to work towards synthesising global rewilding which involves creating a database for rewilding projects and contributing to rewilding guidelines. To aid this process, this chapter will focus on the aims of rewilding and proposes a rewilding framework based on these.

Method

The aims of rewilding outlined in this chapter are based on a 2018 RTG survey of rewilding pioneers. The term 'pioneers' was defined as those who influenced the emergence and evolution

DOI: 10.4324/9781003097822-6

of rewilding concepts, with participants identified through journal and book publications, self-identification, and recommendations from other participants. We received 59 responses, the majority from the USA (26) and western Europe (23, including the United Kingdom, the Netherlands, Spain, Denmark, France, Germany, and Switzerland), with others from Australia (6) and one participant each from Argentina, Cambodia, Canada, Greece, Mauritius, Mexico, and Tanzania (where more than one country of residence were listed, both were counted). Participants were representative of key rewilding organisations or working groups, including the Wildlands Network, Rewilding Institute, Rewilding Europe, Rewilding Britain, and authors from publications such as Wild Earth, Soule and Noss (1998), Donlan et al. (2005), Svenning et al. (2016) and Pettorelli et al. (2018). The survey consisted of 25, mainly open-ended questions resulting in qualitative data and insights into participant's experiences and understanding of rewilding over time. A key theme of 'promoting change' emerged from initial analyses, and further inductive data analysis identified themes under four parent nodes related to change—(1) the drivers of rewilding (change why), (2) aims or intentions of rewilding (change what), (3) rewilding practice or application (change how), and (4) barriers to change. This chapter presents the second of these categories, i.e. the aims of rewilding, and proposes a framework for rewilding, incorporating concepts from the other thematic nodes, where appropriate. The focus on aims, rather than the interventions used in rewilding practice, is because they offer potential to unify rewilding practice under shared, optimistic goals, while remaining broad enough to allow for place-based interpretation.

Aims of rewilding

Participants expressed desire to promote change in three areas—change to a landscape or social-ecological system (i.e. systemic-level change), ecological change and socio-cultural change—and a thematic analysis revealed themes related to change in each of these areas. It should be noted that aims are separated under 'ecological' and 'socio-cultural' based on responses, i.e. the tendency of participants to separate ecological and social processes.

These themes were further categorised as aims and qualities required to achieve these aims (Table 5.1). Also included in Table 5.1 are some of the drivers of rewilding (referring to parent node 1 in the Methods section) which correlate with these aims, highlighting those qualities that are identified as negative for rewilding and to depict the direction of the desired change.

Ecological change

The most common rewilding aims related to ecological restoration or recovery. More specifically, rewilding aims for *'Nature that can be itself'* (i. e., nature that is self-willed or unrestricted by human control or management) and *'Nature that can look after itself'* (i.e., resilient ecosystems that can self-regulate and perpetuate). This does not require nature to stay the same but to have the capacity to adapt to or withstand change.

Qualities identified by participants as contributing to these aims relate to *'Ecological integrity'* or wholeness, meaning systems which contain all the necessary biotic and abiotic processes, such as predation, herbivory, or migration, required for self-regulation. *'Indigeneity'*, relating to indigenous or 'original' ecological composition, was key among many participants as it was perceived that indigenous species have co-evolved, providing guidance on what works in the ecological context and enabling system sustainability and resilience. However, some participants highlighted that rewilding could include the introduction of non-native species as ecological

Direction of rewilding change →

Table 5.1 Aims of rewilding identified from the survey, along with qualities identified as being critical to achieve those aims

Systemic-level aims	Landscapes or social–ecological systems that are sustainable, resilient, 'wild'						
Systemic-level qualities	Coexistence, reciprocity					Societal wellbeing	
Aims (ecological or socio-cultural)	Nature that can be itself	Nature that can look after itself		People accommodating wild nature in their landscapes			
Qualities identified as enhancing rewilding aims	Ecological integrity	Indigeneity	Ecological diversity and heterogeneity	Ability to identify and prevent unsustainable activities	Ecocentrism	Tolerance and adaptability	Appreciating the benefits humans get from nature
Negative (dewilded) landscape features driving rewilding	Human dominated, homogenous landscapes	Degraded ecosystem function, increasing extinction risk	High level of human control, management and overexploitation	Predominant culture anthropocentric or egocentric	Human culture averse to change, risk and unpredictability	People that are hopeless, unfulfilled, discontent, unhealthy	Human-nature detachment, commodification and objectification of nature

surrogates for extinct species, with examples of tortoise introductions in Mauritius and Hawaii, although there is emphasis that this should be done with caution. The final quality under ecological change, *'Ecological diversity and heterogeneity'*, includes biodiversity as well as heterogenous landscape features, and is seen as a quality which contributes to ecological and landscape-scale resilience but also to human interest.

Socio-cultural change

The socio-cultural aims of rewilding are less explicitly discussed in rewilding literature than the ecological aims, but they were clearly understood by participants as being core elements of rewilding—with an understanding that, in order to achieve ecological aims, rewilding requires and should therefore incorporate socio-cultural change. The aim of *'People accommodating wild nature in their landscapes'* is seen as the 'reverse side of the coin' to rewilding's ecological aims and includes the 'letting', 'allowing' and 'giving' inherent in many descriptions of rewilding (e.g. 'allowing nature to be itself' or 'giving nature the space to be itself'). This aim includes reducing control and management (including rewilding or conservation-related management) and negates overexploitation and sustained degradation of ecosystems and their components. While many rewilding projects aim to benefit humans in some way (through economic incentives in particular) some participants highlight that rewilding should not necessarily primarily aim to benefit humans as this is seen as anthropocentric. Many participants expressed the potential for rewilding to have marked effects on *'Societal wellbeing'*, especially contributing to economic sustainability (through ecotourism) and mental health (with increasing optimism and fulfilment through nature connection). It is also assumed that increased societal wellbeing would create further opportunities to achieve rewilding's other aims. Wellbeing is a broad term which incorporates a number of elements contributing to quality of life, perceptions of which differ across different cultures or contexts and therefore need to be locally assessed (Loveridge et al., 2020). Wellbeing can incorporate many different indicators, which of these relate to and interact with other rewilding qualities is a topic that requires further research.

Certain qualities are identified as critical for a system to achieve rewilding's socio-cultural aims. These include the *'Ability to identify and prevent unsustainable activities'* which requires knowledge and monitoring as well as some form of governance or law enforcement. *'Tolerance and adaptability'* relates to people's ability to coexist with the risk and unpredictability inherent in diverse natural systems. The theme of *'Ecocentrism'* incorporates responses which allude to the intrinsic value of non human species or natural processes as well as holism or interconnectivity between species, including humans. Ecocentrism is highlighted as the ethical stance which would most effectively contribute to rewilding aims, preferred over predominant ego- or anthropocentric human values which place the self or humans above all others. This has likely been influenced by rewilding's roots in the Deep Ecology movement which is mentioned by several participants. In the literature, ecocentrism is a complex and debated term but essentially is a stance which acknowledges, like biocentrism, other species' intrinsic value (i.e. that they have value independent of their usefulness to humans) but critically adds an appreciation for holism and the interconnectivities among all species (including humans) in a system (Cocks & Simpson, 2015). In this sense it has been likened to Leopold's (1949: 201) Land Ethic (Conradie, 2019) which further acknowledges the rights of other species and humans' responsibilities to them. Ecocentrism does not negate an *'Appreciation for the benefits humans get from nature'*, which is the final quality identified by participants, and includes knowledge of and appreciation for ecosystem services and natural capital. What these socio-cultural themes also demonstrate, is that rewilding is not necessarily anti-human or anti-culture, as discussed and debated elsewhere (see for example Schulte to Bühne, Pettorelli, & Hoffmann (2021) or Ward (2019)) but is against anthropocentrism, and promotes ecocentric human culture to support coexistence and reciprocity at a landscape scale.

Landscape-level or social-ecological change

The landscape-scale or social-ecological aims of rewilding (resilience, sustainability, and 'wildness') are expressed by participants more broadly and with less clarity or definition that those qualities in the ecological or socio-cultural categories, reflecting the lack of clear definitions for these terms and perhaps limited understanding of how social-ecological interactions play out in a system. This is an important topic for further research and the qualities and aims identified in this chapter could help in providing some metrics against which we can further explore and refine the broader terms of resilience and sustainability in a rewilding context.

Additional qualities which are identified as contributing to these broader aims, along with the ecological and socio-cultural qualities, are those that are not wholly ecological or socio-cultural but shared. '*Reciprocity*' highlights the mutual benefits or symbioses in social-ecological systems—it acknowledges the benefits humans get from nature but critically highlights our impact on other species, emphasising some form of human responsibility to reciprocate, and of course the impacts of other species on one another. '*Coexistence*' encompasses the ability for humans and other species to exist within a landscape at the same time. The socio-cultural quality of '*Tolerance and adaptability*' is especially important for coexistence. '*Coexistence*' and '*Reciprocity*' can be associated with the interrelated concepts of land sharing and land sparing and cores and connectivity, all previously identified as tools for rewilding and conservation (Soule & Noss, 1998; Carver et al., 2021; Finch et al., 2021). Reciprocity in some contexts may involve sparing land for either human activity or nature, such as creating protected areas, while in other instances, higher tolerance for nature could enable coexistence and opportunities for land sharing and/or connectivity. Both approaches would contribute to coexistence at a landscape scale.

The implications of rewilding aims for rewilding practice

The survey data revealed that many drivers of rewilding related to a desire to change the culture and practice within the field of conservation biology. These include a 'doom and gloom', reactionary culture of conservation; increasing awareness based on emerging ecological theories that single-species conservation and protected areas were ineffective; intensive management to create or maintain certain static ecological states (particularly in European cultural or agricultural landscapes and species); and exclusivity and egocentrism in science and conservation. While the socio-cultural aims of rewilding expressed in the previous section are broadly aimed at society as a whole, the implication is that rewilding practitioners must integrate these qualities into rewilding practice. Table 5.2 shows those drivers of rewilding specific to conservation culture and practice, and the desired qualities which participants expressed would enhance the practice of rewilding. Similarities can be drawn between these qualities and those expressed in Table 5.1, for example the aim of societal wellbeing can be linked to a practice that is hopeful, inspiring, and creative and the aim of tolerance and adaptability among human communities links to a rewilding practice that is adaptive to change. The desired qualities of rewilding practice highlighted in Table 5.2 have been integrated into the existing guiding principles for rewilding (Carver et al., 2021). To begin to develop a rewilding framework, we must therefore acknowledge both rewilding aims and the rewilding principles.

Table 5.2 Drivers of rewilding specific to conservation culture and practice, and the desired qualities which participants expressed would enhance the practice of rewilding

Drivers of change related to conservation culture and practice	Related qualities underpinning rewilding practice
'Doom and gloom', reactionary	Innovative, hopeful, inspiring, proactive, creative
Risk averse, high levels of control/management to maintain pre-determined conditions, attempts to define and roll out 'best practice'— i.e. what works here will work everywhere	Adaptive to change, iterative, context-specific approaches, humility over limits to human knowledge/understanding of natural processes ('nature knows best'), thereby aiming to reduce human control and management
Limited focus—e.g. single species or ecosystems	Holistic, landscape-scale, increased acknowledgement of system interconnectivities and complexity, based on knowledge of ecological function and interdependencies, ecocentric
Exclusive, egocentric, lack of engagement with different knowledge or cultural perspectives	Inclusive, collaborative, knowledge democracy, ecocentric

A framework integrating rewilding aims and principles

The data presented here highlights an aspiration for rewilding to transform systems through both ecological and socio-cultural change. It is a long-term process which iteratively creates the potential for a system to achieve sustainability, resilience, and wildness, with the assumption that a system in this state would require no or very little rewilding or conservation management. The reality of rewilding practice is therefore to perform interventions to increase the potential for a system to achieve this state—intentionally making decisions and enacting them to move a system along a continuum from a dewilded to a rewilded state. Given the social-ecological nature of rewilding aims, survey participants highlight many types of interventions that could be used to enhance ecological or socio-cultural qualities of a system, including species reintroductions, habitat restoration, public engagement, and education, encouraging change to policy, human behaviour, or land management. As the most appropriate intervention would depend heavily on the social-ecological context at a particular point in time, instead of specifying types of interventions, any rewilding framework should rather focus on integrating principles into project management. Figure 5.1 proposes such a framework. The stages included in this framework have been developed with reference to the results of the RPS, integrating the need to tackle barriers, such as misinterpretation of the concept of rewilding, and to integrate some of the principles identified in Table 5.2, including the need for context-specific interventions, adaptability, and iterative project management.[1]

Stage 1: The first stage is project self-reflection. This would allow the individuals or organisations planning a rewilding project to assess their own intentions and motivations and to develop their project's aims. Comparing these to defined rewilding aims would allow a project to consider whether it fits within the field of rewilding or whether it is more suited to another field, such as ecological restoration or species reintroduction. An example would be a project which aims to create a small protected area without the long-term aspiration to create connectivity allowing for natural migration, enhancing the ability for nature to be itself and look after itself. This proved to be an issue at the well-known rewilding project,

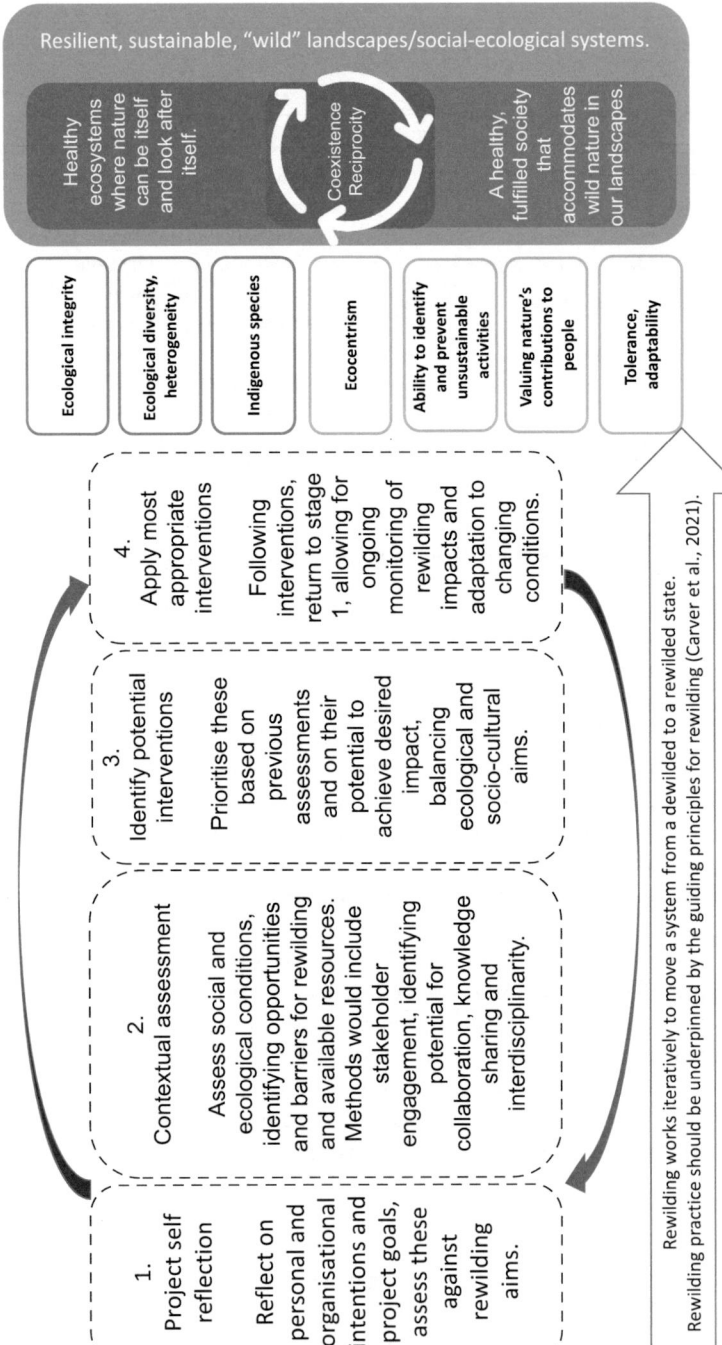

Figure 5.1 Proposed rewilding framework based on results of a survey of rewilding pioneers.

Oostvardersplassen, where limited space and fencing conflicted with project aims of letting nature take its course, as highlighted by Kopnina et al. in Chapter 32 of this book. However, this stage is not intended to be exclusive, but to encourage projects to be creative and optimistic, and aspire to the transformational potential of rewilding. This would allow smaller projects with limited resources and space, or existing projects which have not previously identified as rewilding projects, to embrace rewilding aims if opportunities arise to extend the area and/or impact of their project. There are two very different examples of this stage in the case studies within this book. Firstly, Hilty et al. demonstrate how critical the intention-setting stage was for the Yellowstone to Yukon project, with an influential, broad-reaching vision affecting decisions at local scale. Secondly, Adair and Ashmole demonstrate how even small-scale projects can expand their aims over time, with the Borders Forest Trust approaching local landowners to expand the influence and impacts of the project beyond the area of Carrifran Wildwood. As this example demonstrates, this first stage continues to be necessary in later iterations as the project expands and adapts to changing conditions, barriers and opportunities, such as forming partnerships which would expand project resources and scale, allowing intentional integration of other parties and additional objectives.

Stage 2: To ensure that rewilding interventions suit the local social-ecological context of a project, the second stage requires a thorough assessment of social and ecological conditions. This would include the identification of opportunities or barriers for achieving rewilding aims, necessitating, for example, stakeholder evaluations, knowledge sharing (including local or traditional ecological knowledge), identifying organisations with similar goals, potential for collaboration, and available resources. The methods used for these assessments (and in ongoing monitoring which would be captured in this stage in ongoing iterations) would depend on the scale of the project and the resources available. This stage is crucial for informing rewilding interventions and the first assessment would also provide the baseline for monitoring system change. Methods used for assessments and monitoring can vary dramatically depending on the focus, from less intensive, traditional ecological survey methods such as those undertaken at Carrifran Wildwood (Adair and Ashmole, this book) to intensive, innovative monitoring techniques including remote sensing, eDNA surveys and natural capital accounting approaches as undertaken at Birchfield, enabled through a mutually beneficial partnership with AECOM (White et al., Chapter 20). Inclusion of stakeholders and local knowledge is critical for this stage, as highlighted by Carver et al. (2021, principle 7) and demonstrated in this book by Kealley and Burrows. The question of what to monitor in terms of success are explored in this book by Beyers and Sinclair (ecological) and Root-Bernstein (social), but in reality the methods used and the scope of monitoring will depend heavily on funding and on the skills, knowledge, and commitment of partners and stakeholders.

Stage 3: Based on the above assessments, a list of potential interventions can be created. These would ideally look to take advantage of opportunities and work to overcome barriers. While rewilding theory has in the past emphasised species reintroductions (e.g. Svenning et al., 2016), in reality the interventions used to affect change in rewilding projects can vary greatly, as demonstrated by the case studies in this book which have included habitat restoration (Adair and Ashmole; White et al.); the establishment or restoration of protected areas (Donadio et al.; Pringle and Goncalves; Hilty et al.); prevention of herbivory through culling or fencing (Adair and Ashmole; Kealley and Burrows); controlling non-native species (Kealley and Burrows; Donadio et al.); burning (Kealley and Burrows; Donadio et al.); promoting connectivity through the removal of fences (Donadio et al.); promoting coexistence

(Pringle and Goncalvez); public engagement or joint management programmes (Bhandari and Bhatta; Hilty et al.; Kealley and Burrows); incentives to encourage people to move out of protected areas or to adopt sustainable practices (Pringle and Goncalves; Hilty et al.; Donadio et al.). While this list is not exhaustive, it demonstrates the variety of approaches to overcoming problems identified in stage 2. The initial list of potential interventions must then be prioritised based on current feasibility and their potential to increase rewilding opportunities and success down the line. For example, in Muttawa Kurrara Kurrara in Australia, non-native species have been identified as a major barrier to rewilding and there-fore non-native species management has been prioritised over species reintroductions (Burrows and Kealley). Another example would be using public engagement as a tool to increase the feasibility of future apex predator reintroductions, which may offer a more effi-cient route to ecological restoration than attempting to implement an apex predator reintro-duction where there is very low public acceptance for rewilding (Hawkins et al., 2020).

Stage 4: At this stage, the most appropriate interventions are applied. Following these interventions, the project goes into the next iteration, through stages 1–4, meaning that the system will be reassessed, and plans updated in an adaptive approach based on this ongoing monitoring and feedback. This allows ongoing monitoring of change and effectiveness of interventions which will contribute to the growing rewilding knowledge base. This would include wider impacts which may be directly or indirectly related to the project (e.g. project communications increasing awareness of and support for rewilding and conservation in the region, as demonstrated by Hilty et al., this book).

Discussion and conclusions

The framework outlined above is the first to integrate broad social-ecological rewilding aims based on empirical evidence, but it is important to highlight its limitations, particularly as the framework is developed and tested. Firstly, the qualities identified in Table 5.1 remain broad and will require further interrogation to enhance our understanding of the characteristics that are captured by these themes. For example, 'the ability to identify and prevent unsustainable activities' necessitates knowledge of system boundaries, governance and perhaps commitment to place. In doing so, however, it is also important to maintain the framework's adaptability. So that, for example in the case of identifying and preventing unsustainable activities mentioned above, the type of knowledge and governance is not necessarily prescribed but would be suited to the local culture. A second limitation is that the data used to develop this frame-work is from a limited pool of rewilding experts, mainly from North America and western Europe as this reflects those areas where rewilding concepts and practice were pioneered. The focus on 'pioneers' has provided a very strong base from which to continue developing this framework, but it is essential that further insights into rewilding are gained and that the concepts highlighted within this framework are interrogated in different ecological and cul-tural contexts.

One assumption which requires further interrogation is the separation of human and eco-logical systems expressed by most participants, leading to separation of ecological and socio-cultural aims and qualities identified in Table 5.1. By integrating these under a social-ecological framework we can begin to reconcile the interactions between these qualities and concepts with the intention to address human–nature dichotomy which is discussed as a barrier to and a driver of rewilding practice and evolution in the survey as well as elsewhere in this book (e.g. many chapters in Part IV). Pragmatically, I have kept this separation in the rewilding framework as in terms of policy, practice, education, disciplines, etc., these tend to be separated in the same way. Many cultures may consider this separation of humans and culture from other species and processes as superficial and a false representation of reality (see Fenton and Playdon, Chapter 12) and this may be a key issue that will need to be addressed at some point in the future and may require the proposed framework to reflect changes in understanding or culture, for example if predominant human cultures become more ecocentric or if there is increased understanding of social-ecological interactions and resilience. This and other issues highlighted here point to the need to continue to adapt any rewilding framework to changing contexts, which would include the structure and contents of the framework.

It must also be noted that there are several other proposed rewilding frameworks in the literature that may contribute to the synthesis of rewilding under an accepted framework. Pedersen et al. (2019) propose a framework for trophic rewilding, which focuses specifically on species reintroductions to promote ecological self-regulation, although it does include some interventions to reduce human–wildlife conflict. Torres et al. (2018) use a wider definition of rewilding for a proposed framework which focuses on the ecological aims of recovering eco-logical processes and natural dynamics but also includes reducing management inputs required to maintain the ecological system at its current state. Perino et al. (2019) identify dispersal, trophic complexity, and stochastic disturbance as critical and interacting qualities which con-tribute to the rewilding aim of self-regulating ecosystems, with dispersal-limiting infrastruc-ture, human–wildlife conflicts and environmental risks identified as barriers. Carver's (2014) wilderness continuum suggests that rewilding lies along a continuum of decreasing anthropo-genic influence and increasing ecological integrity and the above frameworks are based on similar conventions. While the evidence presented in this chapter shows that rewilding aims are much broader and more complex than those presented in these existing frameworks, they do provide valuable insight into the qualities of ecological integrity and therefore could be integrated as the proposed framework is developed and interrogated further. More recently, Butler et al. (2021) proposed a social-ecological framework for rewilding which integrates adaptive co-management and social licence to operate, and rewilding aims of enhancing eco-logical conditions and sustainable livelihoods. There are certainly parallels that can be made between these two frameworks, and it highlights increasing acknowledgement within the field of rewilding of the importance of incorporating system science, complexity, and collaboration. The framework proposed in this chapter provides a much broader interpretation of rewilding aims based on empirical evidence, providing further insight into rewilding visions and futures alluded to in Butler et al. (2021) and integrates these into a pragmatic, adaptive, iterative model aimed at rewilding practitioners that may be more accessible for smaller projects with limited resources who aspire to future transformational change. The potential to integrate existing practices such as adaptive co-management into this framework can be further assessed and we should also look to existing frameworks and guidelines from similar fields, such as the IUCN's Restoration Opportunities Assessment Methodology (IUCN, 2016), to draw further insight.

Despite its limitations, the framework presented here provides a useful and evidence-based starting point for unifying rewilding practice under its social-ecological aims and a focal point to enable identification of areas requiring further research or refinement. This would help to broaden our understanding of the qualities identified above and of the methods or interventions which could be used to enhance them. By sharing the findings on rewilding's social-ecological aims, the intention is to encourage the rewilding community to work towards common gaols, to further explore the interconnection between ecological and social systems, and to share experiences and lessons learned related to these aims to enhance the potential for rewilding across systems, cultures and disciplines. Finally, while the framework is aimed at rewilding practitioners who are looking to apply rewilding interventions on the ground, if we truly intend to effect transformational change, we must also look more widely at the systems and institutions in which rewilding research and practice operates. If rewilding is to be a global undertaking, and if it truly has the potential to create transformational change, it must embrace and encourage change across the multiple systems that affect it.

Acknowledgements

This chapter is based on summarised excerpts from the author's PhD thesis, currently in preparation, under the supervision of Ian Convery, Steve Carver, Darrell Smith, and Lisa Fenton.

Note

1 The stages outlined in this framework were refined after discussions during an RTG meeting in November 2021 focusing on developing rewilding guidelines and therefore I acknowledge the additional contributions of Rene Beyers, Steve Carver, Ian Convery, Mick Drury, Adam Eagle, Jessica Rothwell, Alan Watson-Featherstone and thank Emilia and Roger Leese for accommodating us at Birchfield (birchfieldrewilding.org).

References

Butler, J.R.A. et al. (2021) Decision-making for rewilding: an adaptive governance framework for social-ecological complexity, *Frontiers in Conservation Science*, 2. Available at: www.frontiersin.org/article/10.3389/fcosc.2021.681545 (accessed: 8 February 2022).

Carver, S. (2014) Making real space for nature: a continuum approach to UK conservation, *ECOS*, 35(3/4). Available at: www.banc.org.uk/ecos-35-34-making-real-space-for-nature-a-continuum-approach-to-uk-conservation-steve-carver/.

Carver, S. et al. (2021) Guiding principles for rewilding, *Conservation Biology*, n/a (n/a). doi:https://doi.org/10.1111/cobi.13730.

Cocks, S. and Simpson, S. (2015) Anthropocentric and ecocentric: an application of environmental philosophy to outdoor recreation and environmental education, *Journal of Experiential Education*, 38(3): 216–27. doi:10.1177/1053825915571750

Conradie, E.M. (2019) A (South) African land ethic? The viability of an ecocentric approach to environmental ethics and philosophy, in Chemhuru, M. (ed.), *African Environmental Ethics: A Critical Reader*. Cham: Springer International Publishing (The International Library of Environmental, Agricultural and Food Ethics), pp. 127–39. doi:10.1007/978-3-030-18807-8_9

Donlan, J. et al. (2005) Re-wilding North America, *Nature*, 436(7053): 913–14. doi:10.1038/436913a

Finch, T. et al. (2021) Evaluating spatially explicit sharing-sparing scenarios for multiple environmental outcomes, *Journal of Applied Ecology*, 58(3): 655–66. doi:10.1111/1365-2664.13785

Hawkins, S.A. et al. (2020) Community perspectives on the reintroduction of Eurasian lynx (Lynx lynx) to the UK, *Restoration Ecology*, 28(6): 1408–18. doi:10.1111/rec.13243

IUCN (2016) *Restoration Opportunities Assessment Methodology (ROAM)*, IUCN. Available at: www.iucn.org/theme/forests/our-work/forest-landscape-restoration/restoration-opportunities-assessment-methodology-roam (accessed: 9 February 2022).

Leopold, A. (1949) *A Sand County Almanac and Sketches Here and There*. New York: Oxford University Press.

Loveridge, R. et al. (2020) Measuring human wellbeing: A protocol for selecting local indicators, *Environmental Science & Policy*, 114: 461–9. doi:10.1016/j.envsci.2020.09.002

Martin, A. et al. (2021) Taming rewilding—from the ecological to the social: How rewilding discourse in Scotland has come to include people, *Land Use Policy*, 111: 105677. doi:10.1016/j.landusepol.2021.105677

Pedersen, P.B.M. et al. (2019) Trophic Rewilding Advancement in Anthropogenically Impacted Landscapes (TRAAIL): A framework to link conventional conservation management and rewilding, *Ambio* [Preprint]. doi:10.1007/s13280-019-01192-z

Perino, A. et al. (2019) Rewilding complex ecosystems, *Science*, 364(6438): eaav5570. doi:10.1126/science.aav5570

Pettorelli, N. et al. (2018) Making rewilding fit for policy, *Journal of Applied Ecology*, 55(3): 1114–25. doi:10.1111/1365-2664.13082

Schulte to Bühne, H., Pettorelli, N., and Hoffmann, M. (2021) The policy consequences of defining rewilding, *Ambio* [Preprint]. doi:10.1007/s13280-021-01560-8

Soule, M.E. and Noss, R.F. (1998) Rewilding and biodiversity: complementary goals for continental conservation, *Wild Earth*, 8(3): 18–28.

Svenning, J.-C. et al. (2016) Science for a wilder Anthropocene: Synthesis and future directions for trophic rewilding research, *Proceedings of the National Academy of Sciences of the United States of America*, 113(4): 898–906. doi:10.1073/pnas.1502556112

Torres, A. et al. (2018) Measuring rewilding progress, *Philosophical Transactions of the Royal Society B: Biological Sciences*, 373(1761): 20170433. doi:10.1098/rstb.2017.0433

Ward, K. (2019) For wilderness or wildness? Decolonising rewilding. In Pettorelli, N., Durant, S.M., and du Toit, J.T. (eds), *Rewilding*. Cambridge University Press, pp. 34–54. Available at: www.scopus.com/inward/record.uri?eid=2-s2.0-85070630066&partnerID=40&md5=3814873620da51edd8d5c84caafeecfa.

PART II

Theoretical and practical underpinnings of rewilding

6

TROPHIC CASCADES AS A BASIS FOR REWILDING

T.J. Clark-Wolf and Mark Hebblewhite

Introduction

Rewilding is a conservation strategy that is becoming increasingly popular throughout the world. Although the term has multiple meanings, rewilding usually indicates a long-term conservation goal of maintaining or increasing biodiversity, while reducing the impact of present and past human interventions through reintroduction of extirpated species and ecological restoration. Because one of the most important taxonomic groups to have suffered human-caused extirpation are large carnivores (Ripple et al., 2014), they have often become the focus of rewilding strategies. One additional justification for large carnivore rewilding is that they will cause trophic cascades, via trickle-down effects of top predators on plants and other species in an ecosystem. Here, we review the evidence for trophic cascades and the utility of using this strategy to support rewilding of large carnivores.

Most conservationists are familiar with trophic cascades as a basis for rewilding from Yellowstone National Park (YNP) where wolves were reintroduced in 1995. This concept has been popularised by a video called 'How Wolves Change Rivers', which combines footage of Yellowstone with narration by journalist George Monbiot (Sustainable Human, 2014). This video has been viewed > 43 million times, been adopted in school curriculums globally, and has contributed to popularising/mainstreaming the concept that rewilding wolves has caused trophic cascades in Yellowstone. In the video, Monbiot explains that when wolves were reintroduced to the park, they 'radically changed the behavior of the deer [sic]' to avoid areas where they were most vulnerable to predation, and 'bare valley sides quickly became forests of aspen and willow and cottonwood' due to decreased herbivory by deer. The video also states that trophic cascades from wolves helped change the behaviour of rivers and decreased soil erosion in the park. Therefore, 'the wolves … transform not just the ecosystem of the Yellowstone National Park … but also its physical geography'. Although the video is provocative and the idea of trophic cascades by wolves in Yellowstone is highly cited, there is no scientific consensus that these effects have occurred, which we review below. Despite this ambiguity, wolf-induced trophic cascades in Yellowstone are the paradigm of the benefits of large carnivore reintroduction.

In this chapter, we aim to provide some clarity for those who are interested in rewilding to create trophic cascades as a conservation tool. The chapter is in four sections: an introduction

DOI: 10.4324/9781003097822-8

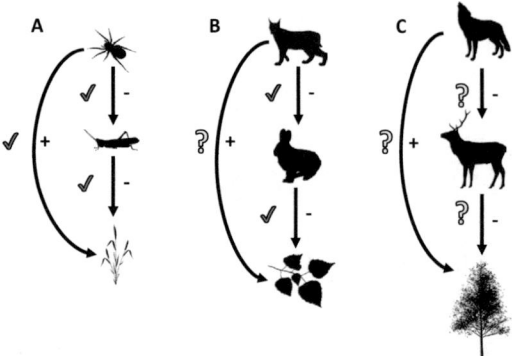

Figure 6.1 Evidence for trophic cascades across ecosystems (**A**), in large carnivores (**B**), and in Yellowstone (**C**). Although **A** and **B** show Connecticut grassland and boreal forest food webs respectively, we intend these silhouettes to be more general and represent trophic cascades across ecosystems and in large carnivores, respectively. Straight lines from predator to herbivore represent the direct effect of predation, which is negative on the herbivore. Straight lines from herbivore to primary producer represent the direct effect of herbivory, which is negative on the primary producer. Curved lines from predator to primary producer represent the indirect effect, which is positive on the primary producer. Checkmarks show that there is substantial positive evidence for the direct or indirect effect, although there may be a few exceptions. Question marks show that there is mixed evidence/a lack of scientific consensus for the effect. We did not find any effects which had a total lack of evidence. (**A**) shows the classic spider-grasshopper-grass/forb trophic cascade (e.g., Schmitz 2008), which here represents evidence for trophic cascades across all ecosystems. (**B**) shows a lynx-snowshoe hare-deciduous shrub trophic cascade, where although direct effects occurred, an indirect effect was not found (Sinclair et al*., 2000). This represents evidence for trophic cascades in large carnivore systems. (**C**) shows a wolf-elk-aspen trophic cascade in Yellowstone, where there is mixed evidence for trophic cascades (Peterson et al., 2014, 2020; MacNulty et al., 2020).

and history of trophic cascades, evidence for large carnivore trophic cascades, evidence for trophic cascades in Yellowstone, and whether we should justify rewilding on the basis of trophic cascades. Throughout, we will refer to Figure 6.1, which summarises the best available evidence on trophic cascades in the first three chapter sections. One thing to expect in the scientific literature is a lack of consensus—where this is the case, we will categorise an interaction as 'maybe'. If we categorise an interaction as 'yes' or 'no', this means that there is majority evidence to suggest that this interaction does or does not occur, but there might be counterexamples one can find in the literature. We will provide evidence for and answer questions surrounding trophic cascades and whether the addition or removal of carnivores will have cascading consequences for ecosystem functions.

Basics of trophic cascades

What is a trophic cascade? A *trophic cascade* is caused by reduced herbivory of plants in response to carnivores' numerical reduction of herbivorous prey, leading to increases in plants' growth, biomass, cover, reproduction, or survival. Therefore, a trophic cascade requires evidence of three interactions: 1) reduction of herbivores by carnivores; 2) reduction of plants by herbivores; and 3) the indirect increase in plants by carnivores via suppression of herbivory (Ford & Goheen, 2015). Three-species cascades are the simplest and most familiar cases (e.g., Yellowstone's wolf-elk-vegetation). The archetypal form of trophic cascades has been generalised and blurred due to the popularisation of the idea to include many other indirect food web interactions (e.g., Ripple et al., 2016). However, we will focus on the trophic cascade that is invoked in rewilding campaigns. Carnivores demographically (consumptive—eating/killing) or behaviourally (non-consumptive—scaring) suppress herbivorous prey, leading to plants' enhanced growth, biomass, reproduction, survival, etc.

Trophic cascades have been known about since at least Darwin's time (Ripple et al., 2016). Many of the early examples illustrate what happens when top predators are eliminated from ecosystems. Aldo Leopold (1949) wrote about the destruction of habitat by deer in the US following wolf extirpation in his *Thinking Like a Mountain*. To quote: 'I have watched the face of many a newly wolfless mountain, and...seen every edible bush and seedling browsed, first to anemic desuetude, and then to death'. Leopold is presaging that the losses of top predators in an ecosystem can abruptly change vegetation and move ecosystems into different states. Later, Hairston, Smith, and Slobodkin (1960) introduced this idea into scientific discourse as the *green world hypothesis*, that is, predators allow plants to flourish by reducing the numbers of herbivores.

One notable example of trophic cascades is in Pacific kelp forests, where sea otters were hunted to near extinction during the maritime fur trade in the 18th–19th centuries. Sea otters predated on herbivorous sea urchins, so the eradication of sea otters cascaded to defoliate kelp forests throughout the Pacific (Estes and Palmisano, 1974). As sea otter populations recover and decline, ecosystems shifted between the states of the kelp-dominated (otters) and urchin-dominated (no otters). Another example of trophic cascades occurred in Venezuela where hydroelectric dams flooded an area in the tropics creating small islands free of predators. This led to 10–100 times greater numbers of rodents, howler monkeys, iguanas, and leaf-cutter ants on these predator-free islands than the mainland, cascading to severely reduce seedlings and saplings of canopy trees (Terborgh et al., 2001).

Trophic cascades occur through two nonexclusive mechanisms—density mediation or behavioural mediation (Ripple et al., 2016). *Density-mediated trophic cascades* occur through a numerical reduction of herbivores by predation, which then lead to increase biomass of primary producers. *Behaviourally-mediated trophic cascades* involve nonlethal, antipredator responses of herbivores to the risk of predation. For example, seagrass productivity increased in response to declined foraging by dugongs and sea turtles in the presence of tiger sharks (Burkholder et al., 2013). There are also *knock-on effects*, where changes in plants via trophic cascades can trigger further changes back up the food chain (Ripple et al., 2016). One controversial example is the trophic cascade of wolves to elk to berry-producing shrubs in Yellowstone, which has been hypothesised to have a knock-on effect to berry-consuming grizzly bears (Barber-Meyer, 2015). For the rest of this chapter, we will primarily focus on trophic cascades, not broader indirect or knock-on effects.

Behaviourally-mediated trophic cascades can be driven by fear of predation, which generate what is known as the *landscape of fear*. This is a mental map that describes the continuous change in predation risk that an animal perceives as it navigates its physical landscape (see Figure 4 in Kohl et al., 2018). The landscape effects of fear can cascade from individuals to ecosystems, including changes in prey physiology, demography, plant growth, and nutrient cycling (Laundré et al., 2001; Laundré et al., 2010; Kohl et al., 2018). However, predator hunting mode and prey antipredator behaviour can modulate the strength of the landscape of fear. For example, active-search hunting predators are expected to have weak/nonexistent fear-mediated effects, whereas sit-and-wait ambush predators should have the strongest effects (Schmitz, 2008). Antipredator vigilance and avoidance of risky places can be negligent in food-limited prey that can easily escape predators without vigilance (Brown, 1999). These behaviours, in addition to predation risk being unevenly distributed across space and time due to highly mobile predators and prey, can lead to highly dynamic landscapes of fear (Kohl et al., 2018). We will revisit these ideas later when we discuss trophic cascades in Yellowstone.

Many of the earliest and best evidence for trophic cascades comes from aquatic ecosystems. This is because whole lakes with contrasting fish communities can be more readily experimented on to generate trophic cascades with the addition or removal of top fish predators. In North American lakes, piscivorous fish (like bass, pike, or salmon) can reduce populations of zooplanktivorous fish (like minnows), which can affect zooplankton communities, which in turn can change phytoplankton levels (Carpenter et al., 1985). Removal of piscivorous fish can therefore change water from clear to green by allowing phytoplankton to flourish. In the Eel River in northern California, steelhead and roach consume fish larvae and predatory insects. These smaller predators prey on midge larvae, which feed on algae. Experimental manipulations of these food webs by removing steelhead and roach led to large increases in algal biomass (Power, 1990). Despite these classic examples, trophic cascades in lake ecosystems are certainly not ubiquitous, with one review finding only 7 of 17 studies of piscivores on algae supporting trophic cascades (Drenner & Hambright, 2002).

Lake ecosystems have had a wide range of experiments and are quite robust, but experiments are a lot harder to do in marine and terrestrial ecosystems. Shurin et al. (2002) ran a meta-analysis of 102 trophic cascade experiments across different ecosystem types and found that the effect of predator removals on plant biomass were stronger in aquatic ecosystems than terrestrial ecosystems. These differences may relate to food web complexity (aquatic webs are simpler and thus cascades are stronger), metabolic rates (large herbivores in terrestrial systems have higher metabolic rates and are less efficient consumers relative to aquatic herbivores), and lack of plant defenses in aquatic systems (reduces strength of herbivory in terrestrial systems), which may lead to weak or nonexistent trophic cascades in terrestrial food webs (Shurin et al., 2002). Yet, many terrestrial predators such as spiders, beetles, lizards, guilds of birds, weasels, ants, and many others have been found to have cascading effects on plants (Terborgh & Estes, 2010).

Trophic cascades in large carnivores

Trophic cascades are difficult to experimentally study in large carnivores, especially due to their wide-ranging behaviour. Because of this, very few trophic cascades have been robustly experimentally studied (Ford & Goheen, 2015). Many studies rely on positive correlations between carnivore and plant abundance, which are not rigorous tests of trophic cascades because they are without experimental control, replications, and multiple competing hypotheses (see Box 1.1; Ford & Goheen, 2015). Because of this, much of the evidence for trophic cascades relies on historical observations of large increases in herbivores accompanied with intensive browsing on plants following large carnivore eradication (see Ripple et al., 2010).

Deer populations exploded throughout North America by the early 1900s following the extirpation of large predators like wolves, cougars, grizzly and black bears. For example, deer biomass was a fifth in areas with wolves than where wolves were absent or rare (Crête & Daigle, 1999). There is also evidence that large carnivores can reduce ungulate populations (see Figure 3A in Ripple & Beschta, 2012 and Figure 2 in Peterson et al., 2014), where each additional predator species results in a stepwise reduction in ungulate abundance. However, in some cases, functional redundancy and interspecific competition may drive compensatory mortality in ungulates. For example, Tallian et al. (2016) found that theft of wolf-killed ungulates by grizzly or brown bears slowed down wolf predation in Yellowstone and Scandinavia, respectively. In sum, there is convincing evidence that predation by large carnivores can have an important top-down influence and reduce ungulate abundance (Figure 6.1), however, counterexamples exist and the role of interspecific competition between predators and human harvest on predator regulation of prey needs to be explored further (Vucetich et al., 2005).

When deer populations exploded throughout North America, many plant communities were reshaped (Ripple et al., 2010). On the Allegheny Plateau in northwestern Pennsylvania, USA, following the extirpation of wolves and mountain lions in the late 1800s, overabundant deer eliminated most northern hardwood saplings like eastern hemlock and deciduous shrubs. A long-term, replicated experiment found that deer reduced sapling density and growth rates, while also favouring browse-tolerant species (Horsley et al., 2003). Deer may be moving the northwest Pennsylvania forests into an alternative state characterised by dense grasses and ferns that inhibit establishment of tree seedlings (Ripple et al., 2010). In sum, intense herbivory by overabundant deer is occurring throughout North America and there is widespread evidence that ungulates impact the structure of ecosystems through herbivory (Côté et al., 2004). But can large carnivore rewilding reverse these continental scale patterns in plant recruitment?

We will now discuss the evidence that large carnivores can cause trophic cascades by discussing one of the most robust natural experiments with wolves in Banff National Park, Canada (Hebblewhite et al., 2005). Wolves naturally recolonised Banff through dispersal in the early 1980s. High human activity surrounding the town of Banff displaced wolves from the region, creating a natural experiment to study the effects of wolves in areas of high human activity ('low wolves') and low human activity ('high wolves'). Elk were a magnitude more numerous, survival and recruitment were higher, and predation rates were lower in the low wolf area. Predation of wolves on elk cascaded to plants, where willow production was 7x greater in high wolf areas due to reduced browsing. High wolf areas also had decreased aspen browsing and increased recruitment (Hebblewhite et al., 2005). The effects on plants led to cascading indirect or knock-on effects for birds and beavers. Songbird diversity and abundance doubled in high-wolf areas due to denser, tall willows available for habitat for nesting and reproduction. The number of active beaver lodges increased due to reduced elk densities and thus, greater abundance of willow via wolf predation (Hebblewhite & Smith, 2010). In sum, the natural experiment in Banff provides compelling evidence that large carnivores like wolves can create trophic cascades that reshape ecosystems.

There are many other examples of purported large carnivore trophic cascades in the literature (Terborgh & Estes, 2010; Estes et al., 2011)—including dingoes in Australia, jaguars on islands in South America, and most notably, wolves in Yellowstone. However, one weakness of many of these studies (which is common in ecology due to difficulties in doing well-controlled experiments), is that they are uncontrolled, unreplicated, and rely on correlations between plant and carnivore abundance (see Box 1.1 and Ford & Goheen, 2015). For example, most of the trophic cascade studies in Yellowstone (e.g., Painter et al., 2015) compare age or size class distributions of plants in areas with and without wolves present, but without a contemporaneous control to ensure such changes were not because of other factors, like weather. The trophic cascade hypothesis predicts that a range of plant age or size classes will be missing when carnivores were extirpated, due to intense elk herbivory. However, this approach rests on the assumptions that: 1) the loss of wolves can account for increases in elk; 2) increased elk led to loss of age/size classes; and 3) that other factors (e.g., weather, disease, other predators) had comparable effects over time on plant and elk abundance, across areas with wolves present and absent (Ford & Goheen, 2015). Unfortunately, these assumptions are rarely tested in the well-known literature in Yellowstone, with many alternative drivers left unconsidered.

In fact, Ford & Goheen (2015) found only two large carnivore experiments out of six where the length of food chains was manipulated (e.g., added/removed predators, herbivores, etc.) and trophic cascades occurred. For example, in the famous Kluane experiment, Sinclair et al.

(2000) found that lynx suppressed hares, and hares suppressed woody plants, but the abundance of woody plants was indistinguishable regardless of whether lynx were present. Similarly, Ford et al. (2015) found that the recovery of African wild dogs in the Laikipia Plateau in central Kenya reduced their primary prey, dik-dik, and dik-dik browsing suppressed tree abundance, but a trophic cascade did not occur. These negative results in the trophic cascade literature point to the complicated nature of the indirect effect of carnivores on plants. Many herbivores are not vulnerable to predation, and many species of herbivores do not suppress plants. Additionally, many plants have high defenses against herbivory. Moreover, food web reticulation, predator hunting strategies and antipredator defences, environmental heterogeneity, and many other factors are known to weaken or eliminate trophic cascades. We suggest that future research should focus on identifying the factors that drive large carnivore trophic cascades, especially with such mixed evidence, instead of just identifying cascades in systems.

Trophic cascades in Yellowstone

Restoration of wolves to Yellowstone was in fact motivated by the intention to restore natural regulation of elk by large carnivores due to overbrowsing in the Yellowstone northern range (Varley & Brewster, 1992). The introduction of wolves into Yellowstone correlated with a precipitous decline in elk populations. Subsequently, many studies inferred that the change in elk numbers or fear of predation decreased herbivory on deciduous shrubs like aspen, willow, and cottonwood. Thus, wolves in Yellowstone could have indirect effects on plants, which could possibly cascade via knock-on or indirect effects to other species, such as beavers. Here, we will review the most current evidence in Yellowstone for trophic cascades, starting with the first logical step of establishing whether wolves indeed caused elk abundance to decline.

There is no scientific consensus on whether wolves were the primary cause of the recent Yellowstone elk population decline. Wolves certainly contributed to the decline in elk, but what is in question is how much of the decline is due to wolves. For example, from 1995 to 2001, wolf populations were small, predation rates were low (3.7%), and they had only a small effect on elk abundance (Peterson et al., 2014; Metz, 2021). Later (2005–2010), elk predation rates increased to around 14% (see Figure 1b in Peterson et al., 2014). However, wolves are low success hunters of a wide range of prey that are typically small, old, in poor health, or rendered vulnerable by landscape features. Predation success rates hardly succeed 20% and are less than 10% in elk. Half of elk killed are calves, and 85% of female adult elk are more than ten years old, indicating that wolves could be primarily compensatory sources of mortality on elk (MacNulty et al., 2020).

There are also a host of alternative hypotheses that drove the decline in elk in Yellowstone (Peterson et al., 2014; MacNulty et al., 2020). The density of grizzly bears in Yellowstone may have tripled between the early 1980s and late 1990s, which correlated with an increase in elk calf mortality. Barber-Meyer et al. (2008) found that bears were the leading cause of elk calf mortality, and that mortality by grizzly and black bears increased from ~23% in 1987–1990 to ~60% in 2003–2005. Similarly, the average number of cougars in northern Yellowstone increased 76% from 1987–1993 to 1998–2004, which could be another predation pressure exerted on elk. Another factor affecting elk abundance is superadditive harvest of female elk by human hunters (Vucetich et al., 2005). Human harvest of female elk, in combination with multiple predation pressures and dwindling numbers, may become a factor reducing elk populations. Bison populations in the park have also been increasing since wolf reintroduction, and dramatically shifted to the northern range where they overlap with elk more. Although early studies indicated that competition between elk and bison was insignificant, these factors may have changed in the present (MacNulty et al., 2020). Other climatic factors such as drought can also modulate the effects of large carnivores and harvest, which is projected to increase in the future with climate warming (Peterson et al., 2014; MacNulty et al., 2020). In summary, the declining number of elk in Yellowstone could still cause a trophic cascade, but not solely due to the reintroduction of wolves into the park.

Despite the lack of consensus on the numerical effects of wolves on elk, many researchers proposed that the landscape of fear was established soon after wolves were reintroduced, causing elk to avoid risky places where wolves could kill them (Peterson et al., 2014). The landscape of fear was inferred from increased vigilance behaviour, shifts in habitat use from nutritious, high-risk open areas to poor-nutrition, low-risk closed forests, decreased diet quality and body fat, and reduced pregnancy rates (see Introduction of Kohl et al., 2018). But other studies have found that elk selected for open areas, maintained body condition and pregnancy rates, and did not change their migratory patterns or winter home ranges in the presence of wolves. Kohl et al. (2018) used extensive GPS data from elk and wolves in Yellowstone to find that the landscape of fear is highly dynamic. Wolves hunted at dawn and dusk, and so elk avoided riskier places when wolves were most active, but safely accessed those places during lulls in wolf activity. Similarly, Kohl et al. (2019) found that in response to predation pressure from both cougars and wolves, elk avoided risky areas in places where both were active, using forested rugged areas during daylight (when cougars are least active) and grassy, open areas at night (when wolves are least active). Lastly, Cusack et al. (2020) found that elk did not change their home ranges in response to wolves. In fact, elk encountered wolves every 7 to 11 days, and frequently survived those encounters. This indicates that changes in temporal use of habitat along with antipredator behaviour during encounters with wolves (e.g., grouping, fighting back, running) help decrease chronic fear-induced effects (MacNulty et al., 2020). In fact, scientific consensus is leaning towards the hypothesis that if any trophic cascade from wolves to plants in Yellowstone is occurring, it is likely a result of a density-mediated, not behaviourally-mediated trophic cascades (Kauffman et al., 2010; Painter et al., 2015; Kohl et al., 2018).

Next, we examine the second causal step in establishing whether wolves caused a trophic cascade, whether there were increased productivity of plants. There is some scientific agreement that reduced densities of elk are driving increased productivity in deciduous plants in Yellowstone, however these changes in growth may be due to multiple factors other than or in interaction with wolf predation, and changes in growth are spatially heterogeneous (Peterson et al., 2020). Aspen, willow, and cottonwood have not grown taller than a height of 1m for decades in northern Yellowstone, and by 2006, aspen saplings have grown to ~1.5m (Peterson et al., 2014). There are large gaps in recruitment of aspen and cottonwood after the 1920s when wolves and other large carnivores were mostly absent (Painter et al., 2015; see Figure 15.1 in Peterson et al., 2020). Other plants had high levels of herbivory from elk, including alder, berry-producing shrubs, and relatively unpalatable conifers (Peterson et al., 2020). However, there are a host of competing hypotheses which could cause the growth of plants like aspen, including temperature, moisture, snow pack, bison herbivory, and competition with conifers.

Plant responses for other woody species are more complex. For example, increased height in willow have occurred with decreased elk herbivory, but the response appears patchy and related to the surrounding water table, beaver flooding, and other factors. Willow was only able to grow tall when protected from elk herbivory in places where the water table was experimentally raised and beavers were restored (Marshall et al., 2013). To emphasise the challenge that interpreting complex trophic cascades cause with uncontrolled, non-experimental observations, consider the beaver. As wolf reintroduction to Yellowstone has been strongly popularised, few are aware that beavers—the ultimate ecosystem engineer (Rosell et al., 2005)—were reintroduced just outside the park from 1986 to 1999 by the US Forest Service (Smith & Tyers, 2012). Thus, combined with Marshall's innovative beaver dam experiments, it is equally as likely that the reintroduction of beavers has been the main trigger of willow responses in Yellowstone. The spatial variability in beaver recovery is also strongly consistent with the spatial variation in willow recovery.

Some point out that the discussion of trophic cascades in Yellowstone has only focused on communities (aspen, willow, and cottonwood) that represent 4% of its total area, and that although these communities are important and biodiverse, many other ecosystem processes occur in the 75% of the area that is comprised by grasslands and coniferous forest which have been unexplored (Peterson et al., 2014). We are only familiar with one study that examined the effects of trophic cascades on grasslands in Yellowstone, which found that reductions in elk decreased grazing intensity and changed biochemical cycling (Frank, 2008). However, this study was carried out during times with relatively insignificant levels of wolf predation (1995–2000), leading some to question these findings (Peterson et al., 2014). Moreover, bison are known to be strong ecosystem engineers of Yellowstone's grasslands (Geremia et al., 2019). However, bison are much harder for wolves to kill than elk, leading wolves in Yellowstone to scavenge bison more often, which implies a lack of direct numerical response of wolves on bison (Tallian et al., 2017; Metz, 2021), and therefore, little likelihood of a trophic cascade.

So in conclusion, we are challenged to answer whether trophic cascades are occurring from wolves to elk to plants in Yellowstone. Much of the evidence for trophic cascades examined age/stage distributions of plants with assumptions that missing age/stage class distributions are due to heightened elk herbivory from missing wolves (Painter et al., 2015; Peterson et al., 2020). As mentioned previously, this method is not experimentally robust and makes some assumptions to arrive at the conclusion of trophic cascades (Ford & Goheen, 2015). Furthermore, these studies, particularly for aspen, are plagued by poor sampling designs and biased field measurements (Brice et al., 2021). Evidence from others (Kauffman et al., 2010; Painter et al., 2015; Kohl et al., 2018) suggests that any indirect effect of wolves on plants is mainly the result of density-mediated trophic cascades. For example, in an experimental test of the behaviourally-mediated trophic cascade hypothesis, Kauffman et al. (2010) found that elk herbivory on aspen demography did not change in areas where elk are at higher risk of predation by wolves. These results lead us to conclude that the evidence for trophic cascades in Yellowstone is mixed. To summarise the most current thinking of effects of elk herbivory on plants (see Peterson et al., 2020): plants have grown taller in some areas, but this is spatially and temporally variable for many reasons including browsing intensity and groundwater avail-ability, and in some areas bison grazing is counteracting reductions in elk herbivory. There is some evidence of knock-on effects of willow growth to songbirds (Baril et al., 2011), but little evidence to show knock-on effects to beavers (Marshall et al., 2013), grizzly bears (Barber-Meyer 2015), or other species. More robust experiments are needed that explore all trophic levels in the system (wolves, elk, plants), instead of just plants, to move beyond this uncertainty (sensu Ford & Goheen, 2015).

Should we rewild large carnivores to produce trophic cascades?

Three important questions to ask when evaluating the potential for rewilding large carnivores to produce trophic cascades are: 1) what do we expect from a trophic cascade; 2) how relevant are general trophic cascades to the specific system in question; and 3) what are other reasons to reintroduce large carnivores? First, introducing a large carnivore to an ecosystem is certainly going to have consequences for an ecological community but predicting the flow and magnitude of the cascade could be exceedingly difficult. As we have shown (Figure 6.1), trophic cascades are not ubiquitous in systems with large carnivores. In some cases, the effects of trophic cascades may be different from what one expects. For example, Schmitz (2006) found that spider predators had no indirect effect on the biomass of a plant community, but they affected plant community composition that increased nitrogen cycling and light penetration. Similarly, although reduced elk herbivory may have increased the growth of aspen sapling height in Yellowstone (Painter et al., 2015; Peterson et al., 2020), sapling growth may be a relatively unimportant factor of aspen demography in clone maintenance (Peterson et al., 2014). This is all to say that reintroducing large carnivores may induce trophic cascades, but whether they matter and whether they can reshape ecosystems will be extremely context specific, depending on the ecological history, complexity of the community to which the large carnivores are restored, primary productivity of the system, and ultimately, the human impacts.

Secondly, many of the places where large carnivore trophic cascades are being reported are in relatively intact areas with small human footprints—like National Parks in the US. However, most ecosystems on Earth are dominated by humans, with huge legacies of destructive land use and harvest. Trophic cascades may be relatively minimal or nonexistent in human-modified landscapes where the effects of humans overshadow any direct or indirect effects of carnivores. For example, the recolonisation of wolves in Sweden led to a change in how humans hunted moose, which reduced numerical effects of wolves on moose (Wikenros et al., 2015). Muhly et al. (2013) found that outside protected areas in Alberta, humans affected vegetation through agriculture and forest modification, which weakened top-down trophic cascades from wolves to elk. To quote the famous wolf biologist David Mech (2012): 'How significant a beneficial effect can wolves have on songbirds compared with the negative effects of logging, grazing, clearing, or farming? How important would wolf effects on trout be where trout are stocked and harvested, streams are polluted, and river banks grazed?' We recommend deeply considering the extensive impacts from human land-use and high human densities which are considerable, prior to reintroducing large carnivores to putatively create trophic cascades.

Lastly, as conservation biologists, it can seem that often the justification for conserving or reintroducing large carnivores is so that trophic cascades can be created (Peterson et al., 2014). As we have shown, the potential for restoring trophic cascades with reintroduced large carnivores needs further research, but could still be an important tool in restoring ecosystems. But why not also argue for carnivore reintroduction on other moral, ethical, or cultural grounds— such as the value of biodiversity or the importance of carnivores to Indigenous Peoples (e.g., Shelley et al., 2011)? We think these considerations could be more productive for planning and prioritising large carnivore reintroduction. To quote Aldo Leopold (1949) again, 'A thing is right when it tends to preserve the integrity, stability, and beauty of the biotic community. It is wrong when it tends otherwise.'

Acknowledgements

TJCW was supported by the National Science Foundation Graduate Research Fellowship (grant 366280). MH's work on trophic cascades in Banff was supported by Parks Canada and NSF LTREB grant #s 1556248 and 2038704.

References

Barber-Meyer, S.M. 2015. Trophic cascades from wolves to grizzly bears or changing abundance of bears and alternate foods? *J Anim Ecol*, 84: 647–51.

Barber-Meyer, S.M, Mech, L.D., and White, P.J. 2008. Elk Calf Survival and Mortality Following Wolf Restoration to Yellowstone National Park.*Wildl Monogr*, 1–30.

Baril, L.M., Hansen, A.J., Renkin, R., and Lawrence, R. 2011. Songbird response to increased willow (Salix spp.) growth in Yellowstone's northern range. *Ecol Appl*, 21: 2283–96.

Brice, E.M., E.J.Larsen, and D.R. MacNulty. 2021. Sampling bias exaggerates a textbook example of a trophic cascade. *Ecology Letters.*

Brown, J.S. 1999. Vigilance, patch use and habitat selection: Foraging under predation risk. *Evol Ecol Res*, 1: 49–71.

Burkholder, D.A., Heithaus, M.R., Fourqurean, J.W., et al. 2013. Patterns of top–down control in a seagrass ecosystem: Could a roving apex predator induce a behaviour-mediated trophic cascade? *J Anim Ecol*, 82: 1192–1202.

Carpenter, S.R., Kitchell, J.F., and Hodgson, J.R. 1985. Cascading trophic interactions and lake productivity. *Bioscience*, **35**: 634–9.

Côté, S.D., Rooney, T.P., Tremblay, J.P., et al. 2004. Ecological impacts of deer overabundance. *Annu Rev Ecol Evol Syst*, 35: 113–47.

Crête, M. and Daigle, C. 1999. Management of indigenous north american deer at the end of the 20th century in relation to large predators and primary production. *Acta Vet. Hungarika*, 47, 1–16.

Cusack, J.J., Kohl, M.T., Metz, M.C., Coulson, T., Stahler, D.R., Smith, D.W., et al. 2020. Weak spatio-temporal response of prey to predation risk in a freely interacting system. *J. Anim. Ecol.*, 89, 120–31.

Drenner, R.W. and Hambright, K.D. 2002. Piscivores, trophic cascades, and lake management. *Sci World*, 2: 284–307.

Estes, J.A. and Palmisano, J.F. 1974. Sea otters: their role in structuring nearshore communities. *Science (80-)*, 185: 1058–60.

Estes, J.A., Terborgh, J., Brashares, J.S., Power, M.E., Berger, J., Bond, W.J., Carpenter, S.R., Essington, T.E., Holt, R.D., Jackson, J.B.C., Marquis, R.J., Paine, R.T., Schoener, T.W., Shurin, J.B., and Sinclair, A.R.E. 2011. Trophic downgrading of Planet Earth. *Science*, 333 (6040): 301–6.

Ford, A.T., Goheen, J.R., Augustine, D.J., Kinnaird, M.F., O'Brien, T.G., Palmer, T.M., et al. 2015. Recovery of African wild dogs suppresses prey but does not trigger a trophic cascade. *Ecology* 96 (10): 2705–14.

Ford, A.T. and Goheen, J.R. 2015. Trophic Cascades by Large Carnivores: A Case for Strong Inference and Mechanism. *Trends Ecol Evol*, **30**: 725–35.

Frank, D.A. 2008. Evidence for top predator control of a grazing ecosystem. *Oikos*, 117: 1718–24.

Geremia, C., Merkle, J.A., Eacker, D.R., Wallen, R.L., White, P.J., Hebblewhite, M., and Kauffman, M.J. 2019. Migrating bison engineer the green wave. *Proceedings of the National Academy of Sciences*, 1–7.

Hairson, N.G., Smith, F.E., and Slobodkin, L.B. 1960. Community structure, population control, and competition. *The American Naturalist*, 94(879): 421–5.

Hebblewhite, M. and Smith, D.W. 2010. Wolf community ecology: ecosystem effects of recovering wolves in Banff and Yellowstone National Parks. In: *The World of Wolves: New Perspectives on Ecology, Behavior, and Policy*. Calgary, Alberta, Canada: University of Calgary Press.

Hebblewhite, M., White, C.A., Nietvelt, C.G., et al. 2005. Human activity mediates a trophic cascade caused by wolves. *Ecology*, 288(8): 2135–44.

Horsley, S.B., Stout, S.L., and DeCalesta, D.S. 2003. White-tailed deer impact on the vegetation dynamics of a northern hardwood forest. *Ecol Appl*, 13: 98–118.

Kauffman, M.J., Brodie, J.F., and Jules, E.S. 2010. Are wolves saving Yellowstone's aspen? a landscape-level test of a behaviorally mediated trophic cascade. *Ecology*, 91: 2742–55.

Kohl, M.T., Stahler, D.R., Metz, M.C., et al. 2018. Diel predator activity drives a dynamic landscape of fear. *Ecol Monogr*, 0: 221440.

Kohl, M.T., Ruth, T.K., Metz, M.C., Stahler, D.R., Smith, D.W., White, P.J., et al. (2019). Do prey select for vacant hunting domains to minimize a multi-predator threat? *Ecol. Lett.*, ele.13319.

Laundré, J.W., Hernández, L., and Altendorf, K.B. 2001. Wolves, elk, and bison: reestablishing the 'landscape of fear' in Yellowstone National Park, U.S.A. *Can. J. Zool.*, 79: 1401–9.

Laundré, J.W., Hernández, L., and Ripple, W.J. 2010. The landscape of fear: ecological implications of being afraid. *The Open Ecology Journal*, 3(1): 1–7.

Leopold A. 1949. *A Sand County Almanac: And Sketches Here and There*. Oxford University Press.

MacNulty, D.R., Stahler, D.R., Wyman, T., et al. 2020. Population dynamics of northern Yellowstone elk after wolf reintroduction. *Yellowstone Wolves Sci Discov World's First Natl Park*: 184–99.

Marshall, K.N., Thompson Hobbs, N., and Cooper, D.J. 2013. Stream hydrology limits recovery of riparian ecosystems after wolf reintroduction. *Proc R Soc B Biol* Sci, 280: 20122977.

Mech, L.D. 2012. Is science in danger of sanctifying the wolf? *Biol Conserv.*, 150 (1): 143–9.

Metz, M. 2021. Estimating wolf predation metrics, patterns and dynamics across time and space in the multiprey system of Yellowstone National Park. Ph.D. Dissertation. University of Montana, Missoula, MT, USA.

Muhly, T.B., Hebblewhite, M., Paton. D., et al. 2013. Humans strengthen bottom-up effects and weaken trophic cascades in a terrestrial food web.*PLoS One*, 8(5): e64311. https://doi.org/10.1371/journal.pone.0064311.

Painter, L.E., Beschta, R.L., Larsen, E.J., and Ripple, W.J. 2015. Recovering aspen follow changing elk dynamics in Yellowstone: Evidence of a trophic cascade? *Ecology*, 96: 252–63.

Peterson, R.O., Beschta, R.L., Cooper, D.J., et al. 2020. Indirect effects of carnivore restoration on vegetation. In: Douglas W. Smith, Daniel R. Stahler and Daniel R. MacNulty (eds), *Yellowstone Wolves: Science and Discovery in the World's First National Park*, Chicago: University of Chicago Press, 2020, pp. 205–10. https://doi.org/10.7208/9780226728483-036.

Peterson, R.O., Vucetich, J.A., Bump, J.M., and Smith, D.W. 2014. Trophic cascades in a multicausal world: Isle Royale and Yellowstone. *Annu Rev Ecol Evol Syst*, 45: 325–45.

Power, M.E. 1990. Effects of fish in river food webs. *Science*, 250 (4982): 811–14.

Ripple, W.J. and Beschta, R.L. 2012. Large predators limit herbivore densities in northern forest ecosystems. *Eur J Wildl Res*, 58: 733–42.

Ripple, W.J., Estes, J.A., Schmitz, O.J., et al. 2016. What is a trophic cascade? *Trends Ecol Evol*, 31: 842–9.

Ripple, W.J., Rooney, T.P., and Beschta, R.L. 2010. Large predators, deer, and trophic cascades in boreal and temperate ecosystems. In: *Trophic Cascades: Predators, Prey, and the Changing Dynamics of Nature*. Washington, D.C., USA: Island Press.

Rosell, F., Boszer, O., Collen, P. et al. 2005. Ecological impacts of beavers Castor fiber and Castor canadensis and their ability to modify ecosystems. *Mammal Review*, 35(3): 248–76.

Schmitz, O.J. 2006. Predators have large effects on ecosystem properties by changing plant diversity, not plant biomass. *Ecology*, 87: 1432–7.

Schmitz, O.J. 2008. Effect of Predator Hunting Mode on Grassland Ecosystem Function. *Science (80-)*: 952–5.

Shelley, V., Treves, A, and Naughton, L. 2011. Attitudes to wolves and wolf policy among Ojibwe tribal members and non-tribal residents of Wisconsin's wolf range. *Hum Dimens Wildl*, 16: 397–413.

Shurin, J.B., Borer, E.T., Seabloom, E.W., et al. 2002. A cross-ecosystem comparison of the strength of trophic cascades. *Ecol Lett*, 5: 785–91.

Sinclair, A.R.E., Krebs, C.J., Fryxell, J.M., et al. 2000. Testing hypotheses of trophic level interactions: A boreal forest ecosystem. *Oikos*, 89: 313–28.

Smith, D.W. and Tyers, D.B. 2012. The history and current status and distribution of beavers in Yellowstone National Park. *Northwest Science*, 86: 276–88.

Sustainable Human. 2014. How wolves change rivers. https://youtu.be/oSBL7Gk_9QU.

Tallian, A., Ordiz, A., Metz, M.C., et al. 2016. Competition between apex predators? Brown bears decrease wolf kill rate on two continents. *Proc R Soc B*, 284.

Tallian, A., Smith, D.W., Stahler, D.R., Metz, M.C., Wallen, R.L., Geremia, C., Ruprecht, J., Wyman, C.T., and MacNulty, D.R. 2017. Predator foraging response to a resurgent dangerous prey. *Functional Ecology*, 31: 1418–29.

Terborgh, J. and Estes, J.A. (eds). 2010. *Trophic Cascades: Predators, Prey, and the Changing Dynamics of Nature*. Washington, D.C., USA: Island Press.

Terborgh, J., Lopez, L., Nuñez, P.V., et al. 2001. Ecological meltdown in predator-free forest fragments. *Science (80-)*, 294: 1923–6.

Varley, J.D. and W.G. Brewster. 1992. Wolves for Yellowstone? A report to the United States Congress, Volume IV, Research and Analysis. National Parks Service, Yellowstone National Park, Wyoming.

Vucetich, J.A., Smith, D.W., and Stahler, D.R. 2005. Influence of harvest, climate and wolf predation on Yellowstone elk, 1961–2004. *Oikos*, 111: 259–70.

Wikenros, C., Sand, H., Bergström, R., et al.2015. Response of moose hunters to predation following wolf return in Sweden. *PLoS One*, 10: 1–21.

7

SPECIES TRANSLOCATIONS, TAXON REPLACEMENTS, AND REWILDING

Mark Stanley-Price

Introduction

This chapter presents the author's reflections on over 30 years of working on species translocations with the IUCN (this includes founding and chairing the IUCN SSC Reintroduction Specialist Group, and chairing a Task Force of the IUCN Reintroduction and Invasive Species Specialist Groups which delivered the 2013 IUCN Guidelines for Reintroductions and other Conservation Translocations). Conservation translocations have grown enormously over the last 40–50 years, evidenced by the number of cases in the IUCN Global Translocation Perspectives, alongside a large increase in translocation-focused journal papers. Alongside reflecting on the development of translocation practice as a scientific discipline, I examine the relationship between species translocations and rewilding. In particular, one unresolved element around rewilding is whether it aims to replace lost taxa to move towards some (undefined) community completeness or complexity, or whether it primarily aims to restore ecological function. We may label them as the compositional and functional approaches. In the latter case, how this is achieved is less important than community intactness and substitutes may be considered to replace (extinct) taxa. Finally, I touch on some of the policy and legal issues surrounding species translocations for the purpose of rewilding.

Translocation and rewilding

Conservation translocation is the human-mediated movement of living organisms from one area, with release in another and is the overarching term to describe *reinforcement* (the intentional movement and release of an organism into an existing population of the same species), *introduction* (the intentional movement and release of an organism outside its indigenous range) and *reintroduction* (the intentional movement and release of an organism inside its indigenous range from which it has disappeared). Reintroductions in general aim to re-establish a viable population of the focal species within its indigenous range' (IUCN, 2013).

Rewilding is a relatively novel term, and has many descriptions or definitions, summarised in Pettorelli (2019). More recently, Carver et al. (2021) defined it as 'the process of rebuilding, following major human disturbance, a natural ecosystem by restoring natural processes and

DOI: 10.4324/9781003097822-9

the complete or near complete food web at all trophic levels as a self-sustaining and resilient ecosystem with biota that would have been present had the disturbance not occurred' (see Chapter 1). Rewilding owes its origins to the classic North American vision of Soulé and Noss (1998), and the origins of rewilding and its different definitions and embodiments have been discussed elsewhere in this book (Chapters 2 and 3).

The theory and practice of translocations is contained in IUCN Guidelines (IUCN, 2008, 2013) for Reintroductions and Other Conservation Translocations, whereas there is, as yet, no comparable guidance for rewilding. The 2013 Guidelines do, however, cover many salient aspects of rewilding, as discussed below.

As already indicated, there has been a significant increase in the number of translocations over the last 40–50 years, evidenced by the number of cases in the IUCN Global Translocation Perspectives, and the number of published papers on translocation, which has increased around eight-fold from the early 1980s to 2005 (Seddon, Armstrong, & Maloney, 2007). In comparison, the number of published papers primarily on rewilding increased approximately ten-times between 2003 and 2017 (Pettorelli & Durant, 2019).

This chapter has two objectives: first, to examine the relationship between species translocations and rewilding, linking this to the use of substitute or replaced taxa where an original form is no longer available (i.e. an introduction); and second, to trace the development of translocation practice as a scientific discipline, with the hope and encouragement that rewilding (with translocation as part of its foundation) will embark on a comparable trajectory. One unresolved element around rewilding is whether it aims to replace lost taxa to move towards community completeness or whether it primarily aims to restore ecological function. In the latter case, how this is achieved is less important than community intactness. We may label them as the compositional and functional approaches.

Regardless of whether an approach is compositional or functional, a common aim of rewilding is the restoration of trophic complexity (Carver et al., 2021). The impacts of restoring wolves (*Canis lupus*) to Yellowstone is a classic example (see for example Ripple & Beschta, 2012), although questions remain as discussed in Chapter 6. Following the return of the wolf in 1995 and 2010 there was a tri-trophic level cascade involving wolves, elk and several riparian tree species heavily used by elk. Further, there were many collateral impacts from the return of the top carnivore cascading down on to the meso-carnivores e.g., coyote, *Canis latrans*. The return of the wolf has also been associated with an increase in bison, scavenging bears, and songbirds in recovering stands of trees. Reduced riverine erosion following from lower numbers of elk has in places altered river morphology (Beschta & Ripple, 2019), allowing an increase in beavers and the consequent biodiversity benefits of their activities.

Lessons learned from early translocations

Well-planned, implemented, and monitored translocation of species have increased greatly both in numbers and the diversity of subject taxa. These are in marked contrast to earlier 'releases' that were often carried out without any assessment of feasibility or suitability. Early animal translocations focused on large and/or charismatic species such as: Golden Lion Tamarin (*Leontopithecus rosalia*), Arabian oryx (*Oryx leucoryx*), Californian condor (*Gymnogyps californianus*), black-footed ferret (*Mustela nigripes*), and several European vulture species. The motivations here were generally to restore species that were known, prized, and missed by the public, and for which there were adequate individuals available, mostly through captive-breeding.

Some generalisations can be made about translocation success based on early translocation efforts:

1. *A thorough understanding of a species' biology and ecology and what is threatening it is more important for translocation success than body size*

   The Arabian oryx is a large species (50–75kg) adapted to an extreme desert environment (Stanley-Price, 1989). It was extinct in the wild in 1972. It is primarily a grazer rather than a browser in habitats where there are probably less than 10 species of grass, all of which it can use as food when released into the desert. It is easy to keep and breed in captivity, has a very high potential rate of increase in the wild; its natural life span might be up to 15 years. Captive-bred oryx have been successfully released at many sites in the Arabian Peninsula, but persistence depends on containing human predation (the original cause of extinction) rather than any technical aspects or the competence of animals to thrive in their natural habitat. As a caution, body size is not necessarily a correlate of ease of translocation. The last wild bison (*Bison bonasus bonasus*), occupying the Bialowieza Primeval Forest, Poland in Europe died in 1919. In 1954 a few bison from captive stock were released back into Bialowieza. On the assumption that the bison was a species of north European temperate forests, the first rewilding efforts in Romania released animals into forests, where they did not thrive. Subsequent research shows that, in fact, the bison prefers mixed landscapes of open grassland and forest, thus the Bialowieza forest was a refuge rather than optimal habitat. The lessons from this are: (1) understand the species' biology and ecology; (2) understand that the first and best place in which to release may not be the site or habitat where the last, original populations existed; and (3) despite individual size, large animals may also be quite delicate in terms of ecological requirements (Kuemmerle et al., 2018).

2. *Scavenging species are easily persecuted*

   Scavengers such as vultures and condors are easy to persecute through illegal poisoning of carcases. With effective captive-breeding, their translocation is similarly helped, or even dependent on, provision of bona-fide clean carcases.

3. *Success of the translocation of keystone species or ecosystem engineers can be measured by the successful establishment of the species and its impacts on the ecosystem*

   For example, the beaver is often termed an 'ecosystem engineer' given its ability to transform landscapes. At the start of the 20th century the European beaver (*Castor fiber*) had been reduced to ~1200 individuals in eight relict populations and had been extinct in the United Kingdom for some 400 years. It was easy to trap and kill the animals for their (at the time) highly valuable products of fur and castoreum (and, because of its tail, the species was deemed to be a fish by the Catholic Church and could hence be eaten on Fridays). Its habits of impounding rivers and its lodges in which it over-winters made direct persecution very effective. On the other hand, the recovery record across Europe shows the beaver to be highly expansionary and it has proved easy to 're-seed' river systems with beavers (Wróbel, 2020). Their immediate environmental engineering is both a quick and easy measurement of their presence and impact (see also Hale & Koprowsk (2018) for a review of the ecosystem-level effects of keystone species translocation).

4. *Simpler life cycles and fewer ecological requirements and interspecies dependencies contribute to translocation success*

   Any translocation strategy must be based on detailed knowledge and an evidence base of factors critical for success. Hence the simpler the life cycle, the easier it will be to re-establish a species' population. In general, large mammals and birds have relatively simple life cycles. But there are exceptions; for example, the black-footed ferret, a smaller mammal (1kg.) has been the subject of rescue for captive-breeding and release in the USA since 1987. The ferret shares it subterranean habitat with its obligate prey, the prairie dog, *Cyonmys spp.*; living in proximity, colonies of both species can be decimated by the shared disease of plague, *Yersinia pestis*. Hence the patchy success record of re-establishment can be largely explained by the ferret's large-scale habitat requirements (Jachowski et al., 2011) and its enforced sharing of space intimately with its obligate prey, when both species suffer from the same disease.

The life cycles of invertebrates may be more challenging for translocation. The Large Blue butterfly, *Maculinea arion*, which has been progressively returned to chalk grasslands in SW England has very complex ecological requirements to breed successfully (Thomas et al., 2009). Its obligate habitat is grassland of around 1.4cm high, the home of seven closely related species of *Myrmica* ants. The Large Blue is a brood-parasite and its larvae depend on these ants for survival. Eggs are laid on *Thymus* spp, which the caterpillars feed on for three weeks; they then drop to the ground and the first *Myrmica* that encounters a caterpillar carries it to its nest mistaking it for one of its own. Survival is greatest if the larva is collected by *M. sabuleti* and less so if by *M.scabrinodis*. Underground survival is determined by soil temperature and for *M.sabuleti* this is best achieved under the 1.4cm sward. Hence, critical to breeding success and successful translocation is having the grass maintained through grazing by sheep or rabbits to this length. If the grass height is below or above this range, the 'wrong' ant species predominate and the prospects for completion of the life cycle are greatly reduced.

The development of translocation science

Translocation guidance and practice has progressed significantly since the early days outlined above. Few contemporary translocations fail to report in terms of defined objectives and timelines, the reasons for success or failure, and lessons learned.

Experiences in Australasia have especially pushed development of translocation science, aiming to integrate science and management (Ewen et al., 2012). Key aspects are choices and alternatives, relevant to welfare and handling in translocations, habitats, and their selection, disease risk and monitoring, genetic issues around small populations, and the benefits of adaptive management when translocation managers are faced with much uncertainty and alternative management options. All contribute to make translocations more rigorous, with lessons and methodologies that can be used across many situations, and that are not merely species- or site-specific.

Since 2012, translocation science has progressed a long way. Considering future conservation introductions, Seddon et al. (2012) write 'We can also envisage an even more extreme conservation introduction whereby populations are established in areas where the species has never been, in species assemblages that have no historical analogues.' In 2013 the scope of the Guidelines for Translocations and Other Conservation Translocations was widened through the inclusion of assisted colonisation of species affected by climate change (Seddon, 2010) and taxon substitution.

There are now also increasing numbers of long-term translocation projects from which we can learn, for example, the reintroduction of the Hihi (*Notiomystis cincta*) in North Island, New Zealand. This programme is characterised by a trans-disciplinary approach, intensive monitoring and adaptive management, emphasising that reintroduction practice involves uncertainty, risk, and ignorance which must be accepted and handled, as well as a strong emphasis on the socio-cultural aspects of the reintroduction. The duration of the Hihi restoration effort, of at least 17 translocations to seven sites starting in 1980, has now led to a best practice guide for Hihi management (Ewen et al., 2018).

The Hihi programme promotes translocation science in many ways, for example, through:

1. the development of new techniques; for example, the application of structured decision making when managers are faced with the issues of multiple options (Panfylova et al., 2019),
2. its own research which is being increasingly granular, with for example, research questions such as 'Can Hihi translocations help restore lost pollination function?' (www.hihiconservat ion.com/), looking at the ecological consequences of re-translocation (rewilding) rather than mere population establishment (species conservation),
3. the sharing of knowledge and techniques; for example, a quantitative risk assessment based on an experimental approach led to a revised management regime for the Mauritius olive white-eye (*Zosterops chloronothos)*, and after four years, the observed outcomes precisely matched the predicted ones (Ferrière et al., 2020).

Despite these developments, and with the inclusion in 2013 of assisted colonisation and taxon replacement in the guidelines, there remain many gaps in guidance and understanding. For example, there is no objective measure of success which can be compared across cases (Seddon & Armstrong, 2019). On the other hand, new modelling approaches can contribute to more informed decision making and hence potentially improve the success rate of translocation programmes (Rayner et al., 2021).

Translocations and rewilding

Below I discuss some key themes which cut across rewilding and translocation science, with the aim of highlighting similarities and the potential for learning across these fields.

Keystone species

One overlap of interests between translocations for species conservation and rewilding in terms of reassembling a species community and restoring ecological functions is the focus on large and/or keystone species. In rewilding, it is often stated that keystone and ecosystem engineer species should be prioritised because they have a cascading effect in the food-web and thus a disproportionally higher impact on the ecosystem than other species. It is notable that the predominantly larger species that have been reintroduced such as bison, vultures and condors and beavers, are also keystone species with important ecological roles. Whether reintroduced for functional or compositional reasons, or for more emotive reasons, large species were also often the subjects of early translocations.

Much rewilding thinking has centred on the role of large species and the imperative to return mega-fauna, which play a disproportionate role in the functioning of ecosystems (Jepson & Blythe, 2020). Across Europe, where grazing lands require restoration or management, a variety of large herbivores (both wild and domestic) have been reintroduced: konik, tauros ((https://rewildingeurope.com/rewilding-in-action/wildlife-comeback/tauros/), bison, water buffalo, roe and fallow deer, elk (https://rewildingeurope.com/areas/), with the intention that their activities will also promote the increase of interdependent species, diversify food chains and encourage the passive return of other species.

One example of this is Knepp in Southern England. This fenced, 1400ha estate has a fac-simile mega-herbivore community (Tree, 2018). Tamworth pigs stand in for wild boar, Exmoor ponies for tarpan, and long-horn cattle substitute for the aurochs, with native roe, red, and fallow deer (and in small numbers, the non-native muntjac). The activity of these herbivores, combined with cessation of conventional agriculture prior to their translocation on to the farm including hedges being no longer maintained and crops not being sown, have had profound effects after 15 years: the very appearance and dynamics of the vegetation are creating a patchwork of woodland-pasture, with scrub. The cows are browsing more than conventional thinking would expect, and the whole area shows signs of herbivore impact—through a browse line, rubbing on trees, dispersal of dung, and hence seeds, disturbed ground from rooting pigs creating micro-habitats, provision of hair for nesting birds (Tree, 2018). However, the level of 'management harvest' of animals from the farm still dictates the evolution of this mosaic with the stated aim being semi-open scrub and wood-pasture landscape.

The cessation of agriculture and removal of farm infrastructure followed by establishment of the substitute (more on this below) herbivore community on the farm are all active measures to establish the potential for reduced management, and in time perhaps, rewilding. The introduced herbivores are considered the building blocks for subsequent passive rewilding. As habitat structure and 'natural' dynamics have taken over, plant species previously unknown on the farm are being recorded, moth and butterfly diversity has increased, with the iconic Purple Emperor (*Apatura iris*) re-colonising by itself, and the annual counts of cuckoos (*Cuculus canorus*) and nightingales (*Luscinia megarhynchos*) increasing (Tree, 2018). Moreover, as ecological complexity increases, further translocation potential opens up, resulting in reintroduced European white stork (*Ciconia ciconia*) breeding at Knepp and dispersing for the first time in 2020, the first such event in England for 600 years.

Rewilding and taxon substitution

The distinction between a translocation purely for compositional purposes and rewilding for returning ecological function is often blurred; this is because ecological function leans heavily on composition and interactions between species that have coevolved. However, contemporary rewilding projects have more explicit goals of restoring function and this opens up the potential for taxon substitution. But even in these cases, the balance between ecological function and composition must be considered. For example, the peregrine falcon was re-introduced to the eastern USA as a cocktail from non-indigenous sources (Barclay & Cade, 1983), thereby replacing the extinct 'duck hawk'. However, to be viable ecological replacements, the birds must have had impacts on their prey populations as well as being part of many other ecological interactions.

Similarly, while returning wild boar (*Sus scrofa*) to English landscapes would constitute reintroducing a lost member of northern Europe's recent fauna, its role in several rewilding projects is taken by certain breeds of domestic pig which could never be claimed a member of the country's post-Pleistocene native fauna. Any released wild boar would meet both the compositional and functional criteria, whereas the domestic pig would be playing merely the ecological function.

An early rewilding publication (Donlan et al., 2006) included the notion of African cheetahs being introduced into the south-west USA to replace the extinct American Cheetah (*Micracinonyx trumani*) as a predator for the swift pronghorn, which has had no effective predator since the end of the Pleistocene Donlan's proposal received some scepticism from the scientific community (Rubenstein et al., 2006), and bemusement from the general public, who may view such proposals for Pleistocene rewilding more as provocations than as serious opportunities. Thus, they may or may not have had an influence on public thinking and the acceptability of rewilding. Pleistocene rewilding with substitute fauna such as cheetah assumes that the current environment is still suitable for these animals to live in and that the introduced species will have comparable effects on the food web as their counterparts did. However, ecosystems evolved in the absence of those species and may have moved into very different systems in which the impact of the introduced species may not be the same as the taxon it replaces and possibly even detrimental for existing native species, including rare or endangered ones (Rubenstein et al., 2006). The biology and ecological requirements of wild or domestic substitutes for Pleistocene fauna or more recent extinct fauna may or may not be similar either. However, the reality is that rewilding will often use substitute species, and some have with great success.

Under 'Conservation introductions' the 2013 guidelines include an element 'Ecological replacement', described as 'Perform an ecological function lost by extinction of an original form'.[1] Species have been substituted at many sites for many reasons. For example, the Arabian ostrich *Struthio camelus syriacus* has been extinct since 1939. Releases in Saudi Arabia used the justified closest relative, the arid-adapted, red-necked ostrich from the adjacent Horn of Africa, *Struthio c. camelus* (Seddon & Soorae, 1999).

On the islands of the south-west Indian Ocean, the dominant herbivore was a number of island-specific giant tortoises, *Cylindraspis* spp. These were driven to extinction, mainly through the actions of mariners taking them on board as a source of fresh meat. Starting with the elimination in the 1970s–80s of the rabbits and goats, introduced by mariners for the same provisioning purpose, the Durrell Wildlife Conservation Trust has been restoring the community of species on Round Island, a few miles off the north end of Mauritius. With habitat protection and planting out of many local plant species, the island's vegetation showed dramatic recovery, but highlighted the absence of any large herbivore. Giant tortoises, <u>Aldabrachelys gigantea</u> were taken from Aldabra (almost 1700km distant) as a substitute and have showed entirely beneficial effects by eliminating non-native grasses, and grazing the native ones to form short lawns, as were recorded in the past (Griffiths et al., 2013).

As discussed, the Knepp community of large herbivores is dominated by species that are substitutes for extinct ones; and these substitutes are domestic species. Management of these species is minimal but consistent with the mandatory requirements of keeping, managing, and marketing domestic species in the farm context. Although wild boar would be available for use in Knepp, their release in England is not permitted because of their role as a reservoir for the highly transmissible African swine fever[2] with an assessed high-risk of escape. While the ecological impacts of the Knepp multi-herbivore community are multiple and striking, the large herbivore community can only be a fraction of the species diversity in the early late Pleistocene before it was progressively reduced. Hence, the impacts of current rewilding may only be a fraction of the ecological complexity resulting from larger, more diverse mega-herbivores. We should remember that reconstructed ecosystems with substitute herbivores may meet many management objectives with historical similarities, but they may be no more than illustrative of earlier Pleistocene ecosystems. Furthermore, Knepp is an example of a novel ecosystem for which there is no natural equivalent or reference. Hence it remains to be seen if a novel system like this will be self-regulating, resilient, and persistent over time, which are all key attributes of accepted rewilding concepts (Carver et al., 2021). These systems might require considerable human intervention long into the future which would defeat the point of rewilding and would be not very different from other forms of nature management. However, projects like this may still be valuable in their own right for biodiversity conservation (as Knepp has shown). Perhaps we should treat them as experiments and use them to test the hypothesis that selected domestic species are good substitutes for extinct ones in terms of their ecological impact, their interactions with other species in the food web and their contribution to self-regulating and resilient ecosystems.

For any taxon substitution, there will always be the obvious risk of such species becoming out of control or invasive. The history of non-native species becoming invasive is very well-documented with a vast number of cases. It is probably true that, so far, there has been no instance in which a species deliberately used as a substitute has become a problem.

Risk may be mitigated by choosing species where an undesirable situation can be reversed; thus, it would be relatively easy to re-capture all giant tortoises on a small island and remove them. This contrasts with the escape in 1993 of hippopotamus *(Hippopotamus amphibius)* in Colombia, which are spreading and increasing; as a major ecosystem engineer, its impacts are already considerable, and may indeed simulate some of those of extinct Pleistocene species.

The IUCN guidelines emphasise the need for adequate risk assessment, but with especial emphasis around taxon substitutions. One can envisage a hierarchy of risk around substitutions: giant tortoises would be at the lower limit. Releases of de-extinct or synthetic species (specifically excluded from this chapter) might be at the extreme end of high risk and low feasibility (Seddon & Armstrong, 2019).

Historical baselines for informing species composition

The debate over many years on the causes of the extinction of large species across different continents has swung between climate change and human persecution as the primary driver. Paradoxically, now that climate change is such a major concern, the verdict seems overwhelming (Sandom et al., 2014) that the arrival of hominins coincided with the start of the extinction patterns seen. For the present purpose, two factors give credence to this:

1. Is it chance that the continent in which *Homo* originated and migrated from, namely Africa, is the sole continent with more than merely the remnants of its Pleistocene fauna? Despite climatic changes in sub-Saharan Africa, it is likely that Africa's large mammal fauna has persisted for so long because it has evolved in parallel with successive evolving hominins (Sandom et al., 2014).

2. The timing of continental extinctions relate very closely to the arrival of humans through the Late Pleistocene (20,000 years BP); further, the pattern of extinctions within continents suggest that the large species were the easiest and earliest to be eliminated (Smith et al., 2015). Given the great impacts that today's restoration of large herbivores can have over geologically mere moments of time, then we can assume that the world's vegetation biomes and their distribution may have been very different from anything we can envisage or see today.

Where rewilding involves the release of species, guidance on justification, planning and design, risk assessment and monitoring is covered by the IUCN Guidelines on Translocations and Other Conservation Translocations (IUCN, 2013). The key test is whether a proposed translocation has conservation benefit, and the prospect of an ecological function being restored should be justification enough. These guidelines raise two further issues:

1. The original guidelines of 1998 (IUCN, 1998) spoke of the requirement to reintroduce within 'historical range'. There is a fundamental problem in defining what is 'historical' because where pre-history ends and history begins is a matter of culture, the means of evidence, human perception, and bias, leading inevitably to arbitrary decisions. Consequently, the 2013 guidelines (IUCN, 2013) replaced historic with 'indigenous range', defined as 'the known or inferred distribution generated from historical (written or verbal) records, or physical evidence of the species' occurrence. Where direct evidence is inadequate to confirm previous occupancy, the existence of suitable habitat within ecologically appropriate proximity to proven range may be taken as adequate evidence of previous occupation.' This overcomes any arbitrary and subjective start of 'historic', such as the arrival of Europeans in North America, where they found large populations of indigenous people. The latter's antecedents presumably were responsible for the extinction of many large species starting with their arrival from Asia some 20,000 years

earlier (Goebel et al., 2008). Given the strength of argument that humans caused the extinction of so many species, should we regard a species' indigenous range as its distribution at the point that *H. sapiens* or an earlier hominin arrived (except for Africa)?

Even if logical, it means it is very difficult after the passage of so much time to say unequivocally whether a species could even be presumed to have existed at a certain site or not, and even less to provide evidence of it.

2. The other problem with this extended time scale is that it makes habitat suitability an impossible challenge, compared to using 400 years BP as the start of a 'historical' reference period.

Until the de-extinction of long-lost mega-herbivores becomes a realistic opportunity (if it ever will), rewilding will depend on extant mega-herbivores, whether wild animals or domestic substitutes; and this will take place either within current or recent range or beyond it where ecological conditions are deemed appropriate. Hence, the aspects of substitution and risk in rewilding are important.

If, logically, we must accept baseline faunas as those existing when *H. sapiens* or earlier hominins arrived in an area, then most of their co-occurring large animal species will no longer exist. Hence, alternatives will have to be found for any rewilding on a functional basis (as discussed in the earlier section on rewilding and taxon substitution).

Policy and law

Translocation involves moving individual plants or animals from one site to another. Active rewilding may also involve the same. Hence, every translocation is bound to have to comply with possibly numerous laws, conventions, or policies (see also Chapter 13, this volume). The range of situations around translocation and rewilding is so great that it is challenging to be specific or comprehensive in the considerations here. However, the following points should be considered in any feasibility or planning exercise:

- The IUCN (2013) Guidelines point out the major issues around international conventions, international and national policies, and laws, with emphasis on animal and plant health, disease spread, and its control,
- Long extinct species, for which translocation is considered or intended, may not be included in a country's lists of species to be given different levels of protection, or none,
- Consequently, such species may have to be treated as extinct, despite intended release within their indigenous range; they might be regarded in law as alien, and many countries have stringent laws and controls on the importation, let alone, release of non-native species.
- This situation is further complicated if different regions or provinces within a country have their own devolved policies about transport or release.
- Complications will arise where released animals naturally cross international boundaries if the release meets requirements of the source country, but the situation is different in the receiving country. This complexity is illustrated by konik horses regularly crossing the river Meuse, the boundary between Belgium and the Netherlands.

This complex set of aspects is potentially more challenging for some rewilding projects:

- Where taxon substitution uses a wild species closely related to a rare or extinct local species, the former may not be allowed to enter the country, based on a concern that the non-native species might out-compete the native one.

- Where large herbivores are an element in rewilding and their numbers have to be controlled, the sale of meat from both wild and domestic species will be subject to stringent health standards; these will probably be even more stringent for domestic herbivores, involving annual round-ups, tagging and health inspections.
- The starting points for development of the new aurochs, the tauros, are ancient cattle breeds. If, with each generation, they are deemed to resemble increasingly the original extinct forms phenotypically and behaviourally, at what point does a domestic species become a wild species? This may become important as national legislations usually have very different laws about the management or use of domestic and wild species. Few legislations allow a species to be wild or domestic depending on the conditions under which it is living.
- To mitigate uncertainty related to translocations, a trial release may be appropriate. This was the response of the Scottish government in the face of an 'unauthorised' release of beavers in Scotland a around 2005. In 2008 the Scottish government started a closely monitored five-year trial, with releases of 16 beavers between 2009 and 2011. Formal evaluation assessed that the ecological benefits of beaver presence were desirable (Gaywood, 2018); hence, the beaver in both populations was given legal protection in Scotland. In 2020, Trees for Life, with legal support from the Lifescape Project, requested a judicial review into the licenced beaver culling in the River Tay catchment in Scotland. The court ruled in 2021 that NatureScot had been issuing beaver culling licenses unlawfully and it revoked all the licences subject to the challenge which remained valid at the date of judgment. (NatureScot now must give written reasons for any culling licence issued, a result which restricts their freedom to license killing of beavers and significantly increases transparency of the licensing process.) Following the judgment, the Scottish government has announced a policy change which, where conflict occurs, will prioritise the translocation of beavers to other areas of Scotland instead of culling being the favoured solution. This ruling promises a future of fewer beavers being culled and active recolonisation of wider areas of Scotland.

There are clearly many aspects around rewilding that will not fit conformably with con-temporary legislation and polices at many levels. From a policy perspective, rewilding needs clear definition of what it is, its purposes, and what outcomes can be expected (Carver et al., 2021). This could shift the emphasis around land use away from reinforcing the compositional approach in favour of the provision of ecosystem services with measurable impacts. This may be a formidable task, but the growing interest in rewilding merits it. It may also provide a defence against a rise in covert 'gorilla' rewilding (Bode, 2020).

Conclusions

Rewilding is explicitly about contributing to community structure, species interactions and ecological function, and translocation is a key tool for pursuing rewilding objectives (Carver et al., 2021). Any future rewilding guidelines are therefore likely to cross-reference IUCN trans-location guidance and other areas of best practice. It is unlikely that rewilding objectives will be met through the translocation of a single species. Cases of linked multi-species translocations in pursuit of rewilding may be rare, but the Ibera project in Argentina (Chapter 16) is a long-term, systematic effort to restore a series of species at different trophic levels.

Responsible translocations now have a substantial record of effort and achievement. As with translocations in the 1980–1990s, leading to the 1998 and 2013 Guidelines, does rewilding need further deployment of a typology beyond the currently clear four and not mutually exclu-sive types: Pleistocene, trophic, ecological, passive (Pettorelli & Durant, 2019, Carver et al.,

2021). Can we develop a holistic paradigm for rewilding? Many of the chapters in this book discuss the potential for this.

What sets rewilding apart, as used by practitioners, from restoration or translocation? Surely, it is appreciation of the distinction between 'making wild again' rather than 'going back to some previous wild'.[3] If this is accepted, then is rewilding uniquely distinguished by the establishment of ecological relationships and activities that have no past analogue? This could be because the species involved have never occurred together, or they comprise a novel assemblage or novel ecosystem under new climate conditions, or closely-related taxa are substitutes, or domestic/semi-wild facsimiles are used for extinct wild forebears? Would this help clarify the boundary between translocation in indigenous range and rewilding, acknowledging that there will be much in common in terms of tools and methods, theoretical and practical? Rewilding is usually considered as a multi-species activity, emphasising interactions, ecological complexity and resilience, attributes that may be merely incidental to single species translocation. Rewilding fauna may be a mixture of indigenous species and suitable substitutes, wild and/or domestic. There may or not be existing reference ecosystems for such a diverse range of possible species compositions. As rewilding experience and ecological change grows over the coming years, rewilded areas might also be accepted as novel ecosystems. For now, the boundary between multiple species translocations and rewilding cannot be crystal clear, for they will share much in terms of tools and methods.

At many sites across western Europe where large herbivores have been released, their impacts and ecological benefits are now well established, and management objectives may be achieved. But the 'hands off, let nature take control' approach leads to many collateral, and pleasant, surprises, such as the recolonisation at Knepp of butterflies, nightingales, and turtle doves, which had never been planned. Moreover, their reappearance provided further insights into what really was preferred habitat.

Thierry and Rogers (2020) provide a spatially explicit method for identifying at fine scale locations where habitat suitability will align with provision of a desired ecosystem function, here seed dispersion on the highly defaunated island of Guam. Their approach is flexible, offering exploration for a wide range of situations, and could help with trade-offs between ecological functions and the species providing these functions, as well as other benefits such as the likelihood of unintentional outcomes.

Current rewilding is not yet ecologically pervasive. Most schemes in western Europe that are explicitly labelled as rewilding have focused on the woodland-pasture system, montane, and riverine or coastal systems. Can rewilding give more attention to other ecosystems, perhaps ones in which mega-herbivores will not be key architects? Does the notion of rewilding have traction on other continents which have also suffered massive defaunation, whether of mega-herbivores or otherwise?

Species translocation has been, and will remain, a central part of the range of translocations for conservation purposes. Rewilding is an extension of this legacy and will hopefully embark on a trajectory of theory and practice as translocation embarked on some 40 years ago. Then perhaps rewilding will grow into the promise of being disruptive, yet a captivating, controversial, 21st-century concept to address ecological degradation in a changing world (Carver et al., 2021).

Notes

1 In the IUCN typology, this is included in the category Conservation Translocation 'outside indigenous range'. Under current rewilding thinking, ecological replacement could be in or outside indigenous range.

2 https://ahdb.org.uk/knowledge-library/africanswine-fever
3 Acknowledging that 'wild' lies on a spectrum of conditions around human intervention either directly on species or on their habitats (Mallon & Stanley-Price, 2013).

References

Barclay, J.H. and Cade, T.J. (1983). Restoration of the peregrine falcon in the eastern United States. *Bird Conservation* 1, 3–40.

Beschta, R.L. and Ripple, W.J. (2019). Can large carnivores change streams via a trophic cascade?. *Ecohydrology*, 12(1): UNSP e2048.

Bode, M. (2020). Covert rewilding: modelling the detection of an unofficial translocation of Tasmanian devils to the Australian mainland. *Conservation* Letters, 2020; e12787.

Carver, S., Convery, I., Hawkins, S., Beyers, R. et al. (2021) Guiding principles for rewilding. 35(6), pp. 1882–93.

Donlan, C.J., Berger, J., Bock, C.E. et al. (2006). Pleistocene rewilding: an optimistic agenda for twenty-first century conservation. *The American Naturalist*, 168(5) (November): 660–81.

Ewen, J.G., Armstrong, D.P., Parker, K.A., and Seddon, P.J. (eds) (2012). *Translocation Biology: Integrating Science and Management*. Blackwell Publishing Ltd.

Ewen, J.G., Armstrong, D.P., McInnes, K., Parker, K.A., Richardson, K.M., Walker, L.K., Makan, T.D., and McCready, M. (2018). *Hihi Best Practice Guide*. Department of Conservation, Wellington, New Zealand.

Ferrière, C., Zuel, N., Ewen, J.G., Jones, C.G., Tatayah, V., and Canessa, S. (2020). Assessing the risks of changing ongoing management of endangered species. *Animal Conservation*, 1–8. https://doi.org/10.1111/acv.12602

Gaywood, M.J. (2018). Reintroducing the Eurasian beaver *Castor fiber* to Scotland. *Mammal Review*, 48, 48–61.

Goebel, T., Waters, M.R., O'Rourke, D.H. (2008). The Late Pleistocene dispersal of modern humans in the Americas. *Science*, 319, 1497–1502.

Griffiths, C., Zuël, N., Jones, C., Ahamud, Z., and Harris, S. (2013). Assessing the potential to restore historic grazing ecosystems with tortoise ecological replacements. *Conservation Biology*, 27(4), 690–700.

Hale, S.L. and Koprowski, J.L. (2018). Ecosystem-level effects of keystone species translocation: a literature review. *Restor. Ecol.* 26, 439–45. doi: 10.1111/rec.12684

Hayward, M.W., Scanlon, R.J., Callen, A. et al. (2019). Reintroducing rewilding to restoration – rejecting the search for novelty. *Biological Conservation*, 233, 255–9.

IUCN (2008). *Guidelines for Translocations*.

IUCN/SSC (2013). *Guidelines for Translocations and Other Conservation Translocations*. Version 1.0. Gland, Switzerland: IUCN Species Survival Commission.

Jachowski, D.S., Gitzen, R.A., Grenier, M.B., Holmes, B., and Millspaugh, J.J. (2011). The importance of thinking big: large-scale prey conservation drives black-footed ferret translocation success. *Biological Conservation*, 144, 1560–6.

Jepson, P. and Blythe, C. (2020). *Rewilding. The Radical New Science of Ecological Recovery*. Hot Science.

Kuemmerle, T., Levers, C., Bleyhl, B. et al. (2018). One size does not fit all: European bison habitat selection across herds and spatial scales. *Landscape Ecology*, 33, 1559–72.

Lawton, J.H., Brotherton, P.N., Brown, V.K., et al. (2010). Making Space for Nature: a review of England's wildlife sites and ecological network. Report to Defra.

Mallon, D.P. and Stanley-Price, M.R. (2013). The fall of the wild. *Oryx* 47 (4), 467–8.

Panfylova, J., Ewen, J.G., and Armstrong, D.P. (2019). Making structured decisions for reintroduced populations in the face of uncertainty. *Conservation and Science and Practice*. https://doi.org/10.1111/csp2.90

Pettorelli, N., Barlow, J., Stephens, P.A., Durant, S.M., Connor, B., Schulte to Bühne, H., Sandom, C.J., Wentworth, J., and du Toit, J.T. (2018). Making rewilding fit for policy. *Appl Ecol.*, 55: 1114–25.

Pettorelli, N. and Durant, S.M. (2019). Rewilding: a captivating, controversial, twenty-first century concept to address ecological degradation in a changing world. In Pettorelli, N., Durant, S.M., du Toit, J.T. (eds), *Rewilding*. (Ecological Reviews) (p. vi). Cambridge University Press.

Rayner, K, Lohr, C.A., Garretson, S., and Speldewinde, P. (2021) Two species, one island: Retrospective analysis of threatened fauna translocations with divergent outcomes. *PLoS ONE*, 16(7): e0253962. https://doi.org/10.1371/journal. pone.0253962

Ripple, W.J. and Beschta, R.L. (2012). Trophic cascades in Yellowstone: the first 15 years after wolf translocation. *Biological Conservation*, 145: 205–13.

Rubenstein, D.R., Rubenstein, D.I., Sherman, P.W., and Gavin, T.A. (2006). Pleistocene park: Does re-wilding North America represent sound conservation for the 21st century?. *Biological Conservation*, 132, 232–8.

Sandom, C., Faurby, S., Sandel, B., and Svenning, J-C. (2014). Global late Quaternary megafauna extinctions linked to humans, not climate change. *Proc. R. Soc. B*, 281(1787): 1–9.

Seddon, P.J. (2010). From translocation to assisted colonization: moving along the conservation translocation spectrum. *Restoration Ecology*, 18(6): 796–802.

Seddon, P.J. and Soorae, P.S. (1999). Guidelines for subspecific substitutions in wildlife restoration projects. *Conservation Biology*, 13: 177–84.

Seddon, P.J., Armstrong, D.P., and Maloney, R. (2007). Developing the science of translocation biology. *Conservation Biology*, 21: 303–12.

Seddon, P.J. and Armstrong, D.P. (2019). The role of translocation in rewilding. In Pettorelli, N., Durant, S.M., du Toit, J.T. (eds), *Rewilding* (Ecological Reviews) (p. vi). Cambridge University Press.

Seddon, P.J., Armstrong, D.P., Parker, K.A., and Ewen, J.G. (2012). Summary. In J.G. Ewen, D.P. Armstrong, K.A. Parker, and P.J. Seddon (eds), *Translocation Biology: Integrating Science and Management*. Blackwell Publishing Ltd.

Smith, F.A., Doughty, C.E., Malhi, Y., Svenning, J-E., and Terborgh, J. (2016). Megafauna in the earth system. *Ecography*, 39: 99–108.

Soulé, M. and Noss, R. (1998). Rewilding and biodiversity: complementary goals for continental conservation. *Wild Earth*, 1–11.

Stanley-Price, M.R. (1989). *Animal Translocations: The Arabian Oryx in Oman*. Cambridge Studies in Applied Ecology and Resource Management, Cambridge University Press.

Thierry, H. and Rogers, H. (2020). Where to rewild? A conceptual framework to spatially optimize ecological function. *Proc. R. Soc. B*, 287: 20193017. http://dx.doi.org/10.1098/rspb.2019.3017

Thomas, J.A., Simcox, D.J., and Clarke, R.T. (2009). Successful conservation of a threatened *Maculinea* butterfly. *Science*, 32580–3.

Tree, I. (2018). *Wilding: The Return of Nature to a British Farm*. Picador,

Wróbel, M. (2020). Population of Eurasian beaver (*Castor fiber*) in Europe. *Global Ecology and Conservation*, 23, e01046.

8

CORES AND CORRIDORS

Natural landscape linkages to rewild protected areas and wildlife refuges

Jonathan Carruthers-Jones, Andrew Gregory, and Adrien Guetté

Connectivity conservation and intact landscapes

Connectivity conservation has been hailed as a key landscape scale strategy to maintain and restore ecologically functional landscapes for the benefit of habitats and species (Worboys et al., 2010). Protected areas are insufficient to conserve biodiversity and maintain landscape connectivity in the face of anthropogenic pressures, such as rapid land use change (Gurrutxaga et al., 2015). Whilst there remains uncertainty about the benefits of connectivity, and specifically corridors, as an approach to mitigating the impacts of global change processes, it has been widely adopted on the grounds that it is hard to imagine any realistic alternative that would be conducive to species persistence (Hilty et al., 2012: 112). Functionally intact networks of protected areas are considered necessary to allow species to move in response to land use and climate change. Networks of protected areas, protecting up to 30% of terrestrial surfaces, are now a feature of international biodiversity conservation targets to 2030, especially within the European Union Biodiversity Strategy to 2030. Additionally, the United States Department of the Interior recently announced a similar plan to conserve 30% of the terrestrial land and water resources within the United States by 2030, but has not yet signed on to the International Convention on Biological Diversity.

Traditionally connectivity conservation has involved targeted conservation actions, such as land acquisition and restoration, focused on the needs of a few focal species. Yet land free of direct anthropogenic disturbance is considered an essential component for achieving conservation outcomes aimed at addressing the twin challenges of climate change and biodiversity loss (Ward et al., 2020). Land free of direct anthropogenic disturbance has not been fragmented by roads or forestry activity and is by definition well connected. Scholars across multiple disciplines have established the importance of intact natural landscapes as key sites for biodiversity (Di Marco et al., 2019), endangered species (Soulé, 2014), and the wider pantheon of ecological processes on which life depends (Chan et al., 2006). Beyond their intrinsic value, these same 'wild' spaces bring additional benefits in terms of human recreation and well-being (Milner-Gulland et al., 2014). Despite growing evidence for their extraordinary value, ecologically intact wild spaces still lack sufficient protection, are rapidly being lost, and require

DOI: 10.4324/9781003097822-10

immediate large-scale connectivity conservation efforts to secure them for future generations (Allan et al., 2020).

Numerous landscape-scale conservation initiatives, such as the Yellowstone to Yukon initiative and the Pan-European ecological network, have focused conservation actions on the identification and protection of intact natural landscapes where there is a low degree of human influence. Natural landscape linkages are a part of those initiatives and offer advantages over species-specific approaches, especially in situations where detailed knowledge of species foraging and dispersal behaviour is lacking (Beier & Brost, 2010). This is especially true where the focus of rewilding is the reintroduction of extinct species, such as the Eurasian Lynx (*Lynx lynx*) in the United Kingdom. In this scenario there is no available information on current behaviour (habitat preferences, foraging behaviour, and dispersal patterns) to accurately determine where the best place to reintroduce them would be and indeed how they would respond to the current landscape. Ecological habitat networks of natural and semi-natural landscape elements, designed to maintain or restore ecological functions, are therefore recognised as a key conservation tool bringing benefits to multiple animal species (Hilty et al., 2019). For example, the Path of the Pronghorn represented a public private partnership to protect critical migration networks for Pronghorn (*Antilocapra americana*) in Wyoming of the United States Intermountain West. The resulting corridor protected and partially rewilded an ~120km long corridor (~70km in federal land ~50km on private land; Berger and Cain, 2014). Beyond protecting the pronghorn, this corridor is also protecting jackrabbits (*Lepus tonwsendii*), wolves (*Canis lupus*), grizzly bear (*Ursus arctos*), and elk (*Cervus elaphus*).

In addition to their value for multiple species, networks of intact natural landscape linkages between islands of stricter protection have significant cost savings. The traditional response to corridor design, requiring land acquisition, active ecological restoration, and species reintroduction, is hugely expensive and often impractical given private land ownership rights in some countries. Given the low levels of funding available, rewilding as a passive ecological restoration strategy is likely to be at the forefront of any connectivity proposals at the scale necessary to meet the ambitious national and international targets for landscape conservation to 2030.

Theory: Research evidence for benefits of creating intact natural landscape linkages using rewilding approaches

The loss of natural habitats inevitably has a direct impact on biodiversity, in part by reducing the number and size of patches or increasing the space among patches that are capable of supporting viable populations. As areas of natural habitat are reduced by anthropogenic fragmentation, the degree to which the remaining fragments are functionally linked by dispersal becomes increasingly important. The Savannah River Site in South Carolina, USA is among the world's best-known experiments studying the impacts of fragmentation and the efficacy of corridors. In the 18 years since establishment of this experiment, corridors and patches cut into the landscape have been largely left alone to natural successional processes, offering experimental insights into passive rewilding processes. As a result of these successional process, since that time, patches were recolonised, resulting in an increase in plant diversity and community complexity (Damschen et al., 2019; see Figure 8.1).

Similarly, numerous arthropods and small mammals have been recolonising and utilising the corridors and patches. Of particular note is that the structural corridor provided immediate benefits for some species, but for most species benefits accrued slowly over time. Experimental systems such as the Savannah River Site demonstrate the great potential that rewilding as a

Figure 8.1 A long-term habitat connectivity experiment. (A) One of ten experimental landscapes each containing a centre fragment that is connected or unconnected (winged and rectangular) to peripheral fragments of open longleaf pine savanna surrounded by dense pine plantations. [Credit: Google Earth 2019] (B) Plant communities within fragments have assembled over nearly two decades and are being restored to their historical ecosystem type using low-intensity fires that mimic historic fire regime, and removal of establishing hardwood species (Damschen et al., 2019).

passive restoration strategy has to create diverse communities. However, to date, relatively few studies have actually addressed the efficacy of passive restoration strategies at applicable regional and national scales and contexts. Of critical importance is the fact that at large scales, rewilded corridors also serve as habitats in their own right. For example, the Stock Route Network (SRN) of New South Wales, Australia was initially established as a protected network to facilitate livestock droving. However, through passive rewilding linked to a decline in use for droving, as well as national level protected area designations and protections, the SRN has emerged as a critical reserve for numerous rare and indigenous bird, mammal, herpetofauna, and marsupials (Lentini et al., 2011).

The degree to which the observed effects at the scale of mesocosm experimental corridor systems also apply to large ecosystem level processes is a complex question (Beier and Gregory, 2014). There are two reasons why this may not be the case. Firstly, mesocosm experimental systems are simplified versions of real-world ecosystems and therefore lack some of the functional redundancies and homeostatic feedback mechanisms of real-world systems. In addition, the simplifying assumptions used to structure the experimental mesocosm can also alter density dependent interactions, and result in changed temporal dependencies and chronologies. For example, over time changes in the composition and diversity of the surface vegetative community from a cultivated field to a grassland and then a shrubland, etc., will result in shifts in the availability of habitats for, for example neotropical migrant birds. These birds will bring with them new seeds that will in turn alter the dynamics of the plant community structure, all of which will be mediated by local temperature and rainfall events, which in turn are influenced by local landcovers. The spatiotemporal scales at which large-scale and long term-changes such as these feedback into each other is largely unknown or explored, and difficult to fully extrapolate form micro- to mesocosm experiments Gardner et al., 2006). Therefore, we do not yet know how results from these experimental systems will scale to provide insights into how rewilding will affect large ecosystems.

Insights into how the results of experimental mesocosms might scale up in space and time can be gleaned from serendipitous natural experiments in passive rewilding. For example, studies of abandoned agricultural fields in the Amazon and Spain have indicated that passive rewilding had an overall net positive impact on native bird and plant diversity taking into account system turnover and time since abandonment (Regos et al., 2016; Rutt et al., 2019). Specifically, presence of invasive generalist species initially drove an overall increase in species richness, but a decline in native or area-sensitive species (Rutt et al., 2019). However, as rewilding progressed, native species reasserted themselves; overall richness fell, but diversity and rarity both increased. Similar patterns have been observed at continental scales in Europe with bird diversity and Natura 2000 reserve areas (Albuquerque and Gregory, 2017) and passive rewilding of the EU Green Belt has resulted in phylogenetically diverse and resilient forest communities (Morel et al., 2019). Despite the fact that such observations generally support what experimental mesocosms have shown us about the potential benefits and timeline of benefits associated with rewilding, there remains a clear need for carefully designed experimental research at multiple scales and across diverse biomes to clearly elucidate the potential benefits of rewilding at applicable spatiotemporal scales.

Practice: Designing rewilding networks

Spatially explicit connectivity modelling at a large landscape scale historically has focused on the 3c's (cores, corridors, and carnivores') approach. Under this system, the landscape is classified in terms of core areas of suitable habitat for a given focal species (often a large or wide-ranging carnivore such as the Iberian Lynx (*Lynx pardinus*) in Europe or the grizzly bear (*Ursus arctos horribilis*) in North America) and the ease of movement for that species, among core areas (Hilty, 2012). A variety of modelling software tools and approaches have been used to define these cores and corridors including least-cost resistance, circuit theory and individual-based movement models. More recently, the focus has begun to shift to look more holistically at the quality of the cores relative to the matrix surrounding them as habitat for a suite of focal species as opposed to a single species model, and defining connectivity corridors for multiple species use (Majka et al., 2007; WHCWG, 2010).

The use of a particular modelling approach depends on the conservation objectives and the scale at which it will be implemented. Nevertheless, all of these approaches use spatial data on land cover as a core input which is then reclassified inside a Geographical Information System (GIS) to reflect the ease of movement for a given species or group of species through that land cover type. The core 'source' habitat areas are then defined using a spatial model, and the final output is a map identifying areas of low landscape 'resistance' among core areas which are then designated as potential corridors of conservation interest for target species (e.g. WHCWG, 2010).

Approaches for modelling intact natural landscape networks that link intact core 'last of the wild' reservoirs are predicated on the idea that as human impacts increase, then a priori 'naturalness' must decrease, an idea often referred to as 'hemeroby' (Grabherr et al., 1998). These models rely on spatial data on naturalness which categorises a landscape in terms of biophysical integrity, human influence, and spatio-temporal intactness (Guetté et al., 2018). Data of this type can be produced at the local, regional, or national level and are weighted by experts to provide locally adapted landcover maps that reflect historical and future climate conditions. This approach outputs a map of potential natural landscape corridors, which link intact core 'last of the wild' reservoirs in the landscape. The underlying logic is that wild animals will choose to move through areas that are natural and will also choose to avoid areas where they come into contact with humans. The preservation of well connected, intact natural areas has many ecological and social benefits such as for human wellbeing, but one of its key functions is to provide habitat for species that are averse to human dominated landscapes, which are often the focus of rewilding projects (Noss, 1991).

When considering connectivity within the wider landscape matrix, the challenge becomes how to prioritise conservation efforts to best maintain the functional integrity of a network of patches (Piquer-Rodríguez et al., 2012). Natural landscape linkage maps are a starting point for network analysis, which can identify both the most important corridors in a network of corridors, and pinch points or barriers in that network that inhibit species movement and may serve as priority areas for passive rewilding (Carruthers-Jones, 2013). Interest in the importance of natural landscape linkages, as well as the complexity and cost of mapping species-specific corridors at national scales, resulted in more sophisticated approaches to connectivity modelling using ecological flow and landscape integrity metrics (Theobald et al., 2012; Dickson et al., 2017). Such approaches are promising for identifying corridors of interest for multiple species, especially long-distance dispersers (Krosby et al., 2015). Careful design of a 'species agnostic' approach is also considered to produce results of greater relevance to large scale implementation of corridor networks in landscape planning (Marrec et al., 2020). Corridor modelling based on recent advances in wilderness mapping provide high-resolution maps of landscape intactness and human influence that are considered of critical value to wild land network development (Cao et al., 2020).

Recent innovations in connectivity modelling methods offer the possibility of modelling the connectivity of the entire landscape, not necessarily tied to core areas or least-cost paths through that landscape. One of these, Omniscape, is a connectivity modelling approach combining a non-species-specific landscape integrity approach with omnidirectional circuit modelling that can be used to map the wider connectivity of natural landscapes at regional scales (McRae et al., 2016). Compared with traditional modelling approaches, an omnidirectional analysis of relevance to multiple species provides greater breadth of insights and understanding into which landscape features are critical to support conservation policy and wider biodiversity goals (Lecours-Tessier et al., 2020). Visualisations of the wider landscape, not linked to specific core areas or corridors, showing the degree of ecological flow, and important areas for rewilding, could help landscape planners to assess the impact of future proposals (See Figure 8.2.)

Paying for the process

One concern with any conservation project is the cost: in the case of rewilding, who will pay for the direct costs associated with rewilding corridors and networks and rewilding in general, and what are the downstream costs in terms of lost revenue likely to be? A rewilding approach to increase landscape connectivity can greatly reduce cost compared to a traditional

Figure 8.2   Omniscape viewer. Allows visualisation of multiple data layers to support spatially explicit decision making. Layer displayed is ecological flow. Yellow areas have high flow potential. © Nature Conservancy 2021, image used with permission.

project-based corridor development. First, since rewilding is a gradual process (Gillson et al., 2011), associated costs can be amortised over a longer period of time, allowing local and regional economies to adjust from a short-term resource extraction-based economy to a long-term sustainable resource and amenity-based economy (Schou, 2021).

Second, many governmental and non-governmental programmes exist that can help pay for the direct costs associated with rewilding and restoration. For example, the United States Farm Bill encompasses a set of programmes to pay subsidies to landowners to rewild their property (e.g., Continuous Enrolment Conservation Reserve Program; Wildlife Habitat Improvement Program). Similar programmes exist within the EU Common Agricultural Policy (CAP) and the World Bank, and funding for ecological restoration of this type is likely to increase in the next decade with global initiatives such as the UN decade on ecological restoration and the Convention on Biological Diversity 30% by 2030.

Third, rewilding can be initiated, and in many cases accomplished, through the integration with a regenerative agricultural practice. Regenerative agriculture is a process driven agriculture framework that focuses on outcomes associated with the triple bottom line of agricultural production: 1) sustainable use and profitability; 2) improved soil and water health/quality; and 3) conservation and wildlife stewardship. For a practice to be adopted it must be shown to sustain or improve all three areas. Recent studies have demonstrated that the integration of natural vegetation areas within agricultural landscapes is considered a potential best practice in regenerative agriculture as it improves both biodiversity conservation outcomes and crop yields (Laguna et al., 2014; Galpern et al., 2020), and in some cases also reduces production costs (Galpern & Gavin, 2020). However, the effectiveness of such practices in a rewilding corridors strategy requires greater spatial planning and integration than currently exists.

Knowledge gaps: Crossing the divide between theory and practice of linking landscapes

Across the spectrum of efforts to enhance connectivity at a large landscape scale, the use of corridors to restore linkages in the landscape has not been without some debate and controversy. Debate over the efficacy and value of corridors has centred around three general questions (Bennet, 1999):

1) Is there sufficient evidence to support the notion that corridors work, or under what specific contexts do corridors benefit wildlife (Gregory & Beier 2014)?
2) Do the benefits of corridors outweigh the potential negative consequences of corridors (Resasco et al., 2014)?
3) Is the establishment of corridors a '*Best Practice*', or might limited conservation resources be better invested in establishing either more or larger core areas (Watling et al., 2020)?

These three questions are intertwined, and an eventual resolution to the first will likely ameliorate concerns associated with the latter two. In this regard, a meta-analysis of established linkages has found that corridors do generally increase connected core diversity, probability of occupancy by focal species, and inter-core movement (Beier & Noss, 2008). However, the majority of corridors in this meta-analysis were short (<100 metres long), were taken from experimental systems that do not necessarily reflect the real-world context in which corridors were meant to function, and examined species known to occupy anthropogenic landscapes (Beier & Gregory, 2012). Thus, the data from and analyses of these systems may not be sufficient to apply to robust and informed linkage development and management. Moreover, current management recommendations from this literature are based on extrapolations from general ecological theory, and not empirical evidence of corridor functionality (Gregory et al., 2020). Consequently, what is needed is a well-designed and replicated analysis of the degree to which real-world, established corridors allow for geneflow and wildlife recolonisation of cores spanning a range of wildlife dispersal abilities and responses to habitat edges (Gregory et al., 2014).

In regard to the second question, the cost-benefit debate over corridors has focused on two issues: the degree to which corridors might serve as a vector of wildlife disease, and the degree to which corridors might facilitate species invasion. To date, the data has suggested that corridors might indeed be vectors for both diseases (Wilschut et al., 2013) and invasive species (Resasco et al., 2014). However, nearly all cases of these negative impacts have come from anthropic landscapes where the corridors represent a degraded subset of the naturally occurring biological community or are themselves an artificial human construct (Haddad et al., 2008). The use of rewilding approaches to develop corridors may actually side-step these concerns since intact natural systems tend to be more robust and resistant to disease and invasion (Lockwood et al., 2007). This tendency of intact systems to contain greater diversity and offer greater stability and resiliency is referred to as the portfolio effect of ecosystems (Schindler et al., 2015). For example, Tilman and Downing (1994) demonstrated that in response to drought, the magnitude of damage to a community was inversely related to species richness. In 1998, Doak et al. demonstrated that the benefits of diversity on reliability was the result of statistical averaging among species that did not vary synchronously through time. Essentially, as some species were lost or declined in abundance, other species increased in abundance to take over the ecological function of lost taxa. This diversification and redundancy in ecological function is difficult to achieve in top-down designed ecological restoration projects, but is often an inherent component of rewilding.

The third question does not focus on whether corridors promote biodiversity conservation *per se*, but rather focuses on whether greater benefit per resource invested might not be achieved by increasing the number or area of cores as opposed to increasing the interconnectedness of existing cores (Watling, 2020). The theory of island biogeography (MacArthur & Wilson, 1967) as applied to terrestrial landscape linkages (Diamond, 1975) states that larger cores, more cores, and/or more connected cores will all sustain greater biodiversity for a longer period of time, and so have greater resiliency than smaller, fewer, or more isolated cores (i.e., portfolio effects; Schindler et al., 2015). Rewilding may also provide a mechanism to resolve this debate, as rewilding has built in economic benefits to help ameliorate the costs associated with corridor implementation.

Inherent to the resolution of concerns two and three about corridor efficacy and biodiversity conservation is the SLOSS (Single Large or Several Small) debate (Primack et al., 2018). In the context of rewilding corridors for connectivity, the question is to what extent does corridor length and width interact with species traits to affect corridor functionality, and similarly to what extent does the corridor functionally increase the area of habitat for a given focal species on the landscape? MacArthur and Wilson (1967) and Diamond (1975) have both argued that linkages among terrestrial reserves will functionally increase habitat area of those reserves and lead to greater conservation outcomes than possible for the reserve areas in isolation. However, more recent simulation studies have questioned the validity of such arguments, pointing out that due to species edge sensitivities, the nested subset of species utilising a series of linked smaller reserves, even if of equal area to a single large reserve, may not fully replicate the pantheon of species and ecological functions contained in a single large reserve area (Ford, 2020; Watling et al., 2020). Similarly, species traits or behavioural eccentricities towards gap crossing or habitat edge sensitivities may also influence the degree to which corridors enhance connectivity or biodiversity or promote the spread of invasive species or vector diseases. The degree to which corridor structural traits address these concerns will greatly affect corridor efficacy (Carroll & Noss, 2021). Unfortunately, these factors are additional unknowns with regards to linkage efficacy and management—we simply do not know how wide corridors need to be to function (Gregory et al., 2020).

A second knowledge gap is the degree to which naturally occurring corridors benefit multiple species. Early studies on experimental systems at the Savanah River Site suggest that corridors will have benefits for multiple species (see Haddad et al., 2003). To date, most empirical studies of naturally occurring linkages have focused on a one or a few closely allied species. For example, a set of naturally occurring corridors in Alberta, Canada provided enhanced movement and occupancy probability of grizzly bears (*Ursus arctos*), grey wolves (*Canis lupus*), and cougars (*Panthera concolor*) (Ford et al., 2019*)*. However, these are all wide-ranging carnivores, and the degree to which these same linkages benefit other taxa (e.g., mule deer (*Odocoileus hemiones*)) is still unexplored. Other reasons for the lack of data on the efficacy of naturally occurring corridor benefits across taxonomic levels might also be due to most naturally occurring linkages being designated (1) by political expediency, (2) on lands that have low economic value, or (3) for aesthetic qualities rather than for their value for animal movement or habitat conservation (Scharf et al., 2018).

Conclusion

Significant evidence exists for the value to multiple species of protecting existing natural landscape linkages and identifying future sites for rewilding to improve the connectivity of the wider landscape. Recognising these values is especially important in restoring connections between isolated and remnant populations of native plant and animal species that have become fragmented by anthropogenic pressures over time. Improving the connectivity of intact natural landscapes using a rewilding approach offers landscape connectivity value to multiple species. This contributes to the ongoing protection and resilience of existing wild spaces, and offers a locally adapted, low-cost and scalable solution for restoring a functional landscape matrix between these islands of wilderness.

To support specific conservation interventions, we clearly need further research to better understand the benefit of corridors for multiple plant and animal species—and especially of the capacity of intact natural landscape linkages to maintain gene flow between populations (Gregory, 2014). From a policy perspective, we need new financial instruments to support rewilding of the wider landscape matrix. These instruments need to be informed in a coherent and spatially explicit way, by targeted multi-scalar modelling that is of relevance at the local regional, national, and international scales.

Acknowledgements

Dr Carruthers-Jones is funded by 'Corridor Talk: Conservation Humanities and the Future of Europe's National Parks', which in turn is funded jointly by the German Research Council (DFG) and the UK's Arts and Humanities Research Council (AHRC).

References

Albuqurque, F.S. and A.J.Gregory. 2017. The geography of hotspots of rarity-weighted richness of birds and their coverage by Natura 2000. *PLoS ONE*. doi.org/10.1371/journal.pone.0174179.

Allan, J.R., Possingham, H.P., Venter, O., Biggs, D., and Watson, J.E. 2020. *The Extraordinary Value of Wilderness Areas in the Anthropocene*. Earth Systems and Environmental Sciences; Elsevier: Amsterdam, The Netherlands.

Beier, P., Majka, D.R., and Spencer, W.D. 2008. Forks in the road: choices in procedures for designing wildland linkages. *Conservation Biology*, *22*(4): 836–51.

Beier, P. and Brost, D. 2010. Use of land facets to plan for climate change: conserving the arenas, not the actors. *Conservation Biology*, 24(3): 701–10.

Bennett, A.F. *Linkages in the Landscape: The Role of Corridors and Connectivity in Wildlife Conservation*. IUCN The World Conservation Union Press.

Berger, J. and Cain, S.L. (2014) Moving Beyond Science to Protect a Mammalian Migration Corridor. *Conservation Biology*, 28(5), 1142–50.

Calabrese, J.M. and Fagan, W.F., 2004. A comparison-shopper's guide to connectivity metrics. *Frontiers in Ecology and the Environment*, 2(10): 529–36.

Cao, Y., Yang, R., and Carver, S. (2020). Linking wilderness mapping and connectivity modelling: A methodological framework for wildland network planning. *Biological Conservation*, 251: 108679.

Carroll, C., and Noss, R. 2021. Rewilding in the face of climate change. *Conservation Biology*, 35(1): 155–67.

Carruthers-Jones, J.S. 2013. The selection, evaluation and testing of methods for the identification of priority connectivity conservation areas: a case study in the Pyrénées. MSc thesis, Centre for Mountain Studies, Perth Scotland.

Chan, K.M.A., Shaw, M.R., Cameron, D.R., Underwood, E.C., and Daily, G.C. (2006). Conservation Planning for Ecosystem Services, *PLoS Biol 4*(11), e379.

Damschen, E.I., L.A.Brudvig, M.A. Burt, R.J. FletcherJr., N.M. Haddad, D.J. Levey, J.L. Orrock, J.Resasco, J.J. Tewsbury. 2019. Ongoing accumulation of plant diversity through habitat connectivity in an 18-year experiment. *Plant Ecology*, 365: 1478–80.

Diamond, J.M. 1975. The island dilemma: Lessons of biogeographic studies for the design of natural reserves. *Biological Conservation*, 7: 129–46.

Dickson, B.G., Albano, C.M., McRae, B.H., Anderson, J.J., Theobald, D.M., Zachmann, L.J., Sisk, T.D., and Dombeck, M.P., 2017. Informing strategic efforts to expand and connect protected areas using a model of ecological flow, with application to the western United States: mapping ecological flow to inform planning. *Conserv. Lett.*, 10: 564–71. https://doi.org/10.1111/conl.12322

Di Marco, M., Ferrier, S., Harwood, T.D., Hoskins, A.J., and Watson, J.E. 2019. Wilderness areas halve the extinction risk of terrestrial biodiversity. *Nature, 573*(7775), 582–5.

Doak, D.F., D. Bigger, E.K.Harding, M.A. Marvier, R.E. O'kMalley, and D.Thomson. 1998. The statistical inevitability of stability–diversity relationships in ecology. *American Naturalist*, 151: 264–76.

Fahrig, L. 2020. Why do several small patches hold more species than a few large patches? *Global Ecology and Biogeography*, 29(4): 615–28.

Ford, A.T., E.J. Sunter, C.Fauvella, J.L.Bradshaw, B. Ford, J. Hutchen, N. Phillipow, and K.J.Teichman. 2020. Effective corridor width: linking the spatial ecology of wildlife with land use policy. *European Journal of Wildlife Research*, 66: 69.

Galpern, P. and M.P. Gavin. 2020. Assessing the potential to increase landscape complexity in Canadian Prairie Croplands: A multi-scale analysis of land use pattern. Frontiers in Environmental Science: doi. org/10.3389/fenvs.2020.00031

Galpern, P., J. Vickbruck, J.H. Devries, and M.P. Gavin. 2020. Landscape complexity is associated with crop yields across a large temperate grassland region. *Agriculture, Ecosystems, and Environment, 290.* doi. org/10.1016/j.agee.2019.106724.

Gardner, R.H., M.W. Kemp, V.S. Kennedy, and J.E. Petersen. 2006. *Scaling Relations in Experimental Ecology.* Colombia University Press.

Gillson, L., R.J. Ladle, and M.B. Araujo. 2011. Baselines, patterns and processes. In R.J.Ladle and R. J.Whittaker (eds), *Conservation Biogeography*. Blackwell Publishing Ltd.

Grabherr, G., G. Koch, H. Kirchmeir, and K. Reiter. 1998b. Hemerobie österreichischer Waldöko-Systeme. *Ver öff. Österr. MAB-Programm*, 17: 493.

Gregory, A.J. and P. Beier. 2012. Desperately seeking stable 50-year-old landscapes with patches and long, wide corridors. *PLoS Biology*. doi.org/10.1371/journal.pbio.1001253.

Gregory, A.J. and P. Beier. 2014. Response variables for evaluation of the effectiveness of conservation corridors. *Conservation Biology*, 28: 689–95.

Gregory, A.J., E. Spence, P. Beier, and E. Garding. 2021. Toward best management practices for ecological corridors. *NAU.* 10(2). doi.org/10.3390/land10020140.

Guetté, A., Carruthers-Jones, J., Godet, L., and Robin, M. 2018. « Naturalité»: concepts et méthodes appliqués à la conservation de la nature. *Cybergeo.* https://doi.org/10.4000/cybergeo.29140

Gurrutxaga, M., Marull, J., Domene, E., and Urrea, J. 2015. Assessing the integration of landscape connectivity into comprehensive spatial planning in Spain. *Landscape Research, 40*(7): 817–33.

Haddad, N.M., D.R.Brown, A. Cunningham, B.J. Danielson, D.J. Levey, S. Sargant, and T. Spira. 2003. Corridor use by diverse taxa. *Ecology*, 84(3): 609–15.

Hilty, J.A., Lidicker Jr, W.Z., and Merenlender, A.M. 2012. *Corridor Ecology: The Science and Practice of Linking Landscapes for Biodiversity Conservation*. Island Press.

Hilty, J.A., Keeley, A.T., Merenlender, A.M., and Lidicker Jr, W.Z., 2019. *Corridor Ecology: Linking Landscapes for Biodiversity Conservation and Climate Adaptation*. Island Press.

Jennings, M.K., K.A. Zeller, and R.L. Lewison. 2020. *Supporting Adaptive Connectivity in Dynamic Landscapes*. Land MDPI. doi.org/10.3390/land9090295.

Krosby, M., Breckheimer, I., John Pierce, D., Singleton, P.H., Hall, S.A., Halupka, K.C., Gaines, W.L., Long, R.A., McRae, B.H., Cosentino, B.L., and Schuett-Hames, J.P., 2015. Focal species and landscape 'naturalness' corridor models offer complementary approaches for connectivity conservation planning. *Landsc. Ecol.* 30: 2121–32. https://doi.org/10.1007/s10980-015-0235-z

Laguna, E., S. Fos, and S. Volis. 2016. Role of micro-reserves in conservation of endemic, rare and endangered plants of the Valencian region (Eastern Spain). *Israel Journal of Plant Sciences*.

Lecours-Tessier, D., Maranger, R., and Poisot, T. 2020. Omnidirectional and omnifunctional connectivity analyses with a diverse species pool. *BioRxiv.*

Lentini, P., J. Fischer, P. Gibbons, D.B. Lindenmayer, and T.G. Martin. 2011. Australia's Stock Route Network: 1. A review of its values and implications for future management. *Ecological Management and Restoration*, 12(2): 119–27.

Lockwood, J.L., M.F. Hoopes, and M.P. Marchetti. 2007. *Invasion Ecology*. Wiley-Blackwell,

MacArthur, R.H. and E.O. Wilson. 1967. *The Theory of Island Biogeography*. Princeton University Press.

Majka, D., Jenness, J., and Beier, P. 2007. CorridorDesigner: ArcGIS tools for designing and evaluating corridors. Available at http://corridordesign.org.

Marrec, R., Moniem, H.E.A., Iravani, M., Hricko, B., Kariyeva, J., and Wagner, H.H., 2020. Conceptual framework and uncertainty analysis for large-scale, species-agnostic modelling of landscape connectivity across Alberta, Canada. *Scientific Reports*, *10*(1): 1–14.

McRae, B.H., K. Popper, A. Jones, M. Schindel, S. Buttrick, K. Hall, R.S. Unnasch, and J. Platt. 2016. Conserving Nature's Stage: Mapping Omnidirectional Connectivity for Resilient Terrestrial Landscapes in the Pacific Northwest. The Nature Conservancy, Portland Oregon. 47 pp. Available online at: http://nature.org/resilienceNW June 30, 2016.

Milner-Gulland, E.J., McGregor, J.A., Agarwala, M., Atkinson, G., Bevan, P., Clements, T., Daw, T., Homewood, K., Kumpel, N., Lewis, J., and Mourato, S. (2014). Accounting for the impact of conservation on human well-being. *Conservation Biology*, *28*(5): 1160–6.

Morel, L., L. Barbe, V. Jung, B. Clement, A. Schnitzler, and F. Ysnel. 2020. *Ecological Applications*, 30(1). doi.org/10.1002/eap.2007.

Noss, R.F. 1991. Sustainability and wilderness. *Conservation Biology*, 5(1): 120–2.

Piquer-Rodríguez, M., Kuemmerle, T., Alcaraz-Segura, D., Zurita-Milla, R., and Cabello, J. 2012. Future land use effects on the connectivity of protected area networks in southeastern Spain. *Journal for Nature Conservation*, *20*(6): 326–36.

Primack, R., Miller-Rushing, A., Corlett, R., Devictor, V., Johns, D., Loyola, R., Maas, B., Pakeman, R., and Pejchar, L. 2018. Biodiversity gains? The debate in local-vs global-scale species richness. Political Science Faculty Publications and Presentations. Portland State.

Regos, A., J. Dominguez, A. Gil-Tena, L. Brotons, M. Ninyerola, and X. Pons. 2016. Rural abandoned landscapes and bird assemblages: winners and losers in the rewilding of a marginal mountain area (NW Spain). *Reg. Environmental Change*, 16: 199–211.

Resasco, J. 2019. Meta-analysis on a decade of testing corridor efficacy: What new have we learned. *Current Landscape Ecology Reports*, 4: 61–9.

Resasco, J., N.M. Haddad, J.L. Orrock, D. Shoemaker, L.A. Brudvig, E.I. Dmaschen, J.J. Tewsbury, and D.J. Levey. 2014. Landscape corridors can increase invasion by an exotic species and reduce diversity of native species. *Ecology*, 95(8): 2033–9.

Rutt, C.L., J.Vitek, M. Cohn-Haft, W.F.Laurence, and P.C. Stouffer. 2019. Avian ecological succession in the Amazon: A long-term case study following experimental deforestation. *Ecology and Evolution*, 9: 13850–61.

Scharf, A.K., J.L.Belant, and D.E.Beyer. 2018. Habitat suitability does not capture the essence if animal-defined corridors. *Movement Ecology*, 6: 18.

Schindler, D.E., J.B. Armstrong, and T.E. Reed. 2015. The portfolio concept in ecology and evolution. *Frontiers in Ecology and the Environment*, 13(5): 257–63.

Schou, J.S., J. Bladt, R. Ejrnaes, M.N. Thomsen, S.E. Vedel, and C. Flfjgaard. 2021. Economic assessment of rewilding versus agri-environmental nature management. *Ambio*, 50(5): 1047–57.

Soulé, M. (2014). The 'New Conservation'. In G, Wuerthner, E, Crist, and T, Butler (eds), *Keeping the Wild*. Island Press.

Theobald, D.M. 2010. Estimating natural landscape changes from 1992 to 2030 in the conterminous US. *Landscape Ecology*, 25(7): 999–1011.

Tilman, D. and J.A. Downing. 1994. Biodiversity and stability in grasslands. *Nature*, 367: 363–5.

Ward, M., Saura, S., Williams, B., Ramírez-Delgado, J.P., Arafeh-Dalmau, N., Allan, J.R., Venter, O., Dubois, G., and Watson, J.E. 2020. Just ten percent of the global terrestrial protected area network is structurally connected via intact land. *Nature Communications*, *11*(1): 1–10.

Watling, J.I., V.Arroyo-Rodriguez, M. Pfeifer, L. Baeten, C. Banks-Leite, L.M. Cisneros, R. Fang, A.C. Hamel-Leigue, T. Lachat, I.R. Leal, L. Lens, H.P. Possingham, D.C. Raheem, D.B. Ribeiro, E.M. Slade, J.N. Urbina-Cardona, E.M. Wood, and L. Fahrig. 2020. Support for the habitat amount hypothesis from a global synthesis of species density studies. *Ecology Letters*. doi.org/10.1111/ele.13471

WHCWG (Washington Wildlife Habitat Connectivity Working Group). 2010. Washington Connected Landscapes Project: Statewide Analysis. Washington Departments of Fish and Wildlife, and Transportation, Olympia, Washington.

Wilschut, L.I., E.A. Addink, H. Heesterbeek, L. Heier, A. Laudisoit, M. Begon, S. Davis, V.M. Dubyanskiy, L.A. Burdelov, and S.M. de Jong. 2013. Potential corridors and barriers for plague spread in central Asia. *International Journal of Health Geography*: doi: 10.1186/1476-072X-12-49

Worboys, G., Francis, W.L. and Lockwood, M. (eds). 2010. *Connectivity Conservation Management: A Global Guide (with Particular Reference to Mountain Connectivity Conservation)*. Earthscan.

9

MAPPING WILDNESS AND OPPORTUNITIES FOR REWILDING

Steve Carver

Introduction

In his essay 'The Green Lagoons', published in the book *A Sand County Almanac*, Aldo Leopold makes specific reference to mapping and the loss of wilderness long before it became a subject for academic study. 'Man always kills the thing he loves, and so we the pioneers have killed our wilderness. Some say we had to. Be that as it may, I am glad I shall never be young without wild country to be young in. Of what avail are forty freedoms without a blank spot on the map?' (Leopold, 1949: 148). This quote belies a conflict that goes beyond the desire for there to remain unmapped portions of the world. While true wilderness is, in essence, the raw landscapes of nature beyond human ken and shaping, there is sadly no longer any such place. The whole world—with perhaps the exception of some of the ocean seabed—has been mapped, at least from the air or from space, and as such might therefore be considered 'known'. This is true even if we know for a fact that no human has ever set foot there, as is the case for some areas of Antarctica that even to this day remain inviolate (Hughes et al., 2011). Humans have roamed for centuries over the other six continents and between a myriad of oceanic islands, all of which have either been visited by explorers or long been inhabited by indigenous people. Like Leopold, the romance associated with those 'Parts Unknown' found on old maps is still within us, and stories of hidden kingdoms and lost worlds fill our collective consciousness: from Shangri-La, Atlantis and Eldorado to my own favourite, La Vallée. With its Pink Floyd soundtrack this film tells the story of the search for a paradise valley said to exist deep within a remote area of Papua New Guinea marked on the map only as 'obscured by cloud'.

While mapping 'reveals' the landscape from behind such veils, be they mythical or real, it robs it of some of its wildness, even if ecologically speaking the area in question remains largely untouched and natural (Griffiths, 2006). But as Leopold recounts we have, of late, had to do it because without such maps it is extremely difficult to protect wilderness from the human exploitation that ultimately destroys their wildness. Despite heart-felt pleas and protestation to the contrary, politicians, planners, and developers only really understand the certainty of numbers and hard lines on a map.

There is a much-quoted paradox in the field of wilderness management that states that while wilderness refers to those lands and ecosystems that lie beyond human modification and

DOI: 10.4324/9781003097822-11

influence, management to protect it from exploitation and over-use risks removing some of the essential wildness found in attributes such as freedom and integrity. The answer to this paradox lies in managing ourselves and our expectations, of human influence over the wilderness itself, and yet even this places restrictions on the personal freedoms we often seek in wild places (Hendee et al., 1990).

This is where mapping comes in. We have mapped the whole surface of the earth, often in minute detail, collecting data about themes as varied as topography, human infrastructure and land cover. Over the last 20–30 years, cartographers and modellers have used these data to create maps of wilderness quality across a range of spatial scales from global to local, revealing the remaining wild bits that are still largely unsettled, unfarmed, and unspoiled by our need for resources such as animals, timber, water, minerals, and energy. Fortunately for Leopold he will never see such a world (he died in 1948 fighting a wildfire on his neighbour's farm) as there now remains no blank spot on the map, no 'parts unknown', no 'obscured by cloud', but thanks—at least in part to such mapping efforts—there is still wild country to be young in.

The world's remaining wilderness is, however, shrinking at an alarming rate with various estimates ranging from a nearly 10% loss between 1993 and 2009 (Watson et al., 2016) to 175km^2 of wilderness lost per day (Theobald et al., 2020). Such alarming rates of attrition comprise a principal threat to biodiversity conservation and UN Sustainable Development Goals (Lu et al., 2015). The post-2020 Global Biodiversity Framework of the Convention on Biological Diversity places 'retaining wilderness areas' as the first of 21 action-oriented targets for 2030. Mapping programmes have provided the numbers and the lines with which politicians and policymakers can gamble on our future, setting targets and deadlines though the CBD, COP, and GBF, so it is essential that they continue to develop and provide the foundational evidence-base needed to underpin our efforts to address the twin crises of climate change and biodiversity loss.

Mapping wild places

While there were a few heroic attempts to map global patterns of wilderness based on the old method of hand-drawn tracing paper overlays (see McCloskey & Spalding, 1989), the mapping of patterns in wilderness quality really came of age with the advent of digital spatial data and Geographical Information Systems (GIS). These essentially did away with the old methods of pen and paper sieve mapping (McHarg, 1969), and opened up a world of possibilities for advanced computer models and algorithms to map landscapes in hitherto unheard-of detail and complexity.

Consider for a moment if you will the problem of mapping remoteness as a key indicator of wildness. In McCloskey and Spalding's original work they mapped—by hand—all areas more than six kilometres from the nearest settlement, road, railway or navigable river using paper copies of 1:2 million scale Jet Navigation Charts. This is a mammoth task in itself, yet remoteness is far more complicated than this. For example, should we consider that all roads are equal? Does a multi-lane highway exert a greater influence than a forest track? Does a hamlet of three or four small dwellings have the same impact as a city of a million people? How does intervening terrain, presenting barriers to movement and resistance to travel, affect their footprint? Does the fact that you can or cannot see the nearest road or settlement from where you stand alter how you think about remoteness? All these factors are difficult if not impossible to map by hand and require digital data and the specialised computer software that GIS provides.

Many of the attributes of wild(er)ness, including those mentioned above, can be measured and mapped along a sliding scale from least to most wild. This recognises that wilderness is a relative concept and where it starts and ends, at least from a perspective of landscape values, depends very much on the background of the observer (e.g. their education, experience, demographic, etc.) and the window of observation (i.e. the region being mapped and studied). Most wilderness mapping is therefore carried out in reference to the concept of a wilderness continuum or environmental modification spectrum. This was first introduced by Roderick Nash in his formative text *Wilderness and the American Mind* (1967) as the 'paved to the primeval' and adapted by Rob Lesslie in his work on mapping wilderness in Australia (e.g. Lesslie & Taylor, 1985). Here, Lesslie and co-workers map four wilderness attributes describing remoteness from mechanised access, remoteness from settlement, apparent naturalness (i.e. how natural if feels based on absence of modern human artefacts such as roads, buildings, and other structures), and biophysical naturalness (i.e. ecological naturalness based on vegetation cover). These are mapped along a continuum from least wild to most wild using GIS methods and then combined by simple weighted summation into a wilderness quality layer. The wilderness continuum adapted from Lesslie for use with rewilding is shown in Figure 9.1.

Subsequent work has largely used continuous data models or map overlay methods (Carver & Fritz, 2016). These have been applied at multiple spatial scales from the global (e.g. McCloskey & Spalding, 1989; Lesslie, 1998; Sanderson et al., 2002), regional or national (e.g. Carver, 1996; Aplet et al., 2000; Kuiters et al., 2013; Müller et al., 2015; Plutzar et al., 2016), or local scales (e.g. Carver et al., 2012; Carver et al., 2013; Orsi et al., 2013; Măntoiu et al., 2016) each mapping wilderness quality within the specified geographical window, from least to most wild. Some studies (e.g. Cao et al., 2019) have compared discrete or continuous mapping approaches by mapping wilderness using both approaches to evaluate the sensitivity of the results to these methodological differences, and identify wilderness boundaries for the purpose of informing protected area networks and their management (see Chapter 6).

One of the problems with these approaches is how scale affects spatial data availability and accuracy. Scotland is perhaps the most intensively mapped country as regards wilderness quality, featuring as it does on global, regional, national, and local level maps. Examination of how

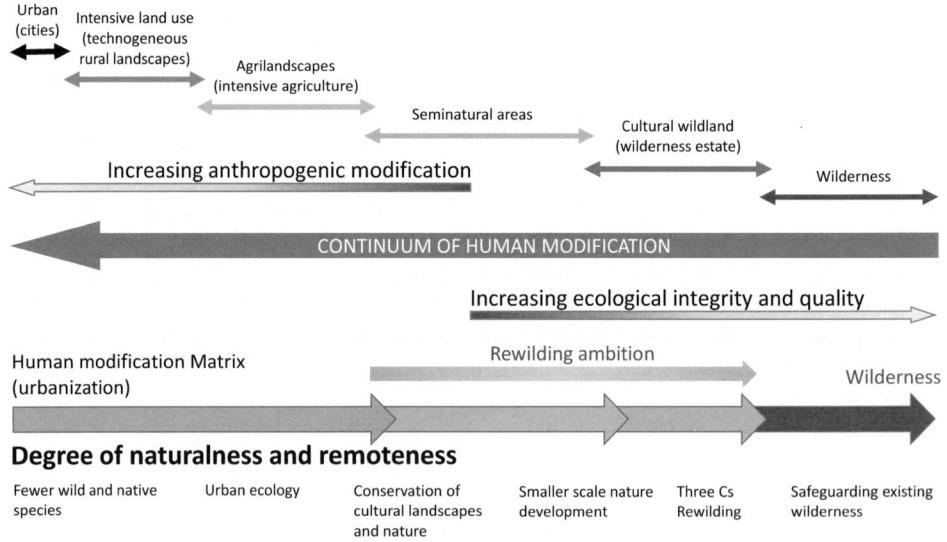

Figure 9.1 Wilderness continuum (after Carver et al., 2021, adapted from Lesslie, 2016).

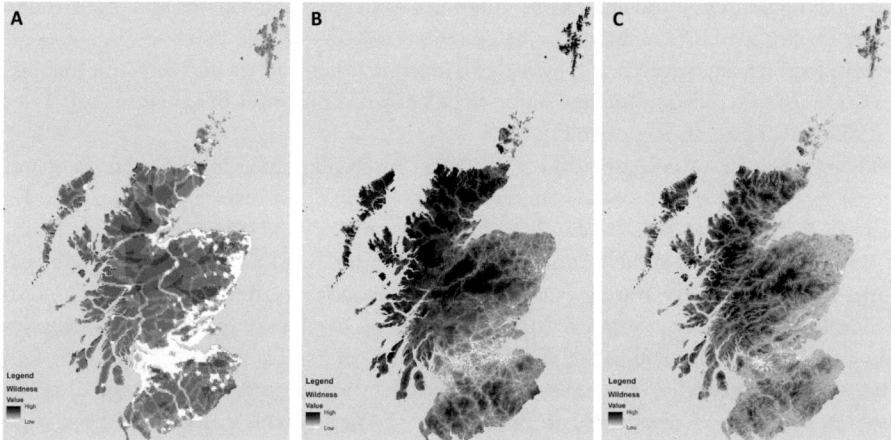

Figure 9.2 Comparison of wilderness quality across Scotland at different spatial scales using the same continuous colour scale: (a) global (after Sanderson et al., 2002), (b) European (after Kuiters et al., 2013), and (c) national (after SNH, 2014).

patterns of wilderness quality change and the location and shape of core areas thus identified across this small, yet geographically varied country, is a good demonstration of the limitations of global and regional-scale mapping efforts. This can be seen in Figure 9.2 which compares data from the Human Footprint (Sanderson et al., 2002) and EU Wilderness Quality Index (Kuiters et al., 2013) both mapped at 1km² resolution, with the national level mapping by Scottish Natural Heritage at 50m resolution, based on methods and techniques developed by Carver et al. (2012).

At the smaller scales, many global spatial datasets are of limited resolution (usually around 1km² pixels) with corresponding limitations in accuracy resulting from generalisation in the source data. Features are missed (e.g. under-recording of roads and other human features), the effect of terrain and landcover on measures of remoteness is ignored, while realistic modelling of walking times and visibility of human artefacts are simply not possible at this scale and resolution. Coupled with the effects of varying data reliability due to basic effects such as the Modifiable Areal Unit Problem (MAUP) and the 'ecological fallacy' in reporting population density (Openshaw, 1984) which state that variation in size and shape of census reporting areas, variation in standards of data recording, data classes, and basic spatial units (BSUs) across national boundaries produces patterns in the data that are really just a function of areas of reporting. This means that global-scale mapping is of limited use in showing anything other than general patterns which then leads to problems in deriving meaningful statistics and cross-border comparisons. Many of these problems can be seen in recent global-scale analyses on loss of wilderness (Watson et al., 2016), protected areas (Allan et al., 2017), marine wilderness (Jones et al., 2018), biodiversity protection (Di Marco et al., 2019), and climate change modelling (Asamoah et al., 2020). While highlighting global patterns, many of these analyses hide regional and local variations which at best restricts their utility in strategic policymaking and decision support at national levels and below, and at worst can lead to misrepresentations and over/underestimates in derived statistics.

Alternatively, large-scale modelling can use national and local datasets that tend to be of much greater integrity and accuracy, and which have been collected using standardised methods and systems geared to local geographies (Carver et al., 2012, 2013; Cao et al., 2019). These data tend to be higher resolution and can be combined with geographically realistic models of wilderness quality attributes such as time-based remoteness models and visible impact assessments that take local terrain, vegetation, physical barriers, and other features into account. Mapping at these scales better conforms to what ought to be known as the '4 Rs' of spatial analysis: Rigour, Reliability, Robustness, Repeatability.

The importance of geographical scale is again nicely illustrated in respect to Scotland by comparing the wildest areas of the country mapped at different spatial scales. When considering the wildest areas at the global scale, the vast majority of Scotland simply doesn't appear in the list (see the 'Last of the Wild' from Sanderson et al., 2002) and yet parts of the Highlands of Scotland do figure in the top 5% wildest areas at a European scale (see Kuiters et al., 2013). Obviously when just considering Scotland on its own, it has its own full spectrum of landscapes from least wild to most wild in the wilderness continuum (see Figure 9.3). This is true for any window of observation, at any scale. Using the wilderness continuum there will always be a spectrum from least to most wild, no matter what the size, extent, or location of the window of observation just so long as the number of pixels used is large enough.

Converting a continuum of wildness into a set of discrete core wild land areas with definite boundaries that are useful for decision and policymaking is both a conceptual and technical problem. On the one hand, defining the start and end point of wilderness along the spectrum shown in Figure 9.1 depends very much on the individual and the wider geographical setting, which includes an acknowledgement of the socio-cultural and historic context of the landscape in question. On the other, it becomes a methodological exercise in dividing a distribution of wilderness quality values along the spectrum into a series of classes such as core, buffer, edge and not wild. Statistics and fuzzy or Boolean methods can be used to define these thresholds in the distribution (see Comber et al., 2010 and Cao et al., 2019).

This was the approach used by Scottish Natural Heritage (now NatureScot) in 2014 to identify the 42 Wild Land Areas (WLAs) across the country shown in Figure 9.3c. They mapped the wilderness continuum based on four wildness attributes: perceived naturalness of land cover, absence of modern human artefacts, remoteness from mechanised access, and the rugged and challenging nature of the terrain which were combined to produce a wildness surface from which wild land boundaries could be drawn (Carver et al., 2012; SNH, 2014).

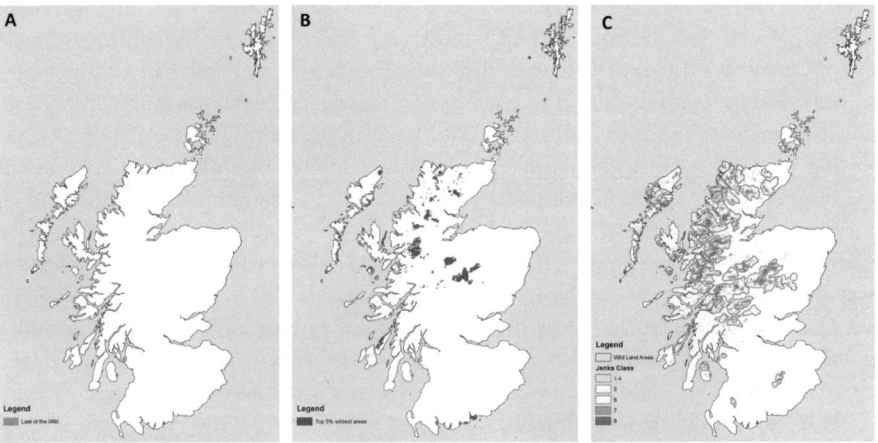

Figure 9.3 Wildest areas in Scotland at different scales: (a) global (after Sanderson et al., 2002), (b) European (after Kuiters et al., 2013), and (c) national (after SNH, 2014).

Opportunity mapping

With wilderness and wild areas mapped at various spatial scales and across different geographical areas, protection is largely a question of designation by applying national and regional nature protection laws under IUCN protected area categories Ia/b and II (IUCN, 2013). More often than not, however, mapping areas that meet all the requirements of wilderness (i.e. both attributes and criteria) can reveal a range of areas that are not yet designated and protected as wilderness even though they are by all measures wild. Areas protected by law as wilderness are referred to as '*de jure*' whilst those areas that fulfil all the criteria for wilderness and yet are not formally protected as such are referred to as '*de facto*' (Bastmeier, 2016). It is these '*de facto*' areas, and those that are close to meeting the required criteria, that protection and nature conservation efforts would do well to best focus their attention for two reasons: first, they are already wilderness areas but just lack formal protection as such, or second, they are close to being wilderness but require some remedial conservation actions (e.g. via rewilding) to improve their ecological condition and restore the wilderness condition.

Opportunity mapping can be used to highlight and target these areas and then focus and direct conservation efforts. GIS methods can be used to identify priority locations to restore ecosystem form and function across a range of scales on the back of wilderness mapping work. In the case of wilderness designation for *de facto* areas, this is simply a question of identifying wilderness areas without formal protection (e.g. by simple map overlay), but for targeting rewilding interventions this relies on more detailed analysis. This firstly requires looking at the attributes used to define wilderness quality across an area of interest, and then secondly using these to identify areas that could possibly meet the requirements for wilderness designation with some additional remedial action or conservation interventions.

For example, consider an area of land that has had four wilderness criteria mapped covering naturalness of land cover, remoteness from mechanised access, absence of modern human artefacts, and topographic ruggedness. If all four attributes are mapped as high, then the criteria for wilderness are met. If one or more of the attributes fall just below the criteria threshold for wilderness, then it is possible that these can be improved through implementation of remedial conservation measures. For instance, if remoteness is reduced by the presence of one or more 4WD tracks, then removal and restoration of those tracks could increase remoteness and push the area into the wilderness category. Similarly, the presence of grazing livestock or timber extraction, might restrict the naturalness of land cover, while removal of these and restoration of natural vegetation patterns would bump up the area into the wilderness category. Likewise, the removal of human features in the area such as the tracks in the remoteness example, or any buildings, dams, power lines, etc. could improve the absence of modern human artefacts criteria. Ruggedness is harder to modify but is very context specific as large flat areas that are not rugged may still be wild depending on the ecosystem in question (for example the Flow Country in northeast Scotland or the sandurs in Iceland's interior highlands). Changes to any of these attributes of wilderness quality, whether by deliberate management interventions or unintentionally via land abandonment, may be considered forms of rewilding, and the use of GIS mapping methodologies to identify areas where such interventions can improve the overall wilderness quality in an area may be used to target those rewilding interventions.

Ecological and landscape connectivity is a key aspect of opportunity mapping and the technology behind the spatial analysis of ecological network analysis has been evolving rapidly over the last few years (Dickson et al., 2019). These tools allow the relatively rapid assessment of habitat suitability/quality for either generic or specific species and the mapping of connecting corridors between core areas. These often rely on assessing multiple possible routes between a finite series of core habitat areas (or patches) across intervening landscapes of lower habitat suitability, by assessing the resistance and barriers to species movement with resistance (or friction) determined by variables such as land cover and human land use, road density and other infrastructure, human population density, and physical barriers such as large areas of open water. Models such as Circuitscape (Hall et al., 2021) use electrical current analogues to mimic how electricity might 'jump' the gap on a circuit board of varying resistance (i.e. the landscape) between positively and negatively charged poles (in this case, suitable habitat patches). One of the advantages of such models over more deterministic least cost path models is that the resulting corridors are of varying width and intensity (similar to current in a circuit board) and can be useful in identifying 'pinch points' in landscapes where movement is concentrated or funnelled to jump across a barrier (e.g. a highway) or squeeze through a gap in the resistance surface (e.g. between two built-up areas).

By combining opportunity mapping and connectivity models it is possible to evaluate and prioritise key gaps in the landscape where remedial measures and rewilding might be most effective in creating more connected ecosystems across a range of scales, from local to continental, thus underpinning the more conceptual models with detailed analytical models. A recent example is CARTNAT where wild(er)ness quality has been mapped across France as 'Haute Naturalité' (High Naturalness) at 80m resolution. Critically, this analysis incorporates an index of ecological connectivity in the model. The outputs are starting to be used at a local level in identifying both unprotected 'de facto' wild lands with the view to expanding protection and connecting these to existing areas of 'Protection Forte' (or high nature protection). This closely mirrors the so-called '3 Cs' model (Cores, Corridors, and Carnivores) developed by Soulé and Noss (1998) wherein wilderness quality indices define the cores (where wilderness quality is high) and resistance to species movement (where wilderness quality is lower) for key stone species (e.g. large carnivores) to move through a landscape following routes of least resistance (i.e. corridors). This is shown in Figure 9.4. An example of this combined mapping at high

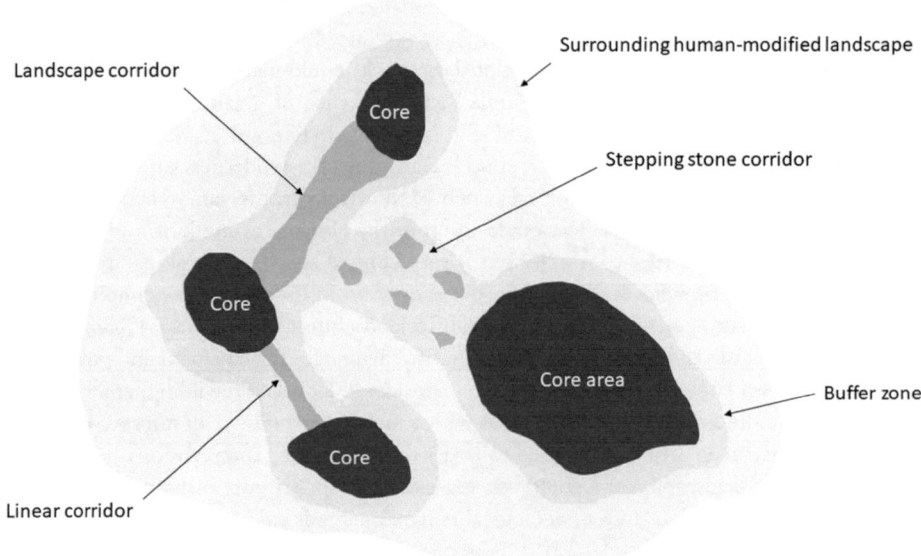

Figure 9.4   The 3 Cs model (after Soulé and Noss, 1998).

resolution is the planning for wildland networks in the Great Taihang region of China where wilderness quality mapping is used as an input to ecological connectivity models (Linkage Mapper and Circuitscape) to identify and assess potential landscape corridors for the north Chinese leopard (*Panthera pardus japonensis*) in this fragmented area (Cao et al., 2020).

Brave new world?

Over the course of the last couple of hundred years cartographers have filled in the last of Leopold's blank spots on the global map. In the space of a little more than 20 years we have moved from broad scale maps and conceptual models of wilderness to detailed mapping at fine levels of resolution powered by satellites, GIS, and digital data. Yet despite these rapid advances in spatial data handling and associated mapping products, there remains a broader need to think more generically as to how these fit into the dialogue on human–nature relationships, particularly around the barriers to rewilding in what can be already crowded, multiple use landscapes. The developing climate and biodiversity crises speak very loudly to this point in that external forcing and associated feedback loops from climate change are beginning to have serious deleterious effects on wilder landscapes and ecosystems. Where natural and semi-natural ecosystems are fragmented and natural habitats are isolated within areas of human-modified land use, many species are less able to migrate as the climate changes, leading to potential biodiversity loss, the simplification of natural ecosystems and the risk of collapse (Gonzalez et al., 2009).

A new 3Cs model focusing on the barriers to change that might hinder a rethinking of the human–nature relationship is therefore required. This needs to take into account geographical as well as community (including society and culture) context and adopt a creativity of thought that is needed to see past traditional models of land as a purely economic resource (see Figure 9.5). Context is important since the geography of a site and its situational setting (landscape, ecosystem, edaphic conditions, human land use, etc.) determines its potential ecosystem form and function as well as the human interaction with it. Community is important since the people owning, managing and living on and around a site or ecosystem have directed, and will continue to direct, the decisions made on its behalf. Many of these decisions are guided and grounded in long histories of culture and traditions that frame the local view of land and

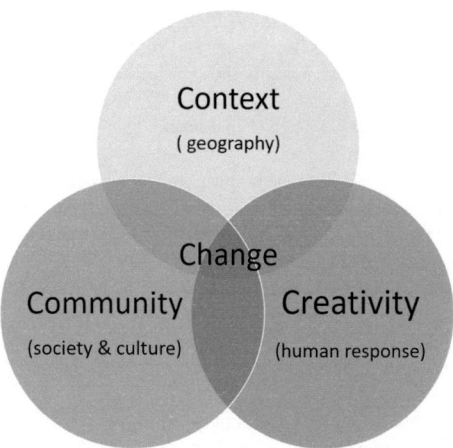

Figure 9.5 Context, Community, and Creativity.

of nature. Finally, creativity is important, because the combined biodiversity and climate crises will force us into new ways of thinking about land and into revising our essential relationship with nature.

And so, we return to where we started and the writings of Aldo Leopold and, in this instance, his thinking about developing a Land Ethic. In Part 3 of his classic book *A Sand County Almanac* Leopold develops the notion that we must extend our social and moral ethics of how we behave to each other in an orderly and just world, to one that includes 'the land' as equal partner in a kind of symbiotic relationship of interdependent mutual benefit. This represents a rebalancing of the human–nature relationship into one where we humans are no longer seen as the prime agent into one where we are equal partners.

Change is at the centre of this model where these three Cs overlap. How we respond to change and act to build more resilient landscapes will dictate how humanity will fare over the next few decades and into the 22nd century. There have already been calls to protect as much as half of the Earth's land surface for nature—see for example, Nature Needs Half programme (Locke, 2014), while the UN CBD proposes the protection of at least 30% of the world's land and sea for nature by 2030. This makes good ecological sense and fits remarkably well with Leopold's Land Ethic. How achievable such an aim is remains to be seen, but geography will determine the pattern of where and just how much land can be preserved for nature in this way. It is for example, easy to see how 50% and more of Canada can be protected, yet there still remains the need to be cognisant of potential traditional cultural uses and ownership of First Nation people. It is much harder to imagine how 50% of a small island nation of farms and cities like the UK could ever meet such a target without a watering-down of just what nature protection means (Starnes et al., 2021). If the global target could be met, then regional and local contributions will vary depending on context, community and creativity, driven by geography and, ultimately, our ability to see the bigger picture in responding to global change at regional and local levels.

References

Allan, J.R., Venter, O., and Watson, J.E. 2017. Temporally inter-comparable maps of terrestrial wilderness and the Last of the Wild. *Scientific Data*, 4(1): 1–8.

Aplet, G., Thomson, J., and Wilbert, M. 2000. Indicators of wildness: Using attributes of the land to assess the context of wilderness. *Wilderness Science in a Time of Change.* USDA Forest Service Rocky Mountain Research Station, RMRS-P-15-VOL-2, Missoula, MT, pp. 89–98.

Asamoah, E., Marco, M.D., Watson, J., Beaumont, L., Venter, O., and Maina, J., 2021. Tracing climate and land-use instability reveals new insights into the future of Earth's remaining wilderness. https://doi.org/10.21203/rs.3.rs-620309/v1

Bastmeijer, K. (ed.). 2016. *Wilderness Protection in Europe: The Role of International, European and National Law.* Cambridge University Press.

Cao, Y., Carver, S., and Yang, R. 2019. Mapping wilderness in China: Comparing and integrating Boolean and WLC approaches. *Landscape and Urban Planning,*192: 103636.

Cao, Y., Yang, R., and Carver, S. 2020. Linking wilderness mapping and connectivity modelling: A methodological framework for wildland network planning. *Biological Conservation*, 251: 108679.

Carver, S. 1996. Wilderness Britain?: Using GIS and Multi-criteria Evaluation Techniques to Map the Wilderness Continuum. School of Geography, University of Leeds.

Carver, S., Comber, A., McMorran, R., and Nutter, S. 2012. A GIS model for mapping spatial patterns and distribution of wild land in Scotland. *Landscape and Urban Planning*, 104(3–4): 395–409.

Carver, S., Convery, I., Hawkins, S., Beyers, R., Eagle, A., Kun, Z., Van Maanen, E., Cao, Y., Fisher, M., Edwards, S.R., and Nelson, C. 2021. Guiding principles for rewilding. *Conservation Biology.*

Carver, S.J. and Fritz, S. 2016. *Mapping Wilderness.* Springer.

Carver, S., Tricker, J., and Landres, P. 2013. Keeping it wild: Mapping wilderness character in the United States. *Journal of Environmental Management*, 131: 239–55.

Comber, A., Carver, S., Fritz, S., McMorran, R., Washtell, J., and Fisher, P. 2010. Different methods, different wilds: Evaluating alternative mappings of wildness using fuzzy MCE and Dempster-Shafer MCE. *Computers, Environment and Urban Systems*, 34(2): 142–52.

Di Marco, M., Ferrier, S., Harwood, T.D., Hoskins, A.J., and Watson, J.E. 2019. Wilderness areas halve the extinction risk of terrestrial biodiversity. *Nature*, 573(7775): 582–5.

Dickson, B.G., Albano, C.M., Anantharaman, R., Beier, P., Fargione, J., Graves, T.A., Gray, M.E., Hall, K.R., Lawler, J.J., Leonard, P.B., and Littlefield, C.E. 2019. Circuit-theory applications to connectivity science and conservation. *Conservation Biology*, 33(2): 239–49.

Gonzalez, A., Mouquet, N., and Loreau, M. 2009. Biodiversity as spatial insurance: the effects of habitat fragmentation and dispersal on ecosystem functioning. *Biodiversity, Ecosystem Functioning and Human Wellbeing*, pp.134–46. Oxford University Press..

Griffiths, J. 2008. *Wild: An Elemental Journey*. Penguin UK.

Hall, K.R., Anantharaman, R., Landau, V.A., Clark, M., Dickson, B.G., Jones, A., Platt, J., Edelman, A., and Shah, V.B. 2021. Circuitscape in Julia: empowering dynamic approaches to connectivity assessment. *Land*, 10(3): 301.

Hendee, J.C., Lucas, R.C., and Stankey, G.H. 1990. *Wilderness Management*. 2nd Edition, North American Press, Golden.

Hughes, K.A., Fretwell, P., Rae, J., Holmes, K., and Fleming, A., 2011. Untouched Antarctica: mapping a finite and diminishing environmental resource. *Antarctic Science*, 23(6): 537–48.

IUCN (2013) *Guidelines for Applying Protected Area Management Categories*. Gland, Switzerland.

Jones, K.R., Klein, C.J., Halpern, B.S., Venter, O., Grantham, H., Kuempel, C.D., Shumway, N., Friedlander, A.M., Possingham, H.P. and Watson, J.E. 2018. The location and protection status of Earth's diminishing marine wilderness. *Current Biology*, 28(15): 2506–12.

Kuiters, A.T., Van Eupen, M., Carver, S., Fisher, M., Kun, Z., and Vancura, V. 2013. Wilderness register and indicator for Europe. Alterra/Wageningen, Wildland Research Institute/Leeds & PAN Parks, 92.

Leopold, A. 1970. *A Sand County Almanac. 1949*. Ballantine.

Lesslie, R.G. (1998). Global Wilderness. UNEP-WCMC: Cambridge, UK. Dataset derived using the Digital Chart of the World 1993 version and methods based on the Australian National Wilderness Inventory (Lesslie, R. and Maslen, M. 1995). www.unep-wcmc.org/resources-and-data/global-wilderness Accessed 09/09/2021

Lesslie, R.G. and Taylor, S.G. 1985. The wilderness continuum concept and its implications for Australian wilderness preservation policy. *Biological Conservation*, 32(4): 309–33.

Locke, H. 2014. Nature needs half: a necessary and hopeful new agenda for protected areas. *Nature New South Wales*, 58(3): 7–17.

Lu, Y., Nakicenovic, N., Visbeck, M., and Stevance, A.S. 2015. Policy: five priorities for the UN sustainable development goals. *Nature News*, 520(7548): 432.

Mǎntoiu, D.Ş., Nistorescu, M.C., Şandric, I.C., Mirea, I.C., Hǎgǎtiş, A., and Stanciu, E., 2016. Wilderness areas in Romania: a case study on the South Western Carpathians. In *Mapping Wilderness* (pp. 145–56). Springer.

McCloskey, J.M. and Spalding, H. 1989. A reconnaissance-level inventory of the amount of wilderness remaining in the world. *Ambio*: 221–7.

McHarg, I.L. 1969. *Design with Nature* (pp. 7–17). New York: American Museum of Natural History.

Müller, A., Bøcher, P.K., and Svenning, J.C. 2015. Where are the wilder parts of anthropogenic landscapes? A mapping case study for Denmark. *Landscape and Urban Planning*, 144: 90–102.

Nash, R.F. 1967. *Wilderness and the American Mind*. Yale University Press.

Openshaw, S. 1984. The modifiable areal unit problem, CATMOG 38. In *Geo Abstracts*, Norwich.

Orsi, F., Geneletti, D., and Borsdorf, A. 2013. Mapping wildness for protected area management: A methodological approach and application to the Dolomites UNESCO World Heritage Site (Italy). *Landscape and Urban Planning*, 120: 1–15.

Plutzar, C., Enzenhofer, K., Hoser, F., Zika, M., and Kohler, B. 2016. Is there something wild in Austria?. In *Mapping Wilderness* (pp. 177–89). Springer.

Sanderson, E.W., Jaiteh, M., Levy, M.A., Redford, K.H., Wannebo, A.V., and Woolmer, G. 2002. The human footprint and the last of the wild: the human footprint is a global map of human influence on the land surface, which suggests that human beings are stewards of nature, whether we like it or not. *BioScience* 52(10): 891–904.

SNH (2014) Landscape policy: wild land. www.nature.scot/professional-advice/landscape/landscape-policy-and-guidance/landscape-policy-wild-land

Soulé, M. and Noss, R. 1998. Rewilding and biodiversity: complementary goals for continental conservation. *Wild Earth*, 8: 18–28.

Theobald, D.M., Kennedy, C., Chen, B., Oakleaf, J., Baruch-Mordo, S., and Kiesecker, J. 2020. Earth transformed: detailed mapping of global human modification from 1990 to 2017. *Earth System Science Data*, 12(3): 1953–72.

Watson, J.E., Shanahan, D.F., Di Marco, M., Allan, J., Laurance, W.F., Sanderson, E.W., Mackey, B., and Venter, O. 2016. Catastrophic declines in wilderness areas undermine global environment targets. *Current Biology*, 26(21): 2929–34.

10

MEASURING SUCCESS IN REWILDING

Ecological overview

Rene Beyers and Anthony R.E. Sinclair

Rewilding and ecological monitoring

Rewilding has been defined as 'the process of rebuilding, following major human disturbance, a natural ecosystem by restoring natural processes and the complete or near complete food-web at all trophic levels as a self-sustaining and resilient ecosystem using biota that would have been present had the disturbance not occurred'. (Carver et al., 2021). 'Rewilded' ecosystems do not require human intervention to persist and they can be seen as optimally adapted to the environment in which they occur. Rewilding often, but not always, takes place at large spatial scales and embraces natural well-connected core areas and the presence of keystone species such as large carnivores. Natural or rewilded food-webs are Trophic Baselines; we define Trophic Baselines as relatively undisturbed, self-sustaining, and persistent ecosystems that serve as a reference for rewilding degraded systems. It is a misconception that trophic baselines are static entities, a pretext for dismissing them (Thomas, 2020). Baselines change all the time through long term natural climate change, the very feature which allows them to control for human exploitation, pollution, and other impacts as well as rewilding to counteract them. Hence, baselines are controls for rewilding manipulations, necessary to detect whether such interference is leading to persistence of the ecosystem or not—without such controls the rewilding agenda is not based on science. These native ecosystems are resilient to short-term natural disturbance and change as a result of evolutionary processes and adaptations to local conditions.

Thus, they are stable, but not static, reference systems of native biota with which we can compare human-disturbed ecosystems. It has been speculated that novel ecosystems which originate in the context of the Anthropocene, will eventually be more resilient than the preceding natural ones as they adapt to the new conditions (Thomas, 2020). However, this premise has to be tested against controls (baselines); they may well be unstable, and lead to an impoverished alternative ecosystem state.

Another misconception is that rewilding is about (re-) creating biodiverse ecosystems irrespective of the provenance of the species. Anthropogenic novel ecosystems with their non-native biota, are often more diverse than the original natural ones and hence meet supposed rewilding objectives (Thomas, 2020). This is an illogical position because it would suggest that

DOI: 10.4324/9781003097822-12

zoos would be the best policy to pursue since they have the greatest diversity of species per unit area. Biodiversity of the native biota, together with their associated ecosystem processes, is what is being aimed for through rewilding, not biodiversity per se. The goal is to restore persistent, complete, interactive food-webs appropriate for the area to be rewilded (using a trophic baseline as reference) (Sinclair & Beyers, 2021).

When Trophic Baselines are no longer available, a model system constructed from historical data and expert opinion (including indigenous knowledge) can be used (see below). From this model, Trophic Baselines can be recreated as self-sustaining, persistent ecosystems. It is often not possible to rebuild a complete food-web due to socio-economic constraints, societal choices, a changed environment, lack of sufficient space, the presence of alien species that cannot be completely removed or the prior extinction of native species. However, we can still set the system on the path to become a more naturally functioning one which we term partial rewilding. Partially rewilded systems usually require some form of continued human management.

A monitoring programme is invaluable to guide and assess the rewilding process—it serves as a form of quality control and validation. The failure of some ecological restoration projects has been linked to a lack of—or poor-quality—monitoring, and a weak connection with ecological science (Lindenmayer, 2020). Monitoring should be discussed in the conceptual stages of a rewilding programme so that appropriate funding, logistics, and personnel can be planned for (Keenleyside et al., 2012). Generally, around 10% or more of the total cost of restoration is estimated to be needed for monitoring (Lindenmayer, 2020). Monitoring is the only way to provide information about the effectiveness and success of rewilding interventions. It is important to know that an ecosystem is on track to an expected desired state (endpoint) and is not moving to or stalling in an unwanted state, such as one that leads to a more degraded system or one that leads to the unexpected extinction of native species. For example, the introduction of a carnivore may have the unforeseen consequence of causing a local extinction of a prey population that may be rare or valuable. This type of information will be helpful for managers to take corrective measures and bring the system back on track. If the latter proves impossible, different targets may be envisioned and goals may have to be adjusted. For example, an ecosystem, through the prior removal of key components and/or long-term changes in the environment, may move to a different state from which full recovery is difficult or impossible.

The ultimate goal of rewilding is to (re-)create a persistent, self-sustaining ecosystem but we may not know whether that is the case unless we monitor it well beyond the initial achievement of rewilding goals. Biological communities and natural processes can take a long time fully to recover from major human disturbance, depending on the organisms involved and the complexity of the community. An extreme example of this is the full recovery of a tropical rainforest after extensive logging, which could take centuries or perhaps millennia.

Long-term monitoring may yield new insights, for example into the effect of the reintroduction of a keystone species after a long absence, and it may generate new questions. Many rewilding programmes are relatively new, the final outcome is uncertain and unexpected events happen. Very few studies have been conducted to assess the effectiveness of rewilding, especially at larger scales. Accelerating human-induced climate change influences the outcome and complicates prediction. A well-designed programme allows us to monitor these effects, assess the direction towards a predicted endpoint and the speed and trajectory of change.

Goals and objectives for monitoring

During the planning stage of monitoring, we need to set clear goals and objectives, and design the programme accordingly. We should ask specific questions that relate to the outcome of a rewilding project. For example, does the re-introduction of a keystone species cause the desired changes in the rest of the community based on what we know from its effects in a reference (Trophic Baseline) ecosystem? Many monitoring programmes do not have questions clearly spelled out and have limited capacity for teasing out ecological processes later on. A 'collect now, think later' (Likens & Lindenmayer, 2018) approach is usually a waste of time and resources. Monitoring is expensive, time consuming, and may involve many people, so it is essential to develop a good framework before collecting data. For example, the mandated monitoring of the abundance of an endangered species allows the detection of trends in the population. However, without correlating these trends to environmental factors and human interventions, the species may be monitored until it becomes extinct because there was no effort to determine the cause of the decline and do something about it.

Monitoring goals may have biological, socio-economic, and other components. They represent the broad vision of what a project aims to achieve. Objectives are more specific and are measurable steps that need to be taken to achieve those goals (see (McDonald et al., 2016, p. 15). The goal of a rewilding project for example is to bring an ecosystem as close as possible to an autonomously functioning target system while successfully addressing socio-economic issues in the area. Objectives may be to restore the full species community, to re-introduce a keystone predator and achieve a viable and ecologically effective population, to restore natural fire regimes, or to limit livestock predation by an introduced carnivore to less than 3% etc.

How do we monitor?

Monitoring response to rewilding interventions

Monitoring the effectiveness of rewilding interventions requires accurate and comprehensive recording of what was done where and when (for example how many animals of a species were re-introduced, how much of an invasive species was removed, etc.). Effort and cost of the interventions should also be documented. To assess the success of rewilding requires a special case of the traditional experimental design of before–after–control–impact, the BACI design (McDonald et al., 2016). There are two questions we need to ask: First, does intervention have an impact? To answer this, we compare the site with the intervention (I) before (B) with after (A) and a similarly disturbed site (C), but without intervention, before (Ct1) with after (Ct2). This will tell us if the intervention had an impact in the experimental site. Second, the far more important question is how successful is the intervention in achieving its endpoint? This requires an undisturbed reference (R) system (see below) with which to compare the intervention (I). Progress is measured by monitoring how I approaches R. Progress cannot be measured by comparing either A or I with B. The BA comparison tells us only that change has taken place, but not by how much or even if it is positive or negative—that comes only from the comparison of I with R. Concurrently, we monitor the disturbed but unmanipulated site (C) and compare it to R. If there is no difference in its approach to R, we don't know if the intervention had any impact on the rewilding as the unmanipulated system achieved the same state as the manipulated system.

So, the special design is BARI with RI being the most important component of the design. To account for long-term change, such as the effects of climate change, we monitor changes in R (see also below). This could be done if we have an existing ecosystem as reference but difficult or impossible with an inferred one. Preferably, we would have several replicates of each site but this may not always be achievable in practice.

End-point and trajectory assessment

To assess the success of rewilding we need an endpoint. We cannot leave the process open-ended as progress cannot be measured and so this approach is not scientifically valid. As we discussed earlier, the endpoint can either be a reference ecosystem or a model informed by historical data and expert opinion. If there is uncertainty about the endpoint, more than one endpoint may be envisioned so that the direction of the rewilding process can be tracked to one of these targets. An endpoint is not necessarily static, it can take into account natural variation, due to for example short term disturbances caused by weather; endpoints can be presented as a range of values or as a mean with appropriate measures of variation.

In terms of achieving the endpoint we can measure recovery completeness by comparing the state of the ecosystem at any time to the level of a reference system in terms of biodiversity and ecosystem functioning. Ruiz-Jaen and Mitchell Aide (2005) wrote a review of articles published in the journal *Ecological Restoration* that address monitoring restoration success. They looked at which ecosystem attributes were assessed and how they were measured and many of these are also relevant for rewilding.

Sinclair et al. (2018) developed a method to monitor both recovery completeness and the trajectory of rewilding (or ecological restoration). It relies on having a defined endpoint for a particular attribute that the ecosystem is expected to recover to. A simple example is a list of species expected to be present at full recovery (endpoint), based on a reference ecosystem. We can determine the state of recovery relative to that endpoint using a Rewilding (or Restoration) Value (RV) which is simply the value of a variable at a specific time during recovery divided by the value of that variable at the endpoint. In our species list example, the RV is the proportion of the full species community present at that particular time during recovery. We define the Rewilding (or Restoration) Index (RI) as the natural logarithm of RV (ln (RV) so that when the endpoint has been achieved (in our example, the species list is complete), the value of RI approaches zero (Figure 10.1a). For the underlying theory and mathematical derivation of the Rewilding (Restoration) Index we refer to Sinclair et al. (2018).

With repeated measurements, the shape of the trajectory towards recovery can be determined. The shape depends on how well species are connected in the food web, i.e.

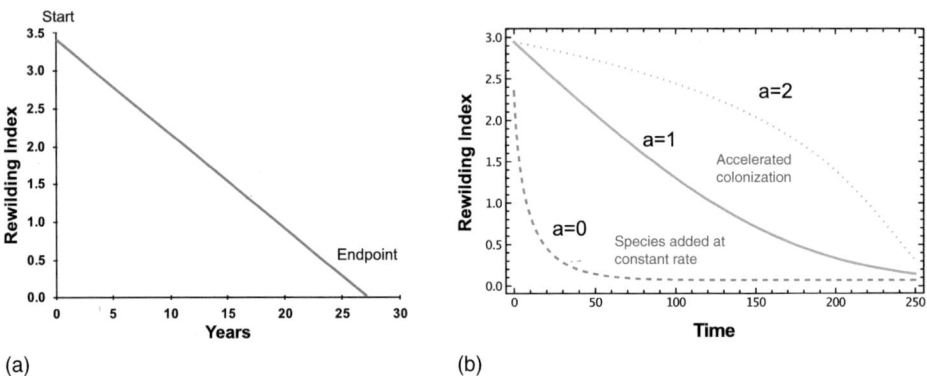

(a) (b)

Figure 10.1 **Rewilding Index and Trajectories** (a) The Rewilding Index progresses to zero when the endpoint is being reached. (b) Rewilding index trajectories for three scenarios with a different colonisation acceleration coefficient (a). When a = 0 colonisation is slow, species are added at a constant rate. Accelerating the rate of colonisation first results in a linear rewilding trajectory (a=1) and becomes convex (a=2) when the colonisation rate accelerates further.

how many links are there between them (Figure 10.1b). If species are well connected and species are added at a constant rate over the course of rewilding, the trajectory will be concave (a=0 in figure 10.1b). However, in many cases recolonisation of species at the beginning is slower and increases as the trophic structure of the system is rebuilt. Populations of species at higher trophic levels take longer to recover. For example, prey need to be present before predators move in. The trajectory will now be linear (a=1). The arrival of certain (keystone) species (for example beaver) facilitates rapid colonisation of other species in which case the trajectory will be convex (a=2) as connections between species increase at an accelerating rate.

A predicted rewilding endpoint may not be reached for several reasons:

- The chosen endpoint is the wrong one
- Climatic conditions have changed so that the reference ecosystem originated in a different climate and cannot be recreated under current conditions
- The conditions that prevent an ecosystem from recovering persist—for example invasive species prevent recolonisation of native ones, continued hunting pressure prevents recovery of (re-introduced) animal populations
- Socio-economic realities or cultural constraints prevent full recovery of an ecosystem
- An ecosystem shifts into an alternate stable state because of a disturbance that has been too intense, too frequent or both and this diminished the resilience of the ecosystem to rebound. This transition may accelerate or be abrupt when a tipping point or threshold has been reached. A rapid change in the shape of the curve that doesn't reach the endpoint (Figure 10.2a) may indicate such a shift. While important, it is hard to predict these tipping points but some progress has been made using simplified models (Jiang et al., 2018).

When the rewilding curve flattens out before RI reaches zero, this is an indication for managers that full recovery is not being achieved. With a good design and the right data, it may be possible to detect the causes and address these. Or, if this is not possible, a different endpoint could be chosen that is more attainable, for example one that takes into account human needs. As we have mentioned earlier, we define this as partial rewilding.

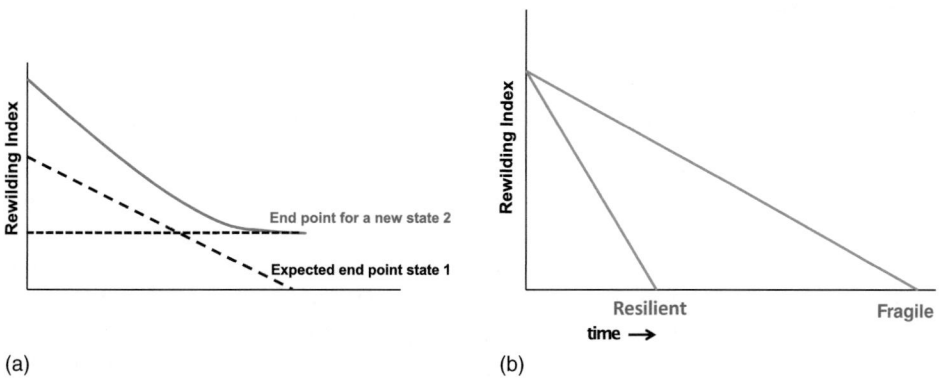

(a)   (b)

Figure 10.2 **Alternative stable state—delayed endpoint** (a) If a system moves into an alternate stable state (state 2), the Rewilding Index does not reach zero (Expected endpoint of state 1) but levels off at a different value (Endpoint for a new state 2). (b) Because a resilient system recovers faster than a fragile system, its Rewilding Index reaches the endpoint (zero) sooner.

If the endpoint is not being achieved because of long-term change, such as climate change, we adapt the monitoring and choose a new endpoint. This new endpoint can be obtained from monitoring the reference ecosystem if available (see above).

We can measure the resilience of an ecosystem by predicting progress to its endpoint. A resilient ecosystem will recover faster than a fragile one (Figure 10.2b). Resilience can be measured for different components of the ecosystem, such as different categories of species. An endpoint may be delayed if for example a disturbance, such as a big fire or flood, sets the system back to an earlier successional stage. Examples of delayed endpoints and differences in resilience in ecosystems can be found in Sinclair et al. (2018).

The Rewilding Index method can not only be used to monitor the recovery of species but also various measures or indicators of biodiversity and ecosystem processes, decreasing human disturbance and any other attributes of interest which we will discuss next.

Space for Time substitution

Space for Time substitution (SFT) methods, also called chrono-sequence studies, are used to study and model temporal phenomena in the future or past using contemporary spatial patterns. Traditionally, they have been used to study plant succession using contemporary ecosystems at different stages of development since a common initial condition to infer a temporal sequence. (Pickett, 1989). So, in theory, these methods could also be applied in rewilding to model the trajectory to an endpoint using ecosystems in various stages of recovery. This method assumes that an ecosystem goes through well-defined chronological states in an obligate order and that similar ecosystems follow the same historical pattern. As far as we know, SFT methods have not been used in a rewilding context but they could conceivably work if multiple sites follow the same trajectory to an identical endpoint. However, several issues may invalidate their application such as: different sites are progressing to different endpoints as is the case with alternative stable states, the return trajectory to a complete food web (endpoint) is different from the outward trajectory that was followed during the unraveling of the ecosystem, long-term climate (or other) change pushes the endpoint to an unknown state which is different from the reference ecosystem. When using SFT methods it is important to validate their central assumption which can be done with long term temporal monitoring as described higher. Even with plant succession, validation of SFT methods by temporal monitoring has been recommended as some classical examples of chrono-sequence studies in succession have been contested (Johnson & Myanishi, 2008).

What attributes do we monitor?

The choice of attributes that we monitor should be driven by the goals and objectives of the monitoring project. For each attribute, a reference value is set at the endpoint. The overarching question is what are the key attributes (or key indicators) that signal the recovery of an ecosystem? This will depend on the type of ecosystem, human impact, and management interventions. Specific questions about attributes should relate to underlying ecological principles of ecosystem functioning and rewilding, such as population regulation through competition and predation, natural disturbance and resilience, community assembly rules, succession, dispersal, landscape ecology, etc. (Lindenmayer, 2020).

Rewilding aims to restore the ecological integrity of a system, which consists of structures (species composition of the community, abiotic elements, etc.) and processes (species interactions, nutrient cycling, disturbance regimes, exchanges with surrounding landscapes, etc.) (Wurtzebach & Schultz, 2016). However, not all conservation that strives for ecological integrity is rewilding. As we have mentioned with partial rewilding, continued human management may be necessary to ensure a high level of ecological integrity, for example in areas that are too small or that have too many outside influences to be self-sustaining (Woodley, 2010). Rewilding however, aims for minimal human interference, so we would not only monitor an increase in ecological integrity but also a decrease of human inputs (Torres et al., 2018).

Table 10.1 is a non-exhaustive list of attributes associated with ecological integrity (biodiversity, ecosystem processes) and human inputs (human stressors and threats, and management

Table 10.1 List of attributes associated with biodiversity, ecological integrity, human forcing, and management

Category	Attribute	Examples–notes
Ecological integrity and ecosystem processes		
Species diversity and abundance	Community composition	Species lists
	Species richness and diversity	Alpha–beta–gamma diversity
	Species abundance	Abundance of keystone and other important species
Trophic complexity / trophic integrity	Food web / Species interactions	Connectance* and other measures of terrestrial, aquatic and soil food web structure
Structural diversity	Vegetation structure	
Spatial structure and diversity	Spatial structure / spatial habitat heterogeneity	Habitat diversity in the landscape and habitat fragmentation
	Spatial connectivity	Connectivity for migration and gene flow
Resilience	Ecological resilience	
Biological processes	Primary productivity	
	Natural vegetation regeneration and succession	
	Species recruitment	
	Dispersal	
	Species movement and migration	
	Pollination	
	Decomposition	Litter production and turnover
Substrate and Abiotic processes	Soil health	Soil mineralisation, soil organic matter turnover
	Water quality	
	Hydrology	
	Nutrient cycling	N / P / K and other nutrients
	Carbon cycling	
Natural disturbance	Natural fire regimes	
	Natural flooding	
	Wind-throw	Forest gap dynamics, storms (frequency, intensity and coverage)

(continued)

Table 10.1 (Cont.)

Category	Attribute	Examples–notes
	Natural avalanches and rock slides	
	Natural disease and pest disturbance	
Local climate	Change in climate variability	Temperature, precipitation, wind
Human stressors		
	Soil health degradation	
	Soil erosion	
	Soil contamination	
	Water pollution	
	Land conversion	Agriculture, tree plantations, urbanisation
	Land degradation	Ranching, logging
	Extractive industries	Mining, oil exploration and exploitation, logging
	Hydrological stressors	Dams and water level regulation, marshland drainage, river channelisation and canalisation
	Infrastructure	Roads, airports, pipelines, canals, ports, etc
	Invasive species–Alien species	Abundance and extent of invasive species
	Hunting, trapping, fishing	
	Harvesting of wild plants and other natural products	
	Removal of dead animals	
	Removal of dead wood	
	Wildlife food subsidies	
Rewilding intervention	Management interventions	Measures of intervention activities
	Resource input	Monetary cost, people, other measures of effort

Note: *Connectance is the proportion of possible links between species that are acquired.

interventions). Very rarely can or should all of these attributes be monitored and they should be carefully selected depending on the ecosystem and the goals and context of the project. Because context and goals will often differ considerably between projects, even in similar ecosystems, it is difficult to standardise the selection of attributes and compare monitoring results between them. Monitoring too many things takes attention away from important questions, and leads to poor quality while trying to do it all, so it is better to focus on fewer relevant variables. For example, we cannot monitor the abundance of all species, but we should look at the important ones, such as keystone species or ecosystem engineers that have a significant influence on other parts of the ecosystem. Indicators that represent a wider community of species or some aspect of ecosystem functioning may also be useful. However, there should be a clear and significant relationship between an indicator and what it is supposed to indicate. In practice, this is often not the case, or the described relationship is vague or non-specific (e.g. species indicators for 'ecosystem health'). Indicators are often specific for particular ecosystems or landscapes and may not be transferable to others.

Different authors use different categories of ecosystem attributes (Wortley et al., 2013). McDonald et al. (2016) for example recognise attributes in six categories for ecological restoration: absence of threats, physical conditions, species composition, structural diversity, ecosystem functionality, and external exchanges. Torres et al. (2018) provide a list of variables and indicators to measure ecological integrity and human input. They combine composite indices of human forcing and ecological integrity to calculate a rewilding score that describes recovery status of the system. Moreno-Mateos et al. (2017) use recovery debt which they define as the 'per annum amount that an ecosystem function or biodiversity is reduced during the recovery process after disturbance ceases'. They monitored annual deficits in biodiversity and functions compared with reference levels in anthropogenically and naturally disturbed ecosystems during recovery. They found that, unsurprisingly, systems disturbed by natural events (hurricanes) recovered faster than human disturbed systems.

Sampling design, methods, and data analysis

A robust statistical design and appropriate data collection protocols are required for any monitoring programme. Sampling techniques, frequency, and timing are dependent on the detectability and behaviour of the species or phenomena being monitored. Changes in vegetation for example can be quantified through time series remote sensing (satellite and aerial images, Lidar), photographs at fixed points in the field, vegetation plots, etc. For animals there is a large number of techniques available such as strip transects and line transects (Distance Sampling) on the ground or from the air and capture-recapture techniques using camera traps, DNA sampling from hair and dung, live traps, etc. It is beyond the scope of this chapter to go into detail here but there is a vast amount of literature available on this topic. We refer to Likens and Lindenmayer (2018) and Spellerberg (2005) for a comprehensive overview and further references.

We highly recommend consulting a statistician, with experience in ecological monitoring, in the planning and design phase of the programme. The statistical power needed to detect significant changes has to be balanced against logistical challenges and costs. Rewilding commonly takes place over big areas and involves different processes operating at different scales which should be taken into account in the design.

It is important to ensure data integrity over the course of the programme and it is imperative to get it right at the start. Changing data collection methods and/or designs half-way through may make comparisons difficult or impossible. For example, one of the authors (RB) was involved in a monitoring programme of elephants and other big mammals in the Okapi reserve in DRC. Several years after a first transect-based survey, the design was changed from a (semi-) random to a systematic design of transect locations. In the transition year, both the transects of the previous survey were repeated and transects of the new design were sampled so that the new transects could be calibrated with the old ones. This costly operation had to be done to ensure integrity and continuity in the data collection.

Data should be collected with a robust error-free system using either paper or digital tools with sufficient backups. It is advisable to store data in a secure data management system where data entry can be validated and data integrity ensured. Data management and analysis costs should be included in the overall budget. Too often, this part of a monitoring programme is poorly funded which may lead to large amounts of unused information that has been very costly to acquire but that in the end does not produce the results that were hoped for. As an example, data management costs for the Long-Term Ecological Research programme account for about 15% of the monitoring project costs (Likens & Lindenmayer, 2018).

Conclusions

Ecological monitoring is an essential component of rewilding. Without it, we are working in the dark. It is important to develop a monitoring programme before the start of any rewilding activities and ensure that there is sufficient and sustainable long-term funding. This should include budgets for design, field logistics, hiring staff and/or training of people (paid staff or volunteers), data management and analysis. Clear goals and objectives should be set based on ecological knowledge about the ecosystem and the desired outcome. An appropriate monitoring programme is an invaluable tool that helps managers to steer and adapt the rewilding process and assess its progress and success.

References

Carver, Steve, Convery, Ian, Hawkins, Sally, Beyers, Rene, Eagle, Adam, Kun, Zoltan, Van Maanen, Erwin, Cao, Yue, Fisher, Mark, Edwards, Stephen R., Nelson, Cara, Gann, George D., Shurter, Steve, Aguilar, Karina, Andrade, Angela, Ripple, William J., Davis, John, Sinclair, Anthony, Bekoff, Marc, Noss, Reed, Foreman, Dave, Pettersson, Hanna, Root-Bernstein, Meredith, Svenning, Jens-Christian, Taylor, Peter, Wynne-Jones, Sophie, Featherstone, Alan Watson, Fløjgaard, Camilla, Stanley-Price, Mark, Navarro, Laetitia M., Aykroyd, Toby, Parfitt, Alison, Soulé, Michael. (2021). Guiding principles for rewilding. *Conservation Biology*, 1–12. https://doi.org/10.1002/cobi.13730

Jiang, J., Huang Zi-Gang, Seager, T.P., Lin, W., Grebogi, C., Hastings, A., Lai Ying-Cheng. (2018). Predicting tipping points in mutualistic networks through dimension reduction. *Proceedings of the National Academy of Sciences*, 115 (4).

Johnson E.A. and Miyanishi, K. (2008) Testing the assumptions of chronosequences in succession. *Ecology Letters*, 11: 419–31.

Keenleyside, K.A., Dudley, N., Cairns, S., Hall, C.M., and Stolton, S. (2012). *Ecological Restoration for Protected Areas: Principles, Guidelines and Best Practices*. Gland, Switzerland: IUCN.

Likens, G.E. (1989) *Long-Term Studies in Ecology Approaches and Alternatives*. Springer.

Likens, G.E. and Lindenmayer, D.B. (2018). *Effective Ecological Monitoring.*CSIRO Publishing. ProQuest Ebook Central. https://ebookcentral.proquest.com/lib/ubc/detail.action?docID=5377555.

Lindenmayer, D.B. (2020). Improving restoration programs through greater connection with ecological theory and better monitoring. *Frontiers in Ecology and Evolution*, 8, Article: 50.

McDonald, T., Gann, G., D., Jonson, J., and Dixon, K.W. (2016). *International Standards for the Practice of Ecological Restoration. Including Principles and Key Concepts*. Society for Ecological Restoration (SER), First Edition.

Moreno-Mateos, D., Barbier, E.B., Jones, P.C., Jones, H.P., Aronson, J., Lopez-Lopez, J.A., McCrackin, M.L., Meli, P., Montoya, D., and Benayas, J.M.R. (2017). Anthropogenic ecosystem disturbance and the recovery debt. *Nature Communications*, 8, Article Number: 14163.

Pickett, S.T.A. (1989) Space-for-time substitution as an alternative to long-term studies, pp. 110–35 in Gene E. Likens (ed.), *Long-Term Studies in Ecology*. New York: Springer-Verlag

Ruiz-Jaen, M.C. and Mitchell Aide, T. (2005). Restoration success: how is it being measured? *Restoration Ecology*, 13(3): 569–77.

Sinclair, A.R.E. and Beyers, R. (2021). *A Place Like No Other. Discovering the Secrets of Serengeti*. Princeton University Press.

Sinclair, A.R.E., Pech, R.P., Fryxell, J.M., McCann, K., Byrom, A.E., Savory, C.J., Brashares, J., Arthur, A.D., Catling, P.C., Triska, M.D., Craig, M.D., Sinclair, T.J.E., McLaren, J.R., Turkington, R., Beyers, R.L., and Harrower, W.L. (2018). Predicting and assessing progress in the restoration of ecosystems. *Conservation Letters*, 11(2): UNSP e12390.

Spellerberg, I.F. (2005). *Monitoring Ecological Change* (2nd ed.). Cambridge University Press.

Thomas, C.D. (2020). The development of Anthropocene biotas. *Phil. Trans. R. Soc. B*, 375: 20190113.

Torres, A., Fernández, N., zu Ermgassen, S., Helmer, W., Revilla, E., Saavedra, D., Perino, A., Mimet, A., Rey-Benayas, J.M., Selva, N., Schepers, F., Svenning, J.-C., and Pereira, H.M. (2018). Measuring rewilding progress. *Philosophical Transactions of the Royal Society B: Biological Sciences*, 373 (1761).

Woodley, S. (2010). Ecological integrity: a framework for ecosystem-based management. In: D.Cole and L.Yung (eds), *Beyond Naturalness: Rethinking Park and Wilderness Stewardship in an Era of Rapid Change*. Island Press,. pp. 106–24.

Wortley, L., Hero, J.-M., and Howes, M. (2013). Evaluating ecological restoration success: a review of the literature. *Restoration Ecology*, 21(5): 537–43.

Wurtzebach, Z. and Schultz, C. (2016). Measuring ecological integrity: history, practical applications, and research opportunities. *BioScience*, 66(6): 446–57.

11

MEASURING SUCCESS IN REWILDING?

Coping with socio-ecological uncertainties in rewilding projects

Meredith Root-Bernstein

Introduction

What does it mean for rewilding to be successful? Many rewilding projects never define short-term fixed goals, making assessment of their progress, and thus assessment of their success or failure, difficult. As I have shown in previous work with colleagues (Root-Bernstein et al., 2018), this obfuscation of success is deliberate, and is critical to maintaining the place-based, adaptive, vision-led, long-term aspects of most rewilding projects. Indeed, it is critical to maintaining these projects at all. Rewilding projects that were integrated directly into government conservation services, and thus depended on short-term measures of success and were less motivated by personal visions, also were more likely to be terminated at the first sign of social conflicts or other difficulties (Root-Bernstein et al., 2018). I posit that there is thus a potential trade-off between developing standard measures of success and promoting and generating long-term social acceptance and resilience in rewilding contexts.

In this chapter I focus on the social elements of rewilded socio-ecological systems, or the 'human dimensions' of rewilding projects. I will examine three kinds of measure of success that can be used to evaluate whether 'human dimensions' of rewilding are moving in the direction that project managers or other stakeholders want and intend. I first examine procedural approaches to enacting success, that is, best-practice processes and recommended steps that are intended to guarantee successful outcomes, since these are the only kinds of success criteria that exist specifically for human dimensions of rewilding or elements of rewilding (e.g. translocation). I then consider the kinds of criteria of success implied or recommended by various approaches to theorising human dimensions of socio-ecosystems or human dimensions of conservation more generally. I divide these into two kinds, approaches that focus on socio-ecological system properties, and approaches that focus on the quality of human–environment, human–wildlife, and human social interactions. Next, I point out that the choice of measure of success from these three broad approaches—procedural, system-level, or interaction-focused—also needs to be evaluated and reflected on. As a second iteration, one may also evaluate and assess the success of decision procedures for selecting success criteria for the human dimensions

DOI: 10.4324/9781003097822-13

of projects. This is pertinent since it would be easy for a powerful social group to select project success criteria that emphasise their own interests to the detriment of other interests. Thus, the roles and use of power within decision-making processes about what success will consist of is an important, if often overlooked, issue. Here again there are multiple approaches to select between in order to assess if power in decision making is being used 'successfully'. Finally, I end with a critique of the need to assess success in rewilding. I suggest that objective, narrow and precise measures of success can ironically lead to project failure, but that qualitative, judgement-based evaluations, especially when embedded in local worldviews that are not fully translatable into scientific or bureaucratic terms, can resist premature declarations of failure. Rather than expecting rewilding projects to achieve some kind of successful state, we might focus on helping people cope with and develop psychological and social resiliency to the unknown and unpredictable aspects of rewilding, and nature itself.

Procedural aspects of enacting success

A variety of theorists of policy and implementation argue that project or programme success is a claim rather than an objective state or outcome (Hoag, 2014; Webber, 2015). Factors affecting how claims of success are formed include the choice of evaluation metrics, which can be selected to ensure attainment of something appearing to be positive, misalignment of different criteria of success for different actors, or assumptions that an ultimate goal necessarily follows from attaining a proximate goal. We should not understand claims of success as necessarily dishonest or as misrepresentations. Rather, they are attempts to approximate the measurement of a state—success—which is actually multidimensional, complex, perspective-dependent, and often temporally unpredictable. For example, following best-practice guidelines is a popular way to help support claims of success. Best-practice guidelines provide measures or tick-boxes for a series of proximate goals, implying that the ultimate goal is likely to have been attained as the outcome, even if it is not or cannot be measured (or even if the outcome appears, by some metric, to be unsatisfactory). The claim of success in the case of following best practice guidelines is thus a claim about the relationship between proximate goals and ultimate goals.

This makes procedural approaches to assessing success particularly apt for rewilding, which often does not have short- or medium-term goals or predicted outcomes. To my knowledge, the only measures of success specifically designed for rewilding are procedural in nature. The IUCN Rewilding Thematic Group published a set of 'universal guidelines for rewilding' (IUCN 2019; see Chapter 18 of this volume). They aim for consensus, and are thus broad, inclusive, and non-prescriptive. Regarding social and economic aspects of rewilding, three of the principals stand out as particularly relevant and a fourth may have additional implications.

6. *Rewilding requires local engagement and support.* This guideline emphasises that rewilding projects should engage stakeholders in ways that are participatory, transparent and inclusive. The goals of engagement should be to take into account concerns around human–wildlife conflicts and the perturbations caused by natural processes, and to educate the public about nature.
7. *Rewilding is informed by both science and indigenous and local knowledge.* Although there are many ways of engaging with indigenous and local knowledge (ILK), the guideline does not prescribe any particular methods. This guideline emphasises mutual learning between local or indigenous groups, scientists, and managers, drawing on natural history and historical ecological knowledge, for the purposes of generation of new knowledge and adaptive management.

10. *Rewilding requires a paradigm shift in the co-existence of humans and nature.* Although quite vague about what paradigm this would actually be, this guideline indicates that rewilding projects should be aware that rewilding implies 'transformative change' towards new ways of coexisting with nature. Beyond implying support for innovative solutions to human-wildlife conflict, it gives the green light to those who wish to integrate rewilding with social and economic change to support sustainability.

4. *Rewilding recognises that ecosystems are dynamic and constantly changing.* While this guideline is focused on calling for management (or lack of management) to promote and allow autogenic and allogenic disturbances, successional processes, and species' range shifts, it would seem to also have implications for social and economic processes in and around rewilding areas. In order to allow rewilded ecosystems to evolve spontaneously, social and economic structures and processes will also need to be flexible.

Additionally, the Human–Wildlife Interactions Working Group of the IUCN SSC Conservation Translocation Specialist Group has published a report on 'Working with people towards conservation solutions' (Consorte-Crea & Bath, 2020). Their recommendations are shown in Box 11.1. These emphasise that good or successful engagement with social aspects of species translocation—a common element of rewilding projects—involve listening and engaging local communities in decision making over the long term, as well as trying to fully consider the complexities of human emotions, attitudes, actions, and social dynamics. Ethnographic methods that produce rich qualitative material over periods of months (at a minimum) are mentioned as useful for this approach.

Box 11.1 Recommendations for 'human dimensions' (HD) work and engaging with people in conservation translocations (Consorte-Crea & Bath 2020)

- Reintroduction requires collaboration between biological and social sciences to address human–wildlife interactions
- Listening to people in different groups of interest, and the general public, in relation to reintroduction issues
- Giving people a voice in decision-making
- Involving different interest groups and local people in a project from the planning stages
- Credibility is built over time, through the development of long-term relationships with groups
- Working with social scientists/HD researchers to understand the attitudes, beliefs and knowledge, behaviour intentions of different interest groups in relation to management decisions and the focus species
- Monitoring changes in attitudes over time, which may result from changes in the size of populations, management decisions, education campaigns, and other changes in circumstances
- Awareness of bias towards scientific knowledge, and against underrepresented groups
- While working towards coexistence and addressing conflicts it is also important to consider the complexity of interactions between people and the focus species, which also involve positive interactions.

System-level perspectives: moving towards optimal socio-ecological systems?

Currently, much emphasis is placed on systems thinking and system-level characteristics like resilience for social and environmental change management (e.g. Laspidou et al., 2020). The IUCN Rewilding Guidelines provide some qualitative descriptive indications of desirable system properties of socio-ecological systems. However, we need to use more specific models or frameworks to define socio-ecosystem characteristics that can be measured to ensure that rewilding is moving towards desirable parts of socio-ecological parameter space. There are several frameworks and models to draw on to find suitable measures.

Resilient trait networks

The ecological literature on ecosystem networks has inspired similar approaches for socio-ecological systems, in which human interactions with other species are modelled as an integrated set of interactions (Sayles et al., 2019). Just as it is unclear in ecological modelling that it is possible for all kinds of functional traits or functional interactions to be optimally resilient at the same time—because this is too complex to assess, and because each set of traits may be expected to be structured independently—the key question seems to be *which traits should be structured in resilient ways* to ensure socio-ecological system stability/ resilience? Although quantification of socio-ecological network structures might be used to assess rewilding success *post hoc*, it is less clear how to actively design or manage socio-ecosystems in order to develop these structures among particular types of traits or interactions, which traits and interactions one should focus on, or what a network that is part-way to developing the correct structure would look like.

Panarchy

The idea that networked interactions are key to understanding socio-ecological dynamics also occurs in other kinds of theory. The panarchy model (Holling & Gunnerson, 2002), which proposes a specific cycle of socio-ecosystem change, also considers 'connectedness' to be one of three variables controlling the movement through this cycle. Unlike network theory, however, Holling & Gunnerson (2002) do not believe that connectedness *causes* resilience, as the two are considered to be two separate, parallel variables. In the panarchy model, resilience is the ability of the system to absorb shock, which declines as biological legacies accumulate and reach some exogenously-determined peak. Biological legacies, in the panarchy context, refer to accumulations of biotic materials such as species, seeds, or soils. Upon reaching the peak the ecosystem is fragilised due to its incapacity to further accumulate biological legacy, leaving it open to collapse from external shocks. The role of connectedness in the panarchy model is to control the level of endogenous vs exogenous control of a given system. Under low connectedness, associated with impoverished and randomly structured socio-ecosystems, endogenous control is low over system dynamics and exogenous control is high. As connectedness increases, increasingly rigid aggregates of components are formed, such that endogenous control over system dynamics is high and exogenous influence is low. This, however, is a relatively negative development because it is correlated with (although does not cause) loss of systemic resilience. Since the panarchy model proposes a constant cycle of socio-ecological change, it would seem that, according to this model, phases in which designers and managers cannot influence system dynamics should be expected. While they may be tempted to regain control by working towards highly complex and integrated socio-ecosystems, these should also be expected to be fragile and rigid in their responses to exogenous shocks.

Telecoupling

The connectedness measure in panarchy seems somewhat contradictory to the claims of the telecoupling literature. Telecoupling is an empirically-driven approach to analysis of environmental and social dynamics, which makes the point that causes of socio-ecological change are often remote to the site of change, especially when economic drivers may affect human behaviour (Liu et al., 2019). Thus endogenous dynamics can be strongly controlled by exogenous connections (Friis & Nielson 2019). Better integration across telecoupling, panarchy, and socio-ecological network modelling would assist assessments of how connectedness across scales affects change in socio-ecological systems.

Land-use allocation

Another set of models of optimal socio-ecological states are not primarily concerned with cross-scalar connections or their network structures. These are land-use allocation models, often but not always drawing on the IPBES Nature's Constribtions to People (ecosystem services) framework (which I discuss more below). In such models, given a set of exogenous variables such as climate or soil, land is allocated to different uses (including nature conservation) in order to maximise specific benefits and reduce specific costs (Deng et al., 2016). In terms of such models, rewilding is successful if it forms part of an optimal allocation of land-uses within a landscape.

Desireable interactions perspective: promoting particular qualities of human–wildlife coexistence

Other perspectives on what makes good or desirable 'human dimensions' of rewilding projects focus not on system-level territorial or network structure, and more on the qualities or values of specific kinds of interactions, often between humans and wildlife but also between humans or between humans and the environment more broadly.

Stakeholder negotiations

One way to define success in a rewilding context would be the negotiation of an agreed management plan supported by a majority of stakeholders. Others might argue that majority support, although a venerable democratic concept, is inadequate as a measure of legitimate or successful negotiation, and would rather prefer to see evidence that principles of equity and justice are being met (Greenberg & Cohen, 2014; Osborne et al., 2021). Lack of success could take the form of social and political conflicts, legal actions, human-wildlife conflict, and violence against humans and property. The IUCN Human-Wildlife Interactions Working Group makes this clear and also provides some recommendations for the methods to achieve the desired management plan outcomes (Consorte-Crea & Bath, 2020). However, one might also argue that avoiding conflict will only lead to stasis rather than transformative change (e.g. Skrimizea et al., 2020). Evidently, the choice of indices of success in terms of social support or agreement and conflict avoidance is highly complex. To further complicate the issue of what to measure, there are alternative framings of what makes a desirable or good set of human social interactions that go beyond social and political concerns to explicitly consider the quality of human interactions with other species.

Service reciprocities

The Nature's Contributions to People (ecosystem services) framework provides a list of kinds of service that nature provides to humanity, including providing air, water, natural resources, cultural inspiration, and so on. These are designed to be measured and subject to assessment. There are different ways to quantify the provision of ecosystem services, and in all cases it is considered optimal to provide the largest quantity of services possible (Díaz et al., 2018). Some researchers have also asked how people make contributions to nature through management (Comberti et al., 2015). While these contributions are not the object of quantification excercises in the Nature's Contributions to People framework, there is no reason that similar measures could not be developed. Similarly to the human contributions to nature concept, other approaches to human ecology have suggested that humans could even be considered to have megafaunal ecosystem processes to contribute to ecosystems, forms of which may currently be reduced or missing in some habitats (Root-Bernstein & Ladle, 2019). Thus, it might be possible to develop an assessment framework in which humans are active endogenous parts of the ecosystem, both receiving and providing ecosystem processes or services. Although this may sound promising as the basis for a 'paradigm shift' in relations to nature (IUCN, 2019), a service tit-for-tat between humans and other species strikes me as intractable and difficult to assess. Reciprocity-based interspecies relationships, in which humans and other species exchange roles, favours, or services, do, of course, exist in many societies. However, many of the reciprocal acts are symbolic and derive their meaning from an ontology (worldview) that permits interpretation, interconversion, and negotiation of interspecies debts in social terms, e.g. in the same terms as between human community members or kin (e.g. Willerslev, 2007). Unfortunately, these symbolic acts are unlikely to travel well outside their worldviews to form universal best-practice solutions (Webber, 2015).

Sustainable development

A sustainability approach focused on the Sustainable Development Goals (see https://sdgs. un.org/goals) could be used to assess the qualities of the human dimension of rewilding projects. The difficulty in using the SDGs for assessment is the large number of goals that should be pursued and optimised simultaneously, some of which are not clearly related to rewilding. At the same time, the SDGs, unlike Nature's Constributions to People, are not closely associated with a methodology of quantification and evaluation, as they are less the subjects of technocratic command-and-control, and more subject to political negotiations.

Coping, tolerance, and resilience

Finally, a promising approach that can deal with some of the challenges of the previously discussed frameworks and models, is to consider the coping strategies and community resilience of stakeholders involved in rewilding. Strangely neglected in the conservation literature, coping mechanisms refer to a range of social and psychological methods for managing, minimising, avoiding, tolerating, compensating for, changing, or accepting the stress, negative emotions, loss, and damage caused by risks (Gogoi, 2018). It is thus broader than risk tolerance, which implies having or expecting minimal psychological or material damage from a risk, and which is often lowest precisely when dealing with the unknown (Carter & Linnel, 2016). Coping strategies may involve tolerance, but can also include ways of living with uncertainty and recovering from harms with significant impacts. The environmental and existential risk literature argues that the best response to risk is to pro-actively develop personal and community-level networks and capacities to learn adaptively from past situations and implement solutions to mitigate or overcome harms (e.g. Paton et al., 2008). Several social characteristics promote community resilience to risks, including social equity and experience with collective action (Carpenter et al., 2012).

An advantage of the coping and risk resilience approaches is that they do not aim for or measure the attainment of some kind of optimal or perfect form of human–wildlife interaction, or socio-ecological state. Rather, life and the world are always characterised by a series of unpredictable and uncontrolled events. This approach seems to fit well with the goals of rewilding: a rewilded landscape is likely to be less predictable and under less anthropogenic control than a conventionally managed one. The unpredictable and unknown, perhaps novel, transformations that rewilding implies disturb some commentators, who thus judge it to be unjustifiable (Rubenstein & Rubenstein, 2016). Some research suggests that tolerance of ambiguity and uncertainty is a personality trait and thus varies between individuals (Furnham & Marks, 2013). Coping strategies also interact with the 'locus of control' or tendency of individuals to think either that they are in control of events, or that events are controlled by outside forces (Thiruchelvi & Supriya 2012). While to my knowledge not studied in relation to environmental change or rewilding, the psychological concepts of tolerance of ambiguity and locus of control seem relevant to how individuals and groups deal with unpredictable species, climate, and socio-ecological change. Developing the coping and resilience capacities of individuals and communities may make more palatable a tradeoff between immediate and salient everyday risks that reduce remote but more severe risks.

Creating social change: Science, communication, and power relations

Frameworks of systemic properties or interaction qualities can be used to argue for and justify particular interventions intended to make rewilding successful. The implementation procedures for arguing for and coming to agreement about rewilding interventions can also be evaluated.

Garmestani et al. (2020) provide an account of using panarchy as a scientific and public communications framework to achieve change in ecological management and policy. The authors sought to enact change to prevent encroachment of a native tree onto native grasslands (Garmestani et al., 2020). In this case, the panarchy model's 'imperative of destruction' was interpreted as justifying using fire management rather than tree felling as a means to control the 'invasive' native trees 'encroaching' on grasslands. Arguably, this reframing of fire as both destructive *and* good is a public communications success. Yet it strikes me that panarchy could just as easily be instrumentalised in defense of woody plant encroachment dynamics, for example by reinterpreting the destruction of prairies and their structural reformulation as woodlands as a natural process of ecosystem transformation. By discussing this case I am not trying to specifically single out panarchy as either excellent or problematic as a framework for developing changes in conservation practice. It is not unusual for some of the most vague, underspecified, or multiply-interpretable theories, frameworks, or ideas, to be widely adopted in guiding human dimensions of conservation interventions (Root-Bernstein, 2020). Some might argue that rewilding itself, and the IUCN guidelines and recommendations discussed in the first section, are also ideas or frameworks that can be reinterpreted in many ways to justify a broad range of very different interventions and goals.

In Root-Bernstein (2020) I argue that the success of some very vague models of human dimensions of conservation can be traced to a lack of formal social theory in conservation on the one hand, and a reluctance to theorise power relations on the other hand. The above-cited tree encroachment case study seems to illustrate this. It is representative of many conservation-related science communication approaches, which frame a certain group of scientists' particular interpretations as 'correct' and all other interpretations and preferences as existing 'because the public has not yet been sufficiently informed'. This frames the evolution of socio-ecological systems as an ontological conflict between different visions of the real and the possible. In this view, scientific managers make successful interventions in human dimensions when they impose their knowledge and values, especially by persuasion (Root-Bernstein, 2020). The persuasive communication methods employed mask the fact that this is a case study in exerting power to shape reality and impose one's will on other people.

Despite the vagueness and open-ended flexibility of rewilding, or of the human dimensions frameworks that it may choose to draw on, it is not inevitable that rewilding projects and programmes need fall into the same trap of not developing or using ideas and analysis of how to best negotiate power, and of thus enacting power in a way that risks being controlling, heavy-handed, or insensitive. Various rewilding projects use co-production and participative-style methods, to explicitly work towards socio-ecological reform and to develop shared visions with stakeholders (Zamboni et al., 2017; Root-Bernstein et al., 2018). There is a large amount of literature arguing for and justifying co-production and participation methods as superior to other modes of working with communities (Turnhout et al., 2020). This literature, while useful and well-meaning, also does not provide a theory of social power, but often contents itself with unrealistic platitudes to the effect that equal and evenly distributed power relations will result in socially and ontologically neutral outcomes. A theory of social power cannot be limited to claims about the conditions under which social power is not supposed to operate. The development of more realistic theories of social power in conservation and rewilding will be useful to guide human dimensions changes in such projects (Turnhout et al., 2020).

Evaluation issues: The design of measures of socio-ecological success

As we have seen, there is an interesting tension and contrast between using vaguely-defined theories to justify rewilding interventions, and using overly prescriptive measures to evaluate their success. The vagueness of certain socio-ecological frameworks or models, and even of the concept of rewilding itself, may be taken advantage of to motivate and justify a wide range of specific, contextual visions. Certain forms of imprecision can be necessary to allow people with different practices, knowledges, and ontologies to work together towards common goals (Root-Bernstein et al., under review). At the same time, this vagueness can be an intellectual weakness that can be turned against projects as a critique, if the intervention is not broadly supported through a shared vision. The specificity of evaluative measures can, by contrast, prevent unpredictable and long processes of adaptive learning and socio-ecological transformation from taking their course. Evaluative measures for success can produce failures that would otherwise be understood not as failures, but as inevitable exposures to environmental risk and unpredictability, conflicts arising from situated inequities, and learning opportunities.

Pragmatically, if we wish to avoid the co-optation of pragmatic guidelines on how to do adaptive learning, coping, or participative management, as restrictive assessment tools, these guidelines should be designed to be unfit for use as standard measures: they should avoid proxy measures readily available in datasets, resist synthesis in quantitative summary statistics, include qualitative analysis, and depend on the judgement of local stakeholders. One way to make them particularly robust to co-optation as evaluative measures may be to embed them in community Indigenous and Local Knowledge (ILK) through co-production approaches, which often defy simple conversion to scientific and bureaucratic data (Whyte et al., 2016), providing an added benefit of working with ILK in rewilding (IUCN, 2019). In addition, an interesting avenue to explore is the establishment (also potentially drawing on participative and co-production approaches) of qualitative modes of evaluation such as those used to decide on things like public purchasing of artworks (Heinich, 2017). What this means is not simply that qualitative evaluations must be negotiated by decision-makers, but that these negotiations can be made on the basis of judgements of quality, emotional attachments, and costs (Heinich, 2017), each of which integrate multiple aspects of the contextual social value of a project. While such negotiations of value are not inherently less biased or more reliable than the use of quantitative measures, they are arguably more integrative and less constraining.

Conclusions

Some forms of rewilding have narrowly-defined ecological visions, while others propose that rewilding is at the core of a complete reform of human relations to nature. Arguably, even rewilding projects that do not specifically intend to influence society or human dimensions of conservation, do so in any case by altering landscapes, affecting experiences, and challenging norms and regulations. The more diverse the impacts that a project has, or wishes to have, on society, the more difficult it is to assess any such changes in terms of success or failure.

Whether any particular socio-ecological change is good or bad is a highly positional, political, moral, and ontological, consideration. While socio-ecological models and frameworks can be pragmatically useful to justify interventions and persuade stakeholders, visions that are jointly developed through participative and co-production approaches can avoid some of the problems of scientific and bureaucratic evaluation, by presenting outcomes as complex socially embedded values. While there are no universal standards of socio-ecological optimisation, a pragmatic and realistic intervention we can implement and measure is helping people to cope with and enact resiliency in the face of uncertainty, risk, and conflict.

References

Carpenter, S.R., Arrow, K.J., Barrett, S., Biggs, R., Brock, W.A., Crépin, A.S., ... and Zeeuw, A.D. (2012). General resilience to cope with extreme events. *Sustainability*, *4*(12), 3248–59.

Carter, N.H. and Linnell, J.D.C. (2016) Co-Adaptation is Key to Coexisting with Large Carnivores. *Trends in Ecology & Evolution*, 31(8), pp. 575–8. doi.org/10.1016/j.tree.2016.05.006.

Comberti, C., Thornton, T.F., de Echeverria, V.W., and Patterson, T. (2015). Ecosystem services or services to ecosystems? Valuing cultivation and reciprocal relationships between humans and ecosystems. *Global Environmental Change*, *34*, 247–62.

Consorte-McCrea, A. and Bath, A. (2020) IUCN-SSC/CTSG Human-Wildlife Interactions Working Group report: Working with people toward conservation solutions. Available from: www.researchg ate.net/publication/344520938_IUCN-SSCCTSG_Human-Wildlife_Interactions_Working_Group_ report_Working_with_people_toward_conservation_solutions (accessed 15 Feb. 2021).

Deng, X., Li, Z., and Gibson, J. (2016). A review on trade-off analysis of ecosystem services for sustainable land-use management. *Journal of Geographical Sciences*, *26*(7), 953–68.

Díaz, S., Pascual, U., Stenseke, M., Martín-López, B., Watson, R. T., Molnár, Z., ... and Shirayama, Y. (2018). Assessing nature's contributions to people. *Science*, *359*(6373): 270–2.

Friis, C., and Nielsen, J. Ø. (eds). (2019). *Telecoupling: Exploring Land-use Change in a Globalised World*. Springer.

Furnham, A., and Marks, J. (2013). Tolerance of ambiguity: A review of the recent literature. *Psychology*, *4*(09): 717–28.

Garmestani, A., Twidwell, D., Angeler, D. G., Sundstrom, S., Barichievy, C., Chaffin, B. C., ... and Allen, C. R. (2020). Panarchy: opportunities and challenges for ecosystem management. *Frontiers in Ecology and the Environment*, *18*(10), 576–83.

Gogoi, M. (2018). Emotional coping among communities affected by wildlife-caused damage in northeast India: opportunities for building tolerance and improving conservation outcomes. *Oryx*, *52*(2), 214–19.

Goulden, S., Erell, E., Garb, Y., and Pearlmutter, D. (2017). Green building standards as socio- questions. *Perspectives in Ecology and Conservation*, *15*(4), 271–81.

Greenberg, J. and Cohen, R.L. (eds). (2014). *Equity and Justice in Social Behavior*. Academic Press.

Heinich, N. (2017). *Des valeurs: une approche sociologique*. Paris: Éditions Gallimard.

Hoag, C. (2014). Dereliction at the South African Department of Home Affairs: time for the anthropology of bureaucracy. *Critique in Anthropology*, *34*(4): 410–28.

Holling, C.S. and Gunderson, L.H. (2002). *Panarchy: Understanding Transformations in Human and Natural Systems*. Washington, DC: Island Press.

IUCN. (2019). *Rewilding Principles*. Available at www.iucn.org/sites/dev/files/content/documents/prin ciples_of_rewilding_cem_rtg.pdf

Laspidou, C.S., Mellios, N.K., Spyropoulou, A.E., Kofinas, D.T., and Papadopoulou, M.P. (2020). Systems thinking on the resource nexus: Modeling and visualisation tools to identify critical interlinkages for resilient and sustainable societies and institutions. *Science of the Total Environment*, *717*: 137264.

Liu, J., Herzberger, A., Kapsar, K., Carlson, A.K., and Connor, T. (2019). What is telecoupling?. In *Telecoupling* (pp. 19–48). Palgrave Macmillan.

Osborne, T., Brock, S., Chazdon, R., Chomba, S., Garen, E., Gutierrez, V., ...and Sundberg, J. (2021). The political ecology playbook for ecosystem restoration: Principles for effective, equitable, and transformative landscapes. *Global Environmental Change*, *70*, 102320.

Paton, D., Smith, L., Daly, M., and Johnston, D. (2008). Risk perception and volcanic hazard mitigation: Individual and social perspectives. *Journal of Volcanology and Geothermal Research*, *172*(3–4), 179–88.

Root-Bernstein, M, Gooden, J., and Boyes, A. (2018). Rewilding in practice and in policy. *Geoforum* 97, 292–304.

Root-Bernstein, M., and Ladle, R. (2019). Ecology of a widespread large omnivore, Homo sapiens, and its impacts on ecosystem processes. *Ecology and evolution*, *9*(19), 10874–94.

Rubenstein, D.R. and Rubenstein, D.I. (2016). From Pleistocene to trophic rewilding: A wolf in sheep's clothing. *Proceedings of the National Academy of Sciences*, *113*(1), E1–E1.

Sayles, J.S., Garcia, M.M., Hamilton, M., Alexander, S.M., Baggio, J.A., Fischer, A. P., ... and Pittman, J. (2019). Social-ecological network analysis for sustainability sciences: a systematic review and innovative research agenda for the future. *Environmental Research Letters*, *14*(9), 093003.

Skrimizea, E., Lecuye, L., Bunnefeld, N., Butler, J. R., Fickel, T., Hodgson, I., ... and Young, J.C. (2020). Sustainable agriculture: Recognizing the potential of conflict as a positive driver for transformative change. *Advances in Ecological Research*, 255–311.

Thiruchelvi, A. and Supriya, M.V. (2012). An investigation on the mediating role of coping strategies on locus of control-wellbeing relationship. *The Spanish Journal of Psychology*, *15*(1), 156.

Turnhout, E., Metze, T., Wyborn, C., Klenk, N., and Louder, E. (2020). The politics of co-production: participation, power, and transformation. *Current Opinion in Environmental Sustainability*, *42*, 15–21.

Webber, S. 2015. Mobile adaptation and sticky experiments: circulating best practices and lessons learned in climate change adaptation. *Geographical Res.*, 53 (1), 26–38.

Whyte, K.P., Brewer, J.P., and Johnson, J.T. 2016. Weaving Indigenous science, protocols and sustainability science. *Sustainability Science*, 11(1): 25–32.

Willerslev, R. (2007). *Soul Hunters: Hunting, Animism, and Personhood among the Siberian Yukaghirs*. Univ of California Press.

Zamboni, T., Di Martino, S., and Jiménez-Pérez, I. (2017). A review of a multispecies reintroduction to restore a large ecosystem: the Iberá Rewilding Program (Argentina). *Perspectives in Ecology and Conservation*, *15*(4), 248–56.

12

REWILDING 'KNOWLEDGES'

Blending science and Indigenous knowledge systems

Lisa Fenton and Zoë Playdon

Robert Macfarlane (2007, pp. 27–8) tells us that 'wild', the root of 'rewilding,' carries 'implications of disorder and irregularity ... an expression of independence from human direction, and wild land can be said to be self-willed land'. Accordingly, 'wildness has been perceived as a dangerous force that confounds the order-bringing pursuits of human culture and agriculture' on the one hand and 'as realms of miracle, diversity, and abundance' on the other. In this reading, in a Western culture whose literature is pervaded by the Abrahamic books of the Judaeo-Christian religions,[1] both definitions evoke a Garden of Eden and a fall from grace. Here, 'rewilding' implies a return to a pre-lapsarian state of innocence, a relinquishment of responsibility to an omnipotent agent, ending conflict and making humans (or at least men) pre-eminent in an Arcadian idyll.

Thus, the linguistic and sentimental etymologies of 'rewilding' have a problematic life in our cultural unconscious, their apparently easy surface understanding concealing difficult, conflicted, below-the-surface identities. The term 'Traditional Indigenous Knowledge' [TIK] is hardly less problematic. As Mary Louise Pratt (2008: 7) pointed out in her groundbreaking work, *Imperial Eyes*, the lens of Eurocentric elitism operated in its contact zones with Indigenous peoples in 'highly asymmetrical relations of domination and subordination'. The Linnaean project epitomises the totalising, classificatory eye that Empire brought to bear on indigeneity, accompanied by a Eurocentric cartography which defined 'conquered' lands as *terra nullius*, nobody's land, and its inhabitants as *homo nullius*, nobodies. Indigenous people's knowledge—their cultural and material practices—was appropriated, where it was useful to Empire's commercial and military programmes, while concurrently the people themselves were repudiated as 'primitive' or 'savage' and certainly, as the infamous tenth edition of the *Systema Naturae* delineated, sub-human: fit for slavery, slaughter, or entertainment in the human zoos that flourished in European capitals in the 19th and early 20th centuries.

TIK was also part of a violent ideological binary manufactured by the European concept of 'scientific knowledge', a project landmarked by Copernicus's publication of his proof of a heliocentric universe in 1543. This new 'natural philosophy' of empirical science was carried forward in the UK by Francis Bacon, Isaac Newton, and the Royal Society that was founded in 1660, against a backdrop of the larger Enlightenment movement. Supported by a growing industrialisation and urbanisation, 19th century British scientific endeavour produced, notably,

DOI: 10.4324/9781003097822-14

Darwin's geological, biological, and anthropological *Voyage of the Beagle* and, notoriously, his cousin Francis Galton's work on human eugenics. As Mary Midgley (2002 p. 58) indicates, such early science was as much a matter of fad as fact: Galileo rejected the idea of gravitational attraction because he was wedded to a 'mechanical model which saw the cosmos as a collection of separate particles interacting only by collision'. It also reflected cultural prejudice, so that Galton's meditations on whether 'nature' or 'nurture' was the strongest determinant of human success culminated in his circular theory of 'hereditary genius', which 'proved' that humanity's finest form was people who already owned or governed everything and everyone. Nor can it be said that modern science is free of similar obfuscation. Much of what lay people take as 'scientific explanation' is in really only description, so that, for example, while medicine can describe the effects of administering a general anaesthetic, it cannot explain how it works. Taxonomical shifts demonstrate how culturally fluid, and sometime regressive, 'scientific knowledge' can be: for example, in the UK trans people lived equally with everyone else until 1970, when they were reclassified from 'intersex' to 'floridly psychotic', lost their human rights, and became the focus of a eugenic project. It was not until 2002 that the British government declared that being trans was not a mental illness, although there are still medical practitioners who act as though it is (Playdon, 2021).

Beyond the uncertainties inherent in these terminologies, however, lay a pressing environmental materiality. Rachel Carson's *Silent Spring* (1962) recorded the poisoning of the biosphere through pesticides, evoking a furiously angry response from chemical companies and agribusiness. A decade later, the Ashby Report (1972: 42) on the control of pollution raised concerns about the greenhouse effect, though whether it would 'increase the earth's temperature as to melt part of the polar icecaps and cause serious flooding' or 'shut out sunlight and trigger off a new ice age' was a matter for speculation. This sense of ecological precarity has only increased in the intervening half-century, increasing the urgency of finding conceptual and material approaches towards its remediation. Acknowledging all the difficulties inherent in the languages, histories, and hermeneutics of rewilding, Indigenous knowledge, and science, therefore, it seems necessary to 'stay with the trouble', as Donna Haraway (2016: 1) puts it, 'not as a vanishing pivot between awful or edenic pasts and apocalyptic or salvic futures, but as mortal critters entwined in myriad unfinished configurations of places, times, matters, meanings'.

Exploring these conceptual complexities is important since how we think about rewilding decides its practice and its popular identity beyond an homogenised, romanticised binary to agribusiness, or a commercial buzzword to improve sales of goods and services. If rewilding is to develop as a liminal, 'blended' practice, drawing from a contact zone between 'scientific' and 'Indigenous' knowledges, then finding an appropriate language to discuss its land-based practices, while avoiding re-conquest and neo-primitivism, is essential to achieving its new cartographies.

Seeking language

Finding language to discuss 'Indigenous knowledge' has proved problematic. Researchers carried out a range of explorations during this century. In 1999, Semali and Kincheloe described the appropriating 'strategy of a language system that is not attached to an ecology or its intelligible essences' while a year later, Marie Battiste and James (Sa'ke'j) Youngblood Henderson pointed to the pejorative superiority implicit in placing 'Indigenous knowledge' in a category separate from Eurocentric colonial 'science'. Responding to some of these concerns, Ellen and Harris (2000: 7) described the concurrent movement of sequestration and repudiation which typified colonial epistemological appropriation: it 'absorbed such pre-existing folk knowledge as was absorbable and, ultimately, consigned what was not to oblivion'. 'Absorbable' knowledge was that which could be instrumentalised for Westernised cultural purposes, so that knowledge and practices gathered under the heading of 'spirituality' were dismissed at best, or at worst, actively destroyed, as, for example, in the Lutheran's burning of Saami drums from the Reformation onwards, or the cultural genocide practised by Christian Residential Schools in Canada from the 1880s to 1996.

Less ideologically burdened terms than 'indigenous', such as 'local', 'traditional', 'ecological', environment', and 'folk', were used by some writers to describe an 'informal' or 'other' kind of knowledge distinct from Western scientific knowledge. But 'traditional' carried the implication of stasis, not the fluidity of lived practice (McGregor, 2008), and while 'local' avoided fossilisation, it was also too vague (Berkes, 2012: 8). When Ingold and Kurttila (2000: 185) pointed out that the idea of local knowledge lay 'in the very activities of *inhabiting* the land, that both bring places into being, and constitute persons as of those places, as local', that raised more problems. As Kapil Raj noted, localities 'constantly reinvent themselves through grounding (that is, appropriating and reconfiguring) objects, skills, ideas, and practices that circulate both within narrow regional or transcontinental—and indeed global—spaces' (cited in Seth, 2009: 378). Consequently, in a local context '(techno-)scientific knowledge and its ways of ordering the world may be both implicated and imbricated in the very Indigenous epistemologies to which it is commonly juxtaposed' (Bonneuil, cited in Seth 2009: 378). Webb (2012) gives the telling example of 'Indian medicine', recognised as a distinct body of knowledge by Western biomedicine in the 19th century but subsequently assimilated into the official European pharmacopoeias. The knowledge arising from an embodied, lived experience and transmitted tradition was thereby depersonalised and devalued as something that would readily be apparent to anyone who lived in that place. As Ellen (2004) had already pointed out, 'once ethnobiological knowledge had been drawn within the orbit of modern science and its origins forgotten, it became difficult to know where to place the boundary between the two'.

Substituting 'ecological' for 'Indigenous' seemed a possible route to take. Recognising the urgent need for decolonisation, Fikret Berkes considered 'traditional ecological knowledge' [TEK] as:

> a cumulative body of knowledge, practice, and belief, evolving by adaptive processes and handed down through generations by cultural transmission, about the relationship of living beings (including humans) with one another and with their environment.
>
> *Berkes, 2012: 7*

Such knowledge was not limited to colonised people, but might be found, for example, in cod fishing in Newfoundland, ranching in Colorado, and the use of Swiss Alpine commons. It was not a universalised knowledge available to everyone, but only to those admitted to its community of practice; it required active practice as well as specific knowledge and an acceptance of the belief system implicit in these ethico-onto-epistemologies; and it was empirical in the sense of being developed from observation and experimentation. As Indigenous scholars Battiste and Henderson had already said, writing back to the Aacademy, 'what is traditional about traditional knowledge is not its antiquity but the way it is acquired and used' (2000, p. 46). Knowing is not separated from doing, so that TEK (McGregor, 2004: 79) 'is expressed as a "way of life"; it is conceived as being something that you *do*'.

But this fruitful site of exploration was not without problems. Berkes highlighted that 'knowledge of the land' was a deeper concept for Indigenous peoples than the usual Western scientific rendering of 'ecological knowledge' (2008, p. 5). As Elders of the Gwich'in people point out, 'spiritual and ethical values have been woven into this knowledge, creating a system that has guided the people and helped them survive' (Gwich'in Elders, 1997: 14). For the Gwich'in, 'knowledge of the land' combines the biophysical and spiritual environments, so that in a Western scientific formulation, 'the closest equivalent of the "Land", taken without its spiritual component, is "ecosystem"' (Gwich'in Elders, 1997, p.14). In this context, 'TEK' seems to be another colonial linguistic imposition on Indigenous peoples, implying that their bio-ecological knowledge can be compartmentalised and separated from their spiritual knowledge, in the familiar appropriative gesture. As Adams (2003: 25) noted, 'the critical branch of science for colonial development was ecology, the "science of Empire"' and the Western scientific formulation of TEK has the taint of 'anti-conquest', the 'strategies of representation whereby European bourgeois subjects seek to secure their innocence in the same moment as they assert European hegemony' (Pratt, 2008: 9).

Wrestling with these problematics, McGregor (2008, pp. 145–6) set out some key tensions:

> From an Aboriginal viewpoint, TEK is conceptualized as both more than and different from Western definitions. Native understandings of TEK tend to focus on relationships between knowledge, people, and all of Creation (the 'natural' world as well as the spiritual). TEK is viewed as the process of participating (a verb) fully and responsibly in such relationships, rather than specifically as the knowledge gained from such experiences. For Aboriginal people, TEK is not just about understanding relationships, it *is* the relationship with Creation. TEK is something one *does*... This means that, at its most fundamental level, one cannot ever really 'acquire' or 'learn' TEK without having undergone the experiences originally involved in doing so.

Consequently, it is only through participation and activity in direct relation with a terrain that Indigenous land-based knowledge can be made, re-made and renewed. As Battiste and Henderson (2000: 41) put it, 'when an Indigenous elder says "I know" it is a temporary reference point ... he or she must respectfully live it and know how to renew it'. Their knowledge is protean, a continually renewing process 'expressed as "way of life"; rather than being just the knowledge of how to live, it is the actual living of that life' (McGregor, 2004: 78).

Land-based practices

Some of these linguistic problems were solved by returning to the lived experience of land-based practices. Bateson (1972: 433) had shown how the action of notching and felling a tree with an axe involved a set of interrelated dynamic relations in the person-axe-tree system, requiring craftsmanship which drew on experience and accumulated knowledge for its success. In 1979, Gibson coined the term 'affordances' to describe the possibilities available for action within a particular environment and the qualities required to utilise them: an ability to read the landscape; adaptability; reflexivity, and responsiveness; and openness to possibility. Put together, these ideas began to form what ecological psychologist James Gibson (1979: 254) called an 'education of attention' a kind of landscape literacy developed through a continual process of engagement with an environment. As Ingold (1996: 178) put it, 'What the practitioner does *to* things, is grounded in an active personal engagement *with* them', an engagement which (Pye, 1968: 22) 'underwrites the qualities of care, judgement, and dexterity that are the essence of skilled workmanship'. What was required, therefore, was an 'ecological approach' to the study of skill, centring the practitioner in a complex web of ecological activity, where individuals learn their own 'ways of doing' (Lave and Wenger, 1991; Ingold, 1996; Ingold and Kurtilla,

2000). Discussing the Saami relationship with the environment, Ingold and Kurtilla (2000) emphasised that this relationship with the natural world is developed and transmitted through cultural practices and traditions, in a knowledge transmission which crucially requires learners' deep engagement with skilled land-based practice. A situated, direct, and applied experience of living with nature and from nature is necessary in order to 'know' nature. Intergenerational transmission of knowledge, therefore, is inseparable from lived, situated experience, developed through active, functional, and local experience. Necessarily, this produces individual variation, which transforms tradition through personal lived experience, as learners grow into their knowledge and adapt it to their particular needs.

Here, learning occurs through the continual process of engagement with a natural environment which forms the basis for cultural practice and traditions. As radical educationist Paulo Freire (1970) put it, people are not containers to be filled with cultural knowledge through a 'banking' system: learning takes place through praxis, a lived experience in which epistemology and ontology are inseparable. This is crucial, for land-based knowledge is necessarily provisional and an acceptance of uncertainty is a defining characteristic of expertise:

> All of a sudden, the most well-known places can 'flip over' and turn strange and hostile, leaving the traveller lost and bewildered. No-one is ever skilled or knowledgeable enough to be able to move in the forest with total confidence: so far as the weather is concerned, one has always to contend with a degree of uncertainty, and it is the recognition of this uncertainty that distinguishes the truly experienced woodsman. Above all, moving in an environment means 'tuning' one's own movement in response to the movements in one's surroundings—other animals, the wind, and so on.
>
> *(Ingold and Kurttila, 2000)*

Helpfully for the decolonisation project, these qualities, expressed as an expansive 'skill', could be found in many locations where expert knowledge was required. Western biomedical practice, for example, moves graduates from the general knowledge and relative certainties learned at medical school to seven years of postgraduate medical education where they learn to specialise and to work within scientific uncertainties and the fluidities of patients of different morphologies, co-morbidities, and cultural imperatives. Similarly, after learning the formal letter of the law, legal graduates learn the informal uncertainties of pleading before different courts, with widely varying clients. Some patients and clients are a 'walk in the park', as the vernacular describes their predictable landscape and others are complex, critical cases, which can shift direction in a moment. Experts in every field, including Berkes's cod fishers, ranchers, and Swiss farmers, and Bateson's tree-fellers, develop skilful practice through practice, through an education of attention, in which the smallest detail could be significant.

Writing about the transmission of environmental knowledge among Inuit men in the Northwest Territories of Canada, Pearce, Write, Notaina, et al. (2011) use the term 'land skills' to gather together the range of knowledge, skills, and experience required for expertise. They describe 'how to set a fish net under the ice in the fall' as an example of the combination of practical knowledge of where to set the net, the rationale underpinning that knowledge, and the hands-on skill necessary to actually set the net. This type of land-based, skilled knowledge can only be accessed, made, or renewed through activity in direct, unmediated relation with 'raw nature', active engagement with the terrain as an intersubjective, interactive component of knowledge. Perceiving an environment, one at the same time perceives what possibilities or opportunities it might afford: this affordance creates a direct link between a knowing agent and their action in that environment. So, the concept of affordance is relational, belonging neither to the perceiving organism nor to the environment, but constituted between both. Its realisation relies on prior knowledge and experience and the capabilities of the actor/agent, so that what a particular object or environment might afford one person (or organism) it might not afford another. In the human sphere, the deciding factor is not only the actor's experience and knowledge but also their goals, plans, values, beliefs, and past experiences (Norman, 1988).

This looks like an encouragingly complete account of a parallel relationship between Indigenous knowledge and Westernised science, and one to which both meta-cultures might subscribe. But unfortunately, both linguistic and land-based skill approaches to understanding the relationship between the two knowledge systems stumble over the same problem: what constitutes knowledge?

Re-conquest and neo-primitivism

The pinch-point is the kind of culturally transmitted knowledge, buttressed by direct personal experience, that constitutes 'belief', and especially the cosmological belief which underpins ideas of the 'sacred'. The European colonial exercise was explicitly theologically-based, a 'crusade' in its now-discredited terminology, an exercise by institutionalised Christianity in brutal indoctrination under the guise of 'compassionate conversion'. Indigenous cosmologies were ignored at best or demonised at worst, and the ceremonies and rituals which celebrated them suppressed. The knowledge they contained and celebrated, about relationality, intersubjectivity, and interdependency was disregarded and frequently the Indigenous languages, which were the only ones capable of communicating these inflected, multifaceted understandings, were lost almost entirely.[2]

Consequently, engagement by Westernised systems with the cosmological aspects of Indigenous knowledges may range from complete detachment, absence, and ignorance, through to a strongly imagined metaphysics, typically expressed as an identification with nature and a return to a romanticised emotional connection with the land (Milton, 2002). Such ideologies include 'green primitivism' (Ellen, 1986), 'earth-based spiritualities,' or 'green religions' (Taylor, 2010), such as 'deep ecology' (Devall and Sessions). However, these approaches may have few points of connection with Indigenous cosmologies, operating rather as 'neo-primitivism', that is, 'a contemporary version of primitivism in which the critical repudiation of earlier primitivist discourse paradoxically enables their re-introduction, under different names and configurations to be sure, as cultural, political, ethical, and aesthetic alternative Western modernity' (Li 2006: ix). Further, as Marie Battiste (2005: 7) points out, under the colonial shadow the concept of Indigenous knowledge as 'sacred' can be problematic:

> Donning the protective cloak of sanctity and religious freedom is an admission that Indigenous people are the hapless victims of biophysical forces that they can endure only as awesome mysteries. In other words, they are as ignorant and superstitious as Eurocentric observers have long maintained.

However well-meant, invoking 'sacred knowledge' runs the risk of falling into re-conquest and neo-primitive narratives as a form of cognitive imperialism.

New cartographies

In 1994, the First World Congress on Transdisciplinarity adopted Nicolescu's *Charter of Transdisciplinarity*, which suggested that:

only a form of intelligence capable of grasping the cosmic dimension of the present conflicts is able to confront the complexity of our world and the present challenge of the spiritual and material self-destruction of the human species.

(Nicolescu, 2002: 147)

Multireferential and multidimensional, transdisciplinarity seeks 'the emergence of new data and new interactions from out of the encounter between disciplines', offering 'a new vision of nature and of reality' through rigorous but open dialogue and discussion. This was a project which had already found expression in the work of feminist philosopher, classical scholar, and depth psychologist Jules Cashford. Using the idea of 'translucency' between different cosmological traditions, she had traced the Western mythic feminine image (Baring and Cashford, 1991), subsequently relating lunar images and rituals to land-based practices (Cashford, 2003) through religious symbolism, fairy tale, and folklore, to interrogate the formation of human consciousnesses.

In 1994, too, feminist theorist Rosa Braidotti (1994: 4) introduced the 'nomadic subject', a critical positioning 'that allows me to think through and move across established categories and levels of experience: blurring boundaries rather than burning bridges'. Nomadism created new cartographies, theoretically-based, and politically informed frames of reference from which to read the present. Such readings are always provisional: they 'need to be redrafted constantly; as such they are structurally opposed to fixity and therefore to rapacious appropriation. The nomad has a sharpened sense of territory but no possessiveness about it' (Braidotti, 1994, pp. 36–7). Speaking two decades later, Braidotti pointed to the new global conditions produced by 'a mutation of advanced capitalism' and the urgency of creating new 'cartographies of power, figurations of the subject, a quest for the present, going out there trying to map out what is happening, because what is happening is new and it cannot be accounted for in the old protocols' (Braidotti, 2014). This new cartography is 'never based on the individual self, it is always relational, it is always in connection to a multiplicity of others, human and nonhuman others, starting from the air we breathe': it is a politics of location, 'a humble call to do something in the place where we are', and it constitutes 'radical immanence', an understanding that the sacred is always present and accessible: there is no transcendent separation between physical and metaphysical. More recently, Braidotti has elaborated these ideas as 'posthuman', seeking, like Haraway, a more inclusive practice of 'becoming-human', a 'qualitative leap based on the need to think in zoe/geo/techno forms' that 'sits alongside a far older tradition of Indigenous philosophy, which likewise understands the power and potentiality of thought as being materially embedded in the geoformations and trans-species influences that shape and define existence in relational terms' (Bignall and Braidotti, 2019, pp. 1–2).

Speaking to this 'far older tradition of Indigenous philosophy', Welch (2017) sets the 'politics of difference' which informs Native American philosophy. Instead of 'an assimilation ideal' based on 'sameness', she foregrounds a 'natural ordering process' of creative chaos, an 'agonism' of purposeful tension and struggle.

Chaos, creativity, and difference are the life forces of difficult yet non-oppressive democratic structures grounded in agonism, since they can simultaneously integrate community members in explorative dispute and attune them to the advantages of complicated but malleable and expressive collaboration.

(Welch, 2017, pp. 371–372)

In this formulation, differences 'are unproblematically sites of inherent intersectionality by virtue of systemic interdependence', requiring a 'respectful wonder' which 'calls on community members to engage imaginatively to try to understand the needs of others' (Welch, 2017, pp. 377, 379).

Turning to material practice, some of the possibilities of these new cartographies are realised in the work of Watson, Linaraki, and Robinson (2021) on their 'Lo-TEK' approach to underwater and intertidal nature-based technologies. Seeking technologies that 'work symbiotically with, rather than against nature', the authors challenge high-tech solutions by drawing on 'traditional Indigenous responses to coastal resilience that instead amplify cultural, ecological, economic, and agricultural resilience' (Watson et al., 2021: 60). Its focus is on 'sustainable values of low energy low impact, and low cost, while producing complex, nature-based innovations that are inherently sustainable' which 'fosters symbiosis between species, while making biodiversity the building block used to construct sustainable technologies' (Watson et al., 2021: 61). The aim is not to use TEK as an appropriated part of Western scientific knowledge but to apply it to a scientific framework, as a 'radical indigenism' which 'argues for a rebuilding of knowledge and explores Indigenous philosophies capable of generating new knowledge' (Watson et al., 2021: 62). The adaptive pathways the project develops in response to sea level rise comprise 'four distinct scenarios: defend, surrender, offend, and retreat', providing 'a holistic design approach of hybrid adaptation strategies that amplify social, ecological, economic, agricultural, and climate resilience for both the environment and all its inhabitants' (Watson et al., 2021, pp. 63, 100).

If these new cartographies provide routes forward for understanding a possible future relationship between the knowledges that are mustered as 'Indigenous' and 'scientific', there remains two issues to be resolved. The first and most obvious is that of process: how to decolonise, what to do to remedy the historical inequities enacted by centuries of colonial military and commercial expansionism? One answer lies in those working in the field adopting the code of practice set out by the United Nations' *Declaration on the Rights of Indigenous Peoples* (2007) and learning from the best practice reports such as that of Canada's Truth and Reconciliation Commission (2015). The point made by both documents is that the lives, experiences, land-based knowledges, and cosmologies of 'Indigenous' people are equally valid, equally to be respected, equally to be valued, to those of Northern and Westernised peoples.

Related to this is the second issue of the location of science in Western knowledge. C.P. Snow's famous lecture, *The Two Cultures* (1959), described a gap between the sciences and the humanities which, he believed, needed to be bridged. Sixty years later, however, the policy of government, the structuring of funding agencies, and the organisation of universities, is still divided between STEM and SHAPE, locating sciences, technologies, engineering, and mathematics as separate from social sciences, humanities, arts, politics, and economics, as though the two never meet in the real world. The present Covid pandemic has brought this division into stark contrast. STEM can develop vaccines and mechanical interventions but only SHAPE can account for 'anti-vaxxers' and 'anti-maskers', not to mention the human, economic, and political consequences of STEM's viral management.

For the disparate movements which comprise 'rewilding' to succeed, therefore, we believe that these urgent questions about knowledge formation must be addressed through transdisciplinarity. An adequate understanding of the challenges and possibilities that attend the theory and the practice of rewilding cannot be achieved by naïve evocations of 'scientific' and 'Indigenous' knowledges. Socio-cultural landscapes, their histories, and contested readings of them are too complex for that. New cartographies are required, with their attendant problematisation and agonism. But as Braidotti and Welch point out, success lies through blurring boundaries, engaging without possessiveness, and of using conceptually rigorous, mutually respectful, essentially posthuman, actively imaginative approaches to create this new, urgent knowledge.

Notes

1 See, for example, Frye, N. (1982) *The Great Code*. London: Routledge and Kegan Paul.
2 This problem, which was especially evident where Residential Schools were created, is epitomised by Macaulay's infamous *Minute on Indian Education* which explicitly aimed to remove the teaching of Sanskrit. Many Indigenous communities, for example, in Canada and the US, are currently working to reclaim their languages and culture.

References

Adams, W.M. (2003) 'Nature and the colonial mind', in Adams, W. and Mulligan, M. (eds), *Decolonising Nature: Strategies for Conservation in a Post-Colonial Era*. Earthscan, pp. 16–50.

Ashby, E. (1972) *Pollution: Nuisance or Nemesis? A report on the Control of Pollution*. HMSO.

Baring, A. and Cashford, J. (1991) *The Myth of the Goddess: Evolution of an Image*. Viking.

Bateson, G. (1972) *Steps to An Ecology of Mind: Collected Essays in Anthropology, Psychiatry, Evolution, and Epistemology*. University of Chicago Press.

Battiste, M. (2005) Indigenous knowledge: foundations for first nations, *WINHEC: International Journal of Indigenous Education Scholarship*, 1: 1–17. https://journals.uvic.ca/index.php/winhec/article/view/19251

Battiste, M.A. and Henderson, J.Y. (2000) *Protecting Indigenous Knowledge and Heritage: A Global Challenge*. Purich Press.

Berkes, F. (2012) *Sacred Ecology*. 3rd ed. Routledge.

Braidotti, R. (1994) *Nomadic Subjects: Embodiment and Sexual Difference in Contemporary Feminist Theory*. Columbia University Press.

Braidotti, R. (2014) 'Thinking as a Nomadic Subject',*ERRANS Project*. Berlin Institute for Cultural Inquiry, 7 November 2014. www.youtube.com/watch?v=HXw8F_Ss3m0

Braidotti, R. and Bignall, S. (2019) *Posthuman Ecologies: Complexity and Process after Deleuze*. Rowland and Littlefield International.

Carson, R. (1962) *Silent Spring*. Houghton Mifflin.

Cashford, J. (2003) *The Moon: Myth and Image*. Four Walls Eight Windows.

Ellen, R. (1986) What Black Elk left unsaid: on the illusory images of green primitivism, *Anthropology Today*, 2(6): 8–12.

Ellen, R.F. (2004) From ethno-science to science, or 'what the Indigenous knowledge debate tells us about how scientists define their project', *Journal of Cognition and Culture*, 4(3–4): 409–50.

Ellen, R.F. and Harris, H. (2000) 'Introduction', in Ellen, R.F., Parkes, P., and Bicker, A. (eds), *Indigenous Environmental Knowledge and Its Transformations: Critical Anthropological Perspectives*. Harwood Academic.

Freire, P. (1970) *Pedagogy of the Oppressed*. Herder and Herder.

Galton, F. (1869) *Hereditary Genius*. Macmillan and Co.

Gibson, J.J. (1979) *The Ecological Approach to Visual Perception*. Houghton Mifflin.

Gwich'in Elders (1997) *Nank'Kak Geenjit Gwich'in Gwinjik: Gwich'in Words about the Land*. Gwich'in Renewable Resource Board.

Haraway, D.J. (2016) *Staying with the Trouble: Making Kin in the Chthulucene*. Duke University Press.

Ingold, T. (1996) Situating action V: the history and evolution of bodily skills, *Ecological Psychology*, 8(2): 171–82.

Ingold, T. and Kurttila, T. (2000) Perceiving the environment in Finnish Lapland, *Body and Society*, 6(3–4): 183–96.

Lave, J. and Wenger, E. (1991) *Situated Learning: Legitimate Peripheral Participation*. Cambridge University Press.

Li, V. (2006) *The Neo-Primitivist Turn: Critical Reflections on Alterity, Culture, and Modernity*. University of Toronto Press.

Macfarlane, R. (2007) *The Wild Places*. Granta.

McGregor, D. (2004) Traditional ecological knowledge and sustainable development: towards co-existence, in Blaser, M., Feit, H.A., and McRae, G. (eds), *In the Way of Development. Indigenous Peoples, Life Projects and Globalization*. Zed Books, pp. 72–91.

McGregor, D. (2008) Linking traditional ecological knowledge and western science: Aboriginal perspectives from the 2000 SOLEC, *Canadian Journal of Native Studies*, 28(1): 139–58.

Midgley, M. (2002) *Science and Poetry*. Routledge.

Milton, K. (2002) *Loving Nature: Towards an Ecology of Emotion*. Psychology Press.

Nicolescu, B. (2002) *The Manifesto of Transdisciplinarity*, translated by K. Claire Voss. State University of New York Press. https://inters.org/Freitas-Morin-Nicolescu-Transdisciplinarity

Norman, D.A. (1988) *The Design of Everyday Things*. Doubleday.

Pearce, T., Wright, H., Notaina, R., Kudlak, A., Smit, B., Ford, J., and Furgal, C. (2011) Transmission of environmental knowledge and land skills among Inuit men in Ulukhaktok, Northwest Territories, Canada, *Human Ecology*, 39(3): 271–88.

Playdon, Z. (2021) *The Hidden Case of Ewan Forbes*. Bloomsbury.

Pratt, M.L. (2008) *Imperial Eyes: Travel Writing and Transculturation*. Second edition. Routledge.

Pye, D. (1968) *The Nature and Art of Workmanship*. Cambridge University Press.

Semali, L.M. and Kincheloe, J.L. (1999) *What Is Indigenous Knowledge?: Voices from the Academy*. Falmer.

Seth, S. (2009) Putting knowledge in its place: science, colonialism, and the postcolonial, *Postcolonial Studies*, 12(4): 373–88.

Snow, C.P. (1959) *The Two Cultures and the Scientific Revolution*. London: Cambridge University Press.

Taylor, B. (2010) *Dark Green Religion: Nature, Spirituality, and the Planetary Future*. Berkeley, CA: University of California Press.

Truth and Reconciliation Commission of Canada. (2015) *Honouring the Truth, Reconciling for the Future*. Winnipeg: National Centre for Truth and Reconciliation.

United Nations General Assembly (2007) *Declaration on the Rights of Indigenous Peoples*. Resolution A/RES/61/295. www.un.org/development/desa/Indigenouspeoples/wp-content/uploads/sites/19/2018/11/UNDRIP_E_web.pdf

Watson, J., Linaraki, D., and Robertson, A. (2021) Lo-TEK: underwater and intertidal nature-based technologies. In Baumeister, J., Bertone, E., and Burton, P. (eds), *Sea Cities*. Springer, pp. 59–105.

Webb, L.A. (2012) On biomedicine, transfers of knowledge, and malaria treatments in eastern North America and tropical Africa. In Gordon, M.D. and Krech, S. (eds), *Indigenous Knowledge and the Environment in Africa and North America*. Ohio University Press, pp. 53–68.

Welch, S. (2017) Native American chaos theory and the politics of difference. In Gary, A., Khader, M.J., and Stone, A. (eds), *The Routledge Companion to Feminist Philosophy*. Routledge, pp. 370–81.

13

REWILDING

A legal perspective

Adam Eagle, Alex Cooper, Rob Espin, Jack Gould, and Elsie Blackshaw-Crosby

Introduction

Rewilding projects can cause geographical, ecological, and social changes on a landscape scale (i.e. on a scale relevant to full ecosystem function) and over long periods, and often require interactions with local and/or indigenous communities and protected species. This gives rise to legal issues which rewilding practitioners should consider. This chapter examines international commitments in biodiversity law, discusses some of the issues which may arise when undertaking a rewilding project, and comments on the direction of travel of law relating to rewilding.

International treaties and conservation law

Much environmental law stems from commitments made by national governments in international treaties on global, regional, and bilateral or multilateral levels (Trouwborst et al., 2017). These treaties are usually implemented through a cascade of instruments moving from general commitments in international treaties[1] to specific obligations in national legislation. For example, commitments in the *Convention on the Conservation of European Wildlife and Natural Habitats* (Council of Europe, 1991, the 'Bern Convention') were implemented in the European Union ('EU') through the Birds Directive and the Habitats Directive, which obliged Member States to enact relevant national legislation (Council Directive 92/43/EEC, 1992 (the 'Habitats Directive'); Directive 2009/147/EC, 2009 (the 'Birds Directive')).

While several treaties provide for signatories to create adequate habitats, the focus of these treaties is on maintaining and protecting these spaces.[2] Rewilding is focused on changing existing land-use; so while these treaties are relevant, there are occasionally tensions between these traditional conservation requirements and rewilding.

The key international treaties in this area are contained in Table 13.1.

Due to the interconnectivity of species, ecosystems, and human activities, practitioners should be aware of the scope of these treaties as well as the particular legislation in their jurisdiction.

DOI: 10.4324/9781003097822-15

Table 13.1 Key international biodiversity treaties (adapted from Trouwborst et al., 2017)

Title	Purpose	Adopted	In force
Convention on Wetlands of International Importance Especially as Waterfowl Habitat (Ramsar Convention)	Contracting parties shall preserve designated wetlands.	1971	1975
UNESCO Convention Concerning the Protection of the World Cultural and Natural Heritage	Defines natural and cultural sites which signatories must preserve as world heritage sites.	1972	1975
Convention on International Trade in Endangered Species of Wild Fauna and Flora (CITIES)	Aims to ensure that international trade in specimens of wild animals and plants does not threaten species' survival.	1973	1975
Convention on the Conservation of European Wildlife and Natural Habitats (Bern Convention)	Aims to conserve wild flora and fauna and their natural habitats.	1979	1983
Convention on Biological Diversity (CBD)	Aims to conserve biological diversity, ensure the sustainable use of the components of biological diversity and ensure the equitable sharing of the benefits arising out of utilisation of genetic resources.	1992	1993

Legal issues regarding rewilding

This section examines some of the legal issues which should be considered when undertaking a rewilding project to ensure compliance with the law and to avoid liability from third parties. The national law of the country in which the rewilding project is situated will be determinative; however, these issues are likely to exist to some degree in most jurisdictions.

Land use: Changing land use

There are a number of legal considerations which should be considered when rewilding a landscape by altering, or ceasing to manage, an area.

Purchasing land: Conveyancing advisors should be instructed and proper due diligence conducted to check whether land is subject to legal restrictions or rights which might constrain rewilding. One particular consideration will be the security of land tenure, which reflects how likely the ownership of the land is to be contested, and reflects legal, social, and political considerations.

Neighbouring landowners: In some jurisdictions, the occupier of land may owe a duty of care towards their neighbour. To comply with this duty, the occupier may need to limit potential hazards to the neighbour to the extent reasonable to do so in the circumstances.[3] Failure to comply with such a duty may give rise to statutory or tortious liability. This is a broad area of law; some pertinent examples are addressed here.

In some countries, legislation compels occupiers to prevent the spread of specified species of plant. In the UK, the *Weeds Act 1959* grants the government powers to serve a notice on an occupier where certain species exist; private parties can also issue complaints under this Act.

As an example, the Knepp estate, a rewilding project in the South-East of England, enacted a policy to ensure compliance with this Act following discussions with neighbours.

Rewilding projects often alter bodies of water, either directly, for example to create a particular habitat or a more 'natural' landscape (which may require licences or permits), or indirectly, for example as the result of the actions of species which have been reintroduced. If water were to flood neighbouring land, it could give rise to tortious[4] or statutory liability.

In practice, one way to help mitigate such risks is to ensure good neighbourly relations from the project's outset, which is also important for reasons beyond management of legal risk.

Planning: Planning law designed to regulate land use is widespread internationally. Permission may be required from a relevant authority when a material change of use of land is planned.

In the EU, Environmental Impact Assessments are required for land uses which are '*likely to have significant effects on the environment by virtue, inter alia, of their nature, size or location*' (Council Directive 2014/52/EU, 2014 I (the 'Environmental Impacts Directive')). These are designed to ensure that the relevant planning authority has knowledge of the environmental impact of a proposal. The requirement would likely apply to afforestation or the creation of wetlands, although may not apply for small-scale projects.

Planning authorities may be required under national law to use their authority to increase biodiversity and consider the ecological impacts of a proposed development.[5]

Access rights: several countries (including Iceland, Finland, Norway, and Scotland) grant their citizens a 'right to roam' (known in Scandinavian countries as 'every man's right') permitting public access to private lands. More commonly, states have legislation allowing members of the public access through land on designated footpaths or 'rights of way', which may restrict the ability of large-scale rewilding projects to enclose land.[6]

A rewilding project seeking to restrict public access to land should consider how to derogate from these rights. In Sweden, the tension between restricting access to a rewilding project for commercial reasons and the Right of Public Access has been noted (Koninx, 2019). In Scotland, the rights granted to the public under the *Land Reform (Scotland) Act 2003* were considered when a rewilding project was granted a licence under the *Dangerous Wild Animals Act 1976* (the 'DWAA') to keep wild boar, which required the habitat to be fenced securely.

Forestry law: Forest and woodland management law will vary across jurisdictions. In countries where forest coverage is low, special licenses or grants may be required to fell trees even where these are non-native species or planted in inappropriate areas (e.g. on peat bogs). In warmer climates, governments may mandate forest management activities such as the removal of unmanaged woodland in order to reduce the risk of forest fires. Community rights to fuel and foraging may also affect a purchaser's ability to maintain a forest.

Land use: Intersectionality

Rewilding projects are often effected on a landscape scale and have significant impacts on the surrounding area; as such they may have consequences for the private legal rights, land rights, or human rights of the population.

As with environmental legislation, national law regarding human rights is often derived from international law.[7] The United Nations ('UN') treaties from which a number of national human rights laws are derived contain the right not to be arbitrarily deprived of property (United Nations, 1948). These can be breached by projects which displace local populations, such as the creation of game parks and conservation areas (Brockington & Igoe, 2006).

Particular care should be exercised regarding indigenous populations who may operate under different legal systems than the dominant system in that country. When establishing national reserves in Tanzania in the 19th century, colonial administrations assumed the jurisdiction of their legal system in interactions with indigenous people, leading to land purchases which one party understood to be legal, but the other did not (Neumann, 2008). In more recent history, compulsory land purchases by the Indian government pursuant to *The Wild Life (Protection) Act 1972* lead to the displacement of native populations, whose rights were later protected by the *Scheduled Tribes and Other Traditional Forest Dwellers (Recognition of Forest Rights) Act 2006* (the 'Forest Rights Act').

The UN Declaration on the Rights of Indigenous Peoples recognises indigenous peoples' rights to their lands, territories, and resources (amongst others) (UN, 2013). While this is not legally binding in itself, some of these principles have been adopted into case law in some nations, and others form part of national legislation, some of which establish specialist tribunals for dealing with disputes.[8] However, disparities may still exist between community-based land rights of indigenous people and the state or private land ownership recognised in national legislation.

Land use: Protected areas

International and domestic legislation may provide for areas to be protected in order to conserve biodiversity. Rewilding projects may seek to attain protected status for sites in order to safeguard them for the future; however, the legislative focus on the conservation of threatened species may pose issues for the ecosystem-focused approach in rewilding.

A number of pieces of international legislation, most notably the *Convention on Wetlands of International Importance Especially as Waterfowl Habitat* (the 'Ramsar Convention') and the Bern Convention, provide a framework for signatories to establish protected land areas. For example, Article 4 of the Bern Convention states that '*[e]ach Contracting Party shall take appropriate and necessary legislative and administrative measures to ensure the conservation of the habitats of the wild flora and fauna species*', which led to the implementation of the Natura 2000 sites in the EU. The Natura 2000 networks consists of Special Areas of Conservation ('SACs') under the Habitats Directive, and Special Protection Areas ('SPAs') under the Birds Directive. These sites include publicly owned nature parks, but the majority of land covered is privately owned.

Protected areas are often good sites for rewilding, and *vice versa*, rewilding sites may seek to become protected areas. Rewilding Europe, a European rewilding organisation, has developed a rewilding network including a number of Natura 2000 sites, and recent work has highlighted the propensity for rewilding activities to meet the requirements of the Habitats and Birds Directives (Fernández et al., 2020). In order to be designated as a Natura 2000 site, a rewilding site would need to be a habitat of a designated species and be nominated to the European Commission by the EU Member State in which it is located; the Member State would then enact domestic legislation to protect these sites from damage, and if necessary introduce positive measures to restore the species or habitat.

States may adopt domestic legislation to provide for protected areas. The majority of protected areas covered by domestic legislation are state-owned, but states may adopt legal protections for privately-owned protected areas as well; these can range from non-binding declarations to voluntary or mandatory restrictions by government. Privately-owned sites may transition to being state-owned parks, such as the creation of the Patagonia National Park in January 2018, previously an area of sheep ranches, which was acquired and rewilded by Tompkins Conservation from 2004 onwards. The National Park was created through a specific decree, *Resolution No. 22/2018 adopted in the Ordinary Session dated October 25, 2018*, following the donation of land and infrastructure from Tompkins Conservation, a private foundation, to the Chilean government.

However, there may be tensions between conservation legislation and rewilding. The Habitats and Birds Directives designate specified species and habitats and oblige national bodies to take steps to maintain their conservation.[9] Rewilding focuses on the interaction of species on an ecological level, which may not promote the conservation of a particular species. For example, the Oostvaardersplassen, a nature reserve in the Netherlands, is the habitat of a number of unmanaged large herbivores. The site was designated as an SPA in 1989; in 1996 it was discovered that the population of spoonbills (a protected species under the Birds Directive) had disappeared, allegedly due to the predation of foxes which were attracted by the carcases of the larger herbivores (Lorimer, 2016). While there is arguably scope within the Habitats and Birds Directives and associated case law for ecological dynamics to be allowed to play out, guidance on rewilding within protected sites could provide useful clarity. Similarly, the management of the Patagonia National Park has attracted criticism from some conservationists, as the removal of cattle may have been a factor in increased predation of the endangered huemul (a native species of deer) in the area (Wittmer et al., 2014).

Land use: Conservation covenants

In addition or alternatively to protected status (which can be limited, varied with respect to enforcement, and inconsistent with rewilding principles), private law mechanisms such as covenants and other land-law mechanisms can be used to preserve rewilding gains for the long-term.

Conservation covenants are voluntary agreements between landowners and 'responsible bodies' (e.g. charities, national governments), which establish obligations binding on current and subsequent owners (permanently or for a specified term). Potential obligations include restrictions (e.g. on damaging/removing conservation value), rights of entry/use (e.g. access for ecological/restorative work), positive covenants (e.g. maintaining land in accordance with specified principles), and enforcement rights (e.g. a 'responsible body' taking action against landowners in respect of breaches).

Conservation covenants, in different forms, can be used under the laws of numerous jurisdictions, including the USA, Chile, Tanzania, Namibia, Australia, New Zealand, Scotland, France, England (shortly, under the Environment Bill), and others. Conservation covenants are already used in rewilding and conservation projects; for example, by 2012 the Adirondack Park of New York State included 3 million acres of private lands with over 781,000 acres under publicly-held conservation easements (Dawson et al., 2015).

However, the utility of statutory conservation covenants is qualified by limitations inherent in their design; for example, positive obligations can be modified/removed by the courts. Further, elements of the statutory conservation covenant scheme rely on government discretion (e.g. the list of organisations able to hold them). Stronger alternatives are sometimes available to safeguard against these limitations. For example, a landowner can transfer the legal interest in their land to a charity, ensuring the long-term endurance of conservation gains. From a protection perspective this approach is ideal, however, it requires the surrender of the asset and is therefore unpalatable to some landowners.

To work around these ownership issues, The Lifescape Project and Wild Europe have developed a 'Legal Mechanism' which facilitates third party long-term protection of rewilding sites, whilst retaining the valuable land interest with the original owner. This involves land-owners exchanging absolute title in their land for long-term interests of near-equivalent value, whilst giving a charity the remainder interest along with rights to enforce ecological protections. Under one version of this Legal Mechanism, the charity will be in a position to enforce legal protections in perpetuity, whilst the landowner will have a legal right to occupy and use the land for 999 years.

Species reintroduction

Reintroduction of species can present the most interesting and challenging part of a rewilding project, but must be carried out in compliance with relevant legal regulations. This section identifies the nature of issues practitioners should address in the short and long term to ensure that the translocation of the specimens is conducted legally, the species is protected, and the project is protected against liability.

Species reintroduction: Reintroducing species

The International Union for Conservation of Nature's *Guidelines on Reintroductions and Other Conservations Translocations* (the 'IUCN Guidelines') recommends that conservation transloca-tion proposals be developed within national and regional conservation infrastructure, including legal and policy frameworks (IUCN, 2013). Two central considerations common to such frameworks are discussed below; however, other issues, such as invasive species restrictions, dis-ease control, and importation restrictions, may arise.

Licensing: many jurisdictions prevent organisations from releasing wild animals without licenses.[10] Rewilding managers should identify which licenses are required to permit the release of the specimens they intend to release.

An example of a licensing regime is the requirement under English and Welsh law for 'dan-gerous' wild animals to be released under the DWAA. The DWAA prohibits a person from possessing a *'dangerous wild animal'*[11] unless they have a licence from the licensing authority. A licence can only be granted where the applicant satisfies several conditions and the applicant is also required to pay certain fees to cover the authority's administration costs in addressing the request.

Transportation: species reintroductions will require project managers to transport live specimens intranationally or internationally. Many jurisdictions regulate the transportation of animals and the IUCN Guidelines emphasise that rewilding managers should take all necessary steps to comply with such regulations to ensure the project complies with applicable law and (more importantly) to safeguard the welfare of the selected animals (IUCN, 2013).

Where the 'source' site for animals is in a different jurisdiction to the 'destination' site, such as the transportation of ungulates from other EU Member States for the Oostvaardersplassen rewilding project, international animal transport regulations may apply. The movement of wild animals within the EU requires compliance with the requirements of the EU Transportation Regulation which stipulates the completion of journey logs, transportation conditions, and max-imum journey lengths (Council Regulation (EC) No 1/2005, 2004 (the 'EU Transportation Regulation')).[12] Further requirements may apply to translocations between unrelated nations; should a project seek to source translocation specimens from outside the EU, the Balai Directive would require rewilding managers to address issues including attaining health certificates from qualified veterinarians before animals can be transported (Council Directive 92/65/EEC, 1992 (the 'Balai Directive')).

Species reintroduction: Continuing obligations

Rewilding stakeholders should also consider their legal obligations after translocated animals have been released at the destination site. This will include assessing wildlife welfare regulations, which may prevent persons killing or disrupting reintroduced animals, and ensuring that the animals do not give rise to legal liability.

Wildlife protection: Rewilding stakeholders should assess whether the laws of the jurisdiction governing the translocation provide sufficient protection to both wild animals and their habitats. There may be positive and negative elements, such as the prohibition on hunting, catching, or killing of wildlife in the People's Republic of China, which is limited to areas of state protection.[13] Since reintroduced animals may roam, the utility of such protection may be limited; conservationists assisting with projects should therefore consider whether released animals would stay within protected areas and therefore receive protection against hunting.

As discussed above, various international treaties protect different types of wildlife and their habitats. Conservationists should consider the implementation of such treaties when assessing whether reintroduced animals would receive adequate legal protection.

Continuing legal obligations: these can relate to both reintroduced animals and their habitats. Many jurisdictions require animal owners to take steps to regulate disease; there are often enhanced obligations in respect of ungulates, as such species can also fall within regulatory frameworks concerning livestock, which can require tagging, veterinary check-ups, and inoculation. Practitioners should comply with these requirements or risk facing civil or criminal penalties. Conversely, in circumstances where released specimens are subject to legal protections, re-capturing, tagging, or otherwise interfering with those specimens may require government-issued licences.

Conservationists should be aware of their responsibilities and the rights of others to the land where translocated animals are released. Landowners may be liable for how they 'use' their land, which could include reintroducing species, and may be required to keep certain species enclosed. Additionally, in some jurisdictions, an owner can be statutorily liable for damage caused by animals which they own.[14]

Direction of travel

Public awareness of the global biodiversity crisis has increased dramatically over the last decade. This may lead to a new wave of legislation and policy; prior to the UN Summit on Biodiversity in September 2020, political leaders representing 64 countries signed a 'Leader's Pledge for Nature', committing to '*meaningful action*' on biodiversity loss, including the development of a framework for biodiversity for adoption at the 15th Conference of Parties of the CBD.

The potential for rewilding to assist in combating climate change is gaining increasing recognition. The interrelation is demonstrated by the Paris Agreement, which notes the importance of biodiversity issues to climate change, states that parties should '*conserve and enhance*' carbon sinks (such as forests), and encourages signatories to implement policy approaches to encourage sustainable management of forests and enhancement of forest stocks.

More widely, the UN Sustainable Development Goals (the 'SDGs'), which the UN Member States adopted in 2015, include goals to protect, restore, and promote terrestrial ecosystems, combat climate change, and protect marine life. It is likely that the SDGs will inform and guide policy of the Member States on these issues.

Finally, developments in laws regulating development may provide additional hope for rewilding, with an emerging trend of requiring developers to include provision for biodiversity gain in their proposed developments, which could unlock substantial funding for rewilding.[15]

Enforcement

Rewilding practitioners should be aware that not all laws relating to the environment or to rewilding specifically will be properly enforced, leaving opportunities to promote rewilding by taking legal or enforcement action. For example, the Lifescape Project and Trees for Life have currently brought a judicial review seeking a ruling that a licence granted by Nature Scot to cull a re-introduced beaver population was unlawful.

Conclusion

In conclusion, it may be helpful to draw together the various issues into a consolidated but non-exhaustive list of considerations.

1. When selecting land, consider whether:
   a. neighbours benefit from rights affecting the land;
   b. planning permissions will be required;
   c. rewilding could affect neighbouring land (e.g. through water damage or encroaching species);
   d. public access to the land is required; and
   e. the land contains legally-protected species which may be affected by rewilding.
2. Consider if and how the project may impact the local population, in particular indigenous populations.
3. Mechanisms exist in private and public law to assist in the long-term conservation of sites; consider whether these help pursue your objectives.
4. When reintroducing a species, consider:
   a. what licences and approvals may be required to release a specimen;
   b. compliance with transportation and disease control regulations;
   c. ongoing protections for the specimens; and
   d. ongoing obligations in respect of the specimens.

Notes

1 For example, Article 8(d) of the Convention on Biological Diversity ('CBD') requires the parties to 'as far as possible and as appropriate [...] promote the protection of ecosystems, natural habitats and the maintenance of viable populations of species in natural surroundings', but does not prescribe the mechanics of this (Convention on Biological Diversity, 1992).
2 For example, Article 8(f) of the CBD provides that signatories shall 'Rehabilitate and restore degraded ecosystems and promote the recovery of threatened species, inter alia, through the development and implementation of plans or other management strategies', but the general focus of Article 8 is on the protection of sites, rather than restoration (Convention on Biological Diversity, 1992).
3 See *Leakey v. The National Trust* [1981] QB 485 for discussions on the origins of this rule in common law jurisdictions.
4 See *Green v. Lord Somerleyton* [2003] EWCA Civ 198 for a summary of this rule in the UK. It is also possible that liability under the rule in *Rylands v. Fletcher* [1868] LR 3 HL 330 may also arise in common law jurisdictions. This occurs where a landowner brings a substance onto their property which escapes by artificial or natural means, causing damage to a neighbour.
5 For example, the Australian *Environment Protection and Biodiversity Conservation Act 1999* states that actions affecting important environmental sites are subject to ministerial approach, which must consider the principle of sustainable ecological development (note that in October 2020, a review into this Act was published and it may be subject to reform). See also the UK National Planning Policy Framework, paragraphs 174–177.

6 For example, in the UK, the *Countryside Rights of Way Act 2000*.
7 For example, the *European Convention on Human Rights* adopted a selection of rights from the United Nations Universal Declaration of Human Rights. Signatories then enacted these rights in their national law.
8 Examples include India's Forest Rights Act; Norway's *Finnmark Act 2005*; and Kenya's *Community Land Act 2016*.
9 For example, Article 6 of the Habitats Directive compels Members States to '*establish the necessary conservation measures [...] which correspond to the ecological requirements of the natural habitat types in Annex I and the species in Annex II present on the sites.*' (emphasis added).
10 E.g. section 14 of the UK *Wildlife and Countryside Act 1981*, which makes it a criminal offence to release wild animals 'not ordinarily resident in Great Britain' without prior authorisation into England and Wales.
11 The DWAA lists species of animal considered dangerous in a schedule which is updated from time to time. This includes species which have been the subject of proposed UK conservation translocation projects, including the Eurasian Lynx.
12 For an in-depth analysis of the Transport Regulation, see Espin, 2018.
13 Article 21 of the *Law of the People's Republic of China on the Protection of Wildlife*.
14 See for example, the UK *Animals Act 1971*, Article 838 of the *German Civil Code* and Article 1243 of the *French Civil Code*.
15 While not yet in effect, the UK *Environment Bill* proposes an amendment to the UK *Town and Country Planning Act 1990* to include biodiversity gain as a condition for the grant of planning permission.

References

Brockington, D. and James, I. (2006). Eviction for conservation: a global overview. *Conservation and Society* [online]. 4 (3): 424–70. (Viewed 20 April 2021.) Available from: www.conservationandsociety.org/text.asp?2006/4/3/424/49276

Dawson, P., Bick, S., D'Luhosch, P. D., Nowak, M., and Kuehn, D. (2013). Conservation easements in the Adirondack Park of New York state.*Science and Stewardship to Protect and Sustain Wilderness Values: Tenth World Wilderness Congress Symposium, October 4-10, 2013, Salamanca, Spain*. Proceedings RMRS-P-74. Fort Collins, CO: U.S. Department of Agriculture, Forest Service, Rocky Mountain Research Station [online]. pp. 150–7. (Viewed 20 April 2021.) Available from: www.fs.fed.us/rm/pubs/rmrs_p074/rmrs_p074_150_157.pdf

Espin, R. (2018). Reintroducing wildlife into the United Kingdom: practical. *UK Journal for Animal Law*, 2 (2): 39.

Fernández, N, Torres, A., et al. (2020). *Boosting Ecological Restoration for a Wilder Europe: Making the Green Deal Work for Nature*. Rewilding Europe [Online]. (Viewed 20 April 2021.) Available from: http://dx.doi.org/10.978.39817938/57

International Union for Conservation of Nature and Natural Resources. (2013). *Guidelines for Reintroductions and Other Conservation Translocations*. http://data.iucn.org/dbtw-wpd/edocs/2013-009.pdf

Koninx, F. (2019). Ecotourism and rewilding: the case of Swedish Lapland, *Journal of Ecotourism* [online].18(4): 332–47. (Viewed 20 April 2021.) Available at: https://doi.org/10.1080/14724049.2018.1538227

Leaders Pledge for Nature. (2020). *Leaders' Pledge for Nature*. www.leaderspledgefornature.org/Leaders_Pledge_for_Nature_27.09.20.pdf

Lorimer, J. (2016). Probiotic legalities: de-domestication and rewilding before the law. In *Animals, Biopolitics, Law: Lively Legalities*. Routledge.

Neumann, R.P. (2008). *Imposing Wilderness: Struggles over Livelihood and Nature Preservation in Africa*. University of California Press.

Trouwborst, A., Blackmore, A., Boitani, L., Bowman, M., Caddell, R., Chapron, G., Cliquet, A., et al. (2017). International wildlife law: understanding and enhancing its role in conservation. *BioScience* [online]. 67(9): 784–90. (Viewed 20 April 2021.) Available at: https://doi. org/10.1093/biosci/bix086

United Nations. (2013). *Indigenous Peoples and the United Nations Human Rights System*. United Nations. Available at: www.ohchr.org/documents/publications/fs9rev.2.pdf

Wittmer, H., Elbroch, L., and Marshall, A. (2014). Conservation of huemul in the future Patagonia National Park: a call for immediate management intervention. *IUCN Deer Specialist Group Newsletter* [online]. 26: 6–13. (Viewed 20 April 2021.) Available at: https://sites.lsa.umich.edu/webbkeane/wp-content/uploads/sites/162/2014/09/Wittmer-et-al.-2014-Conservation-of-huemul-in-the-future-Patagonia-National-Park-a-call-for-immediate-management-intervention.pdf

Cases

Green v. Lord Somerleyton [2003] EWCA Civ 198
Leakey v. The National Trust [1981] QB 485
Rylands v. Fletcher [1868] LR 3 HL 330

Legislation

Animals Act 1971 (UK), c.22. Available at: www.legislation.gov.uk/ukpga/1971/22 (accessed 20 April 2021).

Community Land Act 2016 (Kenya), No. 27 of 2016. Available at: http://kenyalaw.org/kl/fileadmin/pdfdownloads/Acts/CommunityLandAct_27of2016.pdf (accessed 20 April 2021).

Convention on Biological Diversity (1992). [Online]. 1760 UNTS 79, opened for signature 5 June 1992, and entered into force 29 December 1993. (accessed 28 February 2021). Available at: www.cbd.int/convention/text/

Convention on the Conservation of European Wildlife and Natural Habitats (1979). [Online]. ETS No. 104, opened for signature 19 September 1979, and entered into force 1 June 1982 (accessed 28 February 2021). Available at: www.coe.int/en/web/conventions/full-list/-/conventions/rms/0900001680078aff.

Convention on International Trade in Endangered Species of Wild Fauna and Flora (1973). [Online]. 993 UNTS 243, opened for signature 3 March 1973, entered into force 1 July 1975. Available at: https://cites.org/sites/default/files/eng/disc/CITES-Convention-EN.pdf

Convention on Wetlands of International Importance Especially as Waterfowl Habitat (1971). [Online]. 996 UNTS 245, opened for signature 2 February 1971, and entered into force 17 February 1976 (accessed: 28 February 2021). Available from: https://treaties.un.org/doc/Publication/UNTS/Volume%20996/volume-996-I-14583-English.pdf

Council of Europe. (1950). *Convention for the Protection of Human Rights and Fundamental Freedoms*. Rome, 4.XI.1950. Strasbourg, Directorate of Information. Available at: www.echr.coe.int/documents/convention_eng.pdf

Council of the European Union (1992). *Council Directive 92/43/EEC of 21 May 1992 on the conservation of natural habitats and of wild fauna and flora*. *Official Journal of the European Communities* L 206/7, 22 July 1992. Available from: https://eur-lex.europa.eu/legal-content/EN/TXT/PDF/?uri=CELEX:31992L0043&from=EN (accessed 20 April 2021).

Council of the European Union (2009). *Directive 2009/147/EC of the European Parliament and of the Council of 30 November 2009 on the conservation of wild birds*. *Official Journal of the European Union* L 20/7, 26 January 2010. Available from: https://eur-lex.europa.eu/legal-content/EN/TXT/PDF/?uri=CELEX:32009L0147&from=EN (accessed 20 April 2021).

Council of the European Union (2005). *Council Regulation (EC) No 1/2005 of 22 December 2004 on the protection of animals during transport and related operations and amending Directives 64/432/EEC and 93/119/EC and Regulation (EC) No 1255/97. Official Journal of the European Union* L 3/1, 5 January 2005. Available from: https://eur-lex.europa.eu/legal-content/EN/TXT/HTML/?uri=CELEX:32005R0001&from=EN (accessed 20 April 2021).

Council of the European Union (2014). *Directive 2014/52/EU of the European Parliament and of the Council of 16 April 2014 amending Directive 2011/92/Eu on the Assessment of the Effects of Certain Public and Private Projects on the Environment. Official Journal of the European Union*L 124/1, 25 April 2014. Available from: https://eur-lex.europa.eu/legal-content/EN/TXT/PDF/?uri=CELEX:32014L0052&from=EN (accessed 20 April 2021).

Countryside and Rights of Way Act 2000 (UK), c. 37. Available at: www.legislation.gov.uk/ukpga/2000/37/contents (accessed 20 April 2021).

Dangerous Wild Animals Act 1976 (UK), c. 38. Available at: www.legislation.gov.uk/ukpga/1976/38 (accessed 20 April 2021).

Environment Bill. [HL] Bill 16, 2019-20. Available from: https://bills.parliament.uk/bills/2593 (accessed 20 April 2021).

Environment Protection and Biodiversity Conservation Act 1999 (Australia), No. 91 of 1999. Available at: www.legislation.gov.au/Series/C2004A00485 (accessed 20 April 2021).

Finnmark Act 2005 (Norway), Act of 17 June 2005, No. 85.

French Civil Code Livre III: Des différentes manières dont on acquiert la propriété (Articles 711 à 2278). [online]. Available at: www.legifrance.gouv.fr/codes/section_lc/LEGITEXT000006070721/LEGIS CTA000032021488/#LEGISCTA000032021488 (accessed 20 April 2021).

German Civil Code in the version promulgated on 2 January 2002. *Federal Law Gazette* [Bundesgesetzblatt] I page 42, 2909; 2003 I page 738. Available at: www.gesetze-im-internet.de/englisch_bgb/engli sch_bgb.pdf

Great Britain. (2019). *National Planning Policy Framework.*

Law of the People's Republic of China on the Protection of Wildlife (People's Republic of China). Available at: www.china.org.cn/english/environment/34349.htm (accessed 20 April 2021).

Paris Agreement to the United Nations Framework Convention on Climate Change (2015) T.I.A.S. No. 16-1104, 12 December 2015. Available at: https://unfccc.int/sites/default/files/english_paris_agreement.pdf

Resolution No. 22/2018 adopted in the Ordinary Session dated October 25, 2018 (Chile). Nromas Generales, CVE 1508732. *Diario Oficial de la Republica de Chile.* Available at: www.diariooficial.interior.gob.cl/publicaciones/2018/12/11/42226/01/1508732.pdf (accessed 20 April 2021).

Scheduled Tribes and Other Traditional Forest Dwellers (Recognition of Forest Rights) Act, 2006 (India), No. 2 of 2007. Available at: https://legislative.gov.in/sites/default/files/A2007-02.pdf (accessed 20 April 2021).

Title Conditions (Scotland) Act 2003 (UK), asp 9. Available at: www.legislation.gov.uk/asp/2003/9/conte nts (accessed 20 April 2021).

Town and CountryPlanning Act 1990 (UK), c.8. Available at: www.legislation.gov.uk/ukpga/1990/8/conte nts (accessed 20 April 2021).

United Nations (1948). *United Nations Universal Declaration of Human Rights.* Available at: www.un.org/sites/un2.un.org/files/udhr.pdf

United Nations (2015). Transforming our world: the 2030 Agenda for Sustainable Development. *United Nations Sustainable Development.* Available at: https://sdgs.un.org/2030agenda (accessed 20 April 2021).

United Nations (2007). *United Nations Declaration on the Rights of Indigenous Peoples:* adopted by the United Nations General Assembly on 13 September 2007. Available at: www.un.org/development/desa/indigenouspeoples/wp-content/uploads/sites/19/2018/11/UNDRIP_E_web.pdf

Weeds Act 1959(UK), c. 54. Available at: www.legislation.gov.uk/ukpga/Eliz2/7-8/54/contents (accessed 20 April 2021).

Wildlife and Countryside Act 1981 (UK), c. 69. Available at: www.legislation.gov.uk/ukpga/1981/69 (accessed 20 April 2021).

The Wild Life (Protection) Act 1972 (India), No. 53 of 1972. Available at: https://legislative.gov.in/sites/defa ult/files/A1972-53_0.pdf (accessed 20 April 2021).

PART III

Application and impacts
of rewilding

14

REWILDING CASE STUDY

Yellowstone to Yukon

Jodi Hilty, Charles Chester, and Pamela Wright

Y2Y and rewilding: Common origins in the coalescence of conservation biology

Emerging out of the confluence of ideas that became identified as the science of conservation biology, the two concepts of rewilding and Y2Y have run a roughly parallel course. Both have been subject to definitional debate, rewilding with its many interpretations and definitions (Pettorelli et al., 2018) while Y2Y constitutes a vision, a network, an organisation, and a region. Yet it is partly because of the latter—viz., the fact that Y2Y is a place one can outline on a map—that Y2Y has accrued many different forms of conservation. Just as the Y2Y network and organisation have welcomed and worked with hundreds of diverse conservation partners, Y2Y also incorporates a plethora of conservation approaches. Rewilding is one of them, but by no means the whole of what happens under the Y2Y umbrella. Following the development of the equilibrium theory of island biogeography (ETIB) in the 1960s and the ensuing acrimonious debate over the most effective approaches to habitat protection—a debate that is today remembered by the label 'SLOSS' for single-large-or-several-small—*conservation biology* emerged during the 1980s as a coherent and identifiable scientific endeavour (see Triantis & Bhagwat, 2011). It would be simplistic to circumscribe the arena of conservation biology as emanating from a single overarching idea such as ETIB, as many of the myriad scientific underpinnings of conservation biology date to decades earlier. Nonetheless, the need for sufficient habitat conservation was, and remains, at the core of what conservation biology prescribes as a remedy to the loss of biodiversity. For both rewilding and Y2Y, the focus was on the conservation requirements of large carnivores such as grizzly bears (*Ursus arctos*) and wolves (*Canis lupus*), and thus the notion of 'sufficient' habitat necessarily translated into extensive and connected areas.

During the 1990s, the field of conservation biology was influential in the development of a number of innovative conservation applications. One of these was first conceived in 1993 as the 'Yellowstone to Yukon Biodiversity Strategy' by Canadian lawyer Harvey Locke, with the phrase first appearing in print in the 1993–94 winter issue of *Wild Earth* (Locke, 1993/94). The Y2Y concept received support from the Wildlands Project (now Wildlands Network) and the Canadian Parks and Wilderness Society (CPAWS), both of which were involved in organising the 1997 Y2Y Connections Conference in Waterton Lakes National Park in Alberta, Canada. With an overflow attendance of more than 300 conservationists from across the region and

DOI: 10.4324/9781003097822-17

beyond, the Connections Conference is today recalled as the first moment of public engagement for the Y2Y idea, and solidified an active conservation network with two main components: (1) a group of dedicated conservationists who met in person several times per year in various locations throughout the Y2Y region, and (2) a highly participatory emailing list (EML) that came just as many conservationists living and working in the Y2Y region were first connecting to the internet. In a pre-Google era, hundreds of topics were discussed over the EML, often consisting either of requests for information or of requests for support for a specific conservation campaign. Although the prominence of the Y2Y EML would substantially diminish over time, it was critical in forming an extensive, cohesive, and vocal community of individuals— mostly professional conservationists, scientists, and funders—who would promote the Y2Y idea across the region and beyond (Chester, 2006).

For those meeting in person, much early discussion focused on three questions: (1) what was the extent of the Y2Y region, (2) how to apply scientific principles emanating from conservation biology to a binational region as large as Y2Y, and (3) where to place the emphasis of Y2Y's conservation strategy—viz., whether it be on large carnivores, on ecological processes, on economic and social threats, etc. The first question of Y2Y's extent was neither quickly nor easily resolved, but by the end of the first decade had settled into what is now described as the Y2Y region: an area covering 1.3 million square kilometres (502,000 square miles), stretching 3,200 kilometres (1,988 miles) north to south and 500 to 800 kilometres (310 to 496 miles) east to west, and spanning across five American states, two Canadian provinces, two Canadian territories, and the traditional territories of at least 75 Indigenous groups (Figure 14.1; see Y2YCI 2021a). In regard to the second question, much effort was spent during Y2Y's first years in attempting to conduct a scientific assessment of Y2Y as a whole, including efforts to reconcile data inconsistencies not only across the two nations involved, but across state, provincial, and Indigenous (viz. First Nations, Inuit, Métis, and Native American) borders. Over time, this proved to be an expensive and time-consuming endeavor; although efforts persist on this front, ultimately more focus has been placed on examining and working in subregional and local areas, particularly where habitat connectivity or land protection has been identified as critical.

In regard to the third question, it was clear from the 1997 Connections Conference that any conservation strategy under the Y2Y aegis would be widely encompassing in terms of topic matter. Notably, the 138-page document coming out of the conference, informally described as the 'Y2Y Atlas' (Harvey, 1998), covered topics ranging from vegetation cover, Indigenous peoples, and economic trends to wildfire, corridors, and watersheds. Amidst these issues, however, concern over the fate of large carnivores was a consistently emphasised theme, one that would remain a highlight of Y2Y's conservation agenda over the ensuing decades (it is no coincidence that a grizzly bear is pictured on each of the first two pages of the Y2Y Atlas, and it remains the icon for Y2Y to the present day; see Figure 14.2). Furthermore, the dependence of large carnivores on connectivity across extensive landscapes has constituted a key theme from the start; indeed, the title of the Connections Conference was a play on the increasingly used term 'connectivity' in the scientific literature on conservation.

In the first published use of the term rewilding in 1998, Michael Soulé and Reed Noss do not directly mention Y2Y. However, Noss attended the 1997 Connections Conference and authored a chapter in the 1998 Atlas, and Soulé wrote the Preface to the 1998 Y2Y Atlas. So, when they published 'Rewilding and Biodiversity' in that same year, they were fully aware of the Y2Y concept—and not surprisingly, the article emphasises many of the same themes that were highlighted at the 1997 Y2Y Conference and examined in the 1998 Atlas. 'Three major scientific arguments', they wrote, 'constitute the rewilding argument and justify the emphasis on large predators.' The first of these was the importance of 'top-down' ecological and trophic

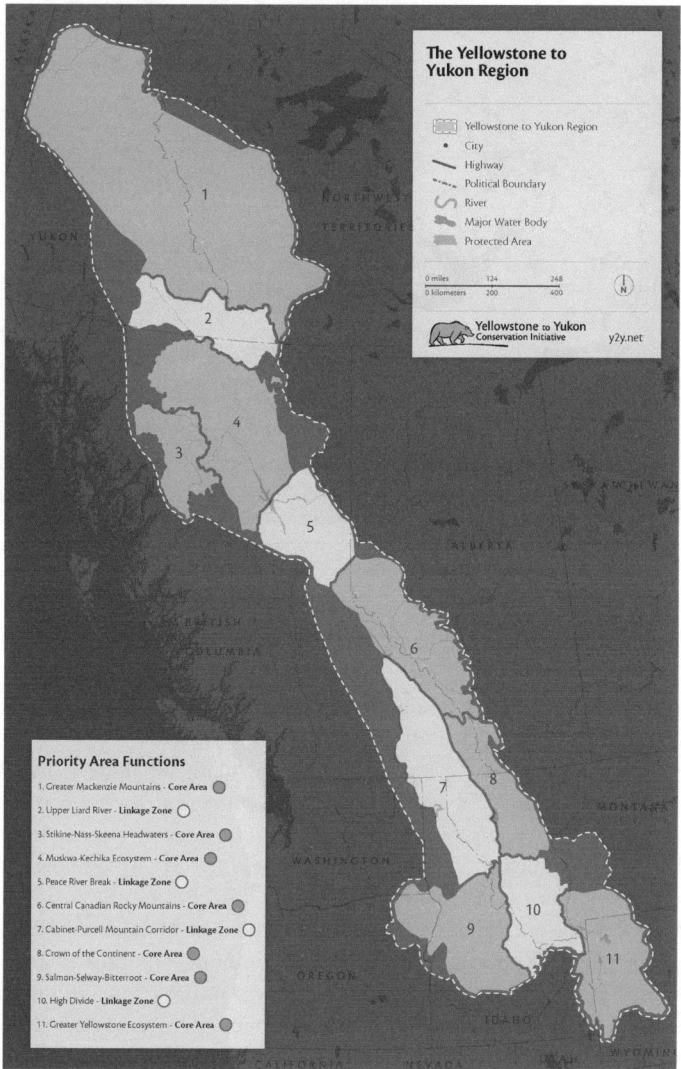

Figure 14.1 The Yellowstone to Yukon (Y2Y) region including Y2Y priority core areas and linkage zones.

interactions, the second the need for large core areas. The third scientific argument focused on connectivity, which was 'also required because core reserves are typically not large enough in most regions; they must be linked to insure long-term viability of wide-ranging species'.

Over the first decade of the 21st century, the in-person and online networks coming out of the 1997 Connections Conference would evolve into an organisation (specifically entitled the Y2Y Conservation Initiative) as well as an inclusive social movement of Y2Y 'partners'. Both of these entities generally emphasised an inclusive, collegial, and generally non-litigious 'big tent' approach to conservation. And while the Y2Y concept would become widely recognised within the applied conservation and conservation biology communities (and beyond) as a vision of landscape conservation at the appropriate and necessary scale for large carnivores—and would

Figure 14.2   Grizzly bear (*Ursus arctos*) sow and cub in Yellowstone National Park. National Park Service/Jim Peaco.

thus remain focused on the core intent and approach of rewilding—many other approaches to conservation would also take place under the Y2Y umbrella.

In other words, even as the science related to rewilding has underscored Y2Y's conservation efforts since the 1990s, Y2Y and its partners have long relied on a wide range of scientific research and knowledge to solve a host of conservation challenges across the region. In broad terms, Y2Y applies this science in three geographic subregions: (1) the northern portion of the Y2Y region, where there is still the opportunity to conserve mostly intact habitat, including places that are key for conservation of culture, carbon, and climate refugia, (2) the middle section of Y2Y, where the focus is on working to minimise the growing threat of fragmentation such as by providing safe wildlife road crossings and restoring and maintaining wildlife corridors, and (3) in the southern area, where the focus is on ecological restoration due to an already relatively large human footprint. It is here, in this southern portion, where the focus on rewilding is most pertinent.

To advance priority projects that rewild the Y2Y region, particularly in areas most important for realising landscape level connectivity, requires science at different scales, diverse collaborative, and a bottom-up approach complemented by a top-down approach to ensure the existence of enabling legislation and policy. Such multi-scale science not only generates an understanding of where the priorities are at the scale of Y2Y, but ensures that more localised science can inform the critical actions within specific regions. In its efforts to advance conservation across the region, Y2Y has been most effective when, early on in any given area, it has brought together most local and regional groups (as well as some national entities) to agree on a shared set of priorities to advance the connect and protect work at local scales (see Cabinet Purcell Mountains case study as an example below). Such shared priorities have also been attractive to funders who see the strengths of the different entities and how their work synergises further conservation. In addition, Y2Y's work at the global and national levels to advance enabling policies on connectivity and protected area conservation provides guidance, impetus, and support for local work (e.g., the establishment of global connectivity guidelines, Hilty et al., 2020). It also builds on the existing 'conservation infrastructure' in the region, bringing positive attention to features such as the world's first national park (Yellowstone), Canada's first national park (Banff) and the world's first international Peace Park on the US/CA border. The following section examines these multi-scale targeted efforts as they relate to enhanced conservation and rewilding.

Evidence and discussion of rewilding

As alluded to above, one of the early critiques of Y2Y was that while the vision was compelling, tangible conservation progress must occur at local scales. The perception was a mismatch between the scale of the Y2Y vision and the ability or authority of conservation actors working in the region to advance tangible actions (Johnson, 2017). The question is whether or not a large vision supports and promotes more local groups of actors to prioritise actions that are important to achieve the large landscape vision.

As one of the earliest large landscape visions that has been actively promoting conservation for more than 25 years, Y2Y is a natural laboratory to address questions about the power of a large landscape vision in advancing conservation, including rewilding. To do this analysis, several aspects of conservation were analysed including: 1) the increase of protected areas within Y2Y as compared to other regions, 2) changes in distribution of the grizzly bear (*Ursus horribilis*, an umbrella species), 3) contributions towards private land conservation, 4) status of safe passage for wildlife across roads including underpasses and overpasses as well as associated critical fencing, and 5) adoption and promotion of the Y2Y vision via scientific and popular media coverage (Hebblewhite et al., 2022). Protected areas increased by more than 80% across the Y2Y region at a time when growth in protected areas was flat in North America. Likewise, the region now has at least 116 designated underpasses and overpasses and associated fencing to help keep wildlife safely connected across busy roads (Hebblewhite et al., 2022). It appears that this region has the most advanced road crossing systems for wildlife of any large landscape. Private lands and co-existence work (see Cabinet Purcell case study below) also have concretely advanced supporting the vision. Likewise, grizzly bears in the US and in Alberta have measurably rebounded in population size. Such advancements provide early evidence that a large landscape vision can inspire local conservation that is important at a large landscape scale.

To date, conservation progress in the Y2Y region has been substantial, and recent announcements indicate that this progress is even building momentum. For example, in the last few years in the Y2Y region, three new and expansive Indigenous Protected and Conserved Areas (IPCAs) in Canada advanced through written agreements between Indigenous Peoples and their provincial, territorial, and federal government counterparts (Y2YCI, 2021b), and these protected areas are expected to be fully designated on-the-ground over the next 18 months (*see* Peace River case study below as an example). Together these would represent more than 56,656km^2 or 14,000,000 new acres of lands in protected status. Likewise, multiple states and provinces have committed to the implementation of new road wildlife crossing structures, which promise to further increase connectivity across busy roadways crisscrossing the Y2Y region. Likewise, in some regions of forest, partners have worked together to decommission and create limitations on new road building.

Case study: Peace River Break and caribou

Within northeastern British Columbia, where the Boreal Plains meet the Northern Boreal Mountains and the Rocky Mountain Hart Range intersects with the Peace River, lies an area referred to as the Peace River Break (PRB). The PRB is a critical pinch-point in the continuity of ecologically intact and functioning landscapes along the north–south extent of the Canadian Rocky Mountains, yet less than 4% of the region has been designated with protected area status (Apps, 2013). The PRB is experiencing industrial-caused disturbances at significant rates: forest harvesting, recreation areas (including heli-ski tenures), mining, agriculture, and seismic lines amount to approximately 27% of the PRB, with the result that half of the PRB is within .5km of roads, reservoirs, and/or oil and gas infrastructure (Mann & Wright, 2018). Human use pressures on this landscape have placed old-forest dependent species such as woodland caribou (*Rangifer tarandus*) in jeopardy. This species is dependent on the food source of arboreal lichens, which in turn requires large areas of continuous tracts of undisturbed alpine and subalpine park-land habitats and old-growth, mid-elevation stands (Johnson et al., 2004). Central and Southern Mountain populations of Caribou are endangered, with most herds at precariously low levels and in decline while others have already been extirpated from the landscape (e.g., the Burnt Pine population). In British Columbia region alone, caribou have dropped from 40,000 to 15,000 animals (Oud 2020).

Over much of the last decade, Indigenous communities, partner organisations such as Y2Y, and researchers have investigated the extent of impacts in the area, and to support conservation initiatives such as the successful Klinse-za Caribou maternity pen initiative led by West

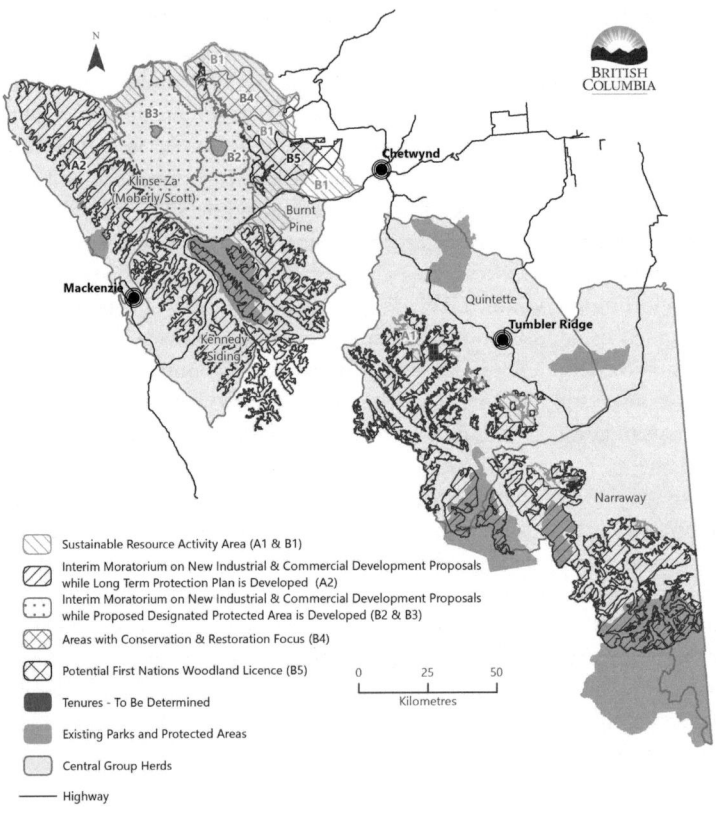

Figure 14.3 A map showing the conservation gains from the Peace River region Caribou Agreement, a signed agreement between the West Moberly and Saulteau First Nations and the British Columbia and Canadian federal government.

Moberly and Saulteau First Nations (Figure 14.3; Apps, 2013; Burkhart, 2018; Curtis, 2018; Mann, 2020). More recently, Y2Y facilitated research and other initiatives that were used to support significant conservation responses in the area. The leadership of the West Moberly and the Saulteau First Nations led to a partnership agreement with the federal and provincial government in 2019 (ECC Canada, 2020), which in turn resulted in the expansion of the Klinse-za Provincial Park (located just south of the Peace Arm) from 2,689ha to 28,000ha in February of 2020 (a further expansion to 206,000ha is planned for the spring of 2021). These expansions are surrounded by other land use agreements where restoration and conservation will be the focus. In addition, the Partnership Agreement includes an interim moratorium on all new tenures and development on a further 550,000ha of high elevation caribou recovery area. Although interim, it can only be lifted if all parties agree—a very unlikely proposition given the long-term caribou recovery goals of the West Moberly and Saulteau First Nations as well as Canada's commitments under the Species at Risk Act (www2.gov.bc.ca/gov/cont ent/environment/plants-animals-ecosystems/wildlife/wildlife-conservation/caribou/partners hip-agreement). These new Indigenous-led conservation efforts represent a remarkable conservation gain for caribou and for climate change resiliency within a critical ecological pinch point in the Peace River Break. Much work remains, however, in rewilding this heavily disturbed landscape.

Case Study: Large carnivore rewilding in the southern region of Y2Y

During the inception of the Y2Y vision in the mid-1990s, many of the individuals at the founding meetings were concerned with the already extensive range loss across North America for both large carnivores as well as hooved animals (Laliberte & Ripple, 2004). As the Laliberte and Ripple maps show, one of the last places where these animals still mostly roam is in the Y2Y region.

Worldwide, applied research has increasingly shown that the loss of large predators leads to a cascade of ecological impacts affecting multiple parts of ecosystems. As one of the more well-studied ecosystems in the world, the Greater Yellowstone Area has been a *de facto* natural laboratory for a plethora of research on such trophic cascades. Specifically, due to the extinction of several large carnivores from all or parts of the ecosystem in the 1900s and the subsequent restoration or rewilding of carnivores back into the system, researchers have been able to understand the role of carnivores by examining differences before and after their restoration.

The body of studies on the pre- and post- restoration of large carnivores in the Greater Yellowstone Ecosystem have covered a diversity of species and topics. While grizzly bears never completely disappeared from the Greater Yellowstone region, less than 150 were thought to persist by the 1970s (USNPS, 2020). Following complete extirpation, wolves (*Canis lupus*) were reintroduced as an experimental population to Yellowstone in 1995. Mountain lions (*Felis concolor*) and wolverines (*Gulo gulo*) were generally assessed as extirpated from the area, and genetics studies suggest that they naturally returned from more northern Canadian populations in the late 20th and into the 21st century (Yellowstone Science, 1994; McKelvey et al., 2014).

Studies of these animal recoveries have shown dramatic effects. For example, the expansion of grizzly bears back to Grand Teton National Park has led to the restoration of willows and an increase in the associated bird communities, as well as a shift in the age structure of a once senescing moose population (Berger et al., 2001). Large carnivores, as talked about more extensively in other chapters of this book, can both reduce hooved animal populations and change their behaviour such as where they spend time in the landscape, resulting in such a cascade of impacts in the ecosystem. Ongoing research details how the restoration of wolves has played an enormous role in shaping ecosystems, ranging from changing riparian vegetation and hydrological processes to altering the abundance of many different species across the park. Notably, the impacts and results are varied across the park where wolves now roam. As one of the world's most well-studied reintroductions, the findings of wolf reintroduction to Yellowstone offers lessons too numerous to expand on here. While there are myriad excellent accounts of Yellowstone wolf reintroduction, we recommend the forthcoming book: *Yellowstone Wolves: Science and Discovery in the World's First National Park* (Smith et al., in press). Overall, the preponderance of evidence supports that the suite of large carnivore species, now restored to Greater Yellowstone, play a significant role in shaping the ecosystem itself.

This important premise of trophic cascades has implications for the Y2Y vision across the Y2Y region. Since large carnivores are absent or in low numbers in other parts of the Y2Y region, including the extensive Idaho wildlands complex where grizzly bear populations were exterminated in 1940s, restoring the full complement of carnivores to such large wild regions will also help to maintain a healthy ecosystem. In addition, experience and scientific research indicate that maintaining such species in any part of Y2Y in isolation can be highly problematic, since the habitat requirements of a viable long-term population often span beyond any individual subregion of Y2Y. Science has clearly demonstrated that even large ecosystems such as the Greater Yellowstone Ecosystem and the Idaho Wildlands complex are too small to sustain some large carnivores, and thus connectivity between these wild regions is also a priority. Efforts by many different non-profits and agencies have been ongoing for decades, racing against an onslaught of human development to try to keep the opportunity for population connectivity open, and today animals like grizzly bears are closer than ever to reconnecting (www.nytimes.com/2017/11/03/science/grizzly-bears-yellowstone-genes.html; https://fwp.mt.gov/conservation/species/bear/management).

Case study: Restoration of habitat and wildlife in the transboundary Cabinet Purcell Mountain region

In the early 1990s, the population of grizzly bears in the transboundary Cabinet Purcell Mountain region of Montana, Idaho, and British Columbia was showing signs of isolating into smaller populations, with one population as low as ten individuals in Montana's Cabinet Yaak Mountains. Grizzly bear science helped to prioritise where key core habitat and connectivity zones should be protected and restored. On the US side, more than 1295km² (>320,000 acres) of habitat were secured through road removal projects on US Forest Service land. But ensuring connectivity among the remaining bear populations required securing private land such as through conservation easements and acquisitions, which significantly increased the security of three identified corridors.

Additionally, the state of Idaho purchased one priority corridor to be restored as a Wildlife Management Area, where the focus was both on recreating wetlands for endemic wildlife such as native bees, native toads and frogs, and other wildlife as well as on increasing connectivity across the broader landscape for bears and other large mammals. More than 20,000 shrubs and trees were planted in recontoured wetlands to help rewild a climate resilient landscape, and a grizzly bear print was found amidst the restoration during the summer of 2020 (J. Grossman, pers. comm.). Other key efforts in the region to support grizzly bear restoration have included the installation of more than 170 electric fences to deter bears from attractants like bee hives, chicken coops, and fruit orchard; educational efforts on preventing human–wildlife conflict, and other projects. With these efforts having built on several decades of work to increase grizzly bear connectivity in the area, recent research on tracked movements between the previously isolated populations and beyond those populations indicates the conservation efforts have already had an impact (Proctor et al., 2018; Hilty et al., 2019).

Case study: Bison restoration

While the collapse of bison (*Bison bison*) across North America is generally well-known, their current status and conservation challenges today are perhaps more complex and less understood by most of the public. Bison flourished across North America until the time of European invasion and the associated slaughter of bison that nearly drove them to extinction in the late 19th century (Sanderson et al., 2007). When a few key individuals realised that bison were on the brink of extinction, they formed the American Bison Society to restore the species in various localities across North America. Bison are a megaherbivore keystone species that literally shape the ecosystem they occupy. By the 1930s, about 20,000 bison had been restored in various conservation herds, a number similar to today (although approximately 500,000 bison are now found in 'ranch' bison that were often cross-bred with cattle). What is less well known is that in some arenas, bison have been and remain anathema—so much so that their status as 'wildlife' was threatened, and still today many states and provinces recognise bison either solely as livestock or in some cases as livestock as well as wildlife. The result is that unlike any other wildlife in the Y2Y region, bison are subject to the unique restriction of confinement to particular areas within their range. They face an additional challenge of various levels of genetic heritage, with ranchers having long sought to interbreed bison and cows to obtain a more hardy but easy to manage animal. Additionally, some conservation herds are small populations that must be managed for inbreeding. A further challenge today is that some populations of bison have acquired diseases transmitted from cattle, most notably brucellosis in the Greater Yellowstone Ecosystem. The presence of disease now presents challenges to relocating bison or allowing bison to roam and mix with cattle due to concerns over disease transmission back to now disease-free cattle (White et al., 2011).

These circumstances meant that for most of the latter half of the 20th century, the population status of bison conservation herds changed little (~20,000 bison), while bison ranching expanded enormously. Conservationists re-awakened to the plight of bison conservation in the late 20th century, and took a fresh look at where restoration of bison at scale could occur in key locations across North America (Sanderson et al., 2007). Three key places where bison restoration is advancing today are in the Y2Y region. The first is on the northern and western edges of Yellowstone National Park, a region where bison leaving Yellowstone National Park were once hazed back into the park or shot. Today due to work of various entities, the state of Montana has now created a buffer zone that allows bison to leave the park within these defined spaces. However, they are still limited in their movements outside the park, and Yellowstone bison are still slaughtered when their numbers are deemed too high to be supported by the habitats they are allowed to access (White et al., 2011; National Park Service 2018).

In the meantime, in Banff National Park in Canada, an experimental population of bison was restored to the northern reaches of the park in 2017 with an initial soft release of 17 bison. Now there are approximately 45 bison that roam freely in the park (Kost, 2020). However, like the situation in Montana and Yellowstone, the confines of jurisdictions that 'allow' for wild bison are still restrictive in Alberta, and these bison are strongly discouraged from leaving the park. Although the Alberta government has created a buffer zone on adjacent non-park public lands, bison are still considered livestock beyond that buffer and currently cannot roam further—although many in the conservation community hope that this will change with time.

Perhaps the most inspiring rewilding project is envisioned by the Blackfoot Confederacy, a transboundary (Montana and Alberta) group of Indigenous Peoples who have been advancing the Iinnii Initiative (Iinnii means bison) since 2009. This effort seeks to conserve traditional lands, maintain Blackfeet culture, and enable bison to return to their lands (Blackfoot Nation, 2020). Recognising that both people and bison are split by political boundaries, this initiative seeks a holistic approach to the restoration of lands, wildlife, and people (Blackfoot Nation, 2020). Some day in the not-too-distant future, bison may once again roam across Blackfoot Confederacy lands including in adjacent national parks such as Glacier and Waterton and per-haps other jurisdictions.

Next steps

Y2Y's mission is to connect, restore, and protect the region so that both people and nature can thrive, and accomplishing it means both protecting extant nature as well as the rewilding and restoration of key ecosystems and species. While the bulk of this work will continue at the grassroots level, such efforts could be accelerated by higher level enabling policies that better recognise and support large landscape conservation. At the global level, a new set of connect-ivity guidelines not only summarise the latest science, but offer concrete and tangible steps for securing ecological corridors and advancing ecological networks across large landscapes such as the Y2Y region (Hilty et al., 2020). These guidelines are meant to help countries as they advance connectivity management and policy in both domestic and transborder contexts. In the USA, wildlife corridor legislation has been repeatedly advanced in the US Congress over the last half dozen years, and while none of these have yet passed, several states, such as California and New Mexico, have passed state level legislation. In Canada, the Pathway to Canada Target I, a federal level initiative to advance biodiversity conservation, includes exploring ecological networks connecting protected areas, but has not yet been enshrined in legislation or practice (*see* Pathway Initiative, 2018). In both countries, the recent interest in and advancement of policy is promising, and if brought to fruition it would help create resources and levers to advance on-the-ground work in the Y2Y region and beyond. Additionally, there are further opportunities to advance policy work on connectivity both in the transportation arena as well as in regard to environmental assessments and funding. Finally, connectivity work across multiple jurisdictions also requires managers of public and private land to understand the cross-border conservation challenges, and to engage with each other in creating solutions to those challenges.

While substantial and important rewilding has been occurring in the Y2Y region, there are still considerable ongoing challenges for conservation. Mountain caribou are found nowhere else on the planet except in the Y2Y region. In 2018, the loss of a transboundary herd between the US and Canada meant that the lower 48 states lost the last of this animal. Many other populations of mountain caribou are suffering major declines in populations. While the Caribou Agreement in the Peace River Region is a model for advancing their recovery, similar measures need to advance in other places. Likewise, in the Yukon, which currently has a rela-tively low human footprint, the federal government committed more than $200 million dollars, to support the maintenance and building of roads. How, and where, any new roads are placed could have a major impact on previously intact big wild areas.

Likewise, although ecosystem fragmentation from the human footprint of built infrastruc-ture and the linear disturbances of roads and other corridors are the single biggest threat to ecological values, the human footprint from recreational use is an emerging issue (Larson et al., 2016). While outdoor recreation provides spectacular opportunities for recreation and to facilitate nature connections there is increasing recognition that we are 'loving' these spaces to death (Simmonds et al., 2018). New and faster technologies such as e-bikes are just accelerating that trend. In the Y2Y region, trails and limited use roads, and the dispersed recreation use associated with them are the primary sources of these impacts. Trails and recreational activity disturb sensitive species (Gaines et al., 2003; Heinemeyer et al., 2019), change interspecies relationships (Ladle et al., 2018), and affect plant species composition, richness, and cover (Barros et al., 2015). Some research indicates that disturbance-sensitive species (e.g., grizzly bears, caribou) avoid heavy-use trails from 500 to 2000 m, irrespective of whether the use is motorised or non-motorised. Unfortunately, there is limited mapping and research of recreational use such that we don't know the true impact and thus cannot manage it effectively (Farr et al., 2018). This means that the intensity of use of public lands, even in parks and core protected areas is increasing, diminishing those spaces for sensitive species in an ever-humanising landscape. The need to understand the cumulative impacts and at the scale that matters for nature is increasingly important, so that we can manage these impacts retaining and restoring key wild places.

Summary

The concept of rewilding and the vision Y2Y were borne of the same ethos. Rewilding remain central to what Y2Y hopes to accomplish over the long term within the most ecologically intact region of North America. With 25-plus years of applied conservation under both, the two paths of rewilding and Y2Y will no doubt continue to intertwine as they lead towards a future of ecological wholeness and restoration.

References

Apps, C. (2013). Assessing cumulative impacts to wide-ranging species across the Peace Break region of northeastern British Columbia. Version 3.0 Yellowstone to Yukon Conservation Initiative, Canmore, AB: Yellowstone to Yukon Conservation Initiative.

Barros, A., Pickering, C., and O. Gudes. (2015). Desktop analysis of potential impacts of visitor use: A case study for the highest park in the Southern Hemisphere. *Journal of Environmental Management*, 150: 179–95.

Berger, J., P.B. Stacey, and M.P. Johnson. (2001). A mammalian predator-prey imbalance: grizzly bear and wolf extinction affect avian neotropical migrants. *Ecological Applicationsm*, 11: 947–60.

Blackfoot Nation (2020). *Iinnii Buffalo Spirit Center.* https://blackfeetnation.com/iinnii-buffalo-spirit-center/.

Burkhart, T. (2018). *Counter-mapping for Conservation: Digital Conservation Atlas Case Study.* [Master's Thesis]. University of Northern British Columbia.

Chester, C. C. (2006). *Conservation across Borders: Biodiversity in an Interdependent World.* Island Press, Washington, DC.

Curtis, I. (2018). *Systematic Conservation Planning in the Wild Harts Study Area* [Master's Thesis]. University of Northern British Columbia.

ECC Canada. (2020). *Environment and Climate Change Canada.* Available at: www.canada.ca/en/environment-climate-change/services/species-risk-public-registry/conservation-agreements/intergovernmental-partnership-conservation-central-southern-mountain-caribou-2020.html

Farr, D.R., Braid, A., and S. Slater. (2018). *Linear Disturbances in the Livingstone-Porcupine Hills of Alberta: Review of Potential Ecological Responses.*Environmental Monitoring and Science Division, Alberta Environment and Parks. BC Ministry of Forests, Lands and Natural Resource Operations. 2013. www.sitesandtrailsbc.ca/about/provincial-trail-strategy.aspx

Gaines, W.L., Singleton, P.H., and R.C.Ross (2003). Assessing the cumulative effects of linear recreation routes on wildlife habitats on the Okanogan and Wenatchee National Forests. *Gen. Tech. Rep. PNW-GTR-586.* Portland, OR: US Department of Agriculture, Forest Service, Pacific Northwest Research Station.

Harvey, A. (1998). *A Sense of Place: Issues, Attitudes and Resources in the Yellowstone to Yukon Ecoregion.* Yellowstone to Yukon Conservation Initiative. April.

Heinemeyer, K., Squires, J., Hebblewhite, M., O'Keefe, J.J., Holbrook, J.D., and J. Copeland. (2019). Wolverines in winter: indirect habitat loss and functional responses to backcountry recreation. *Ecosphere, 10*: e02611.

Hebblewhite, M., J.A. Hilty, S. Williams, H. Locke, C. Chester, D. Johns, G. Kehm, and W.L. Francis. (2022). Can a large-landscape vision contribute to achieving biodiversity targets? *Conservation Science and Practice,* 4(1): e588.

Hilty, J.A., A.L. Jacob, K.G. Trotter, M. Hilty, and H. Young. (2019). Endangered species, wildlife corridors, and climate change in the US West. In *The Environmental Politics and Policy of Western Public Lands.* Wolters, E.L. and B. Steel (eds), Oregon State University.

Hilty, J., Worboys, G., Keeley, A., Woodley, S., Lausche, B., Locke, H., Carr, M., PulsfordI., Pittock, J., White, W., Theobald, D., Levine, J., Reuling, M., Watson, J.E.M., Ament, R., and G. Tabor. (2020). *Guidance for Conserving Connectivity through Ecological Networks and Corridors.* IUCN.

Johnson, S. (2017). *Building a Large Landscape Conservation Community of Practice.* Lincoln Institute of Land Policy, University of Montana, Working Paper WP17SJ1, www.lincolninst.edu/sites/default/files/pubfiles/johnson_wp17sj1.pdf

Kost, H. (2020). Baby boom adds 10 calves to bison population in Banff. CBC News. 6 June. cbc.ca/news/canada/calgary/bison-banff-calves-alberta-1.5601311.

Ladle, A., Steenweg, R., Shepherd, B., and M.S. Boyce. (2018). The role of human outdoor recreation in shaping patterns of grizzly bear-black bear co-occurrence. *PloS one, 13*: e0191730

Laliberte, A. and W. Ripple. (2004). Range contractions of North American carnivores and ungulates. *Bioscience,* 54: 123–38.

Larson, C.L., Reed, S.E., Merenlender, A.M., and K.R. Crooks. (2016). Effects of recreation on animals revealed as widespread through a global systematic review. *PloS one,* 11: e0167259.

Locke, H. (1993/94). Yellowstone to Yukon. *Wild Earth,* 3: 68–72.

Mann, J. (2020). *Climate Change Conscious Systematic Conservation Planning: A case study in the Peace River Break, British Columbia* [Master's Thesis]. University of Northern British Columbia.

Mann J. and P. Wright. (2018). *The Human Footprint in the Peace River Break, British Columbia.* Natural Resources and Environmental Studies Institute. Technical Report Series No. 2, University of Northern British Columbia, Prince George, B.C., Canada.

McKelvey, K.S., K.B. Aubry, B. Keith, N.J. Anderson, J. Neil, A.P. Clevenger, J.P. Copeland, K.S. Heinemeyer, S. Kimberley, R.M. Inman, J.R. Squires, J.S. Waller, K.L. Pilgrim, and M.K. Schwartz. (2014). Recovery of wolverines in the Western United States: recent extirpation and recolonization or range retraction and expansion? *USDA Forest Service / UNL Faculty Publications,* 324. http://digitalcommons.unl.edu/usdafsfacpub/324

National Park Service. 2018. *History of Bison Management in Yellowstone. Yellowstone National Park.* U.S. Department of Interior. www.nps.gov/articles/bison-history-yellowstone.htm [last updated April 2014].

Oud, N. 2020. *First Nations Partner with BC, Canada to protect endangered caribou.* CBC, 21 February 2020. www.cbc.ca/news/canada/british-columbia/partnership-southern-mountain-caribou-1.5471574

Pathway Initiative (2018). *One with Nature: A Renewed Approach to Land and Freshwater Conservation in Canada.* www.conservation2020canada.ca.

Pettorelli, N., J. Barlow, P.A. Stephens, S.M. Durant, B. Connor, H. Schulte to Bühne, C.J. Sandom, J. Wentworthe, and J.T. du Toit. (2018). Making rewilding fit for policy. *Journal of Applied Ecology,* 55: 1114–25.

Proctor, M.F., W.F. Kasworm, K.M. Annis, A. Grant-MacHutchon, J.E. Teisberg, T.G. Radandt, and C. Seervheen. (2018). Conservation of threatened Canada-USA trans-border grizzly bears linked toc comprehensive conflict reduction. *Human-Wildlife Interactions,* 12: 348–72.

Sanderson, E.W., K.H. Redford, B. Weber, K. Aune, D. Baldes, J. Berger, D. Carter, C. Curtin, J. Derr, S. Dobrott, E. Fearn, C. Fleener, S. Forrest, C. Gerlach, C.C. Gates, J. Gross, P. Gogan, S. Grassel, J.A. Hilty, M. Jensen, K. Kunkel, D. Lammers, R. List, K. Minkowski, T. Olson, C. Pague, P.B. Robertson, and B. Stephenson. (2008). The ecological future of the North American bison: conceiving long-term, large-scale conservation of wildlife. *Conservation Biology,* 22: 252–66.

Simmonds, C., McGivney, A., Reilly, P., Maffly, B., Wilkinson, T., Canon, G., and M. Whaley. (2018). Crisis in our national parks: how tourists are loving nature to death. *Guardian,* 20 November 2018.

Smith, D.W., D.R. Stahler, and D.R. Macnulty (eds). In Press. *Yellowstone Wolves: Science and Discovery in the World's First National Park*. University of Chicago Press, Chicago.

Soulé, M.E. and Noss, R. (1998). Rewilding and biodiversity: Complementary goals for continental conservation. *Wild Earth*, 8, 18–28.

Triantis, K.A. and Bhagwat, S.A. (2011). Applied island biogeography. In Ladle, R.J. and Whittaker, R.J., *Conservation Biogeography*. Wiley-Blackwell, 190–223.

USNPS (2020). Grizzly Bears & the Endangered Species Act. US National Park Service. 2 March. ps:www.nps.gov/yell/learn/nature/bearesa.htm.

White, P.J, R.L. Wallen, C. Geremia, J.J. Treanor, and D.W. Blanton. 2011. Management of Yellowstone bison and brucellosis transmission risk—Implications for conservation and restoration. *Biological Conservation*, 144: 1322–34.

Y2YCI (2021a). *The Region: Incredible Variety from Yellowstone to Yukon*. Yellowstone to Yukon Conservation Initiative. http://y2y.net/work/region.

Y2YCI (2021b). *Indigenous Peoples Leading the Way on Conservation in the Yellowstone to Yukon Region and Beyond*. Yellowstone to Yukon Conservation Initiative. https://y2y.net/blog/indigenous-led-conservation-yellowstone-to-yukon-region.

Yellowstone Science (1994). The Yellowstone Lion: *Yellowstone Science*, 2: 8–13. http://npshistory.com/publications/yell/newsletters/yellowstone-science/2-3.pdf.

15

REWILDING CASE STUDY

Carrifran Wildwood

Stuart Adair and Philip Ashmole

Introduction

Carrifran is located in the central Southern Uplands of Scotland. The site extends to 660 hectares and was used for pasturing domestic stock for centuries and quite possibly millennia. The term 'rewilding' was not in currency when we started. We thought in terms of ecological restoration of centuries old sheep walk which had radically altered the original character of the place. We felt that at least some ground in the Southern Uplands should be returned to something closer to its natural state.

Carrifran Glen is a steep sided U-shaped glacial valley, cut through Silurian age greywackes, shales, siltstones and mudstones. The altitudinal range varies from 165m to 821m *a.s.l.* The activity of the ice is witnessed in the steep valley walls, the craggy and ripped cliff faces and the terminal moraine situated at the mouth of the glen. Soils vary from freely draining brown earths disposed over moderate-steep slopes, gleyed brown forest soils in low lying depressions and hollows, podzols on gentler slopes, through flushed peaty mineral soils by springs and mires to blanket peat over high plateaus.

The average annual rainfall for Carrifran is 2250mm. The Met Office records the annual average for daily maximum temperature as 10.8°C and minimum of 3.3°C. There is an annual average of 30.4 of day of lying snow. The longest periods of lying snow are in January and February.

Carrifran forms part of both Moffat Hills Site of Special Scientific Interest (SSSI) and Moffat Hills Special Area of Conservation (SAC). The area is designated primarily due to its botanical interest. The area qualifies as a SAC under the EC Habitats Directive due to the presence of eight Annex I Habitats (habitats of international value). The latter designation also calls for restoration of habitats and reintroductions of species where deemed appropriate. These habitats are rare outside the Highlands and the Moffat Hills and the renowned naturalist, Derek Ratcliffe, considered the Moffat Hills to have some of the best montane habitat remaining in the country south of the Highlands (Ratcliffe, 1959). Primary among these are alpine and boreal heaths; siliceous alpine and boreal grasslands, and hydrophilous tall-herb communities. An example of such a community would be wind-clipped summit moss-heath with plants like the appropriately named Stiff-sedge (*Carex bigelowii*) standing upright against the severe wind buffeting, over the fluffy, grey Woolly Fringe-moss (*Racomitrium lanuginosum*).

DOI: 10.4324/9781003097822-18

Figure 15.1 Carrifran, March 2000.

Origins of the Wildwood Project

In the early 1990s ecological restoration was an idea unfamiliar to most British conservationists, and the term 'rewilding' was unknown. However, pioneering efforts to restore native pinewoods were under way in the Highlands of Scotland, and environmental activists in southern Scotland began to focus on the state of their local uplands, where many centuries of intensive grazing by sheep, cattle, and feral goats had caused ecological degradation, reducing the flora to open grassy-heath and all but completely eradicating any tree and shrub flora, never mind any extensive areas of woodland proper (see figure 15.1). The loss of natural forest was followed, in some areas, by planting of large-scale monocultures of non-native exotic conifers. Members of the Wildwood Group conceived the idea that a grass-roots community group could purchase an entire valley and restore an exemplar 'wildwood' with naturally functioning ecosystems comparable to the broadleaf woodlands and moorlands that had been lost. Their vision was embodied in a Mission Statement:

> The Wildwood project aims to re-create, in the Southern Uplands of Scotland, an extensive tract of mainly forested wilderness with most of the rich diversity of native species present in the area before human activities became dominant. The woodland will not be exploited commercially and the impact of humans will be carefully managed. Access will be open to all, and it is hoped that the Wildwood will be used throughout the next millennium as an inspiration and an educational resource.

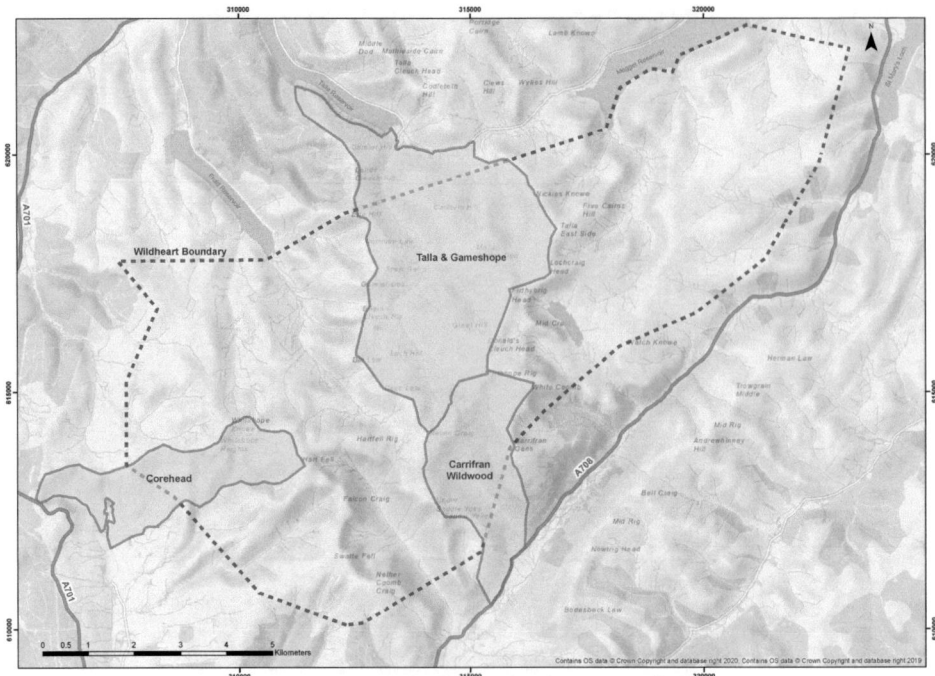

Figure 15.2   The Wild Heart of southern Scotland map © Borders Forest Trust.

In 1995–6, the Wildwood Group saw the funding opportunities offered by the Millennium Forest for Scotland, and joined other native woodland enthusiasts to found Borders Forest Trust. The idea of creating a wildwood had led to the formation of the group, but to be worthwhile, the site had to be a discrete entity with a wide variety of habitats. The group contacted owners of open hill land in the Borderlands by letter and telephone, and eventually found an ideal site for the wildwood some 35 miles south of Edinburgh. Funding was secured by public subscription during a two-year option period, enabling BFT to purchase Carrifran on Millennium Day, 1 January 2000. Major funding for work on the ground was provided by the David Stevenson Trust, which agreed to pay for propagation of all the trees, and by the National Lottery, while the John Muir Trust provided support and wise advice.

The two years of fund-raising were also used to plan the transformation of Carrifran (Newton & Ashmole, 1998). For a grass-roots group with a bold vision, gaining the confidence of relevant authorities is crucial, so planning began with a major conference in Edinburgh in November 1997. Native Woodland Restoration in Southern Scotland: Principles and Practice, organised jointly by Borders Forest Trust and Edinburgh University, attracted more than 150 delegates, including many senior staff of Scottish Natural Heritage (SNH) and the Forestry Commission Scotland (FCS). This mattered, since SNH had a veto on changing the ecology of Carrifran and the FCS could provide grants for native woodland planting.

In 1998 a diverse ecological planning group met monthly in a local pub, convened by local volunteer Adrian Newton (a forest ecologist at Edinburgh University) to discuss and make decisions on all aspects of the work that lay ahead. Late in 1999 the restoration plan for Carrifran was approved by SNH and FCS agreed to fund the planting of 300 hectares of new native woodland.

Woodland establishment

Planting is required to restore native woodland in a landscape such as that in the central Southern Uplands of Scotland, near bereft of existing trees and thus seed-parents for natural regeneration. Over the course of 20 years some 700,000 native tree and shrub species have been established at Carrifran. The planting has been concentrated in the lower part of the valley (<450m *a.s.l.*) with some smaller outliers of high-altitude woodland and montane scrub on the higher ground. Most of the planting was carried out between 2000 and 2008.

When trying to establish native woodland, we chose not to use the 'scattergun' approach. Rather, one looks to match the prevailing natural conditions as far as is possible. At Carrifran, we have used local geology, soils, climate, and existing plant life (in conjunction with what we know of the vegetation history through ancient plant pollen captured and stored in deep peat such as at Rotten Bottom) to determine the most appropriate type of natural woodland to be established in each situation. On the middle and lower slopes three main woodland types have been established: upland Oak–Birch woodland; upland Ash–Elm woodland; and slope/flush Alder–Ash woodland. Reflecting the prevailing nature of the climate, geology and soils, Oak–Birch woodland is the most widespread type with Ash–Elm and Ash–Alder stands confined to areas with suitable moist, flushed, wet, and richer soils. Some wet Birch–Willow woodland has been established on shallow (< 0.5m) peats with, among others, Downy Birch (*Betula pubescens*), Eared (*Salix aurita*), and Grey Willow (*S. cinerea*). In addition, two small groups of Scots Pine (*Pinus sylvestris*), along with Juniper (*Juniperus communis*), have been planted on and around northeast facing crags at 450–500m. Upslope, the latter communities give way to various forms of montane scrub planting with, among others, Juniper, Downy Willow (*S. lapponum*), Tea-leaved Willow (*S. phylicifolia*), Dark-leaved Willow (*S. myrsinifolia*) and Montane Goat Willow (*S. caprea sphacelata*).

Planting has included over 30 species of trees, woody shrubs, and climbers, some of which were already present in tiny numbers. Planted specimens of more than 20 species are already fruiting, implying future diversity much greater than in most native woodland planting schemes. Seeds and planting stock were sought locally by volunteers and propagated by commercial nurseries. For some species, such as sessile oak, we needed to look further afield for appropriate sources. Planting has been conducted in both autumn and spring with little noticeable difference between the two seasons.

Planted broadleaf trees generally require protection from browsing and competition during establishment, but at Carrifran this has been minimised, in an attempt to develop natural looking woodland. Sheep and goats are excluded with a perimeter stock fence, and since natural predators are currently lacking, a deerstalker keeps Roe Deer (*Capreolus capreolus*) numbers low enough to prevent serious browsing damage. Most trees were protected with short vole guards, and 60cm tubes were sometimes used for Oaks; removal of both is now well under way. Pre-planting spot spraying with glyphosate was used in grant-aided areas in the early 2000s but BFT has now terminated herbicide use.

Condition of established woodland

The maturing stands in the lower part of the glen are in good condition. The earliest plantings are now recognisable woodland ecosystems with closed canopy, vertical, and horizontal layering, increasing and dappled shade, leaf litter, and some woody debris on the forest floor, but this is a young woodland, lacking mature trees and deadwood, and thus also lacking many niches for animals. The canopy has a mean height of about 3–5m with occasional specimens reaching 6m.

Downy Birch, Rowan (*Sorbus aucuparia*), and Hazel (*Corylus avellana*) make up the bulk of the planted woodland with lesser amounts of Sessile Oak (*Quercus petraea*), Ash (*Fraxinus excelsior*), Alder (*Alnus glutinosa*), Wych Elm (*Ulmus glabra*), Aspen (*Populus tremula*), Holly (*Ilex aquifolium*), Hawthorn (*Crataegus monogyna*), Juniper, Bird Cherry (*Prunus padus*), and various Willows (*Salix* spp.) and Roses (*Rosa* spp.) Honeysuckle (*Lonicera periclymenum*), and to a lesser extent Ivy (*Hedera helix*), are also quite frequent. The older plantings show good vigour, girth, early crown development and healthy lateral branching. The most mature stands are those that were established on the deep brown earths formerly occupied by Bracken (*Pteridium aquilinum*). The latter are especially suited to Sessile Oak and Hazel and with recovering plant life such as Bluebells (*Hyacinthoides non-scripta*) and Lemon Scented-fern (*Oreopteris limbosperma*), these stands are now taking on the familiar character of Bluebell Oakwood. The montane scrub planting is slower to mature. Nevertheless, Downy Willow scrub is now, nearly 20 years from the original planting, taking on the familiar look of natural formations with Great Wood-rush (*Luzula sylvatica*), Water Avens (*Geum rivale*), and Lady's-mantle (*Alchemilla glabra*) among others. Juniper scrub, albeit slower than the willow scrub, is also starting to take on a more natural look as species like Bilberry (*Vaccinium myrtillus*), Cowberry (*V. vitis-idaea*), Wood-sorrel (*Oxalis acetosella*), Hard-fern (*Blechnum spicant*), and Lemon Scented-fern join the flora.

If the woodland is to prove viable in the longer term, successful natural regeneration is essential. At the moment, and naturally enough bearing in mind the short time period, this is currently scarce but is increasing, most obviously within the pre-existing fragments of woodland. Free from browsing, the few surviving trees, especially Ash, Rowan, and Hazel, once characterised by contorted, irregular, and angular branching have put up multiple vertical shoots and developed better formed crowns. In other situations, Downy Birch and several species of Willows are putting out new, coppice-like growth from low, formerly heavily browsed and suppressed stumps along watercourses, and on ledges and small cliffs. Both extant and planted Honeysuckle have flourished greatly in the new conditions, climbing up trees and sprawling over the ground in quite luxuriant growth. The few surviving Ivy are spreading, but planting of this species is only now getting under way.

Natural regeneration of new plants from seeds of more mature planted and surviving trees is still fairly scarce but increasing as time goes by. Ash, Rowan, Hazel, Alder, Bird Cherry, and Downy Birch are all producing seedlings. Rowan, Hazel, and Bird Cherry seedlings are the most abundant with Rowan especially on heathery banks and more open situations. Hazel is more common under the shade of maturing woodland and Bird Cherry is very common around mature stems of the same species. Alder seedlings are picking out flushed ground. Planted Aspen quickly began to produce suckering shoots which in certain places are becoming very abundant. Downy Birch seed from planted stock is being produced in large quantities but the seed 'rain' is currently mostly falling onto unsuitable dense, rank vegetation and unable to make contact with bare ground and other suitable substrates and Birch seedlings are still very scarce. Sessile Oak is now producing a few seedlings and given the abundance of suitable ground conditions, one would imagine that Oak will follow Hazel and readily regenerate given time. Naturally regenerated Willows are becoming more and more common, especially along watercourses and in wet, sheltered hollows.

Natural plant/vegetation succession
Open ground

Ecological restoration involves much more than native woodland restoration, of course. At Carrifran, long established, poor, sheep-related vegetation such as acid grasslands have been all but completely replaced by heathland and tall-herb vegetation. Bilberry and Great Wood-rush especially have increased their extent enormously—recovering from either suppressed populations within former grassland in the case of the former or expanding out from small scattered stands in the case of the latter. Heather (*Calluna vulgaris*) regeneration is less extensive but nevertheless widespread, especially coming down gullies from existing moorland above and along steep banks by watercourses. On sunny, drier south and southwest facing slopes, the Heather is often joined by Bell Heather (*Erica cinerea*), which has also increased greatly since the time of purchase. Crowberry (*Empetrum nigrum*) and Cowberry have also increased both their range and extent. Ferns, including the relatively rare Mountain Male-fern (*Dryopteris oreades*); small herbs such as Tormentil (*Potentilla erecta*), Common Dog-violet (*Viola riviniana*), and Wood Anemone (*Anemone nemorosa*); grasses including Wavy Hair-grass (*Deschampsia flexuosa*); a suite of typical upland mosses and Reindeer Lichens such as *Cladonia portentosa* and *C. arbuscula* are also on the increase.

Degraded blanket mires are recovering (aided by some remedial works or 're-wetting'—using machines to block drains and re-profile peat hags) with peat builders such as Bog mosses (*Sphagnum* spp.) and Cotton-grasses (*Eriophorum* spp.) colonising formerly bare peats. The regionally rare and notable Northern Bilberry (*V. uliginosum*) has increased its extent. Planted Bog-myrtle (*Myrica gale*) is marking out the water tracks over the surface of mires and wet-heath and now looks very natural.

Among the most important of the upland vegetation types occurring at Carrifran—wind-clipped summit moss-heath—is beginning to show its true status and extent, with the patterns of wind buffeting becoming more obvious as the low growing carpet gives way to taller vegetation wherever the influence of the wind recedes. Characteristic species of this community such as Woolly Fringe-moss (*Racomitrium lanuginosum*) is expanding its range and being joined occasionally by new additions to the summit flora such as Goldenrod (*Solidago virgaurea*).

Springs and associated flush vegetation have changed little in their floristic composition since the removal of stock. The vegetation has become more rank but as yet, all the key species are flourishing and have been joined occasionally by notable species such as Mountain Sorrel (*Oxyria digyna*). The rare Alpine Foxtail (*Alopecurus borealis*) has increased its stature after the withdrawal of stock. In the absence of poaching by stock, cushion mosses such as the distinctive bright yellow-green Fountain Apple-moss (*Philonotis fontana*) have thickened up noticeably.

But perhaps the most eye-catching changes in the flora have come through the spread of previously scarce tall herbs. The latter were confined to inaccessible ledges and cliffs high up in the narrow ravines and small gorges at the time of purchase, when sheep and goats were still present. Since then, the tall herbs have escaped from their previously confined stations and come both downhill and downstream and spread all along both the margins of watercourses and, especially, within formerly bare in-channel gravel bars and shelves. Among the more note-worthy of these plants are Mountain Sorrel, Roseroot (*Sedum rosea*) and Sea Campion (*Silene uniflora*)—species so typical of mountain cliffs and ledges, now coming down as low as less than 200m *a.s.l.*

The list of plants that have colonised these areas includes Great Wood-rush, Water Avens, Lady's Mantle, Early-purple Orchid (*Orchis mascula*), Wood Crane's-bill (*Geranium sylvaticum*), Lesser Birds Foot Trefoil (*Lotus corniculatus*), Common Knapweed (*Centaurea nigra*), Wild Angelica (*Angelica sylvestris*), Common Valerian (*Valeriana officinalis*), Meadowsweet (*Filipendula ulmaria*), Primrose (*Primula vulgaris*), Colt's-foot (*Tussilago farfara*), Marsh Ragwort (*Senecio aquaticus*), Common Milkwort (*Polygala vulgaris*), Ribwort Plantain (*Plantago lanceolata*), Heather and many, many more—all of which can be seen in some places within an area of perhaps no more than *c.* 10 x 10m—very species-rich vegetation indeed by Scottish upland standards (and the situation at Carrifran can be compared directly with the neighbouring Black Hope which is still under a sheep farming regime). In the medium to long term though, such plant communities may prove to be somewhat ephemeral and come and go naturally as heavy catastrophic flooding periodically re-sculptures the scene and the process of gain-loss-gain is repeated in exactly the type of dynamic ecosystem, governed by natural processes, that is so desperately lacking in most of our semi-natural habitats.

Notable plant species

Notable plant species form one of the core qualifying interests of the Moffat Hills SSSI/SAC and the Carrifran portion has its fair share.

At this early stage, it is, of course, still too early to draw any definitive conclusions, but after nearly two decades without grazing by domestic stock or feral goats, not only have the original notable plant species recorded in the glen survived and flourished, expanding their range from cliffs, ledges, and other inaccessible places onto open ground, the list has in fact grown with several new additions recorded since the stock were removed. Mountain Sorrel, Roseroot, Sea Campion, Alpine Meadow-rue (*Thalictrum alpinum*), Pale Forget-me-not (*Myosotis stolonifera*), and Mossy Saxifrage (*Saxifraga hypnoides*) have all expanded their range, in the case of Mountain Sorrel, Roseroot, and Sea Campion quite considerably including down onto the lower ground. Alpine Fox-tail has not yet been recorded at new stations but the original population is stable. The vulnerable species Holly-fern (*Polystichum lonchitis*), Red-listed by the International Union for the Conservation of Nature (IUCN), has one of its only stations in southern Scotland at Carrifran. Other notable plants species occurring at Carrifran include Hoary Whitlow-grass (*Draba incana*), Alpine Willowherb (*Epilobium anagallidifolium*), and Serrated Wintergreen (*Orthilia secunda*).

New additions to the flora since the change of management include Alpine Saw-wort (*Saussurea alpina*), Alpine Cinquefoil (*P. crantzii*) and Lesser Twayblade (*Listera cordata*). The population of the rare fern Oblong Woodsia (*Woodsia ilvensis*) in the Moffat Hills was bolstered by translocations into Carrifran in 2000.

Summary of changes since 2000

The first striking thing (other than the planted trees themselves) that greets visitors to Carrifran is the near complete transition from grassland to heath as the sheep and goats no longer nip the young Bilberry, Heather, Cowberry, and Bell Heather daring to peek their heads above the grassy sward. This alone would have been enough for many but there is more; tall herbs have come out of their refuges and are asking to play with the other plants. Such previously shy, and somewhat elusive, montane beasts as Mountain Sorrel have come down among the commoners and are happy to play on the low ground with the mere Knapweeds of the world. The Wood Anemones have conquered vast new areas previously only available to their sheep-friendly cousins. The sub-arctic and boreal heaths of the Moffat Hills have long been noted for their conservation value. Now, with the pincer-like mouths of sheep and goats removed, the heaths have expanded their extent greatly and are becoming more species-rich—aided by the recent translocation of the rare dwarf-shrub, Bearberry (*Arctostaphylos uva-ursi*). The Bog Mosses grow luxuriantly, free now from the trampling hooves of domesticated herbivores. The deep peats of Rotten Bottom are now growing again for, perhaps, the first time in centuries—building future pollen records with the massive upturn in tree pollen recorded at this very moment but stored for a many a century to come. Pollen itsel fis now a plenty as the plant life, now free from sheep, flowers much more profusely and regularly than when under constant ovine assault. The planted woodland is now starting to resemble natural upland forest and montane scrub and, in many places, could easily be mistaken for natural regeneration (see figure 15.3).

All of the Annex 1 Habitats for which the site is designated as a Special Area of Conservation (and thus of international importance) have either increased their extent or species-richness, or both.

Fauna

The return of birds and other animals to Carrifran

During the planning stages of the Carrifran Wildwood project, before the site was purchased, Peter Gordon (a volunteer member of the Wildwood Group who was working for RSPB) set up a long-term bird survey protocol similar to the British Trust for Ornithology Breeding Bird Survey. In recent years the surveys were organised by the late John Savory. They involve two visits to Carrifran (one in May and one in June) on each of which two observers walk along three parallel linear transects about 200m apart, totalling a distance of 6km. The overall distance walked in each year is thus 24km, effectively surveying most of the valley below 450m.

Woodland planting at Carrifran started in 2000, the same year that feral goats were removed, while sheep were excluded on a phased basis, finishing in 2008. The bird surveys were carried out in 1998, 2000, 2002 and annually thereafter. For comparison, similar surveys have been carried out in roughly alternate years in Black Hope, a topographically similar valley adjacent to Carrifran, where sheep and goats are still present and planting has not been occurring. It is hoped to continue the surveys for a century or more as the woodland and moorland habitats mature.

By late 2019, after 20 years of planting, 52 bird species had been recorded in the May and June surveys on the two sites, of which 26 were in both valleys, 23 were at Carrifran only and three were at Black Hope only (Savory, 2020). During the first eight years of planting only two woodland bird species were recorded in the summer surveys, but from 2008 new woodland species were added rapidly, reaching 19 in 2019. The commonest woodland species are now Willow Warbler (*Phylloscopus trochilus*), Chaffinch (*Fringilla coelebs*), and Lesser Redpoll (*Acanthis cabaret*), followed by Robin (*Erithacus rubecula*), Blackcap (*Sylvia atricapilla*), and Tree Pipit (*Anthus trivialis*) (Savory, 2020).

During the two decades the only significant loss from Carrifran has been Wheatear (*Oenanthe oenanthe*), a species feeding mainly on short grass swards, which have largely disappeared from Carrifran, because the unplanted ground has now developed a heavy sward caused by the removal of sheep; in the grazed valley of Black Hope, Wheatear remains common.

The start of the development of normal species interactions at Carrifran is signalled by recent opportunistic records of a wide variety of raptors including Goshawk (*Accipiter gentilis*), a woodland species, and especially by the sighting of Jay (*Garrulus glandarius*) during the 2019 summer survey, a few years after planted oaks first produced acorns, for which Jays play a key role in dispersal.

Changes in mammal populations have not been formally documented, but Red Squirrel (*Sciurus vulgaris*), another species that caches acorns, has been sighted several times in recent years, and faeces tentatively assigned to Pine Marten (*Martes martes*)—which is expanding its range in southern Scotland—have twice been found on tracks at Carrifran in recent years. Badger (*Meles meles*) trails and disturbance are now very obvious with at least one sett at over 500m. Red Fox (*Vulpes vulpes*) are well fed on the abundant Field Vole (*Microtus agrestis*) population.

Figure 15.3   Carrifran, May 2019.

Some Roe Deer escape culling as intended—provided their numbers do not threaten the trees a little browse is welcome and in-line with ecological restoration principles.

Although data for invertebrate animals is less comprehensive than for birds, moth trapping has shown the expected dramatic increase in diversity as woodland develops. In 2003 68 moth species were recorded, of which only five were considered as typical of woodland habitats. In recent years opportunistic trapping has shown the presence of at least 159 species, of which around 20 are typical of woodland (Singleton, 2020).

A record in 2015 of the micromoth (*Nemophora degeerella*) at Carrifran was notable in being the most northerly in Britain at that time. Of similar interest in 2016 was the finding on planted Bay Willow (*S. pentandra*) of the hemipteran (*Orthotylus virens* (Miridae)) not previously recorded in Scotland. It transpired, however, that the species was also present in a site about 10km to the northeast, suggesting natural colonisation of Carrifran after its habitat became available there (Hewitt, 2020).

The future: The wild heart of southern Scotland

BFT has a vision for the future of the Southern Uplands that includes restoring natural processes to the large upland massif centred around the Moffat/Tweedsmuir Hills: *Reviving the Wild Heart of Southern Scotland*. BFT already owns, and is in the process of restoring, *c.* 31km² of this upland plateau at Carrifran, Corehead, and Talla and Gameshope respectively. Carrifran and Talla and Gameshope are contiguous and cross the Tweed/Annan watershed, Corehead is separated from the latter two sites by a small portion of land (see figure 15.2).

BFT has also entered management agreements with local landowners promoting the restoration of native woodland, especially 'cleuch' (riparian ravine) woodlands. The wider vision includes not only the BFT owned areas in the Moffat/Tweedsmuir Hills, the 'beating heart' as it were, but also other land owners and groups in the area seeking to restore a large core area across the Southern Uplands and forming habitat networks out from the core area down the river valleys and watercourses, the 'arteries' leading out from the beating heart. Such a networked core area could provide habitat connectivity right across the south of Scotland from the North Sea in the east via the River Tweed, the Solway in the southwest via the River Annan and, potentially, the Firth of Clyde, via the River Clyde in the west. Other potential habitat connectivity could be provided via the woods of upper Nithsdale and the wider Galloway Forest Park area and across the Border to Kielder Forest Park. As well as BFT, organisations such as Tweed Forum in the Tweed catchment and Restoring Annan Water (RAW) in the Annan catchment, are carrying out sterling river restoration work, including reinstating meanders, creating flood management ponds, improving riparian and bankside habitat and seeking to achieve integrated, whole catchment management. Together with initiatives being promoted and work being undertaken by other bodies such as SNH, Forestry and Land Scotland (FLS), Royal Society for the Protection of Birds (RSPB), Scottish Wildlife Trust (SWT), and many others, there exists real potential for creating landscape scale, coast to coast, areas of contiguous habitat centred on an ecologically restored central Southern Uplands massif.

Borders Forest Trust have always recognised, however, that it would be impractical to bring back all the large herbivores and predators that once lived in this part of Scotland—at this stage at least. Indeed, culling of roe deer is likely to be necessary for sometime to come yet. The ongoing work of BFT and others offers some hope that the requisite scale may be secured for future generations to contemplate this next move. Meanwhile, a more natural ecosystem is slowly maturing in the south of Scotland.

References

Ashmole, M. and Ashmole, P (eds). (2009) *The Carrifran Wildwood Story; Ecological Restoration from the Grass Roots*. Borders Forest Trust.

Ashmole, P. and Ashmole, M (eds). (2020) *A Journey in Landscape Restoration: Carrifran Wildwood and Beyond*. Whittles Publishing Ltd.

Hewitt, S. in Ashmole, P. and Ashmole, M (eds). (2020) *A Journey in Landscape Restoration: Carrifran Wildwood and Beyond*. Whittles Publishing Ltd.

Newton, A.C. and Ashmole, P (eds). (1998) *Native Woodland Restoration in Southern Scotland: Principles and Practice*. Edinburgh University & Borders Forest Trust.

Ratcliffe, D.A. (1959) The mountain flora of the Moffat Hills. *Transactions of the Botanical Society of Edinburgh*, 37: 257–71.

Savory, J. in Ashmole, P. and Ashmole, M (eds). (2020) *A Journey in Landscape Restoration: Carrifran Wildwood and Beyond*. Whittles Publishing Ltd.

Singleton, R. in Ashmole, P. and Ashmole, M (eds). (2020) *A Journey in Landscape Restoration: Carrifran Wildwood and Beyond*. Whittles Publishing Ltd.

16

REWILDING CASE STUDY

Going wild in Argentina, a multidisciplinary and multispecies reintroduction programme to restore ecological functionality

Emiliano Donadio, Talía Zamboni, and Sebastián Di Martino

Background

Rewilding, the reintroduction of key species to restore ecosystem functionality, is a powerful tool that conservationists are increasingly using to curb the biodiversity crisis (Pettorelli, Durant, & du Toit, 2019). Unlike most of the world, where iconic rewilding projects are being implemented, South America lags when it comes to actively restoring natural areas that have been severely degraded. Reasons for this delay include social, political, scientific, and economic hurdles, which operating either alone or together undermine the successful execution of rewilding programmes. In Argentina, however, an ongoing rewilding programme is overcoming these challenges and effectively implementing rewilding activities (Zamboni, Di Martino, & Jiménez-Pérez, 2017). Most of the programme´s success stems from applying a multidisciplinary and multispecies approach based on a model that we developed in the early stages of the programme. Thus, this model is the result of an adaptive management process, where knowledge gained during the implementation of the programme was used to inform future decisions. This model, named The Full Nature Model, was latter formalised by Jiménez-Pérez (2018).

Basically, the Full Nature Model is supported by four interdependent pillars: protected areas, wildlife, ecotourism, and local communities. Creating and consolidating protected areas is a critical first step. The process starts by identifying private lands deemed to be of high conservation value because they encompass highly threatened ecosystems and biomes poorly represented in the existing system of protected areas. Furthermore, the sociopolitical context should at least suggest that it is feasible to implement the model. Once the land is acquired, the restoration of wildlife species starts through a diverse array of strategies. Simultaneously, infrastructure for public use is built, and the area is promoted as a nature destination, with neighbouring communities becoming involved in the process as local entrepreneurs plan and develop various services related to ecotourism activities. Ultimately, the implementation of the model should result in the creation of legally established protected areas that conserve healthy and fully functional ecosystems, where the observation and enjoyment of wildlife attracts visitors, thus becoming

DOI: 10.4324/9781003097822-19

an engine of economic development for local communities. Here, we summarise the main achievements and challenges, with an emphasis in rewilding activities, as we implemented the Full Nature model in the Iberá wetlands in Argentina.

The Iberá wetlands are located within the Iberá Basin in Corrientes, an 88,200km² province in northeastern Argentina (Figure 16.1). Iberá is home to one of the largest freshwater wetlands in the southern cone of South America, and presents a diverse array of habitats including marshes, lagoons, small rivers, grasslands exposed to seasonal floods, savannas, and subtropical forests. All together, these habitats define a highly dynamic and heterogeneous landscape that in turn supports an exceptional number of species (Giraudo, Bortoluzzi, & Arzamendia, 2006). However, this at first glance healthy system has been impacted by human activities that have resulted in severe processes of habitat degradation and defaunation. During the 20th century the ecosystems of the Iberá wetlands suffered from intensive human exploitation including the establishment of rice farming, cattle ranching, and forestry, the latter represented mostly by extensive plantations of exotic pines. These activities added up to heavy hunting that had already taken a substantial toll on native wildlife, particularly on species of vertebrates with adult body masses ≥ 1kg (large vertebrates from now on). In fact, populations of large vertebrates suffered significant declines within the last century, with several species eventually going extinct (Giraudo, Bortoluzzi, & Arzamendia, 2006). The loss of these species most likely eroded both community structure and ecosystem function across the wetlands.

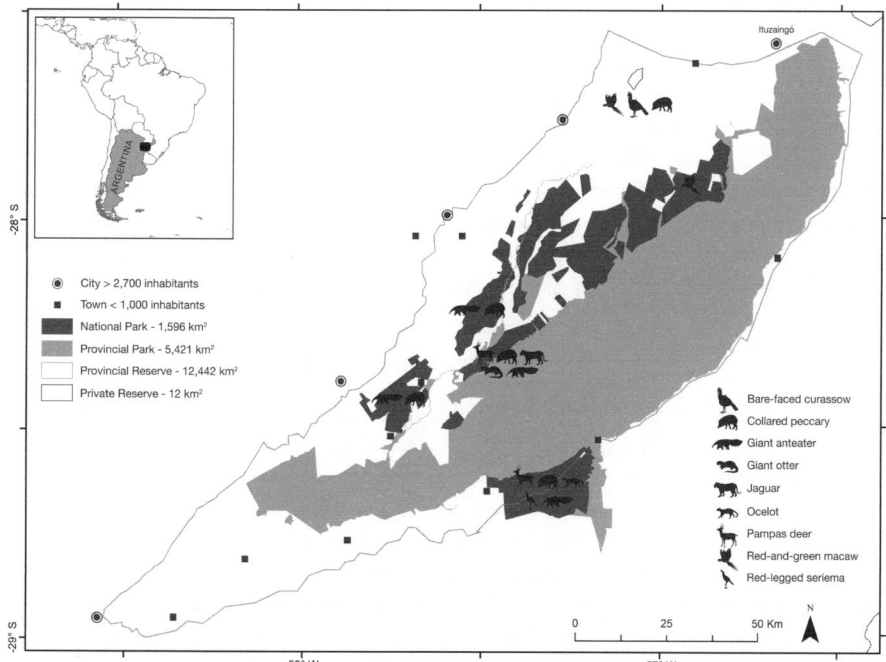

Figure 16.1   Geographical location of the Iberá wetlands and associated urban areas, categories and extension of the protected areas (Iberá Provincial Reserve, Iberá Provincial Park and Iberá National Park) within Iberá, and reintroduction areas for nine (tapirs not shown) species of large vertebrates. Reintroduction projects for red-legged seriemas (*Seriema cristata*) and ocelots (*Leopardus pardalis*), not discussed in the text, are in the planning stages. © Fundación Rewilding Argentina.

First: Secure the land

A multispecies rewilding programme that involves mostly large mammals and birds requires large tracts of protected land to ensure that populations of reintroduced species thrive. The Iberá wetlands featured both attributes, but their conservation status had to be upgraded. Thus, we focused on improving legal protection and law enforcement in the wetlands, while increasing the extent of land strictly devoted to conservation. In 1983, the province of Corrientes designated the Iberá wetlands a provincial reserve, the Iberá Reserve that encompassed ~12,442km^2, of which 5,421 and 7,021km^2 were public and private lands, respectively. Within the boundaries of the Iberá Reserve, multiple land uses, mostly agricultural activities, were allowed, including those that existed at the time the reserve was created. However, some unauthorised livestock operations and poaching persisted, especially in public lands.

Our organisation, Fundación Rewilding Argentina (www.rewildingargentina.org), purchased several private ranches in Iberá, a strategy that resulted in 1,596km^2 of agricultural land being set aside for conservation purposes. We managed these ranches as private reserves where we began executing various conservation actions like the removal of wire fences, the implementation of prescribed burning, the reinforcement, and reintroduction of native wildlife species, and the eradication of livestock, and non-native trees and wildlife. As our territorial presence expanded and consolidated, in 2009 we worked with the government of Corrientes to raise the legal status of the 5,421km^2 of public lands from provincial reserve to provincial park, thus increasing the level of protection of this area because public land allocated to parks cannot be sold by the government. Furthermore, this change in legal status increased frequency of patrolling and law enforcement that, in turn, has led to the almost complete eradication of illegal activities within the provincial park.

In 2018, we donated the 1,596km^2 of land that we had acquired and restored to the federal government, with which the government of Argentina created Iberá National Park, a category of protected area designated by federal law, which thus ensures its conservation in perpetuity. The donation included an agreement that allows us to continue implementing rewilding activities in this now federal land. Overall, our continuous presence on the ground and regular interactions with reserve managers and provincial and federal authorities have led to the overall improvement of the protection and legal status of the protected area. Moreover, the acquisition and subsequent donation of private lands to the Argentine Park Service not only resulted in the creation of a national park, but also increased by 29% the area of the reserve under public domain.

Second: Bring them back

As the consolidation of the Iberá as a complex of protected areas continued, providing the space and the protection for wildlife to thrive, we began planning the reintroduction of species that had vanished from the area, as well as the reinforcement of populations of species at low numbers. We focus on large vertebrates because they are often threatened with extinction and have a high potential to reestablish key ecological processes. Also, as highly charismatic species, they are key to boosting an economy based on wildlife viewing. Our efforts currently involve ten different species, including seven mammal and three bird species. Future reintroductions are planned for lowland pacas (*Cuniculus paca*) and ocelots (*Leopardus pardalis*). Reintroducing these many different species demands the use of a diverse array of tools that range from rehabilitation and captive breeding to the translocation of wild animals.

The importance of apex predators to ecosystems is widely recognised (Estes et al., 2011). Jaguars (*Panthera onca*) and giant river otters (*Pteronura brasiliensis*) went extinct in Iberá; whereas jaguars persist in some regions of Argentina, giant river otters completely vanished from the

Figure 16.2 The ex-situ breeding centre for jaguars, at the core of the Iberá wetlands, is a complex facility that consists of (a) 0.12-ha pens, where individual breeders are kept; (b) 1.5-ha pens, where pregnant and nursing females are kept; and (c) a 30-ha pen, where, before release, jaguars are exposed to most prey and habitats that they will encounter in the park. All pens have 5-metre-high fences. © Fundación Rewilding Argentina.

country (Di Martino et al., 2019). The reintroduction of jaguars began in 2010. It relies on an ex-situ breeding centre built in the core of the Iberá wetland (Figure 16.2); we sourced individuals from national and South American zoos and wildlife shelters (Donadio, Zamboni, & Di Martino, in press; Zamboni, Di Martino, & Jiménez-Pérez, 2017). In 2021, we released three adult females, two of them with their two 4-month-old cubs, and a male. All were fitted with Iridium GPS collars. Releasing females with cubs was critical to anchor females, and the male, to the core area of Iberá and avoid undesirable early dispersal. Current monitoring indicates that in 2022 all four cubs have become independent, and two of the adult females have reproduced in the wild giving birth to two cubs each. Actual population size is 12 jaguars. Cluster searching techniques (Smith et al., 2020) show that jaguars are killing mostly capybaras (*Hydrochoerus hydrochaeris*) and feral hogs (*Sus scrofa domesticus*). We expect to release 8–20 adult jaguars within the next three years.

We source giant river otters from European zoos, which successfully breed the species and have a surplus of individuals. Currently, the first pair of giant river otters reside in a prerelease pen located also in the core of the Iberá. The semi-aquatic pen encloses a section of a lagoon with banks of thick vegetation, where giant river otters have learnt how to catch native live prey. In May 2021, after several unsuccessful attempts, the female has given birth to three cubs, producing the first litter born in the country after more than 30 years. A second pair is currently adjusting to its pen. We keep searching for more individuals in Europe and the United States. Giant river otters are highly social and form family groups that are key to the otters´ reproductive success (Leuchtenberger, Magnusson, & Mourão, 2015). Thus, we plan to release the first family group in 2023, once cubs are > 1 year old. More family groups will be released as new captive-bred individuals join the programme.

Giant anteaters (*Myrmecophaga tridactyla*) also disappeared in Iberá. These specialised insectivores are still poached in other regions of Argentina; when adult females are killed, their pups are kept or sold illegally as pets. Often, authorities seize these pups, which frequently die without proper care. We created a rehabilitation centre to house these orphaned pups and have implemented a strong network of NGOs and governmental agencies that alert us when a pup and, to a lesser extent, an adult, is rescued. Pups are difficult to raise because they are overly sensitive to what they are fed, particularly during the first months of life. Over time, we have developed handling and feeding protocols that have increased first year pup survival from 25% in the early stages of the project to 100% in the last four years. Pups spend an average of 361 days at the centre. Afterwards, we move them to a pre-release pen built where they will be released after an acclimatisation period of ~ 30 days. We freed the first giant anteaters in 2007. Since then, we have established two self-sustaining populations (i.e., populations that are no longer managed; Zamboni, Di Martino, & Jiménez-Pérez, 2017) and two founding groups in Iberá.

Large herbivores are major drivers of terrestrial ecosystems (Danell et al., 2006). In Iberá, the process of herbivory and related mechanisms suffered from the extinction of collared peccaries (*Pecari tajacu*) and lowland tapirs (*Tapirus terrestris*), and dramatic population declines of other large herbivores like the Pampas deer (*Ozotoceros bezoarticus*). The reintroduction of collared peccaries and lowland tapirs began in 2015 and 2016, respectively. We source animals from Argentine and Brazilian zoos and wildlife centres, which are willing to contribute individuals to the project. Rehabilitating and releasing captive-sourced peccaries and tapirs has been particularly challenging. We found that soft releases are the best approach for both species. In short, animals are held in in-situ pre-release pens and fed with a mix of non-native and natural foods until release. At least 21 peccary groups have established in two self-sustaining populations and three founding nuclei (Di Martino et al., 2021).

Conversely, the reintroduction of tapirs faced an unexpected problem. One and a half years after the release and successful adaptation of 11 out of 12 adults, with two females giving birth and at least one other female pregnant, 10 out of 12 tapirs, including two calves, became infected with *Trypanosoma evansi*, a protozoal parasite not known to affect neotropical tapirs. Despite intensive veterinarian care, seven individuals died from the disease. Furthermore, animals that were treated and survived failed to produce antibodies and became reinfected. Faced with this scenario, we decided to recapture the surviving tapirs and temporally halt their reintroduction. We then launched a study at El Impenetrable National Park, located ~520km northwest of Iberá, where we are capturing tapirs and collecting blood samples to evaluate whether wild populations of tapirs develop immune resistance against *T. evansi*. If wild tapirs present antibodies against this protozoan, then the reintroduction of tapirs to Iberá could proceed translocating individuals with a wild origin.

Whereas we mostly work with animals of captive origin, the reintroduction of Pampas deer was entirely dependent upon a wild population. Like in most of its historical range, Pampas deer in Corrientes province underwent a severe retraction in numbers and distribution due to hunting and pathogen transmission from cattle (Merino et al., 2019). Indeed, the last known population of Pampas deer in Corrientes is restricted to its northern portion, where the Iberá Provincial Reserve borders a series of private ranches. Here, native grasslands, prime habitat for the Pampas deer, are being replaced by pine plantations. This scenario generated an opportunity to translocate some individuals from this highly impacted area to the core of the Iberá wetlands, where grasslands had been recovering (Jiménez-Pérez et al., 2009). The first translocation of Pampas deer was implemented in 2009–2012, and the second came in 2015–2019, and involved a total of 37 individuals. To translocate Pampas deer, we darted and immobilised animals, which were then transported by air to a pre-release pen where, upon recovery and a period of acclimatisation, they were released (Jiménez-Pérez et al., 2016; Zamboni, Di Martino, & Jiménez-Pérez, 2017). Density estimates for the population established in 2009–2012 were 3.6 indiv. km^{-2} in 2016 (Ávila, 2017), and 6.1, 10.2, and 7.5 indiv. km^{-2} in 2018–2020. Currently, two self-sustaining populations of Pampas deer have established in Iberá (Zamboni, Di Martino, & Jiménez-Pérez, 2017).

The Red-and-green Macaw (*Ara chloropterus*) and Bare-faced Curassow (*Crax fasciolata*), belong to groups that are important predators and dispersers of large seeds in subtropical and tropical forests and savannas (Blanco, Hiraldo, & Tella, 2018; Brooks, 2006; Galetti et al., 2013). Red-and-green macaws went extinct in Argentina, whereas bare-faced curassows underwent severe populations declines and are extinct in Corrientes province (Chébez, 2008). We source animals from zoos, wildlife shelters, and breeding centres in Argentina and Brazil, the latter only for bare-faced curassows.

Macaws have been the most challenging species to reintroduce because they need training on how to fly and avoid predators and identify native foods. Training occurs in specially designed facilities including a 30×6×6m aviary, where macaws develop their flight musculature and skills by flying along the aviary. Macaws are most vulnerable to predation when foraging on the ground, thus we use a remotely controlled stuffed fox to scare macaws when they land on the ground to feed. Finally, we provide macaws native food, so they learn to recognise it once they are released. Post training, macaws are moved to aviaries built in trees in the locations where they will be released. After release, they are presented with food on their platforms to increase survival and ensure that they become anchored to the release area. In 2015, we had our first experimental release, yet out of the seven individuals released, only one was known to have survived. Poor antipredator skills were regarded as the main cause of failure as ≥ 42% of the individuals were preyed upon (Volpe, Di Giácomo, & Berkunsky, 2017). Lessons from this first release led to an improvement of training techniques (Zamboni, Di Martino, & Jiménez-Pérez, 2017). Between 2016 and 2022, we released 43 (F:M, 21:22) individuals in two locations. Twenty-one (9:12) macaws settled in the release area, became independent and survived, 14 (7:7) died, seven (4:3) became independent, dispersed and their fate remains unknown, and one (1:0) was recaptured. Predation accounted for 64% of the mortality events. Three breeding pairs laid eggs in nest boxes 2.2–3.1 years after release. First reproductive attempts were unsuccessful because the parents damaged the eggs during hatching. Subsequent attempts resulted in all three breeding pairs successfully producing two chicks each (one successful pair in 2021 and two in 2022). Management of chicks included daily monitoring and hand feeding if they showed an empty or half empty crop. This intervention continued until chicks were three months old. Also, the parents were supplemented with additional food presented in a platform next to the nest box. Four chicks, two males and two females, survived. They represent the first red-and-green macaws born in the wild in Argentina in a century. Today, two founding nuclei of red-and-green macaws have been established in Iberá. Releases to reinforce both nuclei will continue for at least three years.

We held bare-faced curassows in a prerelease aviary located in the release area. In 2020, we released nine individuals (5:4). Within the first eight months, four (3:1) individuals were preyed upon, whereas one (0:1) individual dispersed and its fate remains unknown. The two remaining breeding pairs laid eggs 6–10 months after release. First clutches were unsuccessful. One pair had its egg damaged for unknown reasons. The second pair lost its chicks, likely to predators, 24 hours after hatching. Second clutches were laid one month later and were successful. These two breeding pairs will be supplemented with ten (6:4) additional individuals that are presently in quarantine.

Captive-sourced individuals go through exhaustive quarantine and health evaluation processes (Zamboni, Di Martino, & Jiménez-Pérez, 2017). Intensive post-release monitoring and management is based on radio telemetry and involves captive- and wild-sourced individuals (Di Martino et al., 2021; Donadio, Zamboni, & Di Martino, in press; Volpe, Di Giácomo, & Berkunsky, 2017; Zamboni, Di Martino, & Jiménez-Pérez, 2017 and references therein;). After release, we manage individuals for survival and reproductive success. For instance, we recapture and treat, if deemed necessary, animals that are injured or sick; supplement animals that struggle with finding native foods and lose weight; recapture, when possible, and keep animals that have reproduced, but have offspring overly sensitive to predation, in pre-release structures; supplement the diets of animals raising young; provide nest boxes to facilitate reproductive events; and remove feral hogs that kill young collared peccaries and Pampas deer. Further active management includes periodic controlled burns to conserve the grassland habitat and stimulate grass regrowth for pampas deer; and planting 100 pindó palms *Syagrus romanzoffiana* to provide macaws with additional native food resources. Overall, we invest heavily in individuals which are extremely difficult to obtain.

Besides individual monitoring, we implement post-release monitoring at the population and community level. The former is directed to evaluate population trajectories and spatial recolonisation over time, whereas the latter focuses on putative community and ecosystem level effects of reintroducing key species such as apex predators. Monitoring populations has proved to be difficult, given the number of species involved, together with Iberá's diverse array of habitats. We have implemented aerial (Ávila, 2017) and terrestrial transects. Both methods have yielded accurate estimates for Pampas deer and other non-targeted species; but they are impractical in forested habitats and provide information on only one species. Aerial transects are also expensive. We are currently designing a large-scale monitoring scheme using camera traps. The scheme is based on a grid that covers the area of interest and it is subdivided into ~2,5×2,5-km squares. In or nearby the centre of each square, we deploy a camera trap that is operative for 30 days. The number of independent captures (i.e., photos) standardised by sampling effort provides a crude index of abundance per species (Kelly et al., 2012; Meek, Ballard, & Fleming, 2012).

We monitor the community and ecosystem level effects of reintroduced species within the theorical framework of trophic cascades (Terborgh & Estes, 2010). We are evaluating the effects of jaguar reintroduction on the structure and productivity of grass communities using experimental exclosures; the abundance and behaviour of large vertebrate prey, meso predators, and scavengers using camera traps; and prey use by reintroduced jaguars through cluster searching. We will use a similar framework to evaluate the effects of reintroducing giant otters on aquatic systems, especially fish communities, using environmental (Ruppert, Klineab, & Rahman, 2019) and scat (Quéméré et al., 2021) DNA metabarcoding. Likewise, current work is combining DNA metabarcoding and direct observations to evaluate interactions between reintroduced Red-and-green Macaws and local food resources (Volpe et al., 2021). To address these complex questions, we partner with national and international researchers who receive solid logistic support and state-of-the-art equipment (e.g., Iridium GPS collars, camera traps). Collaborative work with independent researchers is a robust strategy to evaluate our own work and validate results before the public and governmental agencies.

Third: Help them thrive

The Iberá is surrounded by several towns with populations ranging from 200 to 20,000 inhabitants. These communities depend largely on exotic tree plantations and cattle ranches that partially operate in the private lands that constitute the reserve. Thus, it has been essential to demonstrate that enhancing land protection and restoring ecosystems would also provide solid economic alternatives. In the long term, the conservation of Iberá and its wildlife will only be effective if local communities are benefitting from their existence.

Consequently, we have directed efforts to develop an economy rooted in ecotourism with wildlife watching as one of its main attractions. To achieve this goal, we have partnered with local and provincial governments and implemented several strategic lines of work. As a result, by increasing the number of entrances to the park from one to ten, public access has been expanded; these entrances have been built in ten different towns. Connectivity between, and infrastructure in towns has been reinforced; over 500km of dirt road was improved, and 88 tourism-related facilities were restored or constructed. Building the local capacity to offer high quality services to tourists has also been essential; thus, 30 workshops dealing with tourism services have been offered in eight towns, reaching 8,200 locals.

Finally, to position Iberá as a first-class destination for ecotourism, we launched a vigorous advertising campaign promoting the destination. This included inviting journalists to visit Iberá so they would publish articles and present their experiences via TV and documentaries. This strategy has yielded outstanding results. Tourist visitation in Iberá increased 70% between 2016 and 2019. This increase has generated employment, which, combined with the upgrading of public infrastructure, has improved the level of wellbeing in local communities. As a result, the government of Corrientes promotes rewilding as an economic activity. Also, local communities and political authorities now appreciate and defend the wild state of the Iberá wetlands and its wildlife, and strongly support rewilding work.

Main challenges and conclusion

Over 15 years, our work in Iberá has encountered many challenges (Jiménez Pérez, 2018; Zamboni, Di Martino, & Jiménez-Pérez, 2017). Initially, public and political distrust were a major issue. Because we acquired private land and worked closely with international NGOs, we were accused of being vehicles of land foreignisation. After the land was donated to create public reserves, relations improved, but we received criticism from some stakeholders for taking land out of production. Overcoming these issues took a transparent communication strategy, numerous meetings with stakeholders and authorities and, above all, results showing that conservation and economic development, in the form of ecotourism, can be effective solutions in battling the current biodiversity crisis and depressed local economies.

One largely unresolved challenge is the unfeasibility to translocate wild-sourced individuals. Both giant river otters and red-and-green macaws are extinct in Argentina. Translocations would be possible if countries with extant wild populations of these species could source individuals. Our experience suggests that this solution can become a bureaucratic ordeal due to local, national, and international regulations between countries. For other species which persist in Argentina, such as jaguars, tapirs, collared peccaries, and Pampas deer, translocation would be possible between protected areas under the same and different jurisdictions. In this instance, opposition results mostly from provincial and federal managers who might not issue the required permits out of excessive caution. Even forward-thinking Argentine academics are hesitant to take on the uncertainties of translocations and often advise against translocations when government officials request their advice.

Rewilding requires the active management of imperiled species. It is a conservation strategy that challenges traditional conservation based on creating reserves and passively monitoring the recovery of populations, communities, and ecosystems. In Argentina, most reserves were created in territories that had already suffered severe processes of defaunation and degradation. Thus, for natural systems to recover key ecological mechanisms, the implementation of rewilding activities is critical. Despite challenges and setbacks, the Iberá programme has managed to secure and upgrade large tracts of land for conservation, begin to restore extinct species, and create an alternative economy, which is in turn boosting local and regional support for rewilding activities. This virtuous cycle leaves behind the false dilemma of conservation versus production, revealing the role that rewilding could have to restore nature and human societies across Latin America.

References

Ávila, B. (2017). *Evaluación de un método de monitoreo aéreo de fauna mediante fotografía en los Esteros del Iberá (Corrientes, Argentina)*. Tesis de Maestría, Centro de Zoología Aplicada, Universidad Nacional de Córdoba.

Blanco, G., Hiraldo, F., and Tella, J.L. (2018) Ecological functions of parrots: an integrative perspective from plant life cycle to ecosystem functioning, *Emu*, 548(118): 36–49.

Brooks, D.M. (ed.) (2006). *Conserving Cracids: The Most Threatened Family of Birds in the Americas*. Houston Museum of Natural Science.

Chébez, J.C. (2008). *Los que se van, fauna argentina amenazada*. Tomo 2. Editorial Albatros.

Danell, K., Bergström, R., Duncan, P., and Pastor, J. (eds) (2006) *Large herbivore ecology, ecosystem dynamics and conservation*. Cambridge University Press.

Di Martino, S., Zamboni, T., Valenzuela, A.E.J., and Gil, G.E. (2019) Pteronura brasiliensis. In SAyDS–SAREM (eds), *Categorización 2019 de los mamíferos de Argentina según su riesgo de extinción. Lista Roja de los mamíferos de Argentina* [online]. Available at: https://cma.sarem.org.ar/es/especie-nativa/pteron ura-brasiliensis (accessed: 1 June 2021).

Di Martino, S., Longo, M., Zamboni, T., et al. (2021) Reintroduction of collared peccary in the Iberá wetland, northeastern Argentina. In Soorae, P.S. (ed.), *Global Reintroduction Perspectives: 2021. Case Studies from Around the Globe*. IUCN SSC Conservation Translocation Specialist Group, Environment Agency–Abu Dhabi and Calgary Zoo, Canada, pp. 246–50.

Donadio, E., Zamboni, T., and Di Martino, S. (in press) Bringing jaguars and their prey base back to the Iberá wetlands, Argentina. In Gaywood, M., Ewen, J., Hollingsworth, P., and Moehrenschlager. A. (eds), *Conservation Translocations*. Cambridge University Press.

Estes, J.A., Terborgh, J., Brashares, J.S., et al. (2011) Trophic downgrading of planet Earth, *Science*, 333(6040): 301–7.

Galetti, M., Guevara, R., Côrtes, M.C., et al. (2013) Functional extinction of birds drives rapid evolutionary changes in seed size, *Science*, 340(6136): 1086–90.

Giraudo, A.R., Bortoluzzi, A., and Arzamendia, V. (2006) Fauna de vertebrados tetrápodos de la reserva y Sitio Ramsar Esteros del Iberá: Análisis de su composición y nuevos registros para especies amenazadas, *Natura Neotropicalis*, 1(37): 1–20.

Jiménez-Pérez, I., Delgado, A., Heinonen, S., and Srur, M. (2009) La conservación del venado de las pampas en Corrientes: amenazas y oportunidades en un paisaje en rápido cambio, *Biológica*, February–March 2009: 28–9.

Jiménez-Pérez, I., Abuin, R., Antúnez, B., et al. (2016) Reintroduction of the pampas deer in Argentina. In Soorae, P.S. (ed.), *Global Reintroduction Perspectives: 2016. Case Studies from Around the globe*. Gland: IUCN/SSC Reintroduction Specialist Group and Abu Dhabi, UAE: EnvironmentAgency-Abu Dhabi, pp. 221–7.

Jiménez-Pérez, I. (2018) *Producción de naturaleza: parques, rewilding y desarrollo local*. Buenos Aires: The Conservation Land Trust Argentina.

Kelly, M.J., Betsch, J., Wultsch, C., Mesa, B., and Mills, S.L. (2012) Noninvasive sampling for carnivores. In Boitani, L. and Powell, R.A. (eds), *Carnivore Ecology and Conservation: A Handbook of Techniques*. Oxford University Press, pp. 47–69.

Leuchtenberger, C., Magnusson, W.E., and Mourão, G. (2015) Territoriality of giant otter groups in an area with seasonal flooding, *PLoS ONE*, 10(5) [online]. Available at: https://journals.plos.org/plos one/article?id=10.1371/journal.pone.0126073 (accessed: 3 June 2021).

Meek, P.D., Ballard, G., and Fleming, P. (2012) *An Introduction to Camera Trapping for Wildlife Surveys in Australia*. PestSmart Toolkit publication, Invasive Animals Cooperative Research Centre, Canberra, Australia.

Merino, M.L., Cirignoli, S., Perez Carusi, L., Varela, D., Kin, M.S., Pautasso, A., Demaría, M., Beade, M.S., and Uhart, M. (2019) Ozotoceros bezoarticus' in SAyDS–SAREM (eds), *Categorización 2019 de los mamíferos de Argentina según su riesgo de extinción. Lista Roja de los mamíferos de Argentina* [online]. Available at: https://cma.sarem.org.ar/es/especie-nativa/ozotoceros-bezoarticus (accessed: 1 June 2021).

Pettorelli, N., Durant, S.M. and du Toit, J.T. (eds) (2019) *Rewilding*. Cambridge: Cambridge University Press.

Quéméré, E., Aucourd, M., Troispoux, V., et al. (2021) Unraveling the dietary diversity of Neotropical top predators using scat DNA metabarcoding: A case study on the elusive Giant Otter, *Environmental DNA*, 00 [online]. Available at: https://onlinelibrary.wiley.com/doi/full/10.1002/edn3.195 (accessed: 3 June 2021).

Ruppert, K.M., Klineab, R.J., and Rahman, S. (2019) Past, present, and future perspectives of environmental DNA (eDNA) metabarcoding: A systematic review in methods, monitoring, and applications of global eDNA, *Global Ecology and Conservation*, 17(00) [online]. Available at: www.sciencedirect.com/science/article/pii/S2351989418303500 (accessed: 28 May 2021).

Smith, J.A., Donadio, E., Bidder, O.R., et al. (2020) Where and when to hunt? Decomposing predation success of an ambush carnivore, *Ecology*, 101(12), e03172.

Terborgh, J. and Estes, J.A. (eds) (2010) *Trophic Cascades: Predator, Prey and the Changing Dynamics of Nature.* Island Press.

Volpe, N.L, Di Giácomo, A.S., and Berkunsky, I. (2017) First experimental release of the red-andgreen macaw *Ara chloropterus* in Corrientes, Argentina, *Conservation Evidence*, 14: 20.

Volpe, N.L., Thalinger, B., Vilacoba, E., et al. (in review) Diet composition of reintroduced Red-and-Green Macaws reflects gradual adaption to life in the wild, *Ornithological Applications*, 124(1): 1–16.

Zamboni, T., Di Martino, S., and Jiménez-Pérez, I. (2017) A review of a multispecies reintroduction to restore a large ecosystem: The Iberá Rewilding Program (Argentina), *Perspectives in Ecology and Conservation*, 15(4): 248–56.

17

REWILDING CASE STUDY

Gorongosa National Park, Mozambique

Robert M. Pringle and Dominique Gonçalves

Introduction

Colonial exploitation, postcolonial depredations, and poverty have created explosive conditions in many of the most biodiverse regions on Earth. Since the mid-20th century, the great majority of armed conflicts have occurred in biodiversity hotspots (Hanson et al., 2009). Similarly, the great majority of mammal and bird species have had conflicts within their ranges (Mendiratta et al., 2021). Yet the ecological impacts of conflict are heterogeneous. On the one hand, wars have devastated many wildlife populations (Daskin & Pringle, 2018); on the other, they are a bulwark against large-scale habitat conversion, which leaves the door open for rewilding.

Mozambique's Gorongosa National Park (GNP) epitomises these conditions. Runaway poaching during the Mozambican Civil War (1977–1992) stripped GNP of >90% of its large-mammal fauna and extirpated several top carnivores (Stalmans et al., 2019). Since 2007, however, an innovative public-private partnership—the Gorongosa Project (GP)—has brought herbivore biomass back to nearly pre-war levels, nurtured the lion (*Panthera leo*) population back to abundance, and reintroduced two locally extinct and globally threatened carnivore species, African wild dog (*Lycaon pictus*) and leopard (*P. pardus*) (Bouley et al., 2018, 2021; Angier, 2021). Meanwhile, GP is working with Mozambique's government to expand the coverage and connectedness of protected areas and has initiated programmes focused on human development (Pringle, 2017).

This portfolio arguably makes GP the most ambitious and successful large-scale rewilding effort anywhere in the world to date. GNP is a model for understanding the ecological effects of defaunation and the trajectory of postwar community reassembly, while GP offers a potentially generalisable model for how diminished protected areas can be upgraded and upsized (Pringle, 2017)—in contrast to the global trend of protected area downgrading and downsizing (Mascia & Pailler, 2011). These models are relevant across large swaths of the Global South where conflict and poverty have destabilised protected areas. In this chapter, we describe GNP's history, the GP framework, the rewilding trajectory, and challenges that still loom.

Biological and historical context

GNP encompasses 4,000km² of lowland and montane savanna, grassland, and forest at the southern tip of the Great Rift Valley in central Mozambique (Figure 17.1). European hunters

DOI: 10.4324/9781003097822-20

marvelled at Gorongosa's wildlife, leading the Portuguese colonial administration to create first a hunting reserve and later, in 1960, a national park. In the early 1900s, some 47 species of large mammal (≥5kg) occurred in GNP, many of which congregated on the productive Rift Valley floodplains around Lake Urema (Tinley, 1977). Although a few species had been extirpated by 1970—white rhinoceros (*Ceratotherium simum*), black rhinoceros (*Diceros bicornis*), roan (*Hippotragus equinus*), tsessebe (*Damaliscus lunatus*)—most were thriving. A 1972 survey recorded 2,500 elephant (*Loxodonta africana*), 3,400 hippo (*Hippopotamus amphibius*), 13,000 buffalo (*Syncerus caffer*), 6,400 wildebeest (*Connochaetes taurinus*), 3,300 zebra (*Equus quagga*), and 3,300 waterbuck (*Kobus ellipsiprymnus*) (Tinley, 1977).

The park was largely unscathed by Mozambique's War of Independence against Portugal (1964–1972) but was throttled by the Mozambican Civil War (1977–1992), an insurgent campaign against the newly independent government. Antigovernment forces were based in Gorongosa, and GNP was the theater for some of the most intense combat. Control of GNP changed hands several times, and combatants shot thousands of animals. Rebel forces are reported to have put special effort into obliterating GNP's infrastructure, and to have traded ivory for weapons with South Africa (Morley & Convery, 2014; Campbell-Staton et al., 2021). A peace accord in 1992 did not relieve GNP's wildlife, as commercial poachers, heavily armed ex-combatants, and a starving populace picked off the remaining game: 'Even small field mice are being unearthed as a meagre source of protein' (Dutton, 1994: 8).

Fixed-wing aerial surveys in 1994 and 1997 indicated near-annihilation of GNP's large mammals. In sum across the two surveys, spotters tallied 8 elephant, 7 hippo, 2 buffalo, 5 zebra, 40 warthog (*Phacochoerus africanus*), 1 bushpig (*Potamochoerus larvatus*), 157 waterbuck, and 148 other antelopes (Stalmans et al., 2019). Although these surveys were limited in coverage (<200km²), several more extensive helicopter counts between 2000–2002 affirmed the general picture. Large-herbivore populations remained extant, but had declined by >90%; a handful of GNP's ~200 lion had survived, but leopard, wild dog, and spotted hyena (*Crocuta crocuta*) had been extirpated (Bouley et al., 2021).

The catastrophic decline of large-mammal populations was associated with significant changes to the landscape. Already in 1994, 'once lawn-like *Cynodon* grasslands… were standing knee-high for lack of grazers,' and 'previously pure grasslands are now invaded by aged woody plants' (Dutton, 1994, pp. 7–8). A study using declassified US spy satellite imagery from 1977 and high-resolution satellite imagery from 2012 found that woody cover had increased by 34% parkwide, by 51–96% in the Rift Valley savannas, and by 134% in the critical Urema floodplain (Daskin, Stalmans, & Pringle, 2016). An invasive woody plant of special concern—the wetland-choking shrub *Mimosa pigra*—proliferated in the floodplain (Guyton et al., 2020).

Phases of rewilding in Gorongosa

The history of rewilding in GNP is deep, intertwined with a history of intensive hunting that long predates Mozambique's 20th-century conflicts. Kenneth Tinley described attempts to (re)introduce several species that were thought to have occurred in GNP historically (Tinley, 1977). White rhino were eliminated by the 1940s; six individuals were imported in 1970. Six cheetah were introduced in 1973, although their historical status in GNP is not entirely clear. Six giraffe (the historical status of which is also unclear) were introduced around 1950 and purportedly eaten by lion.

Tinley was also an early proponent of expanding protected-area coverage from 'mountain to mangroves'—that is, from Mt. Gorongosa to the Zambezi delta—to increase its ecohydrological coherence and secure space for ungulate migrations. In the 1970s, GNP comprised 3,770km² of Rift Valley around Lake Urema along with strips of miombo woodland to the east and west. Tinley envisioned a 30,000km^2 management area encompassing Mt. Gorongosa (an 1,860m massif that generates orographic rainfall that feeds Lake Urema) and extending eastwards to connect with the Marromeu Reserve (Dutton, 1994). This vision remains influential today.

Mozambique emerged from the civil war as one of the world's poorest countries, and its government sought to revitalise protected areas to generate foreign exchange. An initial project in the mid-1990s, in partnership with IUCN and African Development Bank, laid important groundwork by rebuilding basic infrastructure and conceiving a 'more inclusive, less confrontational' approach to law enforcement that brought together rangers who had fought on opposite sides of the conflict (Morley & Convery, 2014: 140). Nonetheless, poaching remained rampant and funding was not commensurate with the scale of the challenge (Morley & Convery, 2014).

GP was conceived in the mid-2000s as a joint venture between Mozambique's government and the Carr Foundation, a US-based non-profit headed by businessman-turned-philanthropist Greg Carr. This public-private partnership coalesced in a 20-year co-management agreement, finalised in December 2007, committing the Carr Foundation to a $24 million investment in exchange for oversight capacity to achieve a long slate of objectives under the headings of conservation, law enforcement, tourism development, community relations, education/training, and infrastructure (Pringle, 2017). This partnership was later extended through 2040.

This long time horizon enables GP to adapt and evolve. The initial focus on ecological restoration and ecotourism has expanded, in coordination with the government, to include complementary objectives in human development for the ~200,000 residents of the park's buffer zone. Initiatives include providing financial and logistical support to district health authorities to extend primary health care services to rural communities; agricultural extension; financial and programmatic support for primary and secondary schools and teachers; and creation of safe spaces where youth can learn life skills, with a particular focus on enabling girls to avoid child marriage and early pregnancy so that they can finish school. We do not have space to detail all of GP's efforts to alleviate poverty and stimulate green economic development; some are described elsewhere (Pringle, 2017; Gorongosa National Park, 2020), but this topic requires separate treatment. GP also supports basic scientific research and a nascent programme in carbon sequestration and climate-change mitigation. The project's original moniker, Gorongosa Restoration Project, contracted to GP in light of this broad and increasingly people-oriented mission. GP's budget has grown in concert with its scope—to roughly $16 million in 2021—with a long list of development aid agencies and NGOs contributing most of the programmatic funding. As of 2019, philanthropic support accounted for nearly two-thirds of GNP's income (although the Carr Foundation was no longer a majority donor) and international aid contributed another quarter; generated revenue accounted for <10% but exhibited the greatest growth of all revenue sources from 2013 to 2019.

The chief rewilding tactic of GP is to facilitate population recovery by curtailing illegal hunting. A revamped ranger squad patrols GNP, removes snares, and apprehends poachers. In addition, over 500 animals of nine species have been translocated into GNP; these translocations are a small fraction of the overall wildlife recovery, but were crucial in reestablishing several species (Stalmans et al., 2019). By gradually expanding protected-area coverage, GP aims to increase connectivity, enable resumption of migration, and buffer the system against climate change. Expansion does not entail evicting people; several communities live inside GNP. Instead, GP seeks to incentivise people to move outside park boundaries, in part by helping them to secure land tenure, which they otherwise lack. This reflects the foundation of GP's strategy (and its original motivation), which is to create fertile socioeconomic conditions for protected-area survival.

Ecologically, GP is open-ended in the sense that it does not intensively manage towards any particular historical baseline. The conditions of 1972 are a convenient reference point because they are well documented (Tinley, 1977), but are not a target per se. The guiding philosophy is rather to ensure that the essential pieces (species) are present, protect the larger system, and let the system reassemble itself.

Trajectory and success of rewilding in Gorongosa

One published framework (Torres et al., 2018) defines rewilding success along two axes (see also Carver et al., 2021). The first is decreased 'human forcing', quantifiable (in theory) as a function of material inputs and outputs. The second is increased ecological integrity, as indexed by the naturalness of disturbance regimes, intactness of communities, connectivity of ecosystems, and complexity of trophic networks.

The former axis is not a useful prism through which to evaluate rewilding success in this social-ecological system. Reducing human forcing in the form of poaching is a major focus; snaring pressure decreased by 65% and lion poaching decreased by 95% from 2015 to 2018 (Bouley et al., 2021). However, curtailing poaching requires massive material investment on multiple fronts—not just law enforcement, but also the community-relations and human-development activities designed to discourage and ultimately obviate illegal hunting. Indeed, human inputs and outputs were incorporated into GP from the outset (Pringle, 2017). The long-term co-management agreement stipulated, among other things, the requirements to maintain 'effective and strict law enforcement'; to 'employ Mozambican nationals and Mozambican firms'; to create 'lodging and tourism activities business'; to build and maintain physical infrastructure; to develop information-technology infrastructure; and to oversee 'animal-reintroduction and... breeding programs'. In short, human inputs are required to enable rewilding, in large part by creating outputs deemed by Mozambique's government to be in the national interest.

Along the latter axis, ecological integrity, GP has been successful by conventional criteria (e.g., Torres et al., 2018; Carver et al., 2021), although the recovery process is ongoing and it may take decades for GNP to settle into a new dynamic equilibrium. We discuss five such criteria (Torres et al., 2018): disturbance regime, species composition, community structure, trophic interactions and ecosystem functions, and ecosystem connectivity.

Disturbance

Fire and flooding are the main abiotic disturbances in GNP. As with human inputs/outputs, the 'disturbance naturalness' criterion (Torres et al., 2018) is nuanced in the context of GP. In Africa, anthropogenic fire has been part of savanna landscapes for as long as *Homo sapiens* has existed. Moreover, fire extent in southern African savannas is determined largely by rainfall (and hence fuel loads), irrespective of management strategy (Van Wilgen et al., 2004). Annual grass fires occurred throughout the park before the civil war (Tinley, 1977), and the same is true today; GP's management approach involves setting patchy, uncontrolled burns early in the dry season to reduce intense fires. While there is need for more research on how GNP's current fire regime compares to historic and prehistoric baselines, there is no directional trend in fire extent over the last 20 years, and the park-wide trend of increasing woody cover from 1977–2012 provides no reason to believe that fire became appreciably more frequent or intense during or after the war (Daskin, Stalmans, & Pringle, 2016). Similarly, hydrological regimes were not actively regulated before or after the war, although subtler anthropogenic effects are possible (Tinley, 1977; Guyton et al., 2020).

Species composition

All large-herbivore populations present in 1972 survived the war, but some were nearly extirpated. For four of these, translocations may have been crucial for persistence. Wildebeest were not detected in postwar aerial surveys until 2007 (n=16 members of the original population); 180 were introduced that same year, and 627 were counted in 2018. Buffalo were sporadically detected in aerial counts from 1994 to 2007 (2–26 individuals); 210 were translocated from 2006 to 2011, and 1,021 were counted in 2018. Eland (*Tragelaphus oryx*) were intermittently detected until 2012 (when 3 individuals were counted); 35 were introduced in 2013, and 142 were tallied in 2018. Zebra were consistently detected, but at extremely low numbers; 15 were imported in 2014, and 44 were counted in 2018. Reintroduction of ungulates extirpated before the war (roan, tsessebe, white and black rhino) is a potential longer-term goal.

Wild dog were successfully reintroduced (Bouley et al., 2021) by importing 45 individuals from genetically distinct source populations between 2018 and 2021; the population now exceeds 100. Intensive camera trapping from 2012 to 2018 failed to detect any leopard. A single male of unknown origin was finally spotted in 2018, and four individuals were introduced in 2021. A hyena was intermittently sighted in 2012, but as of 2021 there was not a viable population; hyena reintroductions began in 2022. Side-striped jackal (*Canis adustus*) occupied GNP in the 1970s but appear to be functionally absent as of 2022; limited reintroductions are planned.

No other species is known to have been lost from GNP in recent decades, although quantitative pre-war data exist only for large mammals and plants. GNP supports dense breeding colonies of waterbirds and more than a dozen globally threatened species of raptors and other birds.

Community structure

By 2018, GNP's total large-herbivore biomass had reached nearly pre-war levels (Figure 17.1), and the lion population was also approaching pre-war abundance (Stalmans et al., 2019). However, relative abundances remain heavily skewed relative to pre-war GNP and intact savannas elsewhere. Medium-sized and solitary-to-moderately-social ungulates (waterbuck, impala, reedbuck, kudu, nyala, warthog) now account for the vast majority of biomass, in contrast to the pre-war fauna dominated by larger-bodied gregarious species (elephant, hippo, buffalo, wildebeest, zebra). In particular, 57,000 waterbuck accounted for >50% of all ungulate individuals in 2018 (Stalmans et al., 2019). This shift in community structure appears to have arisen from differences in traits that affected species' resistance and resilience to poaching during the war. Large size and social groups make animals easy to see and shoot; large size also means slow reproductive rate; and gregarious species rely on herds to detect and avoid predators, including people. Conversely, individuals in smaller groups are more dispersed, making them harder to hunt (resistance), and smaller mammals reproduce faster, enabling swift recovery (resilience). Warthog have shorter gestation and produce three-fold more offspring per litter than GNP's other ungulates. Kudu and especially nyala are woodland-affiliated and cryptic, while reedbuck and especially waterbuck can take refuge in the inaccessible, crocodile-rich floodplain (2,745 *Crocodylus niloticus* were counted in 2020, making it GNP's fifth-most abundant megafauna).

The rapid recovery of mid-sized ungulates (Figure 17.1) has enabled GNP to export animals to support restoration of other defaunated protected areas—a rewilding positive feedback. GNP has exported fourfold more animals than it has imported, including 1619 waterbuck, 200 warthog, 193 reedbuck, 50 oribi (*Ourebia ourebi*), and 48 sable (*Hippotragus niger*). The

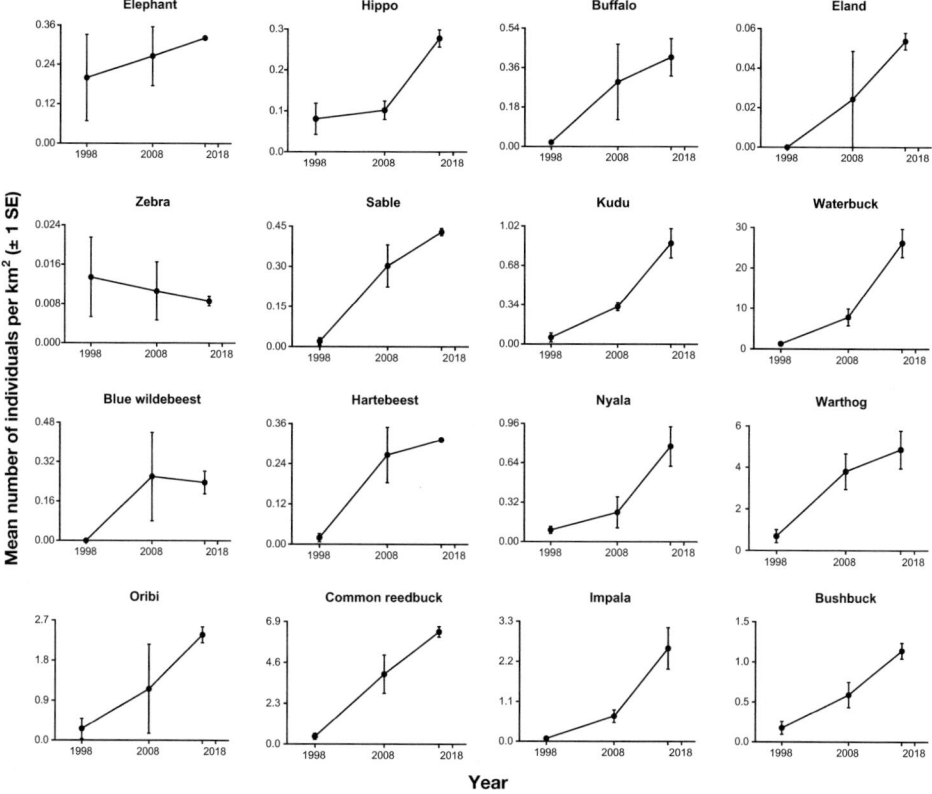

Figure 17.1 Post-war recovery dynamics of 16 large mammalian herbivore species in Gorongosa National Park, Mozambique, 1994–2018 (aerial survey data from Stalmans et al., 2019).

reestablishment of wild dog was so rapid that 5 were exported in 2021. Aside from this selective offtake, GNP's populations are not culled or otherwise actively managed.

Trophic interactions and ecosystem functions

The uneven recovery of large mammals in GNP has produced various ecological anomalies. Intensive ivory poaching caused the evolution of increased frequency of tusklessness in GNP's female elephants, from 18.5% of individuals in the 1970s to 51% in the early 2000s (Campbell-Staton et al., 2021); preliminary data suggest that tuskless females may eat different diets than tusked ones, which might affect their functional role in the ecosystem. Without large scavengers, carcasses decomposed slowly (versus overnight, as typical where hyena are abundant). Vultures have difficulty accessing unmanipulated carcasses, but experimentally creating an incision along the belly led to rapid skeletonisation, suggesting an important role of mammalian scavengers in modulating vulture foraging and nutrient cycling. Similarly, extirpation of large carnivores produced a 'landscape of fearlessness' in which animals did not take typical antipredator precautions. Bushbuck (*Tragelaphus sylvaticus*), ordinarily woodland-restricted antelopes, have increasingly occupied the treeless floodplain, but exhibited strong avoidance of experimentally simulated predator presence (Atkins et al., 2019). The ongoing reassembly of the historical apex-carnivore guild provides an opportunity to test how rapidly such behavioural relaxation reverts. In the first year after wild dog were reintroduced, they subsisted largely on bushbuck (Bouley et al., 2021), suggesting that anomalous behavioural patterns may dissipate rapidly.

Waterbuck have likewise expanded into new habitat, but in the opposite direction (floodplain to savanna) and for a different reason (intraspecific competition instead of risk relaxation). A study from 2015–2019 found that waterbuck were approaching density-dependent limits in the Urema floodplain, depleting preferred food plants, and spilling over into adjoining wood-land habitat where resource concentrations are lower (Becker et al., 2021). As forecasted by that study and a simple logistic-growth model (Stalmans et al., 2019), waterbuck numbers dipped from 2018 to 2020. Waterbuck have unusually high water and protein requirements, and we predict that the population will contract as formerly dominant competitors (e.g., wildebeest, buffalo) and carnivores continue to recover. A parallel scenario played out in Kenya's Lake Nakuru National Park from 1970 to 2004 (Ogutu et al., 2012).

In these no-analog, non-equilibrial conditions, large herbivores exhibited anomalously weak dietary niche differences relative to elsewhere in Africa (Pansu et al., 2022). Abundant, wide-ranging species such as waterbuck ate broad diets that overlapped extensively with other species and included plants that they did not historically eat (Pansu et al., 2019; Pringle & Hutchinson, 2020; Potter et al. 2022). This overlap probably reflects ecological release from interspecific competition and predation risk, which relaxes constraints on where and what ungulates eat. We predict that increasing wildlife densities and community evenness will lead to stronger niche differentiation over the coming decade.

Despite these shifts in community structure, behaviour, and diet, GNP's recovering ungulate population has reestablished at least one key ecosystem function. The invasive shrub *Mimosa pigra* was present in GNP long before the war but was kept in check by herbivores. While the collapse of herbivore populations enabled *M. pigra* to expand, the rapid increase of antelope biomass since 2007 brought the infestation back to pre-war levels by 2019 (Guyton et al., 2020). This finding is significant in the context of trophic rewilding, both because it demonstrates the speed with which ecosystem functions can be recovered even in no-analog communities, and because it eliminates the need for additional human forcing (e.g., chemical or biological con-trol) to mitigate the impacts of biological invasions.

Ecosystem connectivity

At the onset of GP, Mozambique created a 5,400km² buffer zone around GNP. This desig-nation allows sustainable natural-resource use by the ~200,000 inhabitants, but prohibits eco-logically damaging uses of land (e.g., mining) and water (e.g., diversion, pollution). In 2010, the portion of Mt. Gorongosa above 700m elevation was legally annexed to GNP after a thorough consultation process with members of the communities most affected and other stakeholders, conferring legal protection for the chief source of water into Lake Urema along with sustainable-development support to the communities.

Much of the land between GNP and the coastal Marromeu Reserve is occupied by large hunting concessions (coutadas). These reserves are protected areas (IUCN Category VI) with few human inhabitants. In late 2016, GP signed an agreement with the Portuguese company that managed the nearest such reserve, Coutada 12 (C12, 2,000km²), effectively expanding the park by 50% (C12 has not yet been formally appended to GNP, but its addition is anticipated). Importantly, C12's miombo and vleis support small populations of species rare in GNP, such as zebra and leopard (Easter, Bouley, & Carter, 2020).

Moving forward, GP aims to work with the government to create community conservancies that will establish corridor connectivity between the 'core' park, Mt. Gorongosa, and C12; to extend buffer-zone designation in a broad belt between GNP and the coutadas, up to the Zambezi River. These moves may eventually result in a vast contiguous network of protected areas of varying stringency. They would also preserve space for human inhabitants and their livelihoods and create new opportunities for sustainable development.

Challenges

Large animals require large spaces. Large herbivores (and elephant in particular) do not harmonise with agriculture. Large carnivores do not harmonise with livestock. Cultural conservation ethics may be deep (Matos, Barraza, and Ruiz-mall, 2021), but are easily trumped by livelihood needs. Practices that would be sustainable on a given area for a community of 1,000 are not sustainable on the same area for a community of 100,000. Communities are not monolithic and consist of individuals with diverse interests, across a nested set of scales: household, village, district, province, nation. Power asymmetries exist at all of those scales. Mistrust is hard to dissolve, especially among those who have been marginalised by regional, national, and international political and conservation manoeuvres. Resentment of authority and restrictions is a common human trait. Capital has the power to redress past wrongs and also to perpetuate them. The tape of history cannot be rewound. The future is shaped by actions in the present, but durable change requires time, effort, and continuous proof of trust and care. These are simple facts, but they create a great deal of complexity and many challenges for GP.

Community relations

GP's mission is to navigate these challenges in a way that enables large animals and diverse ecosystems to exist in perpetuity amidst a rapidly growing populace. GP's mandate is to execute that mission in a way that adheres to Mozambican law and advances the interests of the nation as perceived by its government. The shadows of colonial dispossession, postcolonial conflict, and regional suspicion of the government all hang over the effort. Inevitably, not all stakeholders express satisfaction; there is a spectrum of attitudes both within and among communities in the buffer zone. Villages closest to GNP, which have greatest access to employment, health, and educational opportunities created by GP, also have some of the most complicated relationships with GNP, owing in part to human–wildlife conflict. Conversely, some members of more distant villages perceive that GP is not doing enough in their communities. Relationships with communities on Mt. Gorongosa are particularly delicate, and ongoing deforestation in that area is a major concern. For most people, benefits accruing at the household level (e.g., agricultural assistance and disaster relief) are more meaningful than those directed towards whole communities (e.g., schools, clinics, wells), yet GP's legal mandate is to support 'projects that benefit the entire Community and not individual actors'. GNP shares 20% of annual revenues with the communities and is legally obligated to channel these funds through community representatives, yet community-level revenue sharing is often overshadowed by individual-level perceptions that representatives misuse funds. Building and maintaining trust is a long-term process and one reason why GP needs a 30-year time horizon, but neighbourly relations require constant maintenance.

Human–wildlife conflict

Residents of villages neighbouring the park suffer frequent crop-raiding and occasional fatalities by elephant and other species, and this threat will grow as wildlife populations increase (Branco et al., 2019). Mozambican law disallows compensation schemes, and fencing the entire perimeter of the park would reduce connectivity (although smaller-scale fencing is under consideration). GP's mitigation efforts have focused on rapid response by rangers (many of whom are posted in the main conflict corridor along the Pungue River), building improved silos (also an asset against rodents), and placing beehive fences at key elephant crossing points (Branco et al., 2020).

Climate change

Models predict that Mozambique will be among the countries worst affected by climate change, with substantial reductions in total rainfall, more frequent drought, higher maximum precipitation and flood risk, and more extreme weather events (Collins et al., 2013). In 2019, one of the worst storms on record in the Southern Hemisphere—Cyclone Idai—directly hit central Mozambique, leading to severe flooding. Whereas the ecological effects on GNP were strong but non-catastrophic, the human toll was immense. GP coordinated local relief efforts, which as of 2021 was the event that people most often noted when asked about the benefits of living near GNP. Impacts of climate change on agricultural livelihoods will influence park-community relations, but in ways that are difficult to predict. Expansion of protected-area coverage may help the coupled human-natural system absorb these impacts (e.g., by buffering floods) but may also strain logistical capacity.

Funding

GP's budget grew from ~$9 million in 2017 to ~$14 million in 2019 to ~$16 million in 2021. This funding is lavish by the standards of African protected areas. Even when GNP is construed as including the entire buffer zone (where much of the budget is spent), the funding per km^2 is an order of magnitude greater than the median for Africa's protected areas (Lindsey et al., 2018). Indeed, GP's current annual budget exceeds the total state-provided protected-area funding for all but four African countries; Mozambique's total contribution in 2017 was ~$2 million (Lindsey et al., 2018). GP aims to catalyse a regional transition out of poverty and create ecologically self-regulating conditions such that GNP requires a much smaller budget. Yet fiscal self-sufficiency for GNP seems distant, with generated revenue accounting for <10% of the budget in 2019. Ecotourism has developed more slowly than initially hoped. Other enterprises such as honey, coffee, cashew, and carbon are emerging, feeding revenues into a trust dedicated to conservation and human-development programmes in Gorongosa.

Instability

Mozambique's peace process is ongoing. In 2019, leaders of the two dominant political parties convened to sign a peace accord in GNP—once the hub of the civil war and now an international symbol of Mozambique's vitality and environmental leadership. Yet instability remains a threat to GNP and Mozambique more broadly. As the national political parties have increasingly embraced peace and demilitarisation, an Islamist insurgency in Cabo Delgado province (~900km from GNP) has intensified since 2017. In 2021, this conflict spread in into Niassa

National Reserve, forcing community members and conservationists to flee. Armed groups with international ties remain a threat to the people and parks of Mozambique, as in 1977.

Conclusion

Gorongosa is proof that socio-ecological systems devastated by conflict can be revived on decadal timescales. The assisted recovery of large-herbivore and carnivore populations has reinstated top-down control and resurrected ecosystem functions—the defining objective of trophic rewilding. GP has achieved this objective through a multifaceted public–private partnership, which envisions GNP as an engine for economic development in a region where three-quarters of people live below the international poverty line. GP's multi-decadal span and broad scope, which contrast sharply with the conservation status quo, is necessary to allow adaptive evolution of targets and tactics, and to build trust needed to resolve the disputes that inevitably arise around any societal institution.

While some ecological scars of GNP's violent history will persist for generations, they will continue to lessen as long as Mozambique remains peaceful. Reintroduction of rhinoceroses, planned for 2040, would represent the healing of scars predating the civil war and would be a capping achievement of GP from a purely ecological perspective. Yet GP's success and viability as a replicable model of protected area upgrading will ultimately be judged in terms of the wellbeing of the people of Sofala Province and the handing back of GNP in a condition that allows its survival in perpetuity.

Acknowledgments

The authors' work in Gorongosa is supported by the Greg Carr and Cameron Schrier Foundations, National Geographic, and the US National Science Foundation (IOS-1656527, DEB-2225088).

References

Angier, N. (2021) How this spot (in Mozambique) got its leopard, *New York Times*, 10 Jan.

Atkins, J.L. et al. (2019) Cascading impacts of large-carnivore extirpation in an African ecosystem, *Science (New York, N.Y.)*, 364(6436): 173–7. doi: 10.1126/science.aau3561.

Becker, J. et al. (2021) Ecological and behavioral mechanisms of density-dependent habitat expansion in a recovering African ungulate population, *Ecological Monographs*, 91: e01476.

Bouley, P. et al. (2018) Post-war recovery of the African lion in response to large-scale ecosystem restoration, *Biological Conservation*. Elsevier, 227 (February), pp. 233–42. doi: 10.1016/j.biocon.2018.08.024.

Bouley, P. et al. (2021) The successful reintroduction of African wild dogs (*Lycaon pictus*) to Gorongosa National, *PLOS ONE*, 16: e0249860.doi: 10.1371/journal.pone.0249860.

Branco, P.S. et al. (2019) Determinants of elephant foraging behaviour in a coupled human-natural system: Is brown the new green?, *Journal of Animal Ecology*, 88(5): 780–92. doi: 10.1111/1365-2656.12971.

Branco, P.S. et al. (2020) An experimental test of community-based strategies for mitigating human–wildlife conflict around protected areas, *Conservation Letters*, 13, p. e12679. doi: 10.1111/conl.12679.

Campbell-Staton, S.C. et al. (2021) Ivory poaching and the rapid evolution of tusklessness in African elephants, *Science*, 374: 483–7. doi: 10.1126/science.abe7389.

Carver, S. et al. (2021) Guiding principles for rewilding, *Conservation Biology*, 35(6): 1882–93. doi: 10.1111/cobi.13730.

Collins, M. et al. (2013) Long-term climate change: Projections, commitments and irreversibility. In Stocker, T.F., D.Qin, G.-K.Plattner, M.Tignor, S.K.Allen, J.Boschung, A.Nauels, Y.Xia, V.B., and P.M.M. (ed.) *Fifth Assessment Report of the Intergovernmental Panel on Climate Change*. Cambridge University Press, pp. 1029–1136. doi: 10.1017/CBO9781107415324.024.

Daskin, J.H. and Pringle, R.M. (2018) Warfare and wildlife declines in Africa's protected areas, *Nature*. Nature Publishing Group, 553(7688): 328–32. doi: 10.1038/nature25194.

Daskin, J.H., Stalmans, M., and Pringle, R.M. (2016) Ecological legacies of civil war: 35-year increase in savanna tree cover following wholesale large-mammal declines, *Journal of Ecology*, 104(1): 79–89. doi: 10.1111/1365-2745.12483.

Dutton, P. (1994) A dream becomes a nightmare, *African Wildlife*, 48: 6–14.

Easter, T., Bouley, P., and Carter, N. (2020) Intraguild dynamics of understudied carnivores in a human-altered landscape, *Ecology and Evolution*, 10(12): 5476–88. doi: 10.1002/ece3.6290.

Gorongosa National Park (2020) Parque Nacional da Gorongosa 2020 Annual Report, https://gorongosa.org/wp-content/uploads/2020/12/12-10-2020-Eng-Highlights-document-reduced-size.pdf.

Guyton, J.A. et al. (2020) Trophic rewilding revives biotic resistance to shrub invasion, *Nature Ecology and Evolution*. Springer US, 4(5): 712–24. doi: 10.1038/s41559-019-1068-y.

Hanson, T. et al. (2009) Warfare in biodiversity hotspots, *Conservation Biology*, 23(3): 578–87. doi: 10.1111/j.1523-1739.2009.01166.x.

Lindsey, P.A. et al. (2018) More than $1 billion needed annually to secure Africa's protected areas with lions, *Proceedings of the National Academy of Sciences of the United States of America*, 115(45): E10788–E10796. doi: 10.1073/pnas.1805048115.

Mascia, M.B. and Pailler, S. (2011) Protected area downgrading, downsizing, and degazettement (PADDD) and its conservation implications, *Conservation Letters*, 4(1): 9–20. doi: 10.1111/j.1755-263X.2010.00147.x.

Matos, A., Barraza, L., and Ruiz-mall, I. (2021) Linking conservation, community knowledge, and adaptation to extreme climatic events: a case study in Gorongosa National Park, *Sustainability*, 13: 6478.

Mendiratta, U. et al. (2021) Mammal and bird species ranges overlap with armed conflicts and associated conservation threats, *Conservation Letters* (December 2020): 1–8. doi: 10.1111/conl.12815.

Morley, R. and Convery, I. (2014) Restoring Gorongosa: some personal reflections. In Convery, I., Corsane, G., and Davis, P. (eds), *Displaced Heritage: Responses to Disaster, Trauma, and Loss*. Boydell Press, pp. 129–42.

Ogutu, J.O. et al. (2012) Dynamics of ungulates in relation to climatic and land use changes in an insularized African savanna ecosystem, *Biodiversity and Conservation*, 21(4): 1033–53. doi: 10.1007/s10531-012-0239-9.

Pansu, J. et al. (2019) Trophic ecology of large herbivores in a reassembling African ecosystem, *Journal of Ecology*, 107(3): 1355–76. doi: 10.1111/1365-2745.13113.

Pansu, J. et al. (2022) The generality of cryptic dietary niche differences in diverse large-herbivore assemblages, Proceedings of the National Academy of Sciences of the United States of America, 119(35): e2204400119. doi: 10.1073/pnas.2204400119.

Potter, A.B. et al. (2022) Mechanisms of dietary resource partitioning in large-herbivore assemblages: A plant-trait-based approach, Journal of Ecology, 110(4): 817–832. doi: 10.1111/1365-2745.13843.

Pringle, R.M. (2017) Upgrading protected areas to conserve wild biodiversity, *Nature*, 546(7656): 91–9. doi: 10.1038/nature22902.

Pringle, R.M. and Hutchinson, M.C. (2020) Resolving food-web structure, *Annual Review of Ecology, Evolution, and Systematics*, 51: 55–80.

Stalmans, M.E. et al. (2019) War-induced collapse and asymmetric recovery of large-mammal populations in Gorongosa National Park, Mozambique, *PLOS ONE*, 14: e0212864. doi: 10.1371/journal.pone.0212864.

Tinley, K. (1977) *Framework of the Gorongosa ecosystem*. D.Sc. Thesis, University of Pretoria, South Africa.

Torres, A. et al. (2018) Measuring rewilding progress, *Philosophical Transactions of the Royal Society B: Biological Sciences*, 373(1761): 20170433. doi: 10.1098/rstb.2017.0433.

Van Wilgen, B.W. et al. (2004) Response of savanna fire regimes to changing fire-management policies in a large African National Park, *Conservation Biology*, 18(6): 1533–40. doi: 10.1111/j.1523-1739.2004.00362.x.

18

REWILDING CASE STUDY

Restoring Western Australia's rangelands: Mutawa/Kurrara Kurrara

Ian Kealley and Neil Burrows

Introduction

Over the last 200 years, some 50% of the world's mammal extinctions have occurred in Australia. Of the 72 species of mammals (excluding bats) known to originally occupy the Australian arid zone, 11 are now extinct, five have disappeared from the mainland and are found only on offshore islands and 15 are now restricted in their range, most becoming absent from the arid zone and persisting in mesic habitats (Short & Smith, 1994; Woinarski et al., 2015). Most extinctions and declines have occurred in the medium size group of mammals, the so called 'critical weight range' mammals (35–5500g), with many ground dwelling bird and reptile species also affected. Altered fire regimes, predation by introduced predators (feral cats and foxes) and habitat degradation by introduced herbivores have been implicated in these declines (Burbidge & McKenzie, 1989).

Much of the Australian arid zone is characterised by the dominance of spinifex, a perennial hummock grass primarily of the genus *Triodia*. Spinifex grasslands occupy about 2.1 million km^2 (about 26%) of the Australian continent. They occur over a diversity of landforms, including sand plains and dune fields, stony plains and rocky hills, in remote and sparsely populated semi-arid and arid regions of Australia. Typical of desert climates, rainfall is highly variable and long periods of drought are common. The summers are long and hot, winters cool and mild. Mature spinifex grasslands are generally structurally simple, with a discontinuous (35–60%) ground cover of *Trioda spp.* (spinifex) hummocks to a height of 50cm. A variety of scattered low shrubs and trees usually grow in association with spinifex. The combination of flammable vegetation and often extreme fire weather make spinifex grasslands highly flammable. Historically, lightning and deliberate burning by Aboriginal people were the main causes of fire (Burrows et al., 2009). Today, most fires are started by lightning, although human-caused ignitions are significant near settlements, along vehicle travel routes and in association with land management practices on Aboriginal and conservation lands. While most *Triodia* species are readily killed by fire, spinifex grassland ecosystems are fire maintained; fires at appropriate temporal and spatial scales are essential for their persistence and health.

Under traditional law and custom, Aboriginal people inherit, exercise and bequeath customary responsibilities to manage their traditional country. The relatively recent displacement

DOI: 10.4324/9781003097822-21

of Aboriginal people from much of central Australia, has coincided with an alarming decline in some native mammal and bird species, and a contraction of some fire sensitive plant communities. Proposed causes of these changes include altered fire regime due to the interruption of traditional Aboriginal burning practices following European colonisation, predation by introduced predators and competition with feral herbivores (Burbidge & McKenzie, 1989). There is evidence of dramatically altered fire regimes in spinifex deserts since the displacement of Aboriginal people, who used fire deliberately and skilfully for a myriad of reasons, but primarily to acquire food. Over millennia, their regular burning, under mostly mild conditions, maintained a fine-grained fire mosaic of patches of vegetation at different seral stages, which restricted the extent of fire and provided habitat diversity. Within 15–20 years of the departure of Aboriginal people and their traditional burning practices, the fire regime in Australian deserts changed to one of very large, high intensity fires, resulting in seral homogeneity (Burrows et al., 2009).

Introduced predators, including the feral cat (*Felis catus*) and the red fox (*Vulpes vulpes*), occur throughout the arid zone. Feral cats were introduced to the Nullarbor coastline by whalers and sealers in the 1830s, and in 1898, hundreds of cats were released near Eucla in an attempt to control introduced rabbits (*Oryctolagus cuniculus*). Cats are well adapted to arid environments and by the early 1900s, they had colonised most of the interior. They do not need to drink water, obtaining moisture from their prey, and they have a wide range of prey species including small mammals, reptiles, birds, and insects. Cats have direct impacts on native fauna through predation but can have an indirect effect by carrying and transmitting diseases such as toxoplasmosis. They have been implicated in the extinction of up to seven species of mainland mammals as well as regional and island extinctions of native mammals and birds. It is well documented that feral cats have caused the failure of many endangered fauna reintroduction (rewilding) attempts in the semi-arid and arid zones and are recognised as the main barrier to reconstructing and protecting faunal assemblages in these environments (Commonwealth of Australia, 2015). The fox was introduced to southeast Australia from England in the 1860s and by the 1930s it had spread across the continent except for the far north. In arid central Australia, fox density is generally low but is temporally and spatially variable. Following good seasons, fox density increases as prey increases, and density is usually higher around playa lakes, riparian zones, breakaways, and other more productive systems in the landscape that provide refugial habitat for rabbits and other prey species. Foxes are considered a threat to 14 species of birds, 48 mammals, 12 reptiles, and 2 amphibians (DEWHA, 2008).

Feral cats and foxes are listed as a key threatening process under the Commonwealth Environment Protection and Biodiversity Conservation Act 1999. The main method of broadscale control is baiting using toxins such as sodium fluoroacetate known as 1080. A cat specific bait, Eradicat®, was developed by DBCA scientists (Algar and Burrows, 2004) and has proven to be efficacious in most, but not all circumstances. Similarly, baiting has been mostly effective at controlling fox populations.

A significant proportion of the Australian arid zone is rangelands subject to grazing (pastoral) leases, with established infrastructure including fences and artificial watering points, the latter enabling much higher densities of native and introduced herbivores than would otherwise be the case. Grazing by introduced herbivores, including domestic stock and feral animals, can cause significant damage to the vegetation and the soil, degrading or destroying native fauna habitat. The extent of environmental damage is proportional to total grazing pressure, or the density of herbivores.

Biophysical description

Rewilding (fauna restoration) is being carried out in the arid rangelands of Western Australia on the former Lorna Glen pastoral lease (now Matuwa) (O'Leary & Kealley, 2016). Comprising some 244,000 hectares, the property straddles the Gascoyne and Murchison Interim Biogeographic Regionalisation for Australia (IBRA) regions (Figure 18.1) and was chosen for this rewilding project because it has the following attributes:

- It is a large area typical of arid zone rangelands ecosystems from which medium size native mammals have declined.
- It contains diverse landform systems and associated diversity of habitats representative of much of the Murchison–Gascoyne rangelands.
- Despite a history of grazing, the vegetation is mostly in good condition with good diversity of plants, reptiles and small mammals.
- There is a good knowledge base including landform system maps, extensive biological survey, sub-fossil and other evidence of mammals that once occurred in the area, fire history, and fire ecology.
- It is accessible and has good infrastructure including buildings for accommodation and storage, and an airstrip.

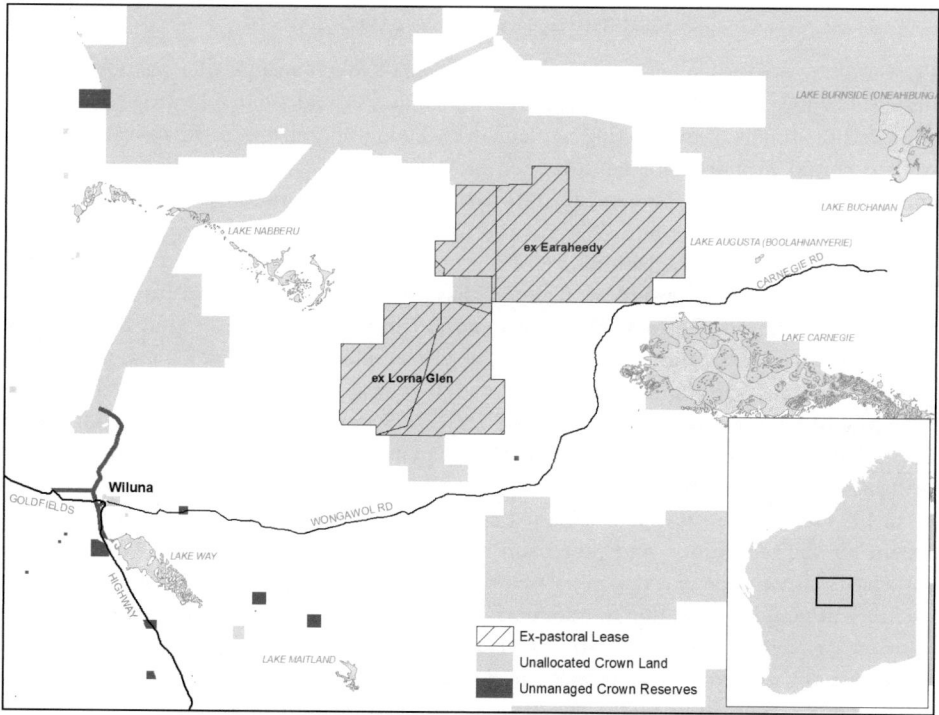

Figure 18.1   Location of the Rangelands Restoration rewilding project in the Western Australian rangelands. © Government of Western Australia, used with permission.

Matuwa is located within the remote central rangelands of Western Australia, towards the eastern edge of pastoral enterprises and close to the geographic centre of the state. A former pastoral lease (Kurrara Kurrara) adjoins Matuwa along the northern boundary and is also managed as an IPA for cultural and conservation purposes. The remaining surrounding land use comprises active pastoral leases (stations or ranches) running cattle. The climate for the area is classified as 'desert', typified by highly irregular, low rainfall with the annual average being about 250mm per year. Most rain falls over the summer months from tropical low-pressure systems and localised thunderstorms. Winters are cool, with the mean daily maximum temperature for July being about 21°C; overnight winter temperatures can fall below 0°C and frosts are not uncommon. Summers are hot with the mean daily temperature for January being about 38°C. The south west of Matuwa is predominantly extensive red sand plains with scattered dunes. Spinifex (*Triodia basedoweii and T. melvillei*), a perennial hummock grass, forms the dominant ground cover, often growing in association with a variety of scattered shrubs and low trees such as mulga (*Acacia aneura*) or eucalypts (*Eucalyptus gongylocarpa, E. kingsmillii*). The north east of the property is characterised by a diversity of landforms including spinifex-dominated red sand plains, calcrete, claypans and ephemeral salt and freshwater claypans, breakaways, low stony hills and broad outwash plains supporting a rich diversity of low scattered woody shrubs, herbs, annuals, and small trees including mulga and gidgee (*Acacia pruinocarpa*). Relatively fine-scale patterning of diverse soils and landforms supports a great diversity of flora and fauna. Typical of the arid region, most medium size mammals have been extirpated from the property in the last 100 years.

Social-cultural-economic background

Prior to European settlement the Matuwa and Kurrara Kurrara area (MKK) was occupied and managed by the Martu Aboriginal people for millennia. Around 1900, following early exploration and then mineral prospecting, settlement by taking up grazing (pastoral) leases progressively occurred with development involving all required infrastructure, buildings, waterpoints, and fences. Grazing enterprises initially involved sheep and wool industries that progressed into mainly cattle by the 1980s and '90s. Aboriginal people continued with traditional activities, often while working on the pastoral stations.

The decline of the sheep and wool pastoral businesses from the 1980s due to economic and environmental reasons led to government inquires and reviews culminating in the Gascoyne Murchison Strategy to rationalise aspects of the pastoral industry and develop a more comprehensive, adequate, and representative conservation reserve system across the rangelands. The Gascoyne Murchison Strategy and National Reserve System funded the purchase in 1999 of Earaheedy (Kurrara Kurrara) and in 2000 of Lorna Glen (Matuwa) pastoral stations as areas to be managed for conservation.

In 1993 the Native Title Act was passed by the Australian Parliament, 'to provide a national system for the recognition and protection of native title and for its co-existence with the national land management system'. Over time this led to granting of exclusive and non-exclusive possession native title to traditional owners, and the development of joint land management arrangements with the state conservation authority.

As part of native title developments and joint management requirements, to achieve management for conservation under the National Reserve System programme, the system of Indigenous Protected Areas (IPA) was introduced in 1997 to formally recognise and fund indigenous land management for conservation and cultural purposes.

The Martu Aboriginal people, traditional owners of the MKK area, mainly based in the nearest town of Wiluna (160km away), were keen to return to Country and be actively involved in land and cultural site management through engagement, field trips, cooperative management, joint management, employment and ranger programmes. This was achieved through a range of programmes and activities from 2003 while the native title claim was progressed up to granting of native title in 2013.

Rangelands restoration objectives

Project objectives and methodologies have evolved and adapted over time, reflecting changing land tenure, land management, and associated changing aspirations, and new knowledge gained along the way. Soon after the pastoral lease was sold back to the Western Australian government, the land reverted to 'unallocated Crown land' under the interim management of the then Department of Conservation and Land Management (CALM). The primary goals at this time (2001) were to:

- Improve the condition (cover, species diversity) of the native vegetation following some 70 years of grazing by domestic stock, large feral herbivores, and artificially high densities of native herbivores (primarily kangaroos).
- Protect extant fauna and reconstruct, as far as practicable, the original suite of medium size native mammal fauna now locally extinct.
- Progress the area to management for conservation and cultural values.

In 2004, a Memorandum of Understanding (MoU) between the Department of Environment and Conservation (DEC—formerly CALM) and the Wiluna native title claimants (Wiluna Martu people) was the formal beginning of co-management. The 2013 determination of exclusive possession native title to Martu over MKK, effectively returned ownership of the land to Martu. It gave native title holders full private property rights, including possession, occupation, use, and enjoyment of the areas, and control over access to and activities carried out on the areas. The declaration of the area as an IPA in 2015 and the implementation of the IPA plan enabled Martu to control the management of these areas, whilst retaining partnership arrangements with the Department of Parks and Wildlife (formerly DEC), and later, the DBCA.

Throughout these changes, the biodiversity and conservation objectives of the Rangelands Restoration Project remained largely much unchanged, with research and management activities coordinated and endorsed by the MKK planning group. The determination of native title and creation of the IPA, provided greater incentive and opportunity to increase the level of engagement of Martu in Rangelands Restoration. In addition to the conservation objectives, for Martu, as outlined in the Tarlka Matuwa Piarku Aboriginal Corporation 2015 IPA Country Management Plan for the area, the MKK 'landscapes' have 'distinctive historical, cultural and ecological associations that are linked to specific outcomes that can provide economic, educational, health and wellbeing and cultural benefits to people and country. Martu view country holistically and see the protection of natural values supporting cultural, educational and economic activities in the IPA area' (TMPAC, 2015).

Managing threatening processes

To prevent further declines in biodiversity, and to reconstruct the original suite of mammals, it was first necessary to mitigate processes that caused their demise. In this region, these processes are primarily habitat damage and degradation by introduced herbivores, predation by introduced predators, and altered fire regimes.

Introduced large herbivores

Artificial waters (bores, wells, dams) were progressively decommissioned over three years and the property was simultaneously de-stocked by 'mustering' (gathering up). Other large feral herbivores, such as horses and camels, were eradicated or controlled by aerial shooting. The entire perimeter of Matuwa, some 225km, was fenced to prevent re-invasion by large herbivores. The progressive decommissioning of artificial water points, in conjunction with commercial culling, reduced the density of native herbivores, such as kangaroos, to more 'natural' levels.

Introduced predators

Feral cats have colonised the vast and remote semi-arid and arid regions of Australia and are a major barrier to rewilding projects (Figure 18.2). They are the most abundant of the introduced predators and are difficult to control, unlike foxes and wild dogs / dingoes, which are at much lower densities and easier to control. This, together with their evasive and largely nocturnal behaviour, and their popularity as a companion animal, makes eradication of this harmful pest unachievable on the mainland. Most common control methods include trapping, poisoning and fenced exclosures. Biological control including immunocontraception, gene drive technology[1] and the introduction of infectious diseases is not currently feasible. Well-designed cat-proof fenced exclosures can be very effective at protecting native animals from predation. Construction and maintenance costs limit the number and size of exclosures that can be built, but where they exist in the arid zone, they have been successful.

Baiting using the toxin sodium fluoroacetate, synthetically produced under the name 1080, has shown most promise as a method for broadscale control of feral cats and other introduced predators (foxes and wild dogs). Being a naturally occurring toxin found in some native plant

Figure 18.2   Feral cats are a major obstacle to rewilding in Australia's arid zone.

species, native animals have developed a high tolerance to it. Introduced species are highly susceptible, with very small doses being lethal. Following the development of a cat bait (Eradicat®), broadscale aerial baiting is now applied in targeted areas of the semi-arid and arid regions of Western Australia.

Baiting with Eradicat® is also lethal to foxes and wild dogs / dingoes should they ingest the baits, but the main target of baiting is feral cats. Baiting aims to significantly reduce the feral cat population and maintain it at low levels, given that eradication in vast open landscapes is unachievable. Feral cats prefer live prey, and when this is abundant, they are reluctant to consume baits, so baiting is less effective. However, they scavenge if they are hungry and the carrion is 'fresh'. There are periods when live prey abundance is low, imposing famine pressure on feral cats. Shortage of prey usually coincides with cold, dry weather, and trials have demonstrated that this is the optimal time to deliver cat baits. At Matuwa, aerial baiting using Eradicat® commenced in 2003 and occurs annually in winter, when 50 baits/km^2 (4.5mg 1080/bait) are delivered from a fixed wing aircraft. Prior to, and several weeks after baiting, introduced predator density surveys are carried out by counting footprints, and more recently, using trail cameras to assess the effectiveness of baiting. Baiting efficacy (knock-down of the pre-bait population) has ranged from 25% to almost 90% (Lohr & Algar, 2020; Figure 18.3).

Although labour intensive, trapping has been an effective method for localised control and 'mop-up' of feral cats provided the area is reasonably accessible by vehicle, there are no significant non-target issues requiring special or complex trap setting methods (Lohr & Algar, 2020). Trapping involves the use of soft-jaw leg-hold traps set in the ground (or elevated if there are non-target issues) and using scent and/or sound lures.

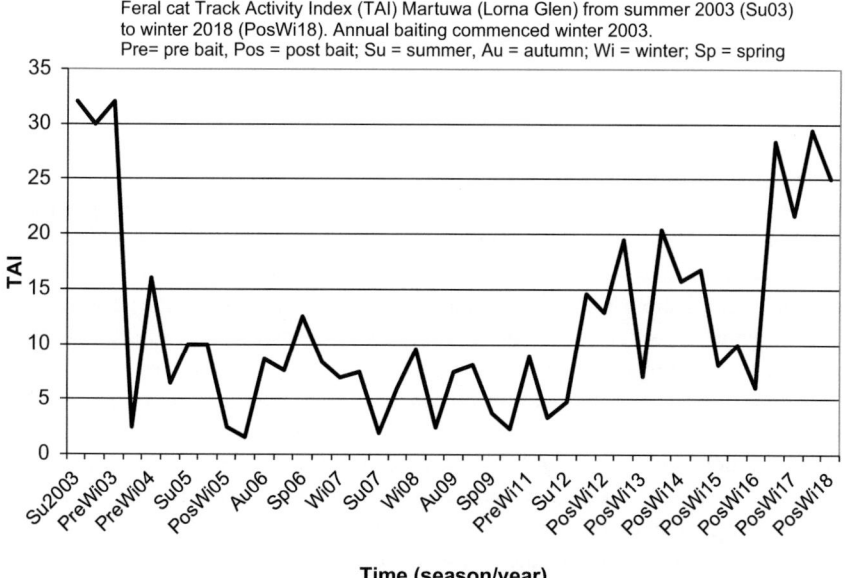

Time (season/year)

Figure 18.3   Variation in feral cat activity track activity index (TAI) based on footprints. Annual baiting with Eradicat® commenced winter 2003 (W03) and was very effective until about 2012. Integrating trapping with baiting can be effective.

Figure 18.4 Traditional Owners (Martu) participating in prescribed burning to mitigate the impacts of bushfires and to create habitat diversity. (Photo Ryan Butler, DBCA).

Fire

Desert Aboriginal people domesticated bushfire many thousands of years ago. Fire has been, and in some areas, continues to be, integral to their physical and spiritual wellbeing. Consequently, the flammable spinifex landscapes were a mosaic of relatively small patches of vegetation at different stages of post-fire development (seral stages) but with most of the landscape in the younger seral stages. This fire-induced diversity provided habitat diversity and the relatively small size of the burnt patches, mostly, <100ha, meant it was efficient for humans (and non-humans) to move through these landscapes to acquire the resources needed for survival.

With the arrival of Europeans, many desert Aboriginal people were displaced or forcibly removed from their homelands. The depopulation of the deserts had many unintended consequences, and among these, was a change in the fire regime. In the absence of regular burning by people, heavy and continuous spinifex fuels developed over vast areas of the desert and, ignited by lightning, high intensity megafires burned great swathes each year. Within 15–20 years of the cessation of traditional burning, the size and intensity of the feral bushfires increased by several orders of magnitude, placing strain on the native wildlife. Some fire sensitive plant communities declined and the combination of mega-fires and introduced predators caused the decline and, in some cases, extinction of medium size mammals and ground dwelling birds. Knowledge of traditional desert Aboriginal burning patterns, together with western science and technology, underpins fire management of Matuwa.

Rangelands restoration—key actions and outcomes

Initially, in 2002, the Rangelands Restoration project had the ambitious goal of re-establishing sustainable populations of 11 arid zone mammal species, including two native rodents, by 2020

Table 18.1 Some mammal species that were known to occur in the central rangelands of Western Australia but which have been extirpated since European settlement. These species, which are extant elsewhere, are candidates for rewilding

Species (Family)	Common name	Conservation status	Source
Macrotis lagotis (**Thylacomyidae**)	Bilby	Vulnerable, locally extinct	Captive bred and from the wild
Trichosaurus vulpecula (**Phalangeridae**)	Brushtail possum	Not threatened, but presumed extinct in the arid zone.	From the wild, south-west Australia
Isoodon auratus auratus (**Peramelidae**)	Golden bandicoot	Vulnerable, presumed extinct in the arid zone	From the wild, island populations, north-west Australia
Lagorchestes hirstus (**Macropodinae**)	Mala	Critically endangered, extinct on the mainland, two island populations.	From the wild, island
Bettongia lesueur (**Potoroidae**)	Boodie	Vulnerable, extinct on the mainland except for 'behind fence populations.	From the wild, island populations north-west Australia.
Myrmecobius fasciatus (**Myrmecobiidae**)	Numbat	Vulnerable, formerly widespread, but now extinct in the arid zone.	Captive bred population.
Perameles bougainville (**Peramelinae**)	Western barred bandicoot	Endangered, extinct on the mainland	From the wild, island populations north west Australia
Pseudomys fieldi (**Muridae**)	Shark Bay mouse	Vulnerable, extinct on the mainland	From the wild, island populations, north-west Australia.
Rattus tunneyi (**Muridae**)	Pale field-rat	Not threatened but severe range contraction.	From the wild.
Dasyurus geoffroii (**Dasyuridae**)	Chuditch	Vulnerable, formerly widespread but now extinct from the arid zone.	From the wild, south-west Australia.

(Table 18.1). These species were chosen because they were known to have occurred in the region from museum records and sub-fossil evidence (Baynes, 2006), and there was suitable habitat on Matuwa. Animals were to be sourced from extant populations in the wild, or from captive bred populations, and flown to Matuwa for release.

The bilby (ground dwelling omnivorous marsupial) was the first species to be re-introduced, with 128 animals being released into the wild at Matuwa over the period 2007–2009. Bilbies were monitored by radio tracking, trapping, analysing DNA extracted from scats, and by observers on foot, all-terrain vehicles, and on horseback, who searched for signs of bilby presence such as scats, digs and burrows. Of the 54 animals fitted with radio collars, over the period 2007–2009, 30 died. Causes of death were feral cat predation (40%); suspected starvation (26%), unknown cause (27%), raptor predation (7%) (Pertuisel, 2010). Despite the high initial mortality rate, the bilbies were breeding, with 75% of adult females trapped over the period 2007–2010 carrying pouched young. Because of their solitary and somewhat nomadic lifestyle (large home range), bilbies were difficult to trap and monitor, especially when the radio transmitter batteries eventually failed. Ongoing monitoring by scat DNA analysis and by observers on

Figure 18.5   Bilby (*Macrotis lagotis*) being released as part of the rewilding project on Matuwa. (Photo Judy Dunlop DBCA).

horseback demonstrated that after the initial high mortality rate, the bilby population steadily increased and dispersed, and by 2014 their population was estimated at 500–600 animals. Over the period 2007–2009, brushtail possums (arboreal herbivorous marsupial) were released into specific habitats with tree bearing hollows. The population initially declined but appears to be persisting. In 2008, mala (marsupial macropod) were released into the wild, but these animals were predated by feral cats within a relatively short time.

While baiting had reduced the population of feral cats, there remained a sufficiently high density to cause concern about the success of future fauna re-introductions into the wild. A decision was made to construct an 1100ha predator-proof fenced compound to acclimatise founder animals, to provide locally bred animals for future releases and to protect species considered to be highly vulnerable to predation. The fenced compound was insurance against an inability to establish free ranging animals due to an inability to adequately control feral cats and other animals. Over the period 2010–2012, boodies, golden bandicoots, mala and Shark Bay mice were released into the compound. Regular trapping and spotlighting demonstrated that in the absence of introduced predators, populations of these species increased, with the exception of Shark Bay mice, which failed to become established because they were likely predated by mulgara (*Dasycercus blythi*), a native carnivorous marsupial. A trial release of boodies from the compound into the wild, failed as the animals were predated by dingoes. Unlike bilbies, boodies live in colonies in burrows and once discovered by predators, the entire colony was quickly predated. In 2012 and 2015, 93 golden bandicoots were released from the compound into the wild. They suffered heavy predation by feral cats and by 2017, only one animal was trapped (Morris et al., 2016 and personal communication.

The boodie population inside the fenced compound has steadily increased, the population estimated to be 452 in 2014. Outside the compound and prior to the commencement of Rangelands Restoration, baseline data about the diversity and density of extant small

Figure 18.6 The 1100ha compound built on Matuwa to protect vulnerable native mammals from predation by feral cats, foxes, and wild dogs.

mammals, reptiles and frogs were obtained from a network of pitfall trap sites installed in the major habitat types across Matuwa. Regular monitoring at these sites was maintained until 2016, when, for logistical and resourcing reasons, they were abandoned. Larger extant species such as mulgara were monitored by trapping and by recording signs of their activity such as footprints and burrows. Notwithstanding the influence of rainfall, the monitoring programme showed a significant increase in the numbers of many extant fauna species, such as mulgara, as a consequence of proactive management, especially introduced predator control (Chapman & Burrows, 2015).

Lessons learnt and future challenges

Of the six native mammals re-introduced to Matuwa, two were successfully established in the wild and four were successfully established in a predator proof compound. Despite considerable control effort, feral cats remain a barrier to rewilding in arid Australia. A key lesson learnt after almost 20 years of operation of Rangelands Restoration is that once natural ecosystems in the Australian arid zone have been degraded, and species have been lost, restoration, and rewilding if even possible, is difficult, time consuming, and expensive. It requires a long-term commitment to proactive interventionist management to control or eliminate threatening processes such as introduced predators and herbivores, and altered fire regimes. Some predator sensitive native birds and mammals cannot be re-established in the wild unless introduced predators such as feral cats are virtually eradicated. Where eradication is unachievable, fenced predator exclosures are an important 'insurance policy' while on-going research and management seeks sustainable solutions. Other species less vulnerable to predation, such as bilbies, possums, and mulgara, can be re-established in the wild if introduced predators are maintained at low densities.

Another key lesson is that large, complex rewilding projects require strong leadership and a dedicated, determined team of people with diverse expertise and knowledge. For this project, the team included applied scientists, conservation practitioners, and importantly, traditional owners. Secure funding from government and the private sector is crucial for the success of rewilding projects. It is also important to constructively communicate with adjoining landholders/neighbours so that any boundary issues can be resolved amicably.

Consistent with adaptive management, monitoring key native fauna populations, and threatening processes, is crucial to evaluating management success, for learning, and for accountability to funding bodies. Monitoring protocols should not be too complex and costly, otherwise they will not endure, but they need to be able to adequately quantify the effectiveness of management. In Australian arid zones, the involvement of Indigenous rangers is a crucial part of ensuring the successful progress of joint management with the traditional owners of Country where rewilding is planned.

Acknowledgments

The Rangelands Restoration project is a collaboration between the Western Australian government (Department of Biodiversity, Conservation and Attractions) and the landowners, the Martu people. Funding for this project was provided by the Western Australian Government and Chevron (as a mining off-set).

Note

1 Gene drives are created using the CRISPR gene editing technique and are designed to spread deleterious genetic effects (e.g. only producing male offspring) throughout a population by reproduction as a tool for population management or eradication (Webber, B.L., Raghu, S., and Edwards, O.R., 2015).

References

Algar, D. and Burrows, N. (2004) Feral cat control research: Western Shield review –February 2003. *Conservation Science W.A.* 5: 131–63.

Baynes, A. (2006) Preliminary assessment of the original mammal fauna of Lorna Glen Station. Department of Environment and Conservation, Perth, unpublished report.

Burbidge, A.A. and McKenzie, N. (1989) Patterns in the modern decline of Western Australia's vertebrate fauna: causes and conservation implications. *Biological Conservation*, 50: 143–98.

Burrows, N., Burbidge, A., Fuller, P., and Behn, G. (2006) Evidence of altered fire regimes in the Western Desert region of Australia. *Conservation Science Western Australia*, 5; 272–84.

Chapman, T. and Burrows, N. (2015) Lorna Glen (Matuwa) small vertebrate fauna monitoring program 2002-2010. Preliminary analysis and review. Department of Parks and Wildlife, Perth, Western Australia.

Commonwealth of Australia (2015). Feral cat threat abatement plan. www.environment.gov.au/biodiversity/threatened/tap-approved.html

Department of the Environment, Water, Heritage and the Arts (DEWHA) (2008) Background document for the threat abatement plan for predation by the European red fox, DEWHA, Canberra. www.environment.gov.au/biodiversity/threatened/tap-approved.html

Lohr, C. and Algar, D. (2020) Managing feral cats through an adaptive framework in an arid landscape. *Science of the Total Environment*, 720; 137631. https://doi.org/10.1016/j.scitotenv.2020.137631

Morris, K., Orell, P., Cowan, M., and Broun, G. (2016) Reconstructing the mammal fauna of Lorna Glen in the Rangelands of Western Australia 2006–2016. Department of Environment and Conservation Perth, unpublished report.

O'Leary, E. and Kealley, I. (2016) A sacred partnership. Managing Matuwa and Kurrara Kurrara. *Landscope*. Department of Environment and Conservation, Perth. Western Australia.

Pertuisel, L. (2010) Modelling the reintroduction of bilbies (Macrotis lagotis) in the rangelands of Western Australia. Department of Environment and Conservation, Perth, Western Australia.

Short, J, and Smith, A. (1994) Mammal decline and recovery in Australia. *Journal of Mammalogy*, 75; 288–97.

TMPAC (2015). Matuwa and Kurrara Kurrara Indigenous Protected Area Country Management Plan 2015-2020. Prepared by Tarlka Matuwa Piarku Aboriginal Corporation and Central Desert Native Title Services, Wiluna WA.

Webber, B.L., Raghu, S., and Edwards, O.R., 2015. Opinion: Is CRISPR-based gene drive a bio-control silver bullet or global conservation threat? *Proceedings of the National Academy of Sciences*, *112*(34): 10565–7.

Woinarski, J., Burbidge, A., and Harrison, P. (2015) Ongoing unravelling of a continental fauna: Decline, and extinction of Australian mammals since European settlement. *PNAS* 112; 4531–40.

19

REWILDING CASE STUDY

Forest restoration: Conservation outcomes and lessons from Terai Arc Landscape, Nepal

Ananta Ram Bhandari and Shiv Raj Bhatta

Introduction: Terai Arc Landscape

The realisation that biodiversity and the integrity of ecosystems cannot be conserved by protected areas alone has led to the concept of rewilding and large landscape approaches to conservation, especially of megafauna and wide-ranging species such as large carnivores. Rewilding has been developed around the three Cs model of the protection of Core areas, establishing and maintaining Connectivity and maintaining or re-establishing healthy populations of Carnivores (or more generally keystone species). The Terai Arc Landscape programme addresses all of the three Cs through a landscape approach that includes protected areas, increasing connectivity between protected areas through forest restoration and increasing the population of keystone species such as tigers, rhinos, and elephants by improving habitat and facilitating their movements. In addition, it highlights and tackles issues of co-existence between humans and wildlife in this densely populated area.

Forests became heavily exploited in the Terai region and a contiguous expanse of dense forests has been fragmented over time. After eradication of malaria in the 1960s, people started migrating from the hills to the Terai, attracted by fertile plains and accessibility. This resulted in declining forest cover in the Terai, reaching annual deforestation rates of 1.3% per annum between 1979 and 1991 (FRA/DFRS, 2014). Rapid forest loss and fragmentation forced megafauna, including tigers, to live in isolated patches and resulted in declining populations. In order to create contiguous habitats connecting natural ecosystems, Nepal declared the area the Terai Arc Landscape in 2001 (MoFSC, 2015).

The transboundary Terai Arc Landscape extends from the Bagmati River, Nepal to the Yamuna River, India covering 5.1 million hectares (MoFSC, 2015). In Nepal, the Terai Arc Landscape extends from the Bagmati River to the Mahakali River covering 2.4 million hectares (Figure 19.1). The Terai Arc Landscape harbours globally important biodiversity as it comprises two global ecoregions viz. the Terai-Duar Savanna and Grasslands, and the Himalayan Sub-tropical Broadleaf Forests (MoFSC, 2004). The Terai Arc Landscape supports one of the most spectacular assemblages of megafauna in Asia, including the Bengal tiger, Asiatic elephant, and Greater one-horned rhinoceros. These are all important

DOI: 10.4324/9781003097822-22

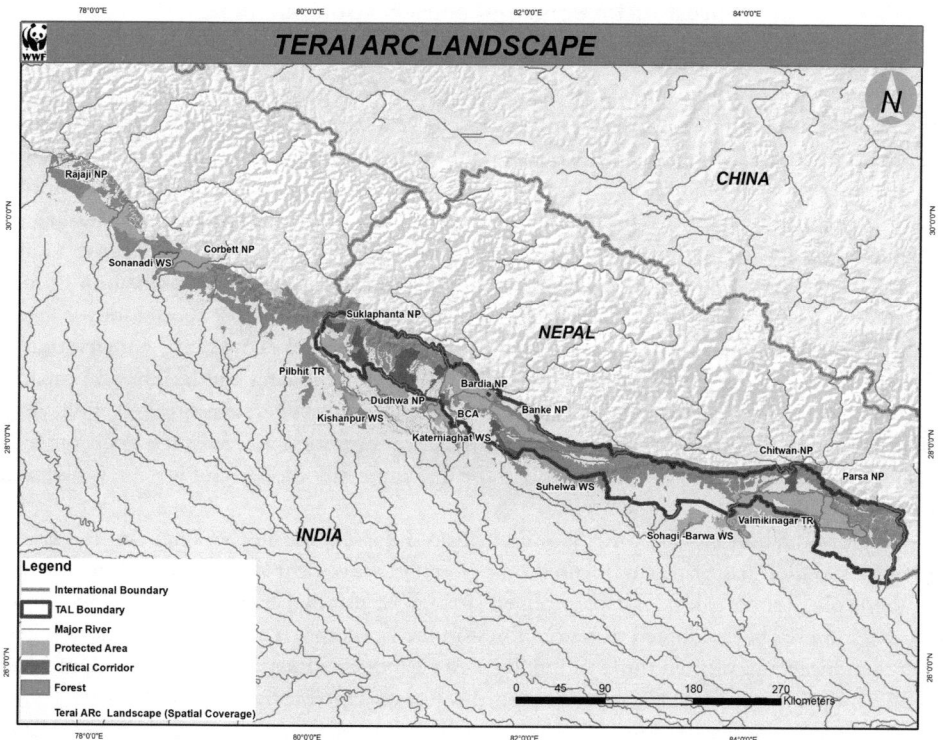

Figure 19.1   Terai Arc Landscape © WWF Nepal.

keystone species, critical components of the ecosystems that they live in. Tropical and sub-tropical broadleaved forests, riverine forests, grasslands, and floodplains of the landscape, represented in six protected areas, three forest conservation areas, and three Ramsar sites, provide habitats for 565 species of birds, 125 fish, 85 mammals, and 47 reptiles (MoFSC, 2015). Moreover, the Terai Arc Landscape has a high human population of over 7.5 million with a rich socio-cultural diversity of indigenous people and ethnic communities (MoFSC, 2015). In recent years, the region has undergone rapid socio-economic and environmental changes characterised by increasing urbanisation, infrastructure development, and in-migration (MoFE, 2018). Therefore addressing the problems of co-existence between humans and wildlife is crucial.

A ten-year strategic plan was developed by the Government of Nepal in 2004 and revised in 2015 envisioning the Terai Arc Landscape as 'a globally unique landscape where biodiversity is conserved, ecological integrity is safeguarded, and sustainable livelihoods of its people are secured' (MoFSC, 2015). The landscape covers various land-uses such as protected areas, forests, wetlands, agricultural lands, and settlements. The landscape strategic plan was developed recognising and complementing the legislative systems of various land uses and engaging multiple stakeholders including local communities, community-based organisations, civil society organisations, and government agencies. WWF has been partnering with the Government of Nepal to design the Terai Arc Landscape, to develop its strategic plans, and to implement conservation strategies on the ground.

Forest restoration and conservation outcomes

In partnership with the Government of Nepal, WWF initiated the Terai Arc Landscape Program in 2001 with the overall goal of conserving and restoring forests and other ecosystems to ensure integrity of ecological, economic, and socio-cultural systems through a multi-stakeholder approach of government leadership, community stewardship, and civil society partnership. The programme aims to mobilise local communities, civil society, and stakeholders for effective land-use management, so that forests and other habitats are conserved and restored creating and maintaining ecological connectivity linking protected areas, forests, and other 'conservation-friendly' land-uses. The major interventions of the programme include conservation of focal species and critical ecosystems; protection and management of forests; restoration of forests and natural landscapes; increased resilience of communities and ecosystems; conservation of freshwater and its ecosystems; improving socio-economic wellbeing and livelihoods; capacity building and institutional strengthening of forestry and protected area institutions; policy and advocacy; conservation awareness; and transboundary cooperation for landscape level conservation. Forest restoration, particularly through the development of forest user groups, is discussed in more detail below.

Forest restoration is one of the WWF's priority interventions in the Terai Arc Landscape. Forest restoration has primarily focused on deforested areas, degraded lands, and flood plains in the ecological corridors. A two-pronged strategy of planting trees and promoting natural regeneration has been adopted to restore forest in the landscape. A total of 3,467,526 seedlings were produced and planted through the Terai Arc Landscape Program. The restoration intervention was intended to increase the original forest cover, to revive wildlife habitats, and to improve ecosystem service flows within the landscape for environmental, social, and economic benefits.

Community forestry, a community-based approach to forest restoration and management, was adopted and engaged local communities in planning and implementing restoration interventions such as planting native trees and fencing off restoration sites to protect plantations and natural regeneration. In community forestry, the local communities who reside adjacent to a forest and depend on them for natural resources, form a forest user group and develop a forest operational plan. Community forestry ensures access and use rights of local communities and indigenous people to forest resources following a specified forest operation plan.

The Terai Arc Landscape Program capacitated and mobilised community forest user groups for restoration management (Figure 19.2a and b). Forest user groups identify restoration

(a) (b)

Figure 19.2 The same landscape shown during restoration by communities in 2003 (a) and in 2011 (b) © WWF Nepal.

sites, plant seedlings and establish fencing to protect the plantation and natural regeneration. Moreover, the Terai Arc Landscape Program supported local communities to create alternative livelihood options through forest-based enterprises, farm and off-farm income generating activities, and nature-based tourism to reduce their dependency on forests. In order to enrich ecological connectivity within the landscape, agroforestry systems have also been practised in various cultivated lands engaging farmers and private landowners.

Major outcomes achieved in the Terai Arc Landscape during the last two decades include an increase in forested land, an increase in the population of endangered species, and an increase in conservation benefits to local communities. The forest area has increased from 1.28 million hectares in 2001 to 1.35 million hectares in 2016 with a net forest gain of 66,800 hectares (MoFSC, 2015). Forest protection, community-based forest management and forest restoration contributed to achieve the gain. A total of 22,791 hectares of degraded forest and degraded land has been restored, whereas 237,050 families have been managing 162,818 hectares of forests as community forests (WWF-Nepal, 2020). Along with protection, increased forest cover in the critical corridors of the landscape have contributed to an increase in the populations of Bengal tiger and Greater one-horned rhinoceros. The Bengal tiger population has nearly doubled in the last ten years, from 121 in 2009 to 235 in 2018 (DNPWC & DoFSC, 2018), while the Greater one-horned rhinoceros population increased from 372 in 2005 to 752 in 2021 (DNPWC, 2021). Anecdotal evidence and community observations indicate that restored forest corridors have a hydrological function as water springs have reappeared (Thapa et al., 2018). The restoration of critical corridors have also contributed to the transboundary movement of wide-roaming megafauna species (Figures 19.3 & 19.4).

The increase in wildlife and movement of wildlife in the corridors, however, has resulted in more human–wildlife conflict. Crop damage, livestock depredation and human casualties by wildlife attacks have all gone up in recent years. Both preventive and curative measures have

Figure 19.3 Transboundary movement (India and Nepal) of collared rhino in Khata corridor, Terai Arc Landscape © WWF Nepal.

Figure 19.4   Collared rhinoceros © WWF Nepal.

been initiated to manage the conflicts. Preventive measures include increasing awareness of wildlife behaviour, predator-resistant pens for cattle, cultivation of deterrent species which can generate income, and alternative livelihood options to reduce local people's exposure to the forest. Curative measures include a relief fund for victims, scholarship fund for victim's children and community-based insurance schemes. These preventive and curative measures have helped to reduce retaliation and negative interactions between humans and wildlife.

The community-based approach of forest restoration and integrated landscape management has resulted in increased conservation benefits to people. Forest restoration and integrated landscape management has created economic opportunities for local communities through forest and farm based green enterprises and nature-based tourism. A total of 2,298 families initiated small and medium scale forest and farm-based enterprises including the production of essential oils, brooms, and nectars (WWF-Nepal, 2020). Similarly, 132 families run home-stays and secure revenues from nature-based tourism.

Despite an increase in human wildlife conflicts, local communities are supportive as they have realised the benefits, which included creating opportunities for livelihoods and income. Local communities have run cooperatives to manage community financial capital which has been created through a revolving fund provided by WWF and their own saving and credit schemes. The saving and credit schemes are managed by the local community groups. They save a certain amount of money on a periodic basis which is used within the groups for work that is needed through self-managed soft loans, with communities deciding on loan arrangements including interest rates and payback period. Approximately USD 5 million in community-owned financial capital has been mobilised so far through 114 cooperatives to support the livelihood and socio-economic wellbeing of communities in the Terai Arc Landscape.

Moreover, forest restoration created an opportunity to Reduce Emissions from Deforestation and Forest Degradation (REDD+) and to mitigate climate change impacts. An Emissions Reduction Program Document of the Terai Arc Landscape was approved by the Forest Carbon Partnership Facility in 2018 (MoFE, 2018). The Government of Nepal has signed an Emission

Reduction Payment Agreement with the World Bank for a result-based payment of USD 45 million for six years sequestering 9.16 million tons of CO_2 equivalent from 2018 to 2024 in the Terai Arc Landscape. The REDD+ actions could also contribute to sustainable financing solutions for landscape level conservation.

Conclusions and lessons

Forest restoration is crucial in integrated landscape management as it creates and maintains connectivity of habitats for the movement of megafauna and for the flow of ecosystem services. Forest restoration in the Terai Arc Landscape has provided a successful example as it has demonstrably increased populations of flagship species such as Bengal tiger and Greater one-horned rhinoceros, even in a densely human-populated landscape.

Multi-stakeholder approaches need to be strengthened in forest restoration and integrated landscape management as it helps to increase ownership and stewardship. A multi-stakeholder approach of government leadership, community stewardship, and civil society partnership was key for the success in this programme. Nepal has recently moved from a unitary system to a federal system of governance, with three government levels viz. federal, provincial, and local, thus, engagement of all tiers of the governments is crucial for success.

Active participation of local communities was also crucial. As most of the local communities depend on forests for their livelihoods, their participation and socio-economic needs have to be integrated into forest restoration and landscape management. Community participation not only enhances social safeguards and natural resource rights of indigenous peoples and local communities, but also contributes to effective implementation on the ground. It also helps to address both the immediate and underlying causes of forest loss and degradation to maintain a functional landscape. Moreover, community forestry being a leading forest management regime in Nepal, technical capacity of forest users, and institutional capacity of forest user groups need to be strengthened to protect and restore forests in the landscape.

As the programme achieved various successful conservation outcomes in the Terai Arc Landscape, forest restoration can be scaled up within the landscape and be scaled out to other landscapes through developing ecological corridors and promoting habitat connectivity to create functional landscapes. Upscaling can be achieved by integrating forest restoration into sub-national, national, regional, and global initiatives.

References

DNPWC. (2021). *Press Release of National Rhino Count Results*. Department of National Parks and Wildlife Conservation, Kathmandu, Nepal

DNPWC and DoFSC. (2018). *Status of Tigers and Prey in Nepal*. Department of National Parks and Wildlife Conservation & Department of Forests and Soil Conservation. Ministry of Forests and Environment, Kathmandu, Nepal.

FRA/DFRS. (2014). *Terai Forests of Nepal (2010-2012): Forest Resource Assessment of Nepal*. Department of Forest Research and Survey, Babarmahal, Kathmandu, Nepal.

MoFE. (2018). *People and Forests- A Sustainable Forest Management-Based Emission Reduction Program in the Terai Arc Landscape, Nepal*. Ministry of Forests and Environment, Kathmandu, Nepal.

MoFSC. (2004). *Strategic Plan 2004–2014, Terai Arc Landscape Nepal*. Ministry of Forests and Soil Conservation, Singha Durbar, Kathmandu, Nepal.

MoFSC. (2015). *Strategy and Action Plan 2015-2025, Terai Arc Landscape Nepal*. Ministry of Forests and Soil Conservation, Singha Durbar, Kathmandu, Nepal.

Thapa, K., Gnyawali, Chaudhary, L., Chaudhary B.D., ChaudharyM., Thapa, G.J., Khanal, C., Thapa, M.K, Dhakal, T., Rai, D.P., Bhatta, S.R., Upadhyay, D., Bhandari, A.R., and Joshi, D. (2018). Linkages among forest, water, and wildlife: A case study from Kalapani community forest in the Lamahi bottle-neck area of Terai Arc Landscape. *International Journal of the Commons, 12(2)* pp.1–20. DOI: http://doi. org/10.18352/ijc.777.

WWF-Nepal. (2020). *Lessons Learnt from 15 Years of Restoring Forest Connectivity: The Terai Arc Landscape in Nepal (unpublished).* WWF.

20

REWILDING CASE STUDY

Monitoring natural capital and rewilding at the
Natural Capital Laboratory, Birchfield, Loch Ness

Chris White, Emilia Leese, Ian Convery, and Philip Rooney

The Natural Capital Laboratory ('NCL') at Birchfield is a partnership venture aiming to 'rewild' an area of land in the Scottish Highlands. This former plantation woodland is located in a remote rural area of Scotland (as defined by the Scottish Government Urban Rural Classification) near to Loch Ness. Site management presents a wide range of restoration challenges, including a shift from commercial/exotic tree species to native woodland, the recovery of degraded peat bogs (caused by tree planting on areas of deep peat and the digging of drainage channels), reintroduction of species missing from the ecosystem, and managing the hydrological flow regime to minimise the impacts of upstream dam construction. To date the focus of the work has been on restoring the native ecosystems on site and engaging local communities with the natural environment. Alongside the rewilding process, an outdoor laboratory has been established to use emerging disciplines such as natural capital accounting and the IUCN Rewilding Principles, as well as emerging technologies to collect data, measure change over time, and communicate the effects of the rewilding process to a broad audience.

The NCL is situated approximately 5km from the south-west shore of Loch Ness and 40km to the west of Inverness (see Figure 20.1). While its 40ha represent a small site in a rewilding context, it is located at a key point between several larger areas of Caledonian Forest ecosystem restoration and rewilding, including the Cairngorms Connect project to the south-east (Cairngorms Connect, 2022), and the Dundreggan estate, managed by Trees for Life, to the north-west (Trees for Life, 2022).

The project was established in 2019 as a joint venture between the landowners Emilia and Roger Leese, engineering firm AECOM, rewilding charity The Lifescape Project, and conservation scientists at the University of Cumbria. The project was an early adopter of the IUCN principles of rewilding (Carver et al., 2021), and in Table 20.1 we outline how the rewilding principles are being applied onsite.

As set out in Table 20.1, Principle 8 concerns the need for an adaptive process to rewilding with high quality monitoring and feedback. Given the expertise of the NCL team in designing and implementing natural capital based monitoring and feedback processes in other situations, there was clearly an opportunity for the project to explore how a 'natural capital' framework could be used to support Principle 8 within a rewilding context. This highly structured and technology heavy approach (which we recognise as fairly unusual for a rewilding project) can

DOI: 10.4324/9781003097822-23

Figure 20.1 Location of the NCL site. Figure produced by AECOM for use in this publication.

provide a quantitative measure of the impacts of the rewilding process over time, and inform how the process can be adapted to better achieve the aims of rewilding. In the following section we outline our activities related to this principle as we feel this is where natural capital can make a significant contribution to rewilding practice.

Natural capital is 'the world's stocks of natural assets which include geology, soil, air, water and all living things. It is from this natural capital that humans derive a wide range of services, often called ecosystem services, which make human life possible' (World Forum on Natural Capital, 2022).

This concept is an extension of the standard economic thinking about capital. Traditionally, economics defines the world in terms of manufactured or financial capital assets such as factories or investments. Over time, and if maintained in good condition, such assets provide a flow of services with value to organisations and wider society, such as manufactured goods or returns on investment. The concept of natural capital extends this notion of the flow of services to natural assets such as woodlands and wetlands. Similar to traditional assets, over time and if maintained in good condition, natural assets will also provide flows of services, such as carbon sequestration and filtration of pollutants (see Figure 20.2), which are valuable to organisations and wider society.

In a now famous paper, Costanza et al. (1997) provided the first meaningful attempt to estimate the value of global natural assets, which they estimated at an average of \$33 trillion per year (by comparison global gross national product was estimated at around US\$18 trillion per year). Much work has followed on from this (e.g. de Groot et al., 2002; Wunder et al., 2008; Fisher et al., 2009) and there is now a huge (and rapidly growing) literature on natural capital. For the casual reader we would recommend Dieter Helm's 2016 book *Natural Capital: Valuing the Planet* as a good starting point.

There is evidence that what might be loosely termed a 'natural capital approach' is being adopted both within the UK and internationally, across a wide range of sectors, and there is a growing number of guidance documents, policies, and tools which are now available.[1]

In the UK, the 2021 'Dasgupta Review' has, according to Sir David Attenborough, finally 'put biodiversity at [the] core of [economics]', and 'by bringing economics and ecology face

Table 20.1 Application of the IUCN Principles of Rewilding (Carver et al., 2021) at the NCL

Principles	Application at the NCL
Principle 1. Rewilding utilises wildlife to restore trophic interactions	Our aim is to restore functioning ecosystems that do not require human intervention to manage them. We will restore habitat using naturally regenerating native vegetation and will intervene where necessary. We will also explore the possibility of reintroducing specific species lost from Birchfield to restore natural functions and a balance between predators and prey.
Principle 2. Rewilding should employ landscape-scale planning that considers core areas, connectivity, and co-existence	Our aim for Birchfield is for it to become a core area of protected native ecosystems within the wider area. Beyond Birchfield's boundaries we will aim to work with other landowners and managers to explore how Birchfield could be connected to other wild areas to provide a corridor for wildlife. We will also seek to understand how rewilding can contribute to the economy and society allowing communities and wildlife to flourish.
Principle 3. Rewilding focuses on the recovery of ecological processes, interactions, and conditions based on reference ecosystems	To guide the rewilding process, we will use the ancient Caledonian Forest as a reference point. We will allow natural processes to recover and we will assist the process as required to ensure that there is a balanced recovery of this ecosystem. This will allow the NCL to connect with other rewilding projects in the area with complementary goals, such as Trees for Life and Cairngorms Connect.
Principle 4. Rewilding recognises that ecosystems are dynamic and change constantly	All plans for the NCL will allow for dynamic, changing processes such as floods, seed dispersal, pollination, movement of species, and changes in climate. We will anticipate these processes occurring and will aim to create the space on the site, and through connectivity with wider areas, to allow these processes to play out. Rather than aiming to manage or control these processes, we will aim to study their implications on people and wildlife in the area.
Principle 5. Rewilding should anticipate the effects of climate change and where possible act as a tool to mitigate impacts	The NCL will explore the potential impacts of climate change on shaping the ecosystems and species in the area and will account for these impacts in the development of the site management plan. Beyond this we will seek to make the NCL carbon negative through restoring forests and peatlands, using renewable energy, adopting a plant based diet, and offsetting any unavoidable emissions through certified carbon offsetting programmes.
Principle 6. Rewilding requires local engagement and support	We recognise rewilding must involve local engagement and support for it to be truly sustainable. We aim to ensure communities have a voice in the implementation of Birchfield's rewilding and to actively encourage involvement where this is possible through engagement events, volunteering opportunities, and feedback on plans and proposals.
Principle 7. Rewilding is informed by science, traditional ecological knowledge (TEK) and other local knowledge	The NCL will act as an ongoing process to collect the data needed to feed into decision making processes and allow informed, science-led decision making. A committee will be appointed to guide and feedback on the decisions made. We will work with local communities and organisations to ensure we use local knowledge to inform project plans. Masters and PhD students will be encouraged and facilitated on site to encourage learning and knowledge generation.

(continued)

Table 20.1 (Cont.)

Principles	Application at the NCL
Principle 8. Rewilding is adaptive and dependent on monitoring and feedback	The NCL will monitor and measure environmental, economic, and social changes on site resulting from the rewilding process. This information will be fed directly into the decision making processes guiding Birchfield's rewilding to allow an adaptive approach. The data will be shared publicly to allow wider engagement and review of the process.
Principle 9. Rewilding recognises the intrinsic value of all species and ecosystems	We recognise animals are not objects and their intrinsic value is independent of their use by humans. We will explore how conservation can engage with the complex ethical questions relating to the intertwined rights and relationships among humans, animals, and plants. We will consider equitable biocentric decision making and evaluate how it informs existing evidence bases for decision making, such as natural capital accounting.
Principle 10. Rewilding requires a paradigm shift in the coexistence of humans and nature	We will use the rewilding process at Birchfield to encourage a shift in the way people currently see, interact, and coexist with the natural world, imagining a different relationship, and explore how people's wellbeing improves as a result. We will provide space for creativity on and off-site inspired by nature and the rewilding process. We will use new technology, such as virtual reality, to showcase what a different future might look like and engage with the complex questions of how we can make that happen for the benefit of all.

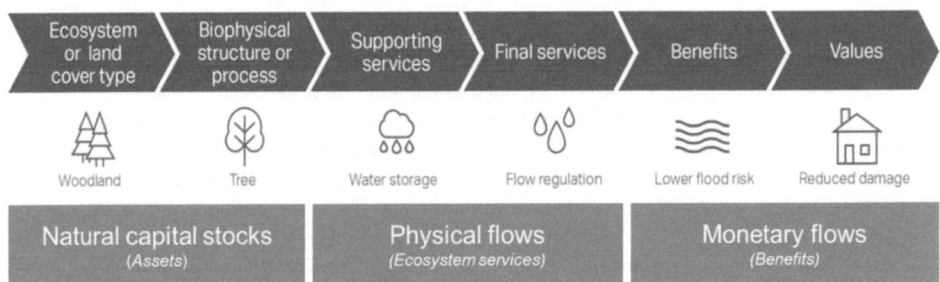

Figure 20.2 Ecosystem cascade diagram setting out the underlying structure of a natural capital approach (after Haines-Young & Potschin, 2010).

to face, we can help to save the natural world and in doing so save ourselves' (Dasgupta, 2021). Broadly speaking, The Dasgupta Review makes the case that a natural capital approach provides a clear and quantitative measure of environmental impacts and can be used to hold organisations to account. Echoing similar arguments made elsewhere, Dasgupta argues that this method ensures the value of the environment is captured within decision making rather than being seen as separate or worthless in financial terms.

However, natural capital is not an uncontroversial approach. Its opponents posit that it could lead to the commoditisation and sale of nature, and reduce the natural environment to a mere number on a balance sheet. Staying with the Dasgupta Review, for example, it has been dismissed as 'the latest attempt at justifying financialisation of nature [and] ignores long-standing problems with capital theory and social cost–benefit analysis…[Dasgupta] offers impossible to achieve valuation, based on old flawed theories and methods, embedded in an unsavoury political economy' (Spash & Hache, 2021).

At the NCL, we have developed a natural capital framework in order to explore how this approach could be applied to a rewilding project and to evaluate the strengths and limitations of the approach. The NCL is also undertaking research in order to better understand how we might improve the efficacy of the approach, with a particular emphasis on understanding how:

- New techniques could be used to improve the collection of data through the use of emerging technologies such as drones, AI (e.g. for camera trap image analysis), robotics, the internet of things, camera traps, remote sensing, AudioMoths, and eDNA analysis;
- The models and frameworks used to analyse such data and provide estimates of the value of natural capital could be improved through developing new models, valuation techniques, and metrics; and
- The complex and technical outputs of a natural capital approach could be made accessible to a broad audience through the use of tools such as digital platforms, video, and virtual reality.

To provide accurate monitoring, one of the first tasks during the first year of the NCL (1 April 2019 to 30 March 2020), was to develop a 'baseline' natural capital account. This baseline account was intended to provide a starting point in time against which changes in the natural environment arising from the rewilding process could be measured, and serve to inform the ongoing assessment of the value (both positive and negative) generated through the rewilding process.

The development of the account drew on the principles set out in the Natural Capital Committees' 'Corporate Natural Capital Accounting' (CNCA) framework (eftec, RSPB, & PwC, 2015). This framework provides a structured format for reporting on natural capital in a manner analogous to the standard financial reporting format undertaken for profit and loss accounts. As set out in Figure 20.3, the framework consists of the following reporting statements and supporting schedules:

- An asset account setting out the extent of natural capital assets within a specific study area and the condition they are in (e.g. the total area of woodland together with a set of metrics indicating the condition that woodland is in)
- A physical flow account setting out the units of each ecosystem service delivered by those natural capital assets each year (e.g. the total tonnes of CO_2e sequestered by that woodland each year)
- A monetary flow account setting out the value of the services provided (e.g. this could include the market value of any carbon sequestered from voluntary carbon markets and/or the avoided social costs of the carbon sequestered)
- A maintenance cost account setting out the costs required to manage the assets within the study area; and
- A balance sheet and statement of change providing a summary of the overall 'profit' and 'loss' provided by the natural capital assets within the study area and how this is changing over time.

The results of the baseline natural capital account prepared for the NCL were published in the first end of year report (AECOM & The Lifescape Project, 2020), and made available online (see Table 20.2). The team also developed a digital natural capital accounting platform

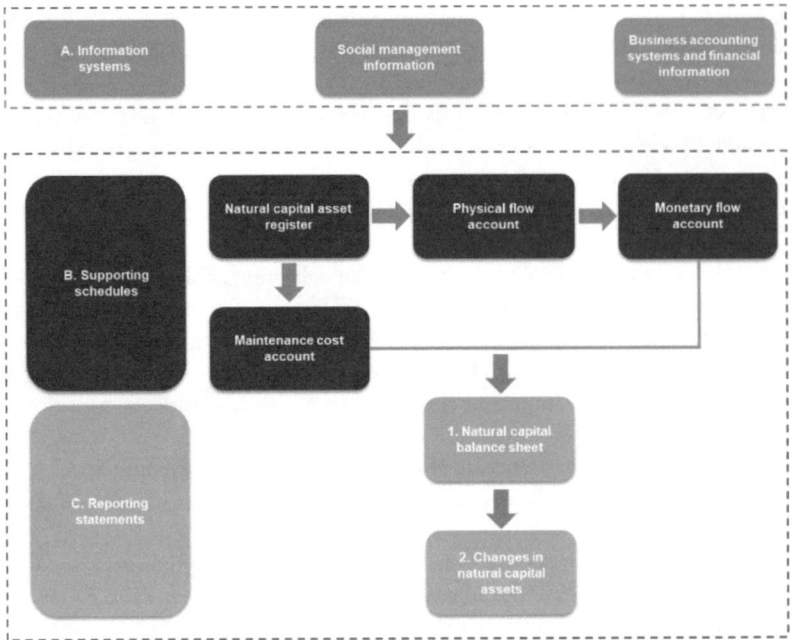

Figure 20.3 Reporting statements and supporting schedules within the CNCA framework. Adapted from eftec, RSPB & PwC (2015).

Table 20.2 Baseline natural capital balance sheet for the NCL (from White et al., 2020)

Natural Capital 2019-20	£ thousands (PV25 years)		
Assets	*Private value*	*External value*	*Total value*
Timber	£ –	£ –	£ –
Energy	£ 13	£ –	£ 13
Water supply	£ 1	£ –	£ 1
Global climate regulation	£ –	£ 260	£ 260
Air quality regulation	£ –	£ 102	£ 102
Flood regulation	£ –	£ 77	£ 77
Water quality regulation	£ –	£ 43	£ 43
Recreation	£ 1	£ –	£ 1
Education	£ –	£ –	£ –
Biodiversity: charismatic species	£ –	£ 36	£ 36
Gross asset value	**£ 15**	**£ 517**	**£ 533**
Liabilities	*Private cost*	*External cost*	*Total value*
NCL costs	-£ 142	£ –	-£ 142
Site costs	-£ 61	£ –	-£ 61
Total liabilities	**-£ 203**	**£ –**	**-£ 203**
Total net value	*Net private value*	*Net external value*	*Total net value*
Total net value	**-£ 188**	**£ 517**	**£ 330**

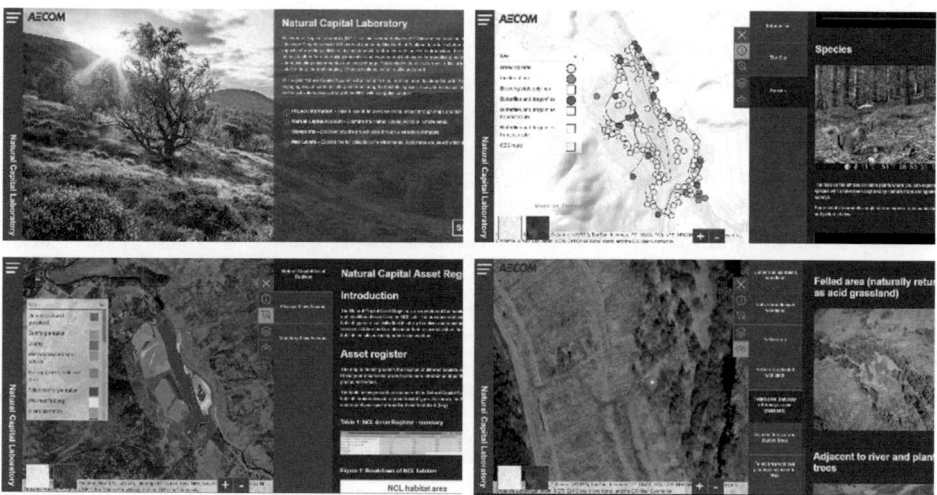

Figure 20.4 Screenshots from the digital natural capital accounting platform (AECOM, 2020) prepared for the NCL. Image produced by AECOM for use in this publication.

for the project (AECOM, 2020) to make this information more accessible to a wider range of stakeholders (see Figure 20.4). This was identified as being important at an early stage given that a key aim of the natural capital approach is to measure environmental impacts and allow people to hold organisations to account for those impacts, something which becomes challenging if the technical nature of the natural capital accounting outputs means they are only accessible to a technical audience. To address this, a publicly accessible version of the accounts was developed in order to allow a non-technical audience to access the results of the accounting process in a visual and engaging manner and be able to hold the project to account for the impacts it delivers.

The initial findings from the baseline natural capital account show a variety of assets and asset conditions on site, including woodland, peatland, and riparian habitats. The current dominant assets are woodland habitats, much of which are in poor condition as assessed by ecologists through a Phase 1, NVC, and Biodiversity Net Gain survey. There is also a potentially significant area of peatland habitat which appears to be degraded due to previous tree planting activities and the installation of drainage channels on site. In addition, a river runs the length of the site which is affected by upstream hydropower activities. In light of this, while there are important natural capital assets on site, there is clear potential to improve their condition and the value of the ecosystem services they provide. It is also important to note that the asset account prepared for the baseline does not provide a complete picture of the extent and condition of all of the natural capital assets on site. While ecosystems and species were well covered, the team identified further information required to cover atmospheric conditions, soil, and water quality. These areas are being explored in the subsequent years of the project and will be included in the updated accounts produced each year.

In terms of understanding the services provided by the natural capital assets on site, the baseline account suggests that the site generates a net value of around £0.3 million over a 25-year period. Regulating services, in particular from carbon sequestration, provide a much greater share of the value than either provisioning (e.g. food, water, and timber) or cultural services (e.g. recreation and aesthetic values). Further, much of the value generated by the site is external value or a public good. This means that the wider public benefits more from the value, rather than the stakeholders who are investing in the site's ownership and management. What's more, while the benefits are mostly public, the costs of managing the site are entirely private and are borne by the NCL stakeholders. Looking purely at the private costs and returns, this means that the net value of the site is around -£0.2 million over the first year.

This issues highlights the financial challenges of investing in such projects. However, this estimate of value does not include other capitals such as social or intellectual capital. The latter is likely to be of significant value. Future versions of the account will extend the accounting framework to cover intellectual, human, financial, and manufactured capital to provide a more complete picture of the returns on investment.

As things stand, the ecosystem service on site with the highest value is global climate regulation. We used high level estimates from studies undertaken elsewhere in the UK to determine the values of carbon sequestration rates for the habitats on site. Because carbon sequestration is a key ecosystem service, further work will be undertaken to refine and enhance the accuracy of these estimates. This is particularly the case for the peatland ecosystem on site which could have significant variance in terms of its carbon balance depending on the specific conditions of the site.

Similarly broad approximations derived from studies carried out elsewhere were also the basis for determining the values for the site's next set of important ecosystem services, namely air quality, water quality, and flood regulation. The literature values may not account for the rural nature of the site, likely lower levels of background pollution, and the limited number of likely beneficiaries of these services. As such, the estimates used might overstate the value of the services provided. A more detailed, site specific model could be developed for the site to provide a better estimate of the extent and value of these services. Data collected via the NCL could be used to refine the models and develop a benefits transfer function, allowing more site specific estimates to be undertaken for both the NCL and other sites.

Other services identified as being of key importance to the site include estimates of the recreational value, which were taken from studies elsewhere and are not unique to the site itself. We could develop a visitor feedback form to collect better data on visitor numbers and help to more accurately quantify the value of each visit using a travel cost method. Visitors also anecdotally report a high aesthetic value of the site. This value is likely to change significantly over time, particularly as blocks of trees are removed and eventually replanted. We could undertake a visual aesthetics survey to quantify the change in the flow of this service. Finally, biodiversity is an important component of the site which is difficult to quantify. A better understanding of the change in species populations over time and the value people place on them through choice modelling surveys would improve this side of the accounts, together with an estimate in the change in biodiversity units generated through the rewilding process over time.

The baseline natural capital account prepared for the NCL is an initial step towards developing a detailed understanding of the environmental, social, and economic impacts of the rewilding process at the Birchfield site. We intend to update this account on an annual basis to capture change over time and to improve and extend the breadth and depth of the accounting process. Meanwhile, we can draw the following initial lessons which may inform this and other rewilding projects considering the potential uses of a natural capital accounting process.

Lesson 1. Measuring change

Natural capital accounting can be used to provide a quantitative measure of environmental change over time. This information could be useful for monitoring the impacts of a project and demonstrating to stakeholders (whether investors, project managers, local communities, or government partners) the progress made towards specific aims or goals. However, it is important to note that by itself, a natural capital approach is inherently a quantitative, numerical approach which can be difficult for non-specialists to understand. And as such, care is needed to develop communication approaches that are tailored to specific stakeholders to help communicate this information.

Lesson 2. Demonstrating value

In addition to monitoring change, natural capital accounting can demonstrate the monetary value of ecosystem services provided by the rewilding process. This may be particularly useful in terms of providing a new perspective on the value of aspects of the natural environment which have previously been overlooked, i.e. demonstrating how native woodland can provide services of equivalent or higher value (e.g. carbon, air quality, water quality, flood regulation, and biodiversity) to plantation woodland habitats (traditionally valued in terms of their timber provisioning services). Highlighting the importance of such values could help to make the benefits of rewilding more transparent to a range of stakeholders, as well as providing transparency about the trade-offs made when significant land use occurs. It should also help when evaluating the efficacy of different rewilding approaches by providing a standardised accounting tool.

Lesson 3. Informing management

Looking beyond reporting on specific impacts and values, natural capital accounting can provide a rich set of data, organised in a formalised structure, which can be useful for day-to-day decision making with regards to site management. On the NCL project, for example, the account identified the presence of peatland habitat on the site and the potential to improve its condition to provide biodiversity and carbon benefits. Prior to the baseline account being prepared, the site management plan aimed to plant this area with trees. The plan has now been updated to account for a peatland restoration component within the rewilding process. The baseline account also identified a significant number of species within the plantation woodland blocks (i.e. red squirrel, pine marten, European badger, and mountain hare) which has led to a shift in the site management plan away from large scale felling of plantation blocks and replanting with native species, towards a slower and more balanced approach to forest management.

While habitats and biodiversity could have been identified without the use of a natural capital accounting framework, the systematic collection and analysis of data on the condition of assets and the presentation of this data to the stakeholders making the land management decisions, helped to facilitate and inform the rewilding process.

Lesson 4. Missing values

While there are potential uses and opportunities for natural capital accounts to be used within a rewilding context, there are also limitations. A key limitation concerns data and evidence needed to reflect the full range of values provided by the rewilding process. A particular issue of concern for the rewilding process is how best to value changes in biodiversity over time. Consistency of approach, correctly identifying causal factors, and data quality are often key challenges.

With regards to data, the remoteness of the site and desire to reduce travel and the project's carbon footprint has led the NCL team to review best practice regarding low cost/low carbon solutions to monitoring change over time. Some of these solutions include the use of emerging new technologies to integrate and automatically analyse different data, including remote sensing to monitor changes in habitat extent and condition and the use of camera traps, AudioMoths, and eDNA to capture and track changes in species diversity.

One promising opportunity is to integrate the use of the Defra Biodiversity Metric 3.0 into the natural capital accounting process to estimate the number of biodiversity units generated and the market value of these, although there are problems with applying the metric to rewilding projects (zu Ermgassen et al., 2021). In the baseline natural capital account, we attempted to value changes in specific species based on estimates of the public's willingness to pay for their protection. However, we need significant additional research to better understand the use and non-use values that the public place on different species. In this respect, we believe using virtual reality in traditional choice modelling techniques may be an interesting avenue to explore.

While there is an increased level of evidence being developed to better understand and value environmental change (biodiversity valuation aside), a broader limitation with the natural capital accounting process is that it does not provide a complete picture of the social and economic impacts of the rewilding process. As numerous rewilding projects have demonstrated elsewhere, it is not enough for a project to demonstrate a positive environmental impact. Long term sustainability requires social and economic issues to also be addressed. For example, building trust with local communities, tackling issues of diversity and inclusivity, generating jobs and livelihoods, and addressing issues around health and safety are all key issues needing consideration, measurement, and monitoring. A move towards a multi capitals accounting framework which covers a broader range of capitals could help fill some of these gaps and provide a more complete picture of impacts (Yorkshire Water, 2021).

In conclusion, the NCL project is a rewilding project in the Highlands of Scotland that is adopting the IUCN Rewilding Principles. A particular emphasis has been placed on Principle 8 around adaptive management and monitoring, exploring how a natural capital framework can be used to measure change in quantitative terms and inform land management decisions. While the implementation of this approach has identified challenges, there are also opportunities in using this approach to make more informed, and more transparent decisions which could help to build longer term support for rewilding projects. The NCL is a multi-year project and future years will aim to build on the initial work, filling in data gaps and developing more robust approaches that can allow a natural capital accounting framework to better support decision making. Further, there is a growing aim to expand the project beyond the initial site at Birchfield, to set up a series of partner sites around the world, implementing rewilding in a practical sense and sharing research and learning to improve the frameworks and approaches for monitoring and managing their implementation.

Note

1 See, for example, within the UK the Defra Enabling a Natural Capital Approach guidance www.gov.uk/guidance/enabling-a-natural-capital-approach-enca, the Natural Environment Valuation Online tool www.leep.exeter.ac.uk/nevo/, and the move towards a natural capital based approach to agricultural subsidies through the new Environmental Land Management scheme www.gov.uk/government/publications/environmental-land-management-schemes-overview

References

AECOM (2020), Natural Capital Laboratory, https://eia.aecom-digital.com/natcap/
AECOM & The Lifescape Project (2020), Natural Capital Laboratory: Year One Report.

Cairngorms Connect (2022), A wild landscape in the making, http://cairngormsconnect.org.uk/

Carver et al. (2021), Guiding principles for rewilding, *Conservation Biology*.

Costanza, R., d'Arge, R., de Groot, R. et al. (1997), The value of the world's ecosystem services and natural capital, *Nature*, 387: 253–60, https://doi.org/10.1038/387253a0

Dasgupta, P. (2021), *The Economics of Biodiversity: The Dasgupta Review*, London: HM Treasury.

de Groot, R.S., Wilson, M.A., and Boumans, M.J. (2002), A typology for the classification, description and valuation of ecosystem functions, goods and services, *Ecol. Econ.*, 41: 393–408.

eftec, RSPB, and PwC (2015), Developing Corporate Natural Capital Accounts, prepared for the Natural Capital Committee. Available at: https://assets.publishing.service.gov.uk/government/uploads/system/uploads/attachment_data/file/516968/ncc-research-cnca-final-report.pdf

Fisher, B., Turner, R.K., and Morling, P. (2009), Defining and classifying ecosystem services for decision making, *Ecol. Econ.*, 68: 643–53.

Haines-Young, R.H. and Potschin, M. (2010), The links between biodiversity, ecosystem services and human well-being. In Raffaelli, D. and Frid, C. (eds) *Ecosystem Ecology: A New Synthesis. BES Ecological Reviews Series*. Cambridge University Press, 110–39.

Spash, C.L. and Frédéric Hache, F. (2021), The Dasgupta Review deconstructed: an exposé of biodiversity economics, *Globalizations*. DOI: 10.1080/14747731.2021.1929007.

Trees for Life (2022), In the heart of the Highlands sits Dundreggan, our flagship example of rewilding in action, https://treesforlife.org.uk/dundreggan/

White, C. et al. (2020), Natural Capital Laboratory: Year 1 Report, https://communications.aecom.com/NCLyear1report

World Forum on Natural Capital (2022), What is natural capital?https://naturalcapitalforum.com/about/

Wunder, S., Engel, S., and Pagiola, S. (2008), Taking stock: a comparative analysis of payments for environmental services programs in developed and developing countries, *Ecol. Econ*, 65: 834–52.

Yorkshire Water (2021), Our Contribution to Yorkshire A review of our impact and public value.

zu Ermgassen, S.O.S.E., Marsh, S., Ryland, K., Church, E., Marsh, R., and Bull, J.W. (2021), Exploring the ecological outcomes of mandatory biodiversity net gain using evidence from early-adopter jurisdictions in England, *Conservation Letters*, 14(6), November/December 2021: e12820.

21

ECO-CIVILISATION PROVIDES NEW OPPORTUNITIES FOR REWILDING IN CHINA

Yue Cao, Shuyu Hou, Steve Carver, Ian Convery,
Zhicong Zhao, and Rui Yang

Introduction

Rewilding takes place along a continuum of scale, connectivity, and human influence and aims to restore ecosystems to achieve self-sustaining nature. International teamwork with IUCN CEM Rewilding Thematic Group (RTG) has developed a global definition and ten universal guiding principles for rewilding. The new adaptation of the '3Cs model' for rewilding emphasises Cores, Connectivity and Co-existence, reflecting the need for transformative change in the relationship between humans and the rest of nature. In this new model, connectivity is not just about the physical/ecological connections between core areas, but also the social and cultural connections between people and nature. The Chinese concept of Ecological Civilisation (eco-civilisation) would appear to offer some further ideas and challenges in this direction.

Eco-civilisation refers to the sum of material, spiritual, and institutional achievements made by human beings for protecting and building a beautiful ecological environment, and it is a social form in which people and nature, environment and economy, and people and society coexist in harmony (Zhou, 2013). Perhaps the earliest published source of the term in Chinese can be traced back to 1985, which refers to eco-civilisation as a method to 'restore the broken unity of human and nature' (Zhang, 1985), partly as a reference to traditional ideas of the unity of nature and humanity (天人合一) in ancient Chinese philosophy. Gare (2012) notes how the concept is related to what he terms the 'creative synthesis of ecological Marxism and the revived ethical and political ideas associated with Chinese traditions of thought' such as Confucianism and Taoism. The formal adoption of the term started in 2007, when the Chinese Central Government published a national strategic document aiming to transform the late 20th to early 21st century industrial period, with its associated history of pollution and environmental damage, into a new historical period of 'Ecological Civilisation' (Wen et al., 2012). At the Third Plenary Session of the 18th Central Committee in 2013, it was agreed that China would implement 'ecological civilisation reforms' in order to mitigate against the impacts of economic development on the environment. Since 2007, more than 4,000 published Chinese articles and books have included eco-civilisation as one of their key words, and more than 170,000

DOI: 10.4324/9781003097822-24

articles in mainstream press-media in China have mentioned the concept (Hansen et al., 2018). Eco-civilisation now has the status of a National Development Strategy and its core concept is embedded in the Civil Code of the People's Republic of China (National People's Congress, 2020) and the 13th and 14th Five-year Plan (CPC Central Committee, 2015; Xinhua Net, 2020) which aims to go beyond the previous GDP-based policy to develop a new, harmonious co-existence between humans and nature, a shift from anthropocentrism to 'ecological holism' (Hu, 2012; Xi, 2018*)*.

The urgent drivers for this policy shift are of global importance. As Pan Yue, the former Vice-minister of China's State Environmental Protection Administration noted in 2005; 'in 2020, there will be 1.5 billion people in China. Cities are growing but desert areas are expanding at the same time; habitable and usable land has been halved over the past 50 years. Acid rain is falling on one-third of the Chinese territory, half of the water in our seven largest rivers is completely useless, while one-fourth of our citizens does not have access to clean drinking water. One-third of the urban population is breathing polluted air, and less than 20 percent of the trash in cities is treated and processed in an environmentally sustainable manner. Finally, five of the ten most polluted cities worldwide are in China… Because air and water are polluted, we are losing between 8% and 15% of our gross domestic product. And that doesn't include the costs for health. Then there's the human suffering: In Beijing alone, 70 to 80% of all deadly cancer cases are related to the environment. Lung cancer has emerged as the No. 1 cause of death' (Pan, 2005, cited in Gare, 2012).

The report of the 19th National Congress of the Communist Party of China (CPC, 18 October 2017) noted that the 'modernization that we pursue is one characterised by harmonious coexistence between man and nature. We should, acting on the principles of prioritising resource conservation and environmental protection and letting nature restore itself, develop spatial layouts, industrial structures, and ways of work and life that help conserve resources and protect the environment' (Xi, 2017). In recent years the eco-civilisation concept appears in the statements that President Xi has frequently made in international conferences. Eco-civilisation has also become the theme of the Convention on Biological Diversity COP 15 as in 'Ecological Civilisation: Building a Shared Future for All Life on Earth' (United Nation, 2019). In practical perspectives, some research shows the ecological condition in China continues to improve over time following the implementation of ecological civilisation construction (ECC) strategy(Gai et al., 2020; Wu et al., 2021).

As of 2020, Gu et al. (2020a) report that five provinces (Fujian, Jiangxi, Guizhou, Qinghai, and Yunnan) and approximately 100 cities and counties have been selected as the first cohort of eco-civilisation demonstration areas and 15 provinces have completed delineation of their ecological conservation redlines (ECRs) (Bai et al., 2018; Jiang et al., 2019). Hansen et al. (2018) argue that eco-civilisation currently constitutes the most significant Chinese state-initiated imaginary of our global future, and that it is therefore crucial to explore what this vision entails. In this chapter we attempt to contribute to this process by presenting a review of eco-civilisation in relation to rewilding, primarily as the starting point for more detailed collaboration as part of ongoing IUCN CEM RTG activities. In such a short chapter we cannot hope to present a comprehensive review of eco-civilisation (for example, see Hansen et al. (2018) and Goron (2018), for a review of the concept and the philosophical and political underpinnings of EC in traditional and contemporary Chinese cultures). Instead, what we aim to do in the following sections is highlight some shared objectives and potential areas of future collaborative research between the rewilding and ecological civilisation communities.

Rewilding and ecological civilisation: A shared vision?

We recognised that China is a mega-wild country. The first identification of wilderness areas at national scale in China has been conducted, which indicates that China contained over 86,000 wilderness patches (with varying relative wilderness qualities), covering approximately 42% of China's terrestrial area (Cao et al., 2019a). Among those areas, about 77% of the existing wilderness patches are not covered by nature reserves, indicating the obvious conservation gaps of China's wilderness areas. Eco-civilisation provides new opportunities for wilderness conservation and rewilding, and the wilderness map could provide basic information and support relevant policy making process. For example, at regional scale, it spatially highlights opportunities to connect existing protected areas and wilderness areas, something which is a core strategy of rewilding. Cao et al. (2020) develops a method to identify pinch-points to create wildland networks and took the Great Taihang region of China as an example to identify key areas for rewilding, which could provide implications for future restoration projects. Besides connectivity conservation, many of the wilderness areas are facing threats and the ecological integrity requires promotion as the integrity of fauna and flora has been damaged due to human activities in the past. Therefore, rewilding of the existing wilderness areas should also be considered and carefully managed.

In fact, rewilding and eco-civilisation would appear to share a number of core values, including a focus on self-willed nature ('let nature renew itself'), the need for a long-term, larger-scale, and holistic conservation perspective,[1] the need for adaptive management based on monitoring and evaluation, alongside a recognition that ecosystems are dynamic and constantly changing (and should anticipate the impact of climate change), and the need for transformative change in the relationship between humans and the rest of nature ('harmonious co-existence').

The concept may evolve in relation to new and emerging policy initiatives, but it is likely that ecological protection and restoration will remain a core principle of eco-civilisation, and we see opportunities for the wider rewilding community to contribute to eco-civilisation and the ecological restoration of China. In the next section situating these ideas in context we explore some core actions of eco-civilisation, and discuss how rewilding might help to shape these projects to achieve better conservation success.

(1) Three types of spaces in the National Territory Spatial Planning System (NTSPS). The guiding ideology of the NTSPS, as outlined by the reporting process following the 18th CPC is that 'we should ensure that the space for production is used intensively and efficiently, that the living space is liveable and proper in size, and that the ecological space is unspoiled and beautiful; and we should leave more space for nature to achieve self-renewal'.[2] We see great opportunity for rewilding to help regulate and restore the three types of spaces (living space, ecological space, and production space). First, the space for production mainly refers to agriculture land. A shift is required from the current intensive system towards more sustainable land-use based on high quality spatial planning that enables connectivity and coexistence across the transition from high-quality farmland through to wild areas. There is also an opportunity to use rural abandonment (which is occurring in China, as elsewhere, in response to economic development and rapid urbanisation) and the resulting 'space for nature' to augment existing wild areas and further enhance connectivity. Second, we see an important role for rewilding processes (even if not formally classified as rewilding) to improve living spaces, simultaneously providing health and wellbeing benefits for humans whilst also creating habitat for nature and a degree of connectivity. Equally, urban wildness has an important and unique value in reconnecting humans with nature, promoting human physical and mental health, conserving biodiversity and maintaining ecosystem services (Cao et al., 2019b). Third, in relation to the conservation of ecological space, rewilding will play an indispensable role in helping shape the connectivity and co-existence across the continuum from highly modified to more natural landscapes and will support the Chinese concept of the ecological conservation redlines (ECRs refer to the area with special important ecological functions and must be strictly protected compulsively within the scope of ecological space) in maintaining the integrity of ecological functions.

(2) National parks and protected areas. China now is in the process of reorganising its protected areas system and establishing a new national parks system, the primary objective of which is to protect the integrity and authenticity of ecosystems (The General Office of the CPC Central Committee and the General Office of the State Council, 2017). A key component of restoring ecological integrity is the restoration of full trophic levels and a recognition of the role provided by highly interactive species. The current proposals for the Amur tiger and Amur leopard National Park in Northeast China provide an excellent opportunity to develop a national park taking rewilding into consideration. The general plan proposes two main strategies for restoring the potential habitat for the Amur tiger and Amur leopard. Firstly, to restore the potential corridors for the species' spreading, the management office will reorganise the forestry stations near the corridors by recovering natural vegetation and change the plantation in a manner closer to natural habitat. Secondly, there will be some rewilding processes in areas that are previously affected by human activities. Some pastures will be slowly recovered as natural grassland to support more ungulates. For croplands, some will be abandoned and recover to natural land, others will be kept only as the food source for ungulates. There are increasing calls for trophic rewilding as part of the new Chinese approach to national parks. Some research has explored the thresholds and potential habitats for the Amur tiger and leopard to identify suitable areas to take actions on conservation and restoration (Gu et al., 2020b; Qi et al., 2020). There are, however, some worrying discussions concerning the removal of people and settlements as part of this approach, and whilst there needs to be careful management (and in some cases removal) of human infrastructure, this needs to be set alongside the growing body of evidence (and IUCN guidance) for community conservation and the role of indigenous and local communities in the management of national parks (Borrini-Feyerabend et al., 2004).

(3) Ecological protection and restoration projects of Mountains-Rivers-Forests-Farmlands-Lakes-Grasslands (MRFFLG). Ecological protection and restoration are core components of ecological civilisation in China. Since 2016, China has developed 25 MRFFLG pilot projects. We see great potential to apply broad rewilding principles to these and other future MRFFLG projects (see Table 21.1). We identify the following similarities between the rewilding practice and the MRFFLG projects in China. First, the principles of MRFFLG form a 'life community' and an emphasis on 'following the laws of nature' are consistent with the rewilding values. Second, both the MRFFLG projects and rewilding projects share a same goal of conserving biodiversity and 'bending the curve' (Mace et al., 2018; Leclère et al., 2020). Third, the MRFFLG projects emphasise 'conservation first, natural restoration first' and 'adopting near-natural methods', which is again consistent with rewilding values of being 'process-orientated' and 'nature led' (Carver et al., 2021). Based on the above similarities, and in order to introduce the lessons that China could learn from the international rewilding communities, Yang and Cao (2019) proposed five action-orientated recommendations for how this might be achieved. These include: conducting baseline investigations of wilderness areas and potential alignment with rewilding; protecting remaining high-value wilderness areas; exploring the rewilding approaches used in different regions; establishing large-scale landscape conservation networks with wilderness and rewilded areas as the core, and promoting ecological experiences and nature education activities based on rewilding. On the other hand,

Table 21.1 List of the 25 MRFFLG pilot projects

Batch	Number	List of the MRFFLG pilot projects
1st	5	Hebei Central Beijing and Tianjin, Jiangxi Ganzhou, Shaanxi Loess Plateau, Qinghai Qilian Mountains, Gansu Qilian Mountains
2nd	6	Changbai Mountain in Jilin, Taishan Mountain in Shandong, Huaying Mountain in Sichuan, Fuxian Lake in Yunnan, Zuojiang River in Guangxi, and Minjiang River in Fujian
3rd	14	Heilongjiang Xiaoxinganling and Three Rivers Plain, Altay Mountain in Xinjiang, Wuliangsuhai in Inner Mongolia, Shizuishan in Ningxia, Xiong'an New District in Hebei, Taihang in Shanxi, Taihang in Henan, South Taihang in Henan, Three Gorges of the Yangtze River in Hubei, Liangjiang in Chongqing, Wumeng Mountain in Guizhou, Lhasa River in Tibet, Xiangjiang River and Dongting Lake in Hunan, Shaoshan in Guangzhou, Qianjiang Source in Zhejiang

whilst the ecological restoration in pilot areas has achieved certain results, conducting more natural recovery strategies is still needed in future actions (Luo et al., 2019), where rewilding could play an important role.

Conclusions

Both rewilding and eco-civilisation are rapidly developing concepts, and both have been criticised in the past for, amongst other things, a perceived lack of cohesive thought and a 'backward-looking' perspective. For example, Hansen et al. (2018) argues that eco-civilisation has developed a 'selective and contested interpretation of China's past'. These are points that require a measured, evidence-based response, and this will form part of the ongoing work of the Rewilding Thematic Group and our collaboration with Chinese colleagues. Criticisms aside, however, we can also see great positivity and ambition in Chinese plans to transform both their environment and their relationship with the natural world. Gare (2012) notes that Chinese proponents of ecological civilisation are justified in their claims that Chinese traditions of thought focused on finding *Dao* (path or way) have much to contribute to working out how to live in the present and how to create a global post-capitalist society.

Eco-civilisation represents a significant commitment on the part of the Chinese government to respond to the growing national and international conservation challenge, and there are important lessons for the UK and mainland Europe in relation to the development of an integrated approach to conservation policy, funding, and communication.

In this chapter, we have highlighted some commonality between the concepts of rewilding and eco-civilisation, and we have also discussed some opportunities for rewilding in China, especially in shaping the ongoing implementation of ecological civilisation. The challenge, as we see it, is to effectively communicate, translate, and integrate the philosophy and science of rewilding and eco-civilisation, whilst also staying true to the ethos and origins of both concepts. There are well documented social, economic, cultural, and ecological benefits of rewilding that align with eco-civilisation and the broader sustainable development agenda. However, this approach does not come without risk; we certainly do not want to dilute rewilding or create 'rewilding lite' and see it become yet another variant of the sustainability industry. The inclusive and collaborative approach we are building with the RTG will be fundamentally important to this process.

Notes

1 http://news.sina.com.cn/o/2016-03-11/doc-ifxqhnev5731713.shtml
2 www.chinadaily.com.cn/china/2012cpc/2012-11/18/content_15939493_9.htm

References

Bai, Y., Wong, C.P, Jiang, B., Hughes, A.C, Wang, M., and Wang, Q. (2018). Developing China's Ecological Redline Policy using ecosystem services assessments for land use planning. *Nature Communications*, 9 (3034).

Borrini-Feyerabend, G., Kothari, A., and Oviedo, G. (2004). *Indigenous and Local Communities and Protected Areas: Towards Equity and Enhanced Conservation*. Best Practice Protected Area Guidelines Series No. 11. IUCN.

Cao, Y., Carver, S., and Yang, R. (2019a). Mapping wilderness in China: Comparing and integrating Boolean and WLC approaches. *Landscape and Urban Planning*, 192: 103636.

Cao, Y., Martin, V.G., and Yang, R. (2019b). Urban wildness: Protection and creation of wild nature in urban areas. *Landscape Architecture*, 26(8): 20–4. [in Chinese with English abstract].

Cao, Y., Yang, R., Carver, S. (2020) Linking wilderness mapping and connectivity modelling: A methodological framework for wildland network planning, *Biological Conservation*, 251.

Carver, S., Convery, I., Hawkins, S., Beyers, R., Eagle, A., Kun, Z., Maanen, E.V, Cao, Y., Fisher, M., Edwards, S.R., Nelson, C., Gann, G.D., Shurter, S., Aguilar, K., Andrade, A., Ripple, B., Davis, J., Sinclair, A., Bekoff, M., Noss, R., Foreman, D., Pettersson, H., Root-Bernstein, M, Svenning, J.C., Taylor, P., Wynne-Jones, S., Featherstone, A.W., Fløjgaard, C., Stanley-Price, M., Navarro, L.M., Aykroyd, T., Parfitt, A., and Soulé, M. (2021). Guiding principles for rewilding. *Conservation Biology*, 35(6): 1882–93.

CPC Central Committee. (2015). *The 13th Five-Year Plan for Economic and Social Development of the People's Republic of China*. Central Compilation & Translation Press. https://en.ndrc.gov.cn/newsrelease_8232/201612/P020191101481868235378.pdf, accessed on 10 April 2021.

Gai, M., Wang, X., and Qi, C. (2020) Spatiotemporal evolution and influencing factors of ecological civilization construction in China. *Complexity*. DOI: 10.1155/2020/8829144.

Gare, A. (2012) China and the struggle for ecological civilization, *Capitalism, Nature, Socialism*, 23:4, 10–26, DOI: 10.1080/10455752.2012.722306.

General Office of the CPC Central Committee (CCCPC) and General Office of the State Council of the PRC (SCPRC). Several Opinions on Delineating and Strictly Protecting the Ecological Conservation Redlines. (in Chinese). 07 February, 2017. http://www.gov.cn/zhengce/2017-02/07/content_5166291.htm (accessed 21 Sep 2022).

Goron, C. (2018) Ecological civilization and the political limits to a Chinese concept of sustainability. *Frontiers of History in China*, 4: 28–52.

Gu, Y., Wu, Y., Liu, J., Xu, M., and Zuo, T. (2020a) Ecological civilization and government administrative system reform in China. *Resour. Conserv. Recycl.*, 155.

Gu, Y., Zhang, F., Liang, X., Liu, C., Xing, S., and Wang, Q. (2020b). Integration of natural reserves based on potential habitat protection of the Amur tiger. [in Chinese with English abstract]. *Chinese Journal of Ecology*. 2020–5.

Hansen, M.H., Li, H., and Svarverud, R. (2018). Ecological civilization: Interpreting the Chinese past, projecting the global future. *Global Environmental Change*, 53: 195–203.

Hu, J. (2012) *A Report to the 18th National Congress of the Communist Party of China*. People's Publishing House, China.

Jiang, B., Bai, Y., Wong, C., Xu, X., and Alatalo, J.M. (2019). China's ecological civilization program—implementing ecological redline policy. *Land Use Pol.*, 81: 111–14. 10.1016/j.landusepol.2018.10.031

Leclère, D., Obersteiner, M., Barrett, M., et al. (2020). Bending the curve of terrestrial biodiversity needs an integrated strategy. *Nature*, 585: 551–6. https://doi.org/10.1038/s41586-020-2705-y

Luo, M., Yu, E., Zhou, Y., Ying, L., Wang, J., and Wu, G. (2019) Distribution and technical strategies of ecological protection and restoration projects for mountains-rivers-forests-farmlands-lakes-grasslands. [in Chinese with English abstract]. *Acta Ecologica Sinica*. DOI:10.5846/stxb201905291108

Mace, G.M., Barrett, M., Burgess, N.D., et al. (2018) Aiming higher to bend the curve of biodiversity loss. *Nat Sustain*, 1: 448–51. https://doi.org/10.1038/s41893-018-0130-0

National People's Congress. (2020). Civil Code of the People's Republic of China. http://english.www. gov.cn/archive/lawsregulations/202012/31/content_WS5fedad98c6d0f72576943005.html, accessed on 10 April 2021.

Qi, J., Holyoak, M., Ning, Y., and Jiang, G. (2020). Ecological thresholds and large carnivores conservation: Implications for the Amur tiger and leopard in China, *Global Ecology and Conservation*, 21: e00837, ISSN 2351-9894, https://doi.org/10.1016/j.gecco.2019.e00837.

United Nations. (2019). Theme announced for landmark 2020 UN Biodiversity Conference: 'Ecological Civilization: Building a Shared Future for All Life on Earth'. www.cbd.int/doc/press/2019/pr-2019-09-05-cop15-en.pdf, accessed on 10 April 2021.

Wen Tiejun, Lau Kin Chi, Cheng Cunwang, Huili He, and Qiu Jiansheng. (2012). Ecological civilization, indigenous culture, and rural reconstruction in China. *Monthly Review*. 2012.02/.

Wu, M., Liu, Y., Xu, Z., Yan, G., Ma, M., Zhou, S., and Qian, Y. (2021). Spatio-temporal dynamics of China's ecological civilization progress after implementing national conservation strategy, *Journal of Cleaner Production*, 285, https://doi.org/10.1016/j.jclepro.2020.124886.

Xi, J. (2017). Report at 19th CPC National Congress. Reported by Xinhuanet. www.chinadaily.com. cn/china/19thcpcnationalcongress/2017-11/04/content_34115212.htm, accessed on April 10, 2021.

Xi, J. (2018). *A Report to the 19th National Congress of the Communist Party of China*. People's Publishing House, China.

Xinhua Net. (2020). China proposes development targets for 14th Five-Year Plan period. www.xinhua net.com/english/2020-10/29/c_139476451.htm, accessed on 10 April 2021.

Yang, R. and Cao, Y. (2019). Rewilding: New ideas for ecological protection and restoration projects of mountains-rivers-forests-farmlands-lakes-grasslands, *Acta Ecologica Sinica* [in Chinese with English abstract].

Zhang, S. (1985). Ways to cultivate individual ecological civilization under the condition of mature socialism [in Chinese]. *Guangming Daily*. 1985.2-57. www.cqvip.com/read/read.aspx?id=49988300, accessed on 10 April 2021.

Zhou, S. (2013). The development and achievements of environmental protection in our country [in Chinese]. www.mee.gov.cn/gkml/sthjbgw/qt/201310/t20131009_261311.htm

22

RESTORING WHAT WE'VE LOST

Lessons from evolutionary history for rewilding and coexisting in landscapes with predators

Joanna E. Lambert and Joel Berger

Introduction

The question of coexistence with wildlife is not a trivial matter. An expanding human population coupled with a vast reduction of wild habitat has consequences for every aspect of our biotic world, not least of which are global increases in human–wildlife conflict (HWC; Nyhus 2016). Although conflict occurs between humans and diverse wildlife, HWC with large-bodied predators is particularly salient because of potential for loss of life and the fear associated with it. Combined with habitat loss, intensive predator control to protect livestock over the past few thousand years, and especially the last ~200, has vanquished apex predators in most landscapes. Today, 64% of Carnivora species are threatened with extinction and 80% are in decline—arguably 'humankind's most pervasive influence on the natural world' with cascading effects on biodiversity, ecological resilience, process, and function (Estes et al., 2011: 301; Ripple et al., 2014; Wolf & Ripple et al., 2018). These facts are at the heart of many rewilding campaigns, hence the emphasis on the repatriation and restoration of apex Carnivora (Foreman, 2004; Berger, 2007; Ripple et al., 2022).

The fact that landscapes have been devoid of predators for centuries means that most humans today have never had direct interaction with them and lack specialised knowledge to facilitate coexistence (Røskaft et al., 2003). Such circumstances yield fearful and negative attitudes towards sharing landscapes with species such as grey wolves (*Canis lupus*) and grizzly bear (*Ursus arctos*) and have important consequences for HWC and rewilding success. Yet, we have coexisted with predators for >99% of our species' evolutionary history. How did we do this?

The following excerpts are taken from our own field notes:

> We were on an impromptu hike deep in grizzly habitat, moving quietly and quickly off-trail in a densely vegetated area. We had binoculars, but no bear spray or backpacks containing first aid kits, and certainly no weapons. The lack of gear was in part an oversight (no bear spray?!) and partly based on the rationalization that this was just a 20-minute jaunt, not an expedition. At least we had knowledge. It was the calls

DOI: 10.4324/9781003097822-25

of several ravens that alerted us first, followed by undeniable odor of a dead animal. As seasoned field biologists, our neocortex quickly retrieved information—we knew the signs of a kill site and the likelihood for encountering a predator, potentially a grizzly bear. Another quick realization (processed by our hippocampus and prefrontal cortex): we were lacking critical tools. It was then that our primitive limbic brain activated response based on fear (fight–flight–freeze).In our case we did not stay to fight, but initially froze, and then walked away from the scene.

(The authors in the Greater Yellowstone Ecosystem, June 2019)

During a visit to Baffin Island in the Canadian arctic, I had a frank conversation with a young Inuit (Nunatsiarmiut) inhabitant. As we walked together in an area of high polar bear abundance, I queried the young man as to how he lived in this landscape with so many bears. He answered by saying 'We are scared, but we are used to it.'

(Lambert in Nunavut, August 2014)

We use these quotes to illustrate the central points of our chapter: coexisting with large-bodied predators is facilitated by having enough experience and knowledge to mitigate fear/naivety, appropriately assess risk, and make informed decisions that foster sharing of landscapes. We suggest that understanding neurobiology and human–predator coexistence in the past can provide tenets for coexistence and successful rewilding that rests on a deep understanding of *habituation, learning,* and associated *tools* for coexistence.

Though our discussion has some relevance to the rich literature regarding the ecology of fear, landscapes of fear, and landscapes of coexistence, addressing these concepts directly are beyond the scope of this chapter (Brown et al., 1999: Laundré et al., 2001; Berger, 2007b; Oriol-Cotterill et al., 2015; Miller & Schmitz, 2019). Instead, after defining key terms and scope, we establish why carnivoran predators represent a special case in HWC. We then situate current circumstances of human–predator conflict in light of our understanding of how humans share(d) landscapes with Carnivora in deep time (Pleistocene), recent history (Holocene), and now (Anthropocene). We provide insight into the underpinning biology of fear and offer suggestions for addressing HWC based on habituation and the transmission of learned behaviour.

Defining terms and scope

Before proceeding we take this opportunity to be explicit about the animals of focus in this discussion—and the terms to reference them (Table 22.1). 'Predator' and 'carnivore' are often conflated and used interchangeably in the literature. However, while they have overlapping meanings, they are not synonymous. 'Predator' refers to any organism that consumes another organism. Technically, this can include herbivores (1° consumers) who are quite literally predators of plants (producers). 'Carnivore' has a narrower definition than that of predator as it relates to any animal in the third or higher trophic level of a food web (animals that eat other animals) and does not include predators of plants (i.e., herbivores). 'Carnivore' is frequently used to refer to species belonging to the Order Carnivora; however, these animals are most accurately referred to as 'carnivorans' and not 'carnivores'. In short: not all predators are carnivores, not all carnivorans are carnivores (e.g., the frugivorous maned wolf, *Chrysocyon brachyurus,* and kinkajou, *Potos flavus*), and not all carnivores are carnivorans.

Because we centre our discussion on species that belong to Order Carnivora, we use the term 'carnivoran' rather than the more commonly used 'carnivore' (Van Valkenburg, 2010). We do not

Table 22.1 Definitions of key terms used in this chapter

Term	Definition
Predator	An organism that consumes another organism. Includes herbivores that prey on plants (i.e., plant predators).
Carnivore	An animal that eats other animals; i.e., a carnivorous animal. An ecological term referencing animals that are in the 3rd trophic level or higher. Is not constrained to taxon; many species not classified within Order Carnivora eat other animals and are technically carnivores/carnivorous.
Order Carnivora	A monophyletic group of species in the Class Mammalia. Species are classified based on the presence of a set of specialised teeth (*carnassials*) adapted for slicing the flesh of other animals. Includes the last common ancestor of all dog/dog-like and cat/cat-like animals and their descendants. Not all species within Order Carnivora are carnivores.
Carnivoran	The abbreviated name of those species belonging to the Order Carnivora. A taxonomic name with no reference to ecological positioning in a food web.
Food Web	A graphical depiction of the ecology of who eats whom. Food webs are organised into trophic levels. Sometimes referred to as a consumer-resource system or trophic pyramid.
Producers	Organisms classified in the 1st trophic level (plants).
1° Consumer	Animals that eat plants, classified within the 2nd trophic level. Also called 'herbivores'.
2° Consumer	Animals that eat animals (carnivorous), classified within the 3rd trophic level. Also called 'carnivores'.
1st Trophic Level	First level of a food web. Includes all plants (producers).
2nd Trophic Level	Second level of a food web. Includes all herbivores (1° consumers).
3rd Trophic Level	Third level of a food web. Also known as carnivores (2° consumers).
Apex Predator	A predator at the top of a food web not typically preyed on by any other species in the system. Examples include grizzly bears (*Ursus arctos*) and grey wolves (*Canis lupus*). Note that whether a species is deemed 'apex' varies by ecosystem. In the Russian Far East, for example, tigers (*Panthera tigris*) kill grizzly bears.
Meso-predator:	A mid-ranked predator in a food web, often (though not always) omnivorous. Prey selection tends to be of smaller animals. Increasingly, meso-predators are becoming apex predators as they are released from competition due to removal of previous apex predators. Examples include coyote (*Canis latrans*) and raccoon (*Procyon lotor*).
Fear	An emotional response controlled by the brain's limbic system and the Hypothalamus–Pituitary–Adrenal axis (HPA). Activated by a perceived threat to life ('risk').
Human-Wildlife Conflict (HWC)	Conflict that arises when animals pose a direct and recurring threat to the livelihood and safety of people, leading to the persecution of that species and conflict among stakeholders about what should be done about the conflict (IUCN, 2020).
Coexistence	Dynamic but sustainable state in which humans and large carnivores co-adapt to living in shared landscapes where human interactions with carnivores are governed by effective institutions that ensure long-term carnivore population persistence, social legitimacy, and tolerable levels of risk (Carter and Linnell, 2016).
Rewilding	The process to restore ecosystem to achieve a self-sustaining autonomous nature (Carver et al., 2021). Includes all aspects of ecological structure, function, process, and resilience.

include herbivores (as plant predators) in our use of the term 'predator' but instead to any carnivoran with the capacity to prey on humans, their livestock, or pets and hence have high potential to be engaged in conflictual encounters. This can include mesopredators—a mid-ranking predator in the middle of a trophic level (e.g., coyote, *Canis latrans*) and apex predators—top-ranking predators that are consumed by no other predator in the species assemblage (e.g., grizzly bear, *Ursus arctos*). Though the term mesopredator can technically include omnivorous non-carnivorans (e.g., baboon, *Papio anubis*) we will restrict our discussion to members of the Order Carnivora.

We also clarify our use of 'fear', 'conflict', 'rewilding', and 'coexistence' (Table 22.1). Fear is an emotional response to perceived risk to life; as we discuss in detail below, this response is managed by the brain's limbic system and Hypothalamus-Pituitary-Adrenal axis (HPA) and is shaped by experience and knowledge. 'Conflict', as defined by the IUCN Human-Wildlife Conflict Task Force, occurs when animals pose a direct and recurring threat to the livelihood and safety of people, leading to the persecution of that species and conflict among stakeholders about what should be done about the conflict (IUCN, 2020). 'Rewilding' is the process by which ecosystems are restored (Carver et al., 2021).

'Coexistence' among species can have several meanings. In ecological terms, coexistence *sensu stricto* references the state of two of more species living in the same place at the same time (sympatry), with no explicit indication as to whether the species are in either direct or indirect interaction or whether they have any impact (negative or positive) on the population abundance of the other species (Holt, 2013). Here, we employ a narrower and inherently more anthropocentric definition that explicitly refers to humans sharing landscapes in ways that facilitate long-term carnivore persistence of carnivores and human safety (Carter & Linnell, 2016). Note that this definition does not mean absence of risk; it means tolerance of an acceptable level of risk (Carter & Linnell, 2018). Finally, while our discussion is relevant to settings around the world, we focus on circumstances of human-predator interactions in the American West in settings outside of protected areas.

The special case of carnivoran predators = perfect storm for HWC

Globally, the scale and scope of HWC is intensifying, with serious consequences both for conservation of biodiversity as well as human well-being and livelihood; carnivoran predators are certainly not the only taxa involved in deleterious interactions with humans (Nyhus, 2016). For example, in the 20th century birds were responsible for the loss of over 450 human lives due to collision with aircraft. In the US alone Hymenoptera (including bees, wasps, and hornets) caused the death of 1,109 people between 2000 and 2017 (Thorpe, 2016; MMWR, 2019). Birds and bees notwithstanding, several features of our relationship with carnivoran predators render them an especially challenging case in HWC:

- *First,* although quite rare, carnivoran predators can and do attack humans, livestock, and pets.
- *Second,* humans are both predator and prey. As a prey species throughout our evolutionary history our neurobiology is wired to fear predators. Carnivoran predators are frightening. It is appropriate that we are fearful of them.
- *Third,* while HWC occurs with diverse taxa, including relatively benign species (e.g., grey squirrels, *Sciurus carolinensis*, that deplete bird feeders or damage roofs), a negative interaction with carnivoran predators may potentially result in the loss of life or serious injury to humans, their livestock, or pets. This means that sharing landscapes with animals such as grizzly bears, grey wolves, and mountain lions (*Puma concolor*) requires a diverse tool kit and specialised knowledge beyond that required for life with wild animals in general.

- *Fourth,* because of concerted efforts by all levels of US governance throughout the 19th and much of the 20th centuries, predators—especially grizzly bears, grey wolves, and mountain lions—were all but completely extirpated throughout their ranges in the American West.
- *Fifth,* because of this successful extermination campaign, encountering large carnivoran predators became increasingly rare as the 20th century unfolded. Along with a loss of the animals themselves came a loss of the special knowledge required to coexist with them.
- *Sixth,* virtually all predators involved in HWC in the American West are increasing in numbers, including black bear (*Ursus americanus*), grizzly bear, mountain lion, grey wolves, and coyotes. Some species (e.g., black bear, coyote, mountain lion) are adapting to anthropogenic landscapes, while others (e.g., grey wolves, grizzly bears) are benefitting from their protected status and restoration from rewilding campaigns.

This potent combination of factors compounds to create a perfect storm of human–predator conflict that is informed by fear. Yet, for most of our species' existence we coexisted with predators; only recently has this relationship shifted. We thus situate the remainder of our discussion within the broader context of our relationships with predators in deep time and in recent history.

Humans and predators in deep time and more recent history

Answers to when and how anatomically modern *Homo sapiens* (hereafter 'humans') evolved are as diverse and complex as any in the paleontological record (Bräuer & Smith, 2020). A fragmented fossil record, issues including (but not restricted to) dating inconsistencies, and disagreement as to the primacy of genetic data over fossil information have resulted in a bewildering array of phylogenetic interpretations. We do not address these matters here. Suffice it to say that a 'best fit' interpretation using both fossil and genetic evidence suggests that humans evolved in Pleistocene Africa some 300,000 to 190,000 kya—a time of a considerably more diverse, abundant, and physically larger guild of carnivoran predators than there are today (van Valkenburgh et al., 2015; Galway-Witham & Stringer, 2018).

The Pleistocene epoch (2.6 mya–11.7 kya) was a period of extremes in virtually all ways that mattered to the fitness and survivorship of our direct ancestors, not least of which were the predators with whom they shared habitat. The significance of a terrestrial, carnivoran predator guild in shaping extinct ecosystems tends to be underestimated because there is literally no modern analogue of such a guild anywhere on the planet today (van Valkenburgh et al., 2015). Prior to the advent of the Holocene (11.7 kya), mammalian predator guilds were more species-rich and abundant than extant assemblages. Moreover, carnivoran predator guilds were, on average, 47% larger and contained more (5–6) large (>100kg) carnivorans per community than any intact assemblage now (Figure 22.1; van Valkenburgh et al., 2015).

Our ancestors were thus sympatric with diverse large-bodied, fierce predators and likely in competition for similar large prey species. Though eventually humans innovated a sedentary horticultural lifestyle, most of our lineage's existence was characterised by mobility and specialised big-game hunting (Speth, 2010). Starting at around 1.5 million years ago, the direct forebearer of modern humans—*Homo erectus*—embellished upon an earlier (far simpler) chopping tool technology (Oldowan) to a more sophisticated tool technology (Acheulean) comprising hand axes, cleavers, picks, and flakes for processing carcasses of large mammals (Muller et al., 2022). The use of spears and ambush tactics, in combination with a diverse tool kit facilitated predation that did not require contact with the prey itself (Bunn, 2019). In this way, early humans occupied a unique niche within the broader predator guild consuming ungulate prey.

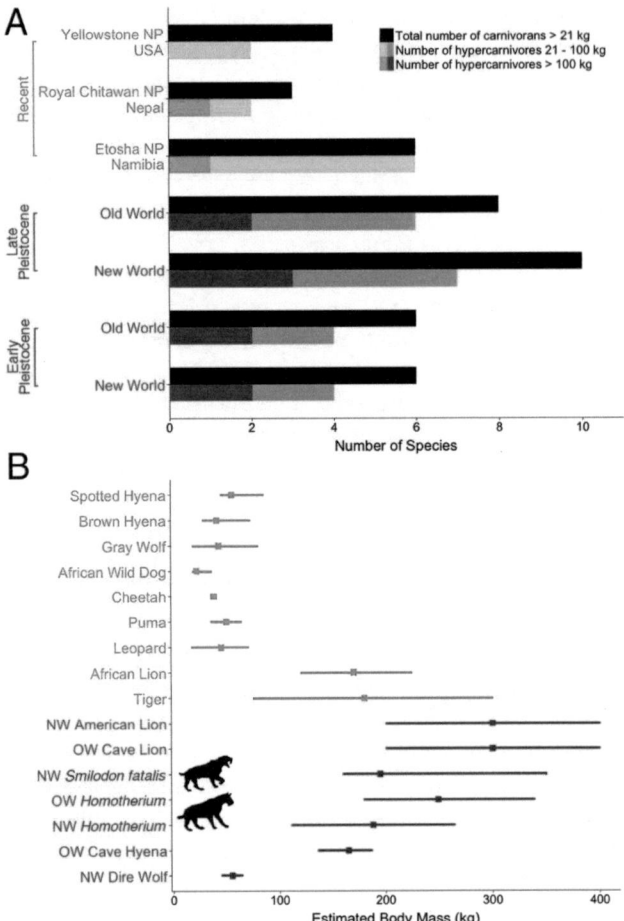

Figure 22.1 Predator guild composition for four Pleistocene (red) and three extant (blue) communities. Indicated for each guild are the total number of species of carnivorans (hypercarnivores and omnivores, e.g., ursids) with masses>21kg (black), the subset of these that are hypercarnivores (two-toned bar), and the subset of these that are hypercarnivores with masses >100kg (dark blue or red). (B) Estimated body masses (mean and range) of extant (blue) and extinct (red) hypercarnivores. Silhouettes are provided only for the sabertooth cats because they lack modern analogs. Reproduced with permission from Van Valkenburg, B., Hayward, M.W., Ripple, W.J., and Roth, L. (2015). The impact of large terrestrial carnivores on Pleistocene ecosystems. *PNAS*, 113(4): 862–7.

Importantly, we have also been prey throughout our evolutionary history (Treves & Naughton-Treves, 1999). Fossil and archaeological evidence from throughout the hominin fossil record suggests that at any given time anywhere from 6–10% of humans and human ancestors were preyed on. This evidence comes in the form of tooth marks on bones, holes in crania, and the taphonomy of assemblages (Sussman & Hart, 2005). This is in line with the approximated percentage of predator caused mortality on terrestrial primates today (e.g., *Cercopithecus aethiops*, *Papio anubis*). In short, humans and our hominin ancestors have always had a profoundly important relationship with carnivoran predators as both competitor and prey, begging the question:

How does a day-active ape, with a weak musculature and regressed climbing abilities, with poor to mediocre night vision, incapable of hard biting, disarmed and harmless, a fat and historically tasty morsel for predators, survive on the ground with many night-active predators? And survive night after night for decades, despite menstruation, births, the crying of babies, snoring, or the scent of wounds acquired accidentally in the course of daily activity?

(Geist, 2016: 3)

How our Pleistocene ancestors survived: fear and habituation, learning and tools

How our ancestors coexisted with a fierce and abundant predator guild is explained by our neurobiology, cultural innovation, and reliance on learned behaviour. As we describe, though like most vertebrate species humans are wired for fear, a combination of habituation, experience, and knowledge facilitated risk-assessment and decision making. Furthermore, Pleistocene humans had the tools both to procure prey and prevent predation and, as a social species, this learned knowledge was shared among kin and other conspecifics.

Coexisting with large-bodied carnivorans such as saber-toothed cats (e.g., *Homotherium* spp) and cave bears (*Ursus spelaeus*) meant that early humans and our hominin forebears dealt with the potential threat of predation on a 24hr basis. Under these circumstances fear would have been an omniscient reality. The neurobiology of fear as emotion dates to well over 350 million years ago and is shared by all tetrapods (Laberge et al., 2006). Fear is managed by a set of structures (e.g., amygdala, hypothalamus) in the brain known as the limbic system that prepare the body for action (Figure 22.2; Gross & Canteras 2012). These responses occur so rapidly that the cascading effects occur even before the neocortex (e.g., prefrontal cortex, hippocampus) has completely processed the events—we can react without conscious cognitive processing (Figure 22.2).

However, no organism can persist in a constant state of fight-flight-freeze. An activated HPA response is metabolically expensive and comes at the expense of other life-sustaining behaviours; multiple studies have demonstrated that scared prey species spend less time in feeding, produce fewer offspring, and have lowered survivorship overall (Reisland et al., 2021; Ogden, 2016; Zanette et al., 2011; Surachi et al., 2016). Fear also impacts gene expression and hijacks the brain's ability to learn and store memories critical for survival (Ogden, 2016; Feng et al., 2015; Lima and Dill, 1990; Mathis et al., 2008). Thus, while natural selection has outfitted organisms to respond to extreme threat, it is not adaptive to remain in this state.

The reality is that coexisting with predators does not mean constant imminent threat of attack. While an individual must be sensitised to the threat of predation, they must also be able to evaluate signals such that they do not continually over-respond to that threat. This is accomplished via habituation (a diminished response to stimuli through exposure) and learning (Eisenstein et al., 2001). Optimal behaviour in a landscape of fear is thus a balance between sensitisation and habituation—i.e., of *knowing* when to be afraid of predators and when not to be. This is true not just for humans, but for all prey species and is highly conserved across taxa (Eisenstein et al., 2001; Berger, 2007; Carthey & Blumstein, 2018).

While behavioural response to some predators (e.g., snakes) appears to be instinctive (Isbell, 2011), knowledge of most predators and how to behave around them is acquired through experience and learning and would have started at a very young age. Existing as they did in a landscape with diverse large-bodied carnivoran predators, Pleistocene humans would have had rich knowledge of other species' behaviour and habitat requirements to successfully predict when and where encounters might occur and thus be able to assess risk appropriately.

Tools were part of this learned repertoire and were central to our ancestor's survival and innovations during the Pleistocene (e.g., hand axes and spears) were key to maintaining distance from predators in combat (Figure 22.3). The addition of dogs may have facilitated our survival

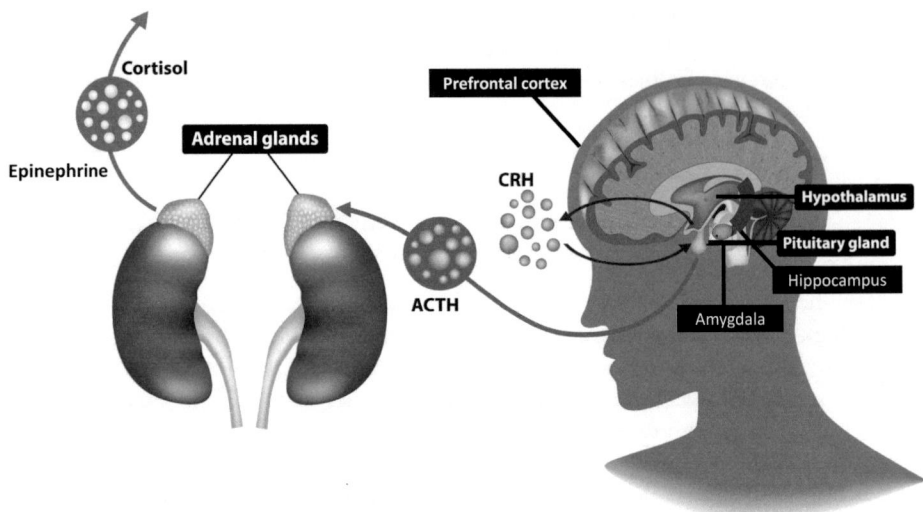

Figure 22.2 Illustration of limbic and Hypothalamus-Pituitary-Adrenal (HPA) axis. Upon exposure to a threatening stimulus, the amygdala signals the hypothalamus, which activates the sympathetic nervous system and delivers signals to the adrenal glands. The adrenals emit epinephrine that stimulates physiological response throughout the body in preparation for either fight, flight, or freeze. These responses occur so rapidly that the cascading effects occur even before the neocortex has completely processed the events—we can react without conscious cognitive processing. As the initial cascade of epinephrine subsides, the hypothalamus signals the HPA axis which maintains the activated stress response through activation of corticotropin-releasing hormone (CRH), which triggers adrenocorticotropic hormone (ACTH) from the pituitary, which travels to the adrenals to yield cortisol. These processes occur so rapidly that physiological response can precede processing by the neocortex (e.g., hippocampus, prefrontal cortex).

during the Pleistocene as well, both by thwarting predation and by providing a competitive edge against other predators, including other hominin species (e.g., Neanderthals; Shipman 2015a, 2015b). As early humans radiated out of Africa and colonised Eurasia some 50–70,000 ya, they encountered multiple wolf species, including grey wolves (*Canis lupus*). A prevailing hypothesis of dog domestication is that ca. 30,0000–40,000 years ago, some of these grey wolves overcame their fear of humans and initiated a process of self-domestication to take advantage of a new food source in the form of scavenged meat from human kill sites or perhaps by consuming our feces which must have been widespread around encampments (Larson et al., 2012).

Humans, in turn, learned the utility of allowing these calmer, less aggressive versions of wolves to remain in proximity. Such proto-dogs would have served as excellent sentries and provided predator protection, and were essentially an additional tool for humans to successfully exist in the glaciated and fiercely competitive landscapes of Pleistocene Eurasia (Shipman, 2015a).

The Holocene and Anthropocene: Loss of predators and the knowledge of how to coexist

Arguably one of the most important shifts in human evolutionary history occurred in the transition from a hunter-gatherer lifestyle to that of agriculture and permanent settlement ~10,000 years ago (Childe, 1936; Weisdorf, 2005). Hallmarks of this transition include domestication of plants and animals and a shift from living in temporary shelters to permanent

(a)   (b)

Figure 22.3 Artist reconstructions of Pleistocene predator guilds. Illustrating tool kit of anatomically modern *Homo sapiens* ('humans'). (a) early humans interacting with cave bear (*Ursus spelaeus*) in Pleistocene Eurasia. Note use of ambush tactics, spears, and other tools that do not require direct contact with predator. (b) A pack of early dogs ('proto-dog') working cooperatively in a hunt of woolly mammoth (*Mammuthus primigenius*). Hypotheses regarding dog domestication have suggested that early modern humans were competitively successful because they used early dogs as a tool for increasing vigilance against predation and for maximising big-game hunting success. Both images reproduced with permission from Alamy.

settlements. However, the Neolithic Revolution was more than just a transition in food acquisition. It had profound consequences for myriad aspects of our relationship with the natural world including radical modification of landscapes for food crop cultivation with associated irrigation and deforestation. This allowed for production of surplus food that buoyed increasing human populations during periods of food scarcity.

The Neolithic Revolution also transformed our relationships with wildlife (Russell, 2015). Both crops and domesticated animals would have been an abundant and readily procured source of food by wild animals, thus the onset of human–wildlife conflict as we understand it today along with predator removal that has continued since. Prior to domestication and life in permanent settlements, it is highly unlikely that humans would have killed predators to reduce competition for prey species.

The cultural–cognitive distinction between animals that were 'wild' versus 'not wild' (i.e., domesticated) would have formed as humans settled into village life, sustaining themselves with livestock and crops, and protecting those livestock and crops from unwanted exploitation from other species. Distinct from 'not wild', wild animals would have further been designated as those to be hunted for food versus those to be killed for self-protection or to thwart predation on livestock. Sharing landscapes with unwanted wild animals meant keeping those animals at a distance—accomplished by using livestock guardian dogs and the innovation of fencing and walls which would have served the dual purpose of preventing the intrusion of unwanted human competitors as well as competing or predatory animals (Hayward & Kerley, 2009).

Over the subsequent 10,000 years, predator control has been a reality of human existence. State mandated removal of wolves from ancient Greece, for example, was underway in earnest by 500 BC. In Europe, policies for predator reduction have been in place for hundreds of years. Indeed, 'control to the point of extirpation was seen as social duty, for the economic benefit of the community, with sponsorship (e.g., by feudal patronage or state bounty schemes) as an incentive, and often with penalties for shortcomings' (Reynolds & Tapper, 1996: 128). When and wherever humans achieved a high population density, predator control was highly effective. European lions (*Panthera leo*) were extirpated in Macedonia around 100 AD and in Armenia and Georgia approximately 1000 AD (Masseti and Mazza 2013). By the 18th century there were no brown bears, grey wolves, or Eurasian lynx (*Lynx lynx*) in Great Britain.

Significantly, though these animals were physically absent, they loomed large in fairy tales and mythology—a means by which to keep fear alive and present in the absence of the threat itself. The pedagogical significance of fairy tales is a much-discussed topic, with arguments both for and against their utility in teaching appropriate levels of fear to children who do not have direct experience with the source of threat (Tatar, 1993). Little Red Riding Hood in particular has been imbued with heightened meaning in the controversary surrounding rewilding with wolves in the American West (Figure 22.4; Lappalainen, 2019).

It must have been a shock, then, when European colonisers arrived upon the eastern shores of what is now the United States and encountered species known primarily from tales told by the likes of the Brothers Grimm and Aesop. What ensued was a campaign against predators that was as fierce if not more so than it had been in Europe. It was certainly more intensive an effort—the decimation of predator populations that took thousands of years in western Europe occurred over a span of a mere 100–200 years in the United State. The addition of various lethal poisons (e.g., cyanide, strychnine, sodium fluoroacetate) to the repertoire of more typical forms of killing such as shooting and trapping made the US campaign against predators particularly effective (Gulliford, 2015). Species such as grey wolves and grizzly bear were viewed as a threat to the American way of life, an assault against Manifest Destiny (Lopez, 1979). Apex species were targeted, but no predator—including smaller mesopredators—were spared (Flores, 2016). Attitudes towards these species were quite literally described in terms of fear and loathing, such that even the so-called conservation president Theodore Roosevelt described grey wolves as 'beasts of waste and desolation' and mountain lions as 'bloodthirsty and cowardly' (Johnston, 2002; Roosevelt, 1889). By the mid- to late-19th century, resolutions to remove predators were in place at all levels of governance from local to state to territory to Federal. In 1915, Congress appropriated Federal funds in direct support of predator removal, and it worked (de Calesta, 1976). By the mid-1960s greys wolves were reduced to a handful (~400) in the upper Midwest, and grizzly bear distribution reduced by 98%.

The significance of the extirpation programme of predator removal in Europe and the United States has implications beyond the loss of the animals themselves. It means that, for many generations (at least three in the American West), humans have not had to deal with the

Figure 22.4 Sign held in protest against grey wolf reintroduction. Source: *Ruidoso News*, 14 February 2017.

exigencies of life in sympatry with carnivoran predators. As numbers of grizzly bears, wolves, and mountain lion declined so too did knowledge of their behaviour and habitat use and an ability to appropriately assess risk (Bruskotter et al., 2017). More than just knowledge of the behaviour of predators, information regarding the use of tools (e.g., livestock guardian dogs, hazing devices) is not commonly held amongst all, but instead has become the domain of specialised non-profits (e.g., Defenders of Wildlife, Working Circle).

In addition to a loss of knowledge and the tools to facilitate coexistence, humans are no longer habituated to the experience of having predators on the landscape. A lack of experience with predators means an uninformed understanding of the likelihood of attack in various contexts (Herberholz & Marquart, 2012). We are, essentially, naïve prey as has been shown in other species in the absence of predators. Within just a few generations of the absence of predators, ungulate prey (e.g., moose, *Alces alces,* elk, *Cervus elaphus,* caribou, *Rangifer tarandus*) fail to respond appropriately (e.g., clustering, flight) to stimuli mimicking predator attack (Berger et al., 2001, 2007).

A lack of knowledge or experience with frightening stimuli such as predators not only results in naïve behavioural responses, but also an activated limbic response (fear), impairing our ability to engage in rationale decision-making by interrupting pre-frontal cortex processing (Olsson & Phelps, 2007). Furthermore, fear itself induces a feeling of dread which, in turn, shapes tolerance towards the feared object (Fischhoff et al., 1978). This translates to interactions with wildlife (Prokop et al., 2013; Dickman 2010). Work on perceptions towards predators indicates that fear underpins hostility towards wildlife, while experience with those species reduces negative affect towards that wildlife (Røskaft et al., 2003; Prokop et al., 2013).

Fear can occur even with no direct experience of the threat itself—the same neural structures underpinning fear from direct exposure to a threatening stimulus are involved in learned fear in the absence of that stimulus. Recent work in neuroscience and psychology has demonstrated the profound importance of both observational and social learning of fear, including the fear of predators (Lima & Dill, 1990; Olsson & Phelps, 2007; Ogden, 2016; Gross & Canteras, 2012). Humans can learn to be fearful of something if they observe others being fearful of that thing; this is activated by mirror neurons which fire when an individual observes an action and then mirrors the behaviour of the other. Fear is also socially learned by humans and other non-human primates (Kilner & Lemon, 2013). Social learning of fear (SLoF) is an indirect process of learning sometimes referred to as vicarious learning that occurs through listening to others—or reading the written words of others—conveying fear about an object (Olssen & Phelps, 2007). SLoF is typically measured in the absence of the source of the socially learned fear. In this way, it is distinct from so-called contagious fear responses such as when horses may spook because they detect other animals in the herd running (Olsson, 2012).

In sum, humans are not born with an innate fear of predators such as grey wolves (Figure 22.5). Fear of such species has been enculturated from lessons derived from fairy tales and myths, which is then reinforced by interacting with others (socialisation of fear) who also fear those animals. A lack of knowledge in combination with enculturated and socially learned fear yields dread and hostility to the source of the dread. If the source of fear is an animal that may impact your livelihood, this only adds to the emotionally charged milieu of conflict and decreased tolerance for sharing landscape with carnivoran predators.

Implications for HWC and rewilding

These issues have important and practical implications for rewilding campaigns that involve the reintroduction or relocation of apex predators. Although predators are not the only taxon of concern in rewilding initiatives, they feature large in discussions of biodiversity and ecosystem

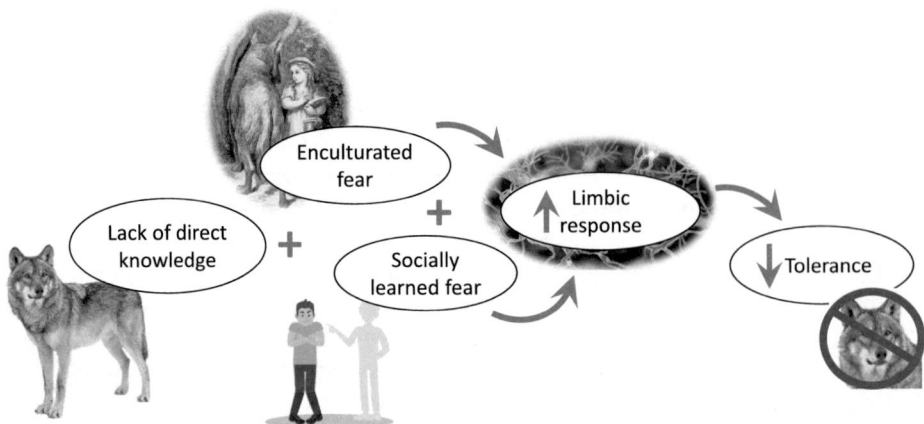

Figure 22.5 Simplified model illustrating the interaction of a lack of knowledge about a fearful object, enculturated fear, socially learned fear, a limbic response, and tolerance as discussed in text. In sum, we suggest that a lack of knowledge and experience about something that is potentially dangerous, in combination with enculturated fear about that thing that comes in the form of well-known stories passed through generations, along with fear that we are taught directly in social settings can result in a highly charged limbic response to that thing in its absence and without any direct experience. The result of these circumstances can lead to reduced tolerance.

restoration. Apex predators have been removed from virtually every ecosystem on Earth with marked implications for all other trophic levels in those systems ('trophic downgrading', Estes et al., 2011: 301). Considerable efforts around the world, including in the American West, have centred on the restoration of apex predators into systems.

At issue is that in rewilding campaigns predators are commonly returned to areas where humans have not had to deal with them for several generations, if not centuries. Moreover, these initiatives are often directed by conservation biologists (who have deep understanding of animal behaviour) who may make assumptions that most people are generally aware of threats presented by wildlife. But this is not the case (Dickman, 2010).

In fact, mismatch between perception of risk from predators and actual risk can be extreme (Nyhus, 2016). Compounding this complexity is the fact that humans are vastly (>1000x) more likely to accommodate risk if it is voluntary rather than imposed—such as in statutory reintroductions of predators (Starr, 1969; Dickman, 2010; Bruskotter & Wilson, 2014; Bruskotter et al., 2017). The example of grey wolf reintroduction to Yellowstone National Park in 1995 is a case in point. In the Greater Yellowstone Ecosystem much of the contention surrounding wolf repatriation centred on resentment towards US Fish & Wildlife and the Federal Government for imposing wolves in a region that had not had wolves for ~70 years (Wilson, 1996). Similarly, central to the current controversy surrounding wolf reintroduction in Colorado is resentment towards human populations on the Front Range (eastern slope) of the Rocky Mountains where >85% of Colorado residents reside (e.g., Denver metropolitan area) imposing a vote in favour of wolves that would result in restoration of wolves in the much less populated, rural landscapes of the western slope (Niemiec et al., 2020).

Nonetheless, there are tangible and lost-tested strategies and tactics to life in landscapes with large-bodied carnivoran predators that data back thousands if not millions of years that centre on *habituation, learning,* and *tools* (including dogs):

Habituation and Learning: In any rewilding campaign that involves restoration of carnivoran predators, discussions should occur sooner rather than later to habituate local human inhabitants to the experience of life with animals such as grizzly bear. This should include video and as much imagery as possible (e.g., see Episode 28 of the Wyoming Game & Fish Department's Work in the Wild series on preparation for predator attacks www.youtube.com/watch?v=yTR2ZYfX wbI). Re-enactment scenarios in which human subjects are trained how to respond in encounters with predators can be particularly useful. Such trainings should occur more than just once and capitalise on repetition to habituate the subject to the stimulus, tantamount to military training tactics around the world that emphasise task repetition and its value for reducing limbic response and maximising appropriate neocortical assessment of risk (Hagman, 1980). A repetition training tactic is also undertaken by professional firefighters and law enforcement in preparation for real encounters. We identify this as a tactic not so much from the perspective that humans will be attacked by predators (the reality is that throughout the American West, dangerous encounters with predators are extremely rare). Instead, we raise this as an important means by which to expose individuals to the stimulus and then—importantly—to contextualise the exposure in terms of learning. The combination of habituation and learning is particularly powerful.

Animal prey also need to be trained if they have not been habituated to life with predators or lack critical learned tactics. This is the case both in rewilding programmes that aim to reintroduce naïve captive-bred or translocated animals (Griffin et al., 2000; Berger et al., 2001; Carthey & Blumstein, 2018) as well as the domesticated animals that live in the landscape receiving the reintroduced animals. Despite intensive selective pressure on organisms for avoiding predation, learning and experience are essential to developing appropriate behavioural response to a predator and involves optimisation of multiple elements of nervous system function including sensory assessment of threat stimuli, rapid neurological transmission of information, and fast and accurate motor activation (Herberholz & Marquart, 2012). Training campaigns for teaching predator-naïve cattle and other livestock have recently sprung up throughout the northern Rocky Mountain West post grey wolf restoration and grizzly bear range expansion (e.g., www. rancherpredatorawareness.com; Coats, 2021).

Habituation and learning also occurs through time: the longer humans have lived with the object of fear (e.g., grey wolves, grizzly) the more commonplace the experience becomes. As a stimulus becomes increasingly unremarkable, emotional reaction is down-regulated by the parasympathetic nervous system. With increasing familiarity comes increased tolerance. For example, in work conducted in rural communities in Montana and Washington, Young and colleagues (2015) demonstrate that fear of—and attitudes towards—carnivoran predators becomes lessened and less hostile the longer livestock producers have lived with them. Beyond just habituation, tolerance towards predators is shaped by learning. Throughout regions of Europe, Asia, and Africa where high abundance of carnivoran predators have persisted, coexistence is achieved because of long-lived practices related to animal husbandry and thwarting predators (Linnell et al., 2001; Young et al., 2015; Chapron et al., 2014; Athreya et al., 2013; Woodroffe, 2000)

Tools: There is a vast literature on the tools that can be used in mitigating conflict and living with predators and a description of this array is beyond the scope of this discussion. There are many excellent sources (see for example: Sowka, 2012; Shivic, 2014; Jaicks, 2022; Living with Wildlife Foundation Resources guides https://lwwf.org/resource-guides; Colorado State University Extension Resources: https://extension.colostate.edu/topic-areas/people-predat ors). Among these tools are included:

- The use of *livestock guardian dogs and other animals*. Humans have been in association with proto-dog and fully domesticated dogs for over 30,000 years. As discussed, in the earliest phases of this association, it is likely that a primary reason for humans tolerating early proto-dog (wolves) in proximity would have been their utility in protecting against other predators. This practice has continued through history. Today, breeds such as Kangal, Komodor, Akbash, Maremma, Tatra sheepdog, Karakachan, and Anatolian shepherds are particularly effective. Other animals such as donkeys, mules, and llamas can be used as well.

- *Altering livestock husbandry and management* to decrease likelihood of predation. These practices run the gamut of training livestock to engage in anti-predator behaviour (e.g., huddling, standing ground instead of running), maintaining cohesiveness of herds, bringing animals into enclosures at night, etc.
- *Increase human vigilance:* Having humans on the ground—either on horseback or vehicle—both increases detection rates and can serve to frighten off predation.
- *Creating barriers against predation.* The use of barriers such as walls, fences, and trenches has been in place for at least 10,000 years with the advent of domesticated livestock and are highly effective.
- *Modifying predator behaviour through hazing:* Diverse hazing devices exist from fladry, to light and noise-making devices.

We also suggest that an additional tool in the toolkit of human-wildlife coexistence is *compassion* (Figure 22.6). As with fear (midbrain, particularly amygdala) and learning (neocortex, particularly hippocampus and prefrontal) compassion has its roots in neurobiology but must be practiced (Chierchia & Singer, 2017). Though compassion as a tactic in conversation has been forwarded with respect to our ethics regarding non-human species (e.g., Ramp & Bekoff, 2015), it has only very rarely been proposed as a tool for facilitating human–wildlife coexistence in its application towards other humans. In the Kavango-Zambezi Transfrontier Conservation Area, Namibia, training in non-violent communication was shown to be highly effective in conversations related to the threats of life in landscapes with large predators for diffusing human-human conflict (Kanksy & Maassarani, 2021). As conservation practitioners who may have vastly different experiences with wildlife than those with whom we are working to impose repatriated species, being compassionate about the fact that fear is a universal human experience can go a long way in shaping our approaches at the outset.

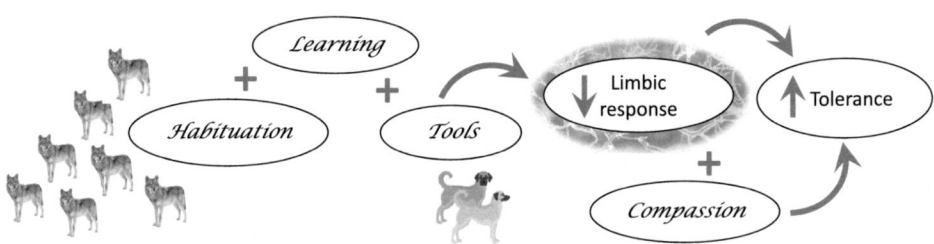

Figure 22.6 Simplified model illustrating the interactions among habituation, learning, and access to tools on limbic response and tolerance. As discussed in the text, habituation to a stimulus occurs through repeated exposure. Habituation about a potentially dangerous stimulus (in this case predators) is enhanced through learning that can come from either observing the predator with no deleterious consequences or from direct teaching about how to live with predators in workshops and other trainings. Tools such as using livestock guardian dogs and hazing devices are part of this learning. Habituation and learning result in a lowered limbic response which then allows higher level cognitive processing to occur (risk-assessment, decision-making) and increased tolerance. We also suggest that compassion and understanding about the state of fear in others and recognising and validating these emotions can increase tolerance in rewilding tactics.

A final word

We recognise that myriad variables impact human perceptions of, emotions about, and attitudes towards wildlife (Knight, 2000; Manfredo et al., 2009; Dickman, 2010; Bruskotter & Wilson, 2014; Convery & Davis, 2016; Nyhus, 2016; Angelici, 2016; Frank et al., 2019). We have presented only a partial picture of this complexity and have generalised across culture, economics, education, psychology, personal experience, values, and even political party affiliation (Knight, 2000; Treves & Karanth, 2003; Teel et al., 2007; Manfredo, 2008; Manfredo et al., 2009; Dickman, 2010; Bruskotter & Wilson, 2014; Treves & Bruskotter, 2014; Lute et al., 2016; Nyhus, 2016; Pooley, 2016; Frank et al., 2019). For example, affiliation with either Republican or Democratic political parties dominated other variables (e.g., ages, sex, private-land ownership, education) in one recent evaluation of attitudes towards grey wolves in eastern Oregon (Hamilton et al., 2020). We also are cognisant that we have presented a picture best understood through the lens of western European sociohistory. It is certainly not the case that everyone in the American West is afraid and intolerant of predators (Young et al., 2015;

Figure 22.7 Examples of Large Carnivore Species that Inhabit (either Persisting or Recovering) Multi-Use Landscapes outside Protected Areas. These examples are nonexhaustive but illustrate the fact that a range of carnivore species are currently sharing landscapes with humans around the world. Negative effects and conflicts associated with these carnivores vary greatly in each of these landscapes, but can sometimes be severe. The importance of shared landscapes for global carnivore recovery efforts necessitates a more holistic conceptualisation of human–carnivore coexistence that can be operationalised on the ground. Graphic and legend reproduced with permission from Carter, N.H., Linnell, J.D.C. (2016). Co-adaptation is key to coexisting with large carnivores. *Trends in Ecology and Evolution*, 31: 575–8.

Manfredo et al., 2021). Finally, in no way do we mean to deny individual agency in decision-making by suggesting humans are the hapless victims of their own biology and evolutionary history.

Caveats aside, in the end our message is simple: humans have coexisted with predators for virtually all of our history. In regions of the world where carnivoran predators have persisted or have returned, coexistence is still occurring (albeit with varying degrees of associated conflict), including in the American West (Figure 22.7; Carter & Linnell, 2016). Recognising that habituation to their restoration—either by geographical expansion of their populations or by reintroduction—is an achievable objective facilitated by learning is fundamental to our mission of restoring wildlife and ecosystem function. We are all wired for fear. Being reminded of this can facilitate compassion and foster communication with those whose lives are most directly impacted by the predators themselves.

References

Angelici, F.M. (2016). *Problematic Wildlife: A Cross-Disciplinary Approach*. Springer Publishers.

Athreya, V., Odden, M., Linnell, J.D., Krishnaswamy, J., and Karanth, U. (2013). Big cats in our backyards: Persistence of large carnivores in a human dominated landscape in India. *PloS One*, 8: e57872.

Berger, J. (2001). Carnivore repatriation and Holarctic prey: narrowing the deficit in ecological effectiveness. *Conservation Biology*, 21: 1105–17.

Berger, J. (2007). Carnivore repatriation and Holarctic prey: Narrowing the deficit in ecological effectiveness. *Conservation Biology*, *21*(1): 1105–16.

Berger, J. (2007b). Fear, human shields and the redistribution of prey and predators in protected areas. *Biological Letters*, 3: 620–3.

Berger, J. (2008). *The Better to Eat You With: Fear in the Animal World*. University of Chicago Press.

Bräuer, G. and Smith, F.H. (2020). *Continuity or Replacement: Controversies in Homo Sapiens Evolution*. Routledge.

Brown, J.S., Laundre, J.W., and Gurung, M. (1999). The ecology of fear: optimal foraging, game theory, and trophic interactions, *Journal of Mammalogy*, 80: 385–99.

Bruskotter, J.T., Vucetich, J.A., Manfredo, M.J., Karns, G.R., Wolf, C., Ard, K., Carter, N.H., Lopez-Bao, J., Chapron, G., Gehrt, S.D., and Ripple, W.J. (2017). Modernization, risk, and conservation of the world's largest carnivores. *BioScience*, 67: 646–55.

Bruskotter, J.T. and Wilson, R.S. (2014). Determining where the wild things will be: using psychological theory to find tolerance for large carnivores. *Conservation Letters*, 7: 321–32.

Bunn, H.T. (2019). Large ungulate mortality profiles and ambush hunting by Acheulean-age hominins. *Journal of Archaeological Science*, 107: 40–9.

Carter, N.H. and Linnell, J.D.C. (2016). Co-adaptation is key to coexisting with large carnivores. *Trends in Ecology and Evolution*, 31: 575–8.

Carter, N.H., Shrestha, B.K., Karki, J.B., Pradhan, N.M.B., and Liu, J. (2012). Coexistence between wildlife and humans at fine spatial scales. *Proceedings of the National Academy of Sciences*, 109: 15360–5.

Carthey, A.J.R. and Blumstein, D.T. (2018). Predicting predator recognition in a changing worlds. *Trends in Ecology and Evolution*, 33: 106–15.

Carver, S., Convery, I., Hawkins, S., Beyers, R., Eagle, A., Kun, Z., van Maanen, E., Cao, Y., Fisher, M., Edwards, S.R., Nelson, C., Gann, S... (2021). Guiding principles for rewilding. *Conservation Biology*, 35(6): 1882–93.

Chapron, G., Kaczensky, P. Linnell, J.D.C., von Arx, M., Huber, D., Andrén, H., ... Boitani, L. (2014). Recovery of large carnivores in Europe's modern human-dominated landscapes. *Science*, 346: 1517–19.

Chierchia, G. and Singer, T. (2017). The neuroscience of compassion and empathy and their link to prosocial motivation and behavior. In J-C. Dreher and L. Tremblay (eds), *Decision Neuroscience: An Integrative Approach*. Academic Press, pp. 247–57.

Childe, G. (1936). *Man Makes Himself*. Watts and Company.

Coats, M.L. (2021). Training cattle with predator awareness. Unpublished training manual, Rancher Predator Awareness, www.rancherpredatorawareness.com/pdfs/training_cattle.pdf

Convery, I. and Davis, P. (2016) *Changing Perceptions of Nature*. Boydell & Brewer.

deCalesta, D.S. (1976). Predator control: history and policies. Extension Circular 710, Oregon State University Extension Service.

Dickman, A.J. (2010). Complexities of conflict: the importance of considering social factors for effectively resolving human-wildlife conflict. *Animal Conservation*, 13: 458–66.

Eisenstein, E.M., Eisenstein, D., and Smith, J.C. (2001). The evolutionary significance of habituation and sensitization across phylogeny: A behavioral homeostasis model. *Integrative Physiological and Behavioral Science*, 36: 251–65.

Estes, J.A., Terborgh, J., Brashares, J.S., Power, M.E., Berger, J., Bond, W.J., Carpenter, S.R., Essington, T.E., Holt, R.D., Jackson, J.B.C., Marquis, R.J., Oksanen, L., Oksanen, T., Paine, R.T., Pikitch, E.K., Ripple, W.J., Sandin, S.A., Scheffer, M., Schoener, T.W., Shurin, J.B., Sinclair, A.R.E., Soule, M.E., Viraanen, R., and Wardle, D.A. (2011). Trophic downgrading of planet Earth. *Science*, 333: 301–6.

Feng, S., McGhee, K.M., and Bell, A.M. (2015). Maternal predator-exposure and offspring ability to generalize a learned colour-reward association in sticklebacks. *Animal Behaviour*, 107: 61–9. doi:10.1016/j.anbehav.2015.05.024

Fischhoff, B., Slovic, P., and Lichtenstein, S. (1978). Fault trees: Sensitivity of estimated failure probabilities to problem representation. *Journal of Experimental Psychology: Human Perception and Performance*, i, 330–44.

Flores, D. (2016). *Coyote America: A Natural and Supernatural History*. Basic Books.

Foreman, D. (2004). *Rewilding North America: A Vision for Conservation in the 21st Century*. Island Press.

Frank, B., Glikman, J.A., and Marchini, S. (2019). *Human–Wildlife Interactions: Turning Conflict into Coexistence*. Cambridge University Press.

Galway-Witham, J. and Stringer, C. (2018). How did *Homo sapiens* evolve?, *Science*, 360: 1296–8.

Geist, V. (2016). A brief history of human-predator conflicts and potent lessons. Keynote Address, *Proceedings 27th Vertebrate Pest Conference, University of California, Davis*, pp 3–12.

Griffin AS, Blumstein DT, Evans CS (2000) Training Captive-Bred or Translocated Animals to Avoid Predators. Conservation Biology 14(5) 1317–1326.

Gross, C.T. and Canteras, N.S. (2012). The many paths to fear. *Nature Reviews Neuroscience*, 13: 651–8.

Gulliford, A. (2015). Looking back on a century of poisoning predators. *High Country News*, 24 November.

Hagman, J.D. (1980). Effects of training task repetition on retention and transfer of maintenance skill. US Army Research Institute for the Behavioral and Social Sciences Research Report 1271, Alexandria, VA.

Hamilton, L.C., Lambert, J.E., Lawhon, L.A. et al. (2020). Wolves are back: Sociopolitical identity and opinions on management of Canis lupus. *Conservation Science and Practice*, 2(7): e213.

Hayward, M.W. and Kerley, G.I.H. (2009). Fencing for conservation: Restriction of evolutionary potential or a riposte to threatening processes? *Biological Conservation*, 142(1): 1–13.

Herberholz, J. and Marquart, G.D. (2012). Decision making and behavioral choice during predator avoidance. *Frontiers in Neuroscience*, 6: 125.

Holt, R. (2013). Species coexistence. S.A. Levin (ed.), *Encyclopedia of Biodiversity* (second ed.), Academic Press, pp. 667–8.

Isbell, L.A. (2011). *The Fruit, the Tree, and the Serpent*. Harvard University Press.

IUCN (2020). *IUCN SSC Position Statement on the Management of Human–Wildlife Conflict*. IUCN Species Survival Commission (SSC) Human-Wildlife Conflict Task Force. Available at: www.iucn.org/theme/species/publications/policies-and-position-statements

Jaicks, H. (2022). *The Atlas of Conflict Reduction: A Montana Field Guide to Sharing Ranching Landscapes with Wildlife*. Anthem Press.

Johnston, J. (2002). Preserving the beasts of waste and desolation: Theodore Roosevelt and predator control in Yellowstone. *Yellowstone Science*, Spring Issue: 14–21.

Kansky, R. and Maassarani, T. (2022). Teaching nonviolent communication to increase empathy between people and toward wildlife to promote human-wildlife coexistence. *Conservation Letters*, 15: e12862.

Kilner, J.M. and Lemon, R.N. (2013). What we know currently about mirror neurons. *Current Biology*. 23(23): R1057–R1062.

Knight, J. (2000). *Natural enemies: people–wildlife conflicts in Anthropological Perspective*. London: Routledge.

Laberge, F., Muhlenbrock-Lenter, S., Grunwald, W., and Roth, G. (2006). Evolution of the amygdala: new insights from studies in amphibians. *Brain, Behavior, and Evolution*, 67: 177–87.

Lappalainen, K. (2019). Recall of the fairy-tale wolf: 'Little Red Riding Hood' in the dialogic tension of contemporary wolf politics in the US West. *Interdisciplinary Studies in Literature and Envionment*, 26(3): 744–67.

Larson, G., Karlsson, E.K., and Lindblad-Toh, K. (2012). Rethinking dog domestication by integrative genetics, archeology, and biogeography. *Proceedings of the National Academy of Sciences*, 109(23): 8878–83.

Laundré, J.W. et al. (2001). Wolves, elk and bison: re-establishing the 'landscape of fear' in Yellowstone National Park, USA. *Can. J. Zool.*, 79: 1401–9.

Lima, S.L. and Dill, L.M. (1990). Behavioral decisions made under the risk of predation: A review and prospectus. *Canadian Journal of Zoology*, 68: 619–40. doi:10.1139/z90-092

Linnell, J.D.C., Swenson, J.E., and Anderson, R. (2001). Predator and people: Conservation of large carnivores is possible at high human densities if management policy is favorable. *Animal Conservation*, 4: 345–9.

Lopez, B. (1979). *Of Wolves and Men*. Charles Scribner's Sons Press.

Lute, M.L., Navarrete, Nelson, M.P., Gore, M.L. (2016). Moral dimensions of human–wildlife conflict. *Conservation Biology*, 6: 1200–11.

Manfredo, M. (2008). *Who Cares About Wildlife? Social Science Concepts for Exploring Human-Wildlife Relationships and Conservation Issues*. Springer Publishers.

Manfredo, M.J., Berl, R.E.W., Teel, T.L., and Bruskotter, J.T. (2021). Bringing social values to wildlife conservation decisions. *Frontiers in Ecology and the Environment*, 19: 355–62.

Manfredo, M.J., Teel, T.L., Berl, R.E.W., Bruskotter, J.T., and Kitayama, S. (2021b). Social value shift in favor of biodiversity conservation in the United States. *Nature Sustainability*, 4: 80–80.

Manfredo, M.J., Vaske, J.J., Decker, D.J., Duke, E.A., and Brown, P.J. (2009). *Wildlife and Society: The Science of Human Dimensions*. Island Press.

Masseti, M. and Mazza, P.P.A. (2013). Western European Quaternary lions: new working hypothesis. *Biological Journal of the Linnean Society*, 109: 66–77.

Mathis, A., Ferrari, M, Windel, N., Messier, F., and Chivers, D. (2008). Learning by embryos and the ghost of predation future. *Proceedings of the Royal Society B*, 275: 2603–7. doi:10.1098/rspb.2008.0754

Miller, J.R.B. and Schmitz, O.J. (2019). Landscape of fear and human-predator coexistence: applying spatial predator-prey interaction theory to understand and reduce carnivore-livestock conflict. *Biological Conservation*, 236: 464–473.

MMWR (2019). QuickStats: Number of death from hornet, wasp, and bee stings, among males and females—National Vital Statistics Systems, United States, 2000-2017. *Morbidity and Motality Weekly Report*, 68: 649. doi: http://dx.doi.org/10.15585/mmwr.mm6829a5external icon

Muller, A., Shipton, C., and Clarkson, C. (2022). Stone toolmaking difficulty and the evolution of hominin technological skills. *Scientific Reports*, 12: 5883. https://doi.org/10.1038/s41598-022-09914-2

Niemiec RM, Berl REW, Gonzalez M, Teel T, Camara C, Collins M, Salerno J, Crooks K, Schultz S, Breck S, Hoag D. (2020). Public Perspectives and Media Reporting of Wolf Reintroduction in Colorado. Peer J. https://peerj.com/articles/9074/

Nyhus, P.J. (2016). Human–wildlife conflict and coexistence. *Annual Review of Environment and Resources*, 41: 143–71.

Ogden, L.E. (2016). The surprising consequences of being scared. *BioScience*, 66: 625–31.

Olsson, A. (2012). Social learning of fear. In: Seel, N.M. (eds), *Encyclopedia of the Sciences of Learning*. Springer. https://doi.org/10.1007/978-1-4419-1428-6_862

Olsson, A. and Phelps, E.A. (2007). Social learning of fear. *Nature Neuroscience*, 10: 1095–1102.

Oriol-Cotterill, A., Valeix, M., Franck, L.G., Riginos, C., and Macdonald, D.W. (2015). Landscapes of coexistence for terrestrial carnivores: the ecological consequences of being downgraded from ultimate to penultimate predator by humans. *Oikos*, 124: 1263–73.

Pellman, B.A. and Kim, J.J. (2016). What can ethobehavioral studies tell us about the brain's fear system? *Trends in Neurosciences*, 39: 420–31. doi:10.1016/j.tins.2016.04.001

Pooley, S., Barua, M., Beinart, W., Dickman, A., Holmes, G., Lorimer, J., Loveridge, A.J., Macdonald, Marvin, G., Redpath, S., Sillero-Zubiri, C., Zimmermann, A., and Milner-Gullan, E.J. (2016). An interdisciplinary review of current and future approaches to improving human-predator relations. *Conservation Biology*, 31: 513–23.

Prokop, P., & Fančovičová, J. (2013). Self-protection versus disease avoidance: The perceived physical condition is associated with fear of predators in humans. *Journal of Individual Differences*, 34(1), 15–23. https://doi.org/10.1027/1614-0001/a000092

Ramp, D. and Bekoff, M. (2015). Compassion as a practical and evolved ethic for conservation. *BioScience*, 65(3): 323–7.

Reisland, M.A., Malone, N., and Lambert, J.E. (2021). Endangered apes—can their behaviors be used to index fear and disturbance in anthropogenic landscapes? *Diversity*, 13(12): 660. https://doi.org/10.3390/d13120660

Reynolds, J.C. and Tapper, S.C. (1996). Control of mammlaina predators in game management and conservation. *Mammal Review*, 26(2/3): 127–56.

Ripple, W.J., Estes, J.A., Beschta, R.L., Wilmbers, C.C., Ritchie, E.G., Hebblewhite, M., Berger, J., Elmhagen, B., Letnic, M., Nelson, M.P., Schmitz, O.J., Smith, D.W., Wallach, A.D., and Wirsing, A.J. (2014). Status and ecological effects of the world's largest carnivores. *Science*, 343(6167): DOI: 10.1126/science.1241484

Ripple, W.J., Wolf, C., Newsome, T.M., Hoffmann, M., Wirsing, A.J., and McCauley, D.J. (2018). Both the largest and smallest vertebrates have elevated extinction risk. *PNAS*, 115(26): E5847–E5848.

Ripple WJ, Wolf C, Phillips Mk, Beschta Rl, Vucetich JA, Kauffman JB, Law BE, Wirsing AJ, Lambert JE, Leslie E, Vynne C, Dinerstein E, Noss R, Wuerthner G, Dellasala DA, Bruskotter JT, Nelson MP, Eileen C, Darimont, Ashe DM (2022) Rewilding the American West. BioScience biac069, https://doi.org/10.1093/biosci/biac069

Roosevelt, T. (1889) *The Wilderness Hunter*. GP Putnam's Son Publishers.

Røskaft, E., Bjerke, T., Kaltenborn, B., Linnell, J.D.C., and Andersen, R. (2003). Patterns of self-reported fear towards large carnivores among the Norwegian public. *Evolution and Human Behavior*, 24: 184–98.

Russell, N. (2015). Neolithic human–animal relations. *Historisch Tijdschrift Groniek*. 175: 206–7.

Shipman, P. (2015a). *The Invaders: How Humans and Their Dogs Drove Neanderthals to Extinction*. Harvard University Press.

Shipman, P. (2015b). How do you kill 86 mammoths? Taphonomic investigations of mammoth megasites. *Quaternary International*, 359–60: 38–46.

Shivik J (2014) The Predator Paradox: Ending the War with Wolves, Bears, Cougars, and Coyotes. Beacon Press. Boston, Massachusetts 208 pp

Sowka, P. (2012). Living with Predatros Resource Guide Series. Produced by the Living with Wildlife Foundation in cooperation with Montan Fish, Wildlife and Parks. 2012 edition. SwanValley, Montana.

Speth, J.D. (2010). *The Paleoanthropology and Archaeology of Big-Game Hunting: Protein, Fat, or Politics?* Springer Press.

Starr, C. (1969). Social benefits vs. technological risks. *Science*, 165: 1232–8.

Suraci, J.P., Clinchy, M., Dill, L.M., Roberts, D., and Zanette, L.Y. (2016). Fear of large carnivores causes a trophic cascade. *Nature Communications*, 7 (art. 10698). doi:10.1038/ ncomms10698

Sussman, R. and Hart, D. (2005). *Man the Hunted: Primates, Predators and Human Evolution*. Basic Books Publishers.

Tatar, M. (1993). *Off with Their Heads! Fairy Tales and Culture of Childhood*. Princeton University Press.

Teel, T.L., Manfredo, M.J., and Stinchfield, H.M. (2007). The need and theoretical basisi for exploring wildlife value orientations cross-culturally. *Human Dimensions of Wildlife*, 12: 297–305.

Thorpe, J. (2016). Conflict of wings: Birds versus aircraft. In: Angelici, F. (eds), *Problematic Wildlife*. Springer. https://doi.org/10.1007/978-3-319-22246-2_21

Treves, A. and Bruskotter, J.T. (2014). Tolerance for predatory wildlife. *Science*, 344: 476–7.

Treves, A. and Karanth, K.U. (2003). Human-carnivore conflict and perspectives on carnivore management worldwide. *Conservation Biology*, 17: 1491–9.

Treves, A. and Naughton-Treves, L. (1999). Risk and opportunity for humans coexisting with large carnivores. *Journal of Human Evolution*, 36: 275–82.

Van Valkenburgh, B. (2010). Carnivores. *Current Biology*. 20: R915–R919.

Van Valkenburg, B., Hayward, M.W., Ripple, W.J., and Roth, L. (2015). The impact of large terrestrial carnivores on Pleistocene ecosystems. *PNAS*, 113(4): 862–7.

Weisdorf, J.L. (2005). From foraging to farming: explaining the Neolithic Revolution. *Journal of Economic Surveys*, 19(4): 561–86.

Wilson, M.A. (1996). The wolf in Yellowstone: science, symbol, or politics? Deconstructing the conflict between environmentalism and wise use. *Society and Natural Resources*, 10: 453–68.

Wolf, C. and Ripple, W.J. (2018). Rewilding the world's large carnivores. *Royal Society Open Science*, 5: 172235.

Woodroffe, R. (2000). Predators and people: Using human densities to interpret declines of large carnivores. *Animal Conservation*, 3: 165–73.

Young JK, Ma Z, Laudati A, Berger J (2015) Human–Carnivore Interactions: Lessons Learned from Communities in the American West. Human Dimensions of Wildlife 20 (4) 349–366.

Zanette, L.Y., White, A.F., Allen, M.C., and Clinchy, M. (2011). Perceived predation risk reduces the number of offspring songbirds produce per year. *Science*, 334: 1398–1401. doi:10.1126/science.1210908.

23

REWILDING AND FARMING

Could the relationship be improved through adopting a three compartment approach to land use?

Julia Aglionby and Hannah Field

What would a positive relationship between rewilding and farming look like in England? Do terms matter? When is extensive farming better categorised as rewilding and when is rewilding really farming with nature? Can farming and rewilding co-exist and complement each other to meet national ambitions to produce sufficient high-quality food, and address the interconnected biodiversity, climate, and health crises?

This chapter explores these questions digging beneath the binary polarisation that has in recent years characterised much conversation, social media, and writing in this space. Barriers to land use change are explored and ideas to address them provided. We focus on England, which is currently facing a number of policy and funding challenges and opportunities, but we recognise that many of these issues will have resonance in other countries and contexts. The analytical framework adopted is the three-compartment approach to land use as recommended by the National Food Strategy (Dimbleby, 2021). Could this framework better enable the co-existence of farming and rewilding when partnered with appropriate government levers and delivery mechanisms including the Environmental Land Management (ELM) schemes?

Policy context

In September 2020 following the Convention on Biodiversity recommendations the UK Government agreed to protect 30% of England by 2030 (often known as the 30x30 commitment) (Gov.uk, 2020). The Prime Minister noted only a further 4% of England will need to be protected to achieve this ambition as National Parks and AONBs already cover 26% of England, though others question whether these areas being IUCN Category V areas qualify as protected areas for nature (Starnes et al., 2021; Bailey et al., 2022). A stronger legal level of protection for nature in England is provided for land designated as a Site of Special Scientific Interest (SSSI) though these are more limited in extent at eight percent of England or approximately one million hectares. Furthermore, many SSSI designated features of interest are reliant on continuing farming operations or grazing i.e., they are managed habitats.

DOI: 10.4324/9781003097822-26

In January 2022, George Eustice announced this Government's ambition to bring over half our SSSIs into favourable condition by 2042 and utilise the new ELM schemes to create or restore up to 300,000 ha of habitat by 2042 (Gov.uk, 2022a). It is not clear if the 300,000ha is in addition to the SSSI land achieving favourable condition. Given 300,000ha is 2.3% of England, 50% of SSSIs is 4% of England and 2042 is 12 years after 2030 it remains unclear how the government is proposing to achieve its 30x30 target. Perhaps it is a paper exercise in designating protected areas rather than an ambition to deliver outcomes for nature. It is worth noting as context that Rewilding Britain, the nation's leading champion for rewilding, has an ambition to rewild 5% of Britain and for a further 25% to be returned to nature friendly uses e.g. high nature value (HNV) farming and native woodlands (Rewilding Britain, 2021). Their ambition increases to rewild 10% in National Parks and AONBs. Overall, in England 70% of land is currently in agricultural use (Defra, 2019). Overlaying these ambitions for nature, in 2019 the UK government enshrined in law a commitment to achieve net zero carbon emissions by 2050 (*The Climate Change Act 2008 (2050 Target Amendment) Order 2019*, 2019).

The figures above demonstrate how muddled are England's targets and terminology around food production, climate resilience, and nature conservation. This makes it harder to have constructive conversations as people come to the table with different understanding and definitions of nature conservation, nature recovery, farming, and rewilding. All too quickly defensive positions are adopted as land managers, conservationists, and farmers separately feel threatened. So how do we move forward productively in a relatively small country when pressures to increase food security are growing due to global instability, the cost-of-living crisis, and less predictable yields due to climate change.

One way to structure a conversation about the interface between farming and rewilding is through the lens of a land use framework, which can be at a local level, a strategic national vision, or both. To date, government has shied away from land use frameworks noting land is predominately privately owned and the ultimate decision about land use rests with landowners, tenants, and commoners. Understandably farmers and landowners resist the imposition of external plans even though their choices in practice are heavily influenced by Government setting the levers and incentives. These include taxes, financial support for farming and the environment, legislation on nature, planning, access, pollution, food, and animal welfare standards. Defra's aversion to creating a land use framework has led to a scatter gun approach to regulations and incentives rather than an outcomes-based approach. If we are going to successfully produce sufficient food and water, protect nature, address the climate emergency, and provide opportunities for public health and well-being then purposeful place-based land use is critical (FFCC, 2022, 2019; Seppelt et al., 2018).

The National Food Strategy, an independent review commissioned by the UK Government and led by Henry Dimbleby reported in 2021 through 'The Plan'. This presents evidence-based recommendations for future land use recognising the interlinkages between food production, biodiversity, climate, and well-being. They recommend a land use framework based on a three compartment model of semi-natural land, low yield farmland, and high yield farmland. This is a rejection of the dichotomy of land sharing or land sparing that has too often served to polarise human land managers and is based on ecological evidence that demonstrates that nature is enhanced by a diverse range of place-based land management approaches. We also move further away from dichotomy and compartments by suggesting a land use continuum, highlighting that those in land management need to make decisions from their current context, supporting a place-based approach. The 'places' could be referred to as Social-Ecological Systems (SES) which highlights the importance of social and ecological relationships and dynamic processes which need to be supported for adaptability (Berkes, 2017; Folke et al., 2003).

A striking piece of evidence from 'The Plan' is that 20% of our land produces 3% of our calories and on that 20% of land there is also a high opportunity cost in terms of missed opportunities for carbon sequestration and biodiversity enhancement. In terms of contribution to nutrition, the AHDB inform us that in the UK lamb constitutes 2% and beef 6% of adult protein intake (AHDB, 2022). The Plan recommends 5–8% of land in England is released completely from agricultural production, i.e. becomes semi-natural, while the balance of less productive land is encouraged, through adequate financial rewards, to adopt low intensity farming practices. An additional 2.6% of England's farmland is considered to be required for new housing. The net result of the recommendations proposed by the National Food Strategy is a reduction of land being farmed from 70% to about 63%.

Dimbleby stresses the three compartment model is not a manifesto for clearing farming communities from the uplands rather explicitly demands sufficient public money to enable upland farmers to transition to extensive HNV farming, recognising the cultural and aesthetic value that sheep farming and agro-pastoralism has in England, as well as encouraging more use of rare breed cattle. The three compartment model encourages a rebalancing to integrate nature recovery into and alongside food and timber production rather than a wholesale change. Furthermore, it explicitly recommends that nature recovery in England should not be at the expense of exporting the environmental impact of our food production to low-income countries and the Global South. Additionally, there is an ongoing debate around sustainable intensification in the High Yield compartment, whether intensification can be sustainable given the high level of inputs (fertiliser, feed, pesticides). As technology develops and net zero solutions are implemented then the technocrats predict we can reach a higher output with lower input model than at present (Pretty, 2018).

So how can 'Rewilding' fit into the future of a diverse range of land uses of England as imagined by the three compartment model? The National Food Strategy has few direct references to rewilding; perhaps recognising the unfortunate current toxicity of the word; instead it adopts the term 'semi-natural land'. This concept of the semi-natural is consistent with thoughtful proponents of rewilding. For instance, in 'What is Rewilding?' Jepson and Schepers usefully discuss the difference between wilderness as a state and rewilding as a process. They comment: 'Rewilding is not a state; it is a process. It is about moving up a scale of wildness and giving the ecosystems a functional "up-grade" whatever their nature, scale, and location.' They, and others, emphasise that restricting the use of the term rewilding to the creation of wilderness (9–10 on their scale) would result in many missed opportunities and note the term can also be used to include moving from 2 to 3 or 5 to 7 (Jepson & Schepers, 2016; Jepson et al., 2018; Jepson & Blythe, 2020). This approach of progressive rewilding is also supported by Torres et al. (2018) and Hawkins (Chapter 5 in this book). Whilst we recognise that not all 'rewilders' share Jepson and Schepers process-focused approach, most if not all sites that identify as rewilding in England are semi-natural as it would be hard to find any true wilderness in England (Carver et al., 2021).

This acknowledgement of a process of wilding combined with the three compartment model suggests a continuum approach has merit in diffusing tensions and enhancing outcomes. In Figure 23.1 we present a continuum of land use from intensive farming to wilderness in order of intensity reflecting the approach in the wilderness continuum (Carver et al., 2021). This inevitably, in common with all models, oversimplifies reality and it is recognised that a change in land use can be made from one state directly to another without adopting the intermediate phases but shows the plurality of approaches that can be adopted and the process of wilding can happen at several stages.

A LAND USE CONTINUUM

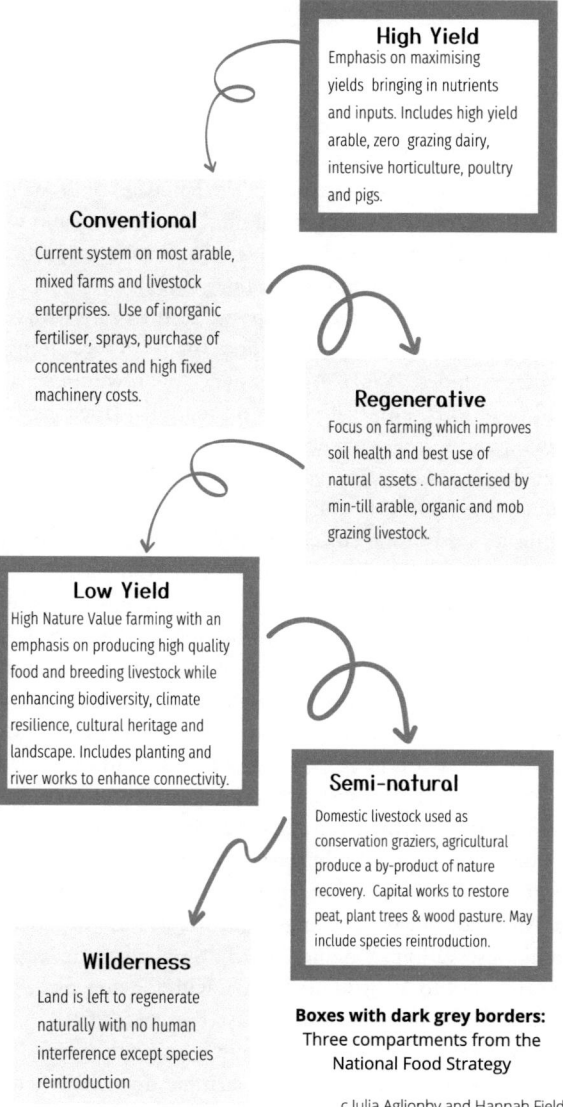

High Yield
Emphasis on maximising yields bringing in nutrients and inputs. Includes high yield arable, zero grazing dairy, intensive horticulture, poultry and pigs.

Conventional
Current system on most arable, mixed farms and livestock enterprises. Use of inorganic fertiliser, sprays, purchase of concentrates and high fixed machinery costs.

Regenerative
Focus on farming which improves soil health and best use of natural assets. Characterised by min-till arable, organic and mob grazing livestock.

Low Yield
High Nature Value farming with an emphasis on producing high quality food and breeding livestock while enhancing biodiversity, climate resilience, cultural heritage and landscape. Includes planting and river works to enhance connectivity.

Semi-natural
Domestic livestock used as conservation graziers, agricultural produce a by-product of nature recovery. Capital works to restore peat, plant trees & wood pasture. May include species reintroduction.

Wilderness
Land is left to regenerate naturally with no human interference except species reintroduction

Boxes with dark grey borders:
Three compartments from the National Food Strategy

c Julia Aglionby and Hannah Field

Figure 23.1　A 'Land Use Continuum'.

Barriers to land use change

As revealed by a 2018 workshop on Rewilding and the English Uplands there is a complex web of factors that are barriers to rewilding (Sandom et al., 2019). In the rest of this chapter we consider how to overcome four barriers to land use change and facilitate a progression to a less adversarial conversation about the appropriate intensity of farming. The first barrier is identity and language, the second the finances of delivering public goods, the third land tenure, and the fourth advice and facilitation.

Language and identity

Currently there is palpable and increasing concern among many livestock farmers that their sense of value, identity, and purpose as food producers is being eroded by narratives in both the mainstream media and on social media espoused by a range of environmental activists, including vegans, rewilders, and climate change campaigners. Tim Bonner of the Countryside Alliance summarises concerns about rewilding; 'the language of "rewilding" and its very name assumes that there is no place for farming within it. I have argued and written for years that the concept of "rewilding" will never be accepted by the farming community as long as its name specifically describes an end to agriculture' (Bonner, 2019).

Such sentiments were also expressed in response to the twitter survey by Steve Carver and Ian Convery in February 2022 asking; 'What does the term "rewilding" mean to you (in one sentence)?' Responses included, 'Sadly, my experience of "rewilding" has been that it means the vilification of food producers, especially those on marginal lands, by those being led by poor or irrelevant research...and further concentration of land ownership in the hands of billionaires. I hope that changes.' And some expressed these views more forcefully e.g.; 'Same as it did in Nazi Germany: an excuse for elitist "enlightened" crusaders with rural idyll fantasies to invade & steal from existing stewards using conservation as an excuse for land grab & genocide of the culture; expect fierce resistance.' Others, while still expressing concern, were more nuanced in their answers e.g. 'I'm also not a farmer but I do know some who are offended by the implication that #rewilding and farming are by definition opposed to each other. I hope your state of play article will include how to keep this conversation open and respectful.'

Farmers are passionate about their role feeding the nation and stewarding the countryside. This brings a sense of identity and purpose while delivering public goods is not seen as core to identity and rewilding is considered by many to be antithetical to their sense of identity (Reboah, 2019). Nature conservationists are often equally committed to a vocation to deliver nature recovery. The question then for conservationists, noting 70% of England is farmed, is how can they best achieve their ambitions for nature at scale in a farmed or semi-natural landscape. Some approaches have proved unproductive. For instance, rubbishing the sheep sector, as George Monbiot did with his reference to 'woolly maggots', has served to alienate many farmers from conservation (Monbiot, 2013). Similarly poor engagement in Wales in the Summit to Sea project led to Rewilding Britain withdrawing from that project following a breakdown in trust with the farming community (Forgrave, 2019). There are different, more successful approaches; for example, Ben Goldsmith in using the term, 'a wilder way of farming' proffers a more inclusive paradigm presenting farming and wilding as collaborative partners (Goldsmith, 2021).

Other articles such as by Ian Convery et al. sought to bring together multiple interests collaboratively but the authors' passion for a particular outcome meant they missed the mark resulting in further alienation of farmers from nature conservation in the Lake District (Convery et al., 2020). This extract illustrates the bias, 'The current LDNP management approach, with a focus on a partnership model where some voices lobby against change in order to protect upland farming and common lands, is not conducive to the current wildlife crisis, climate breakdown, or indeed presenting a meaningful response to the Glover Report. Common lands are important, but they do not necessarily need to contain sheep'. This article came just a year after the Lake District National Park Partnership has endorsed a statement on 'Co-operation not Conflict' seeking a way forward to simultaneously enhance natural and cultural capital in a World Heritage Site inscribed, in part, for it agro-pastoral sheep systems. The statement was brokered between key farming and nature conservation organisations by the Foundation for Common Land and this approach was recommended in the Glover Review as a way forward for land use in contested spaces (Glover, 2019). Commons are complex and inherently place-based, so top-down assumptions about what can or should happen need to be avoided. Instead, collaboration between landowners, commoners, and other stakeholders including public value of commons needs to be facilitated to ensure multiple benefits from commons can be delivered, accounting for the diverse range of perspectives

This approach is consistent with the Convention on Biological Diversity (CBD), which in its meetings in 2021, stressed the need to adopt a Theory of Change and in Target 9 of the framework includes 'customary sustainable use by indigenous peoples and local communities' (*First Draft of the Post 2020 Global Biodiversity Framework*, 2021). Nations are urged to include local communities customary land use as part of the solution, not characterise it as the problem. Farmers all over the world wish to identify as farmers and that requires continuing farming. Farming in UK legislation is defined as 'the occupation of land wholly or mainly for the purposes of husbandry' and husbandry while not defined in statute has in case law been given the meaning 'to include all forms of tillage of soil and use of land by livestock held for its produce or for food' (HMRC, n.d.; Anon, 1972).The term rewilding is a newer term and while not yet defined in law (see Eagle et al.,Chapter 13), the Houses of Parliament Post Note states rewilding usually refers to 'reinstating natural processes that would have occurred in the absence of human activity' (Post Note No 537, 2016).

We posit that explicitly acknowledging rewilding as a process, not a state, with in tandem considering land use choices as a continuum, as set out in Figure 23.1, i.e. adopting a plural dynamic rather than binary static state, would allow for more constructive conversations. Adapting language for instance from rewilding to perhaps 'wilding' may also assist. This alongside the use of softer terms such ecological or nature enhancement and regeneration could create a more enabling and less threatening environment for farmers to adapt and potentially compartmentalise their land use, especially those in areas well suited to the compartment; 'low yield farmland', often known as High Nature Value (HNV) farming.

Financing the delivery of public goods

Fear arising from uncertainty often translates to defensiveness and currently uncertainty is being amplified on England's hill farms due to financial stress arising from their dependence on Basic Payment Scheme (BPS) which is due to be phased out by 2027. Newcastle University via the Farm Business Survey provides an annual analysis of hill farms (Harvey & Scott, 2022). Figures from 2019/2020 are summarised in Table 23.1 showing the huge challenge facing these micro businesses.

Farmers are therefore considering how they can, in the post BPS world, run a business that gives a fair return on labour and capital spent on the farm. The full details of ELM are yet to be released by Defra but many upland farms have limited options as they already are in agri-environment schemes; in 2020/21 agri-environment income averaged £13,045 per annum on LFA farms. In most cases land already in CS cannot be in ELM due to the requirement to avoid double funding the same public good provision. This means, unless payment rates increase substantially, farms will have to look to other changes to enhance profit margins either through altering their farming system, on farm diversification or working off farm. The work of Chris

Table 23.1 Farm Business Survey LFA Data 2020/21

Upland Farm Business Income (FBI)	£33,360
Value of Unpaid Family Labour	£28,284
Farm Corporate Value	£5,076
BPS	£30,445
FBI excluding BPS and accounting for unpaid family labour	-£25,369

Note: * Average for 202 surveyed farms from FBS Hill Farming Report, Harvey and Scott (2022).

Clark of Nethergill Associates in 'Less is More' demonstrates, that most farms do not optimise their farming to produce the maximum sustainable output (MSO), i.e. they are subsidising agricultural production with BPS (Clark & Scanlon, 2019). Profitability could be improved by adjusting (often reducing) livestock numbers as well as reducing inputs and fixed costs such as investment in machinery and bought in feed. A move to MSO also in most cases increases opportunities for earning more from delivering public goods via Countryside Stewardship, and in the future ELM, so enabling farms to deliver more for nature and climate alongside food production; the middle 'compartment' of low yield farming.

Farmers are familiar with selling agricultural produce in auction marts and to traders, as well as being rewarded for delivering for nature through stewardship schemes. What farmers are not so familiar with is treating the delivery of public goods as an enterprise. This inevitably has resulted in farmers considering land being turned over to a low yield or semi-natural use as 'wasted' and a risk to national food security as well as an erosion of culture, local economies, and communities. One mechanism for reducing tension between rewilding and farming is an effective system for funding the delivery of public goods from land. The work of Little and colleagues at the University of Sheffield emphasises the need to engage farmers and land managers and enable informed choices and long-term plans to be made ((Hurley et al., 2022); see *Advice and facilitation* in this chapter).

Three approaches to addressing this barrier are recommended. Firstly, treat schemes for delivering public goods as a business venture whether paid for by the state or a private firm. Effective contracts that are attractive, clear, long lasting, and enforced are required. Being attractive means paying sufficient rates for nature recovery to be a credible land use choice for a farmer.

Secondly, directly fund capital works for nature recovery. Nature recovery and rewilding in many instances require capital investment up front to kick start delivery e.g., for peat restoration, river wiggling, and tree planting. Farm businesses often lack the financial reserves to bank roll these works as is required by current grants. While some rewilders espouse the benefits of allowing nature to take its course (Carver, 2019), and this approach avoids the need for capital expenditure, it does delay the growth in flow of ecosystem goods and services so many rewilders, as well as more traditional conservationists, acknowledge the benefits of intervention, often termed active rewilding.

Thirdly, risk for farmers needs to be managed as the time scales for delivery of public goods are substantially longer than agricultural produce with a higher level of uncertainty about the quality and quantity of what will be produced. For example, you cannot guarantee curlew or snipe will start nesting, but you can create habitat that is attractive to them. If farmers consider achieving the indicators of success as too high risk and failure could result in funds having to be repaid they will not sign up.

Land tenure

Passive rewilding in allowing natural processes to take their course requires long term management control of the land and a commitment to cease farming which is less likely to happen if the property rights are split among more than one party and especially if the land is let under an agricultural tenancy where agricultural husbandry is the primary land use purpose. Thirty-eight per cent of land in England is tenanted and in the uplands the proportion of tenanted land is considerably higher at 52% (Defra, 2019) and often the duration of the tenancy is less than ten years. Furthermore, 38% of the moorland region is common land with complex and split property rights, with the land owned by one person but layered with rights of common usually for livestock.

Even where the end point is not wilderness but a desire to enhance nature and carbon sequestration within a farmed setting, the split of property rights bundles (on tenanted and common land can be a barrier (Ostrom, 1990). This is because tenant's rights are for farming and commoners' rights for grazing domestic livestock while natural assets such as trees and carbon are usually reserved to the landlord or landowner. The incentive for a tenant or commoner to enhance natural assets is therefore limited given much of the income from ELM will be clawed back in rent or via internal agreements. The transaction costs of negotiations and governance regulating these separate interests currently holds back nature recovery in the uplands, despite the significant potential due to the land being less agriculturally productive so less critical for securing national food security.

There are several ways these barriers can be overcome if the relationship between tenants and landlords and commoners and landowners is seen as a partnership rather than feudal. This relates to a Social-Ecological Systems (SES) view of commons, where relationships and trust are essential for longevity (see *Advice and facilitation* in this chapter). Unfortunately, some landlords instead see the potential income from public goods as an opportunity to take land in hand from tenants and rewilding with its limited management requirements means a tenant is no longer required (Tidman, 2022). Advice on the business of rewilding is popping up emphasising rewilding as a new business opportunity with Savills estimating a net return of £562 per hectare for conversion to rewilding compared with £263 per hectare for an arable Farm Business Tenancy (*The Business of Rewilding; Spotlight Savills Research*, 2022).

Property rights are a significant factor in determining the delivery of public goods; at the moment the political and legislative landscape in England is laissez faire as to the impact on the tenanted sector of enhancing public goods. Defra's review into the tenanted sector by Baroness Rock is an opportunity for social justice to travel hand in hand with environmental justice (Gov.uk, 2022b).

A further issue is about the duration of agreements as nature recovery often requires many years to achieve and can be undone in an instance. Long term agreements for providing public goods either as contracts or more legally binding conservation covenants can mitigate this risk and encourage uptake of schemes.

Advice and facilitation

Advisory services in farming are essential to many businesses in enabling navigation through complex regulations, policies, markets, and financial opportunities. Government (local and national) provision of such services was usual before the mid-1980s, since becoming broadly privatised (Garforth et al., 2003; Sutherland et al., 2013). Access to high-quality advice and consultancy from trusted sources is a major barrier to decision-making due to a lack of skilled advisors and the cost of such services. Additionally, credibility, legitimacy, trust, longevity, and expertise are found to be core to effective advisory services in farming, which take time to develop (Sutherland et al., 2013; Hurley et al., 2022). This lack of access to advice is reinforcing uncertainty about what decisions to make within the context of financial and tenure considerations outlined earlier in this chapter. It is also causing farmers to turn more to each other to share experiences and knowledge, and turn away from seeking 'experts' due to the lack of trust created over many years (Rust et al., 2022). Considering how land managers want to access information is key to enabling place-based decision-making processes.

There is a lack of access to clear information around place-specific opportunities (farm-level, catchment/landscape-level) that deliver multiple benefits from land management practices. This lack of information is paired with a lack of advice, increasing the administrative burden on land managers to source what they need for decision making. As an example, commons (specifically common land in England) are Common Property Resources and the porous boundaries between them (physically and socially) require levels of agreement and collaboration between landowners and commoners to deliver multiple benefits (Ostrom, 1990, 2000). Commons provide the opportunity to join up across landscapes, but this requires facilitation and advice that is place-based and dynamic with appropriate governance structures that enable adaptation and self-organisation (Ostrom, 2000; Ostrom et al., 1999; Cinner et al., 2019).

Another example is the development of carbon markets, which has accelerated in the last few years and are viewed as an unregulated 'wild west', where there is no clarity around or mechanism for 'selling' carbon credits (Brill, 2021). This is having various impacts on land management, including a new form of land grabbing which was recently highlighted in Wales, where private investors are seeking to buy farms to plant trees for carbon off-setting requirements. This could undermine local communities and lead to inappropriate land use for the area (Jones, 2022). Private finance has the potential to support the delivery of public goods, but, similarly to public finance, more clarity, mechanisms, and associated advice would enable this.

Facilitation is also becoming a core requirement to support landscape or catchment scale delivery for nature and people due to the need for collaboration. In the Local Nature Recovery and Landscape Recovery ELM schemes, collaboration between neighbours—land managers, landowners, and farmers—is often required to achieve landscape-scale ambitions. This can be hugely challenging as building good relationships between neighbours and social capital can take time and can require facilitation by a trusted facilitator (Arnott et al., 2021), causing further financial and time barriers to accessing opportunities. This could also favour large landowners who own large tracts of land in directing land use decision-making, with fewer people involved and therefore less collaborative effort required. Facilitating and enabling collaboration and peer to peer learning is vital, as farmers turn to each other particularly if there is more perceived financial risk of change (Vigani et al., 2021). Enabling farmers to learn about opportunities from other farms which are similar in size and context via farm visits and social media, for instance, could help to support change along the Land Use Continuum.

Advice and facilitation could help explore the multiple benefits of differing land management practices more collaboratively along a Land Use Continuum. There needs to be more trained advisors and facilitators to enable services to be personal to the farmers and land managers, use clear language specific to audiences, be place-based/localised, and be credible. This would enable relationships to be built and decision-making so farmers to choose where along the continuum from wilderness to intensively farmed they manage their land, and to adjust that position over time. Returning land to a semi-natural state, farming less intensively, high nature farming, or wilder farming become options rather than threats.

Conclusion

On the surface, there are substantial barriers to rewilding and farming becoming comfortable bedfellows and extracts from the press and social media tend to support this conclusion. In exploring the barriers, it becomes clearer that this conclusion is a result of pitting two land uses as polar opposites and mutually incompatible practically, psychologically, and philosophically. While this is one world view, fortunately other views are available and deliverable, as illustrated in England by wilder farming at Geltsdale in Cumbria, high nature value farming at Hill Top Farm in the Yorkshire Dales, regenerative farming and rewilding at Knepp, and wood pasture including on common land, whether long standing as in the New Forest or as proposed at Torver and Coniston. These are all in addition to more ambitious rewilding projects in the UK and overseas which are recognised as having barriers in common (Jepson & Blythe, 2020).

Intellectually and practically, co-existence or abutting of contrasting forms of land management is eminently possible. This would create core areas for nature supported by nature connectivity within and across farmland at a landscape scale alongside maintaining the cultural heritage of our farmed landscapes as well the capacity to secure high quality food provision in line with Lawton's vision (Lawton, 2010)

This is the good news. The less encouraging news is that current successful initiatives are the exception rather than the rule and the barriers that have been created as described above will remain and frustrate efforts to do more at scale for nature, climate, and sustainable food production unless addressed. The Government appear well intentioned to deliver on their 25-year Environment Plan and Defra as a department make all the right noises about enabling change to enable farmers and landowners to deliver public goods but the current appraisals of new policies are less than encouraging when the detail is examined.

The conclusions of a number of reports on ELM and Post Brexit Agricultural Policy in 2021 and 2022 from the Public Accounts Committee, Efra Select Committee, National Audit Office, and Institute of Government range from the concerned to the critical and on occasion, the damning (Public Accounts Committee, 2021; Efra Committee, 2021; Marshall et al., 2022; NAO, 2021). A common theme the reports emphasise is that clarity by Government on desired broad outcomes is not in itself sufficient to address the climate and biodiversity crises alongside producing sustainable and sufficient food. Implementation requires current and future incentives and levers to be designed, modeled, and stress tested to be confident of achieving the desired outcomes. And once that has been undertaken efficient delivery adopting a theory of change approach is required. A consistent criticism from all these reports was the lack of clarity to enable farmers to forward plan.

The departure from the EU provides a potential opportunity with a relatively blank drawing board following exit from the Common Agricultural Policy. Regretfully the above inquiries conclude the current government risks wasting this opportunity in England through a lack of direction and a failure to demonstrate the link between ambitions, schemes, funding, and delivery in a complex world where land tenure complicates how benefits and responsibilities are shared and where government appears not yet ready to reward the full cost of delivering public goods.

Diffusing the stand-off between farming and rewilding is eminently achievable if government chose to break down the four barriers described above through a combination of properly valuing public goods through adequate payment levels, legislating to ensure landowners, tenants, and commoners have a fair transition and providing good quality advice. Overarching this is the need to recognise that land use is a continuum with competent land managers all along that continuum having separate but equally important and interconnected roles. A land use framework as envisaged in the National Food Strategy's three compartment model would be an excellent first step for designing support schemes to give effect to the changes sought at a holding, sub catchment, county, and national scale. The evidence would suggest a continuum approach, as set out in Figure 23.1, rather than distinct compartment would better reflect the reality of land use and demonstrate more explicitly the plurality of options both between and within land holdings as well as supporting change towards wilder and more nature rich farming systems.

While these conclusions arise from evidence in the English context, similar issues are found internationally where socio-ecological systems need to enhance their adaptability and dynamism as place-based situations if they are to support both social and ecological health. This is best enabled through place-based institutions and clear effective governance which shapes decision making in policy development and implementation, working in a joined-up way across all aspects of land-use planning and management. Achieving successful top-down/bottom-up integration is easier to write about than to achieve; it will require the rewilding and farming communities (with the multiple supporting stakeholders) to actively and meaningfully engage with each other. Such a willingness would enable more purposeful land use for the greater benefit of people, businesses, and ecology.

References

AHDB (2022) *Simply Beef and Lamb*. [Online] Available from: www.simplybeefandlamb.co.uk/nutrition/ (accessed 18 April 2022).

Alison, J. and Wentworth, J. (2016) Post Note Number 537: Rewilding and Ecosystem Services. London: Houses of Parliament Parliamentary Office of Science and Technology..

Anon (1972) *Lowe v. J W Ashmore Ltd*. 46 TC 597.

Anon (2019) *The Climate Change Act 2008 (2050 Target Amendment) Order 2019*.

Anon (2021) *First Draft of the Post 2020 Global Biodiversity Framework*.

Anon (2022) *The Business of Rewilding; Spotlight Savills Research*.

Arnott, D., Chadwick, D.R., Wynne-Jones, S., Dandy, N., and Jones, D.L. (2021) Importance of building bridging and linking social capital in adapting to changes in UK agricultural policy, *Journal of Rural Studies*, 83: 1–10.

Bailey J. J., Cunningham, C. A., Griffin, D. C., Hoppit, G., Metcalfe, C. A., Schéré, C. M., Travers, T. J. P., Turner, R. K., Hill J. K., Sinnadurai, P., Stafford R., Allen D., Isaac N., Ross B., Russi D., Chamberlain B., Harvey Sky N., McKain S. (2022). Protected Areas and Nature Recovery. Achieving the goal to protect 30% of UK land and seas for nature by 2030. London, UK. Available at: www.britishecologicalsociety.org/protectedareas.

Berkes, F. (2017) Environmental governance for the Anthropocene? Social-ecological systems, resilience, and collaborative learning, *Sustainability (Switzerland)*, 9(7): 1232.

Bonner, T. (2019) Why rewilding is toxic in the countryside. [Online]. Available from: www.countryside-alliance.org/news/2019/10/why-rewilding-is-toxic-in-the-countryside (accessed 4 April 2022).

Brill, S. (2021) A story of its own: creating singular gift-commodities for voluntary carbon markets, *Journal of Cultural Economy*, 14(3): 332–43.

Carver, S. (2019) Rewilding through land abandonment, in Nathalie Pettorelli, Sarah M. Durant, and Johan T. du Toit (eds), *Rewilding*. [Online]. Cambridge University Press. pp. 99–122.

Carver, S., Convery, I., Hawkins, S., Beyers, R., Eagle, A., Kun, Z., van Maanen, E., Cao, Y., Fisher, M., Edwards, S.R., Nelson, C., Gann, G.D., Shurter, S., Aguilar, K., Andrade, A., Ripple, W.J., Davis, J., Sinclair, A., Bekoff, M., et al. (2021) Guiding principles for rewilding, *Conservation Biology*, 35(6): 1882–93.

Cinner, J.E., Lau, J.D., Bauman, A.G., Feary, D.A., Januchowski-Hartley, F.A., Rojas, C.A., Barnes, M.L., Bergseth, B.J., Shum, E., Lahari, R., Ben, J. and Graham, N.A.J. (2019) Sixteen years of social and ecological dynamics reveal challenges and opportunities for adaptive management in sustaining the commons, *Proceedings of the National Academy of Sciences of the United States of America*, 116(52): 26474–83.

Clark, C. &Scanlon, B. (2019) *Less is more: Improving profitability and the natural environment in hill and other marginal farming systems.*

Convery, I., Stainer, S., Gere, C. &Lloyd, K. (2020) 'Reimagining the Lake District.'The Eologist

Defra (2019) *The Future Farming and Environment Evidence Compendium.*

Dimbleby, H. (2021) 'National Food Strategy. The Plan,'*National Food Strategy,*

Efra Committee (2021) *Environmental Land Management and the Agricultural Transition Second Report of Session 2021–22 Report.*

FFCC (2019) *Our Future in the Land.*

FFCC (2022) *The Case for a Land Use Framework.*

Folke, C., Colding, J., and Birkes, F. (2003) Synthesis: building resilience and adaptive capacity in social-ecological systems, in F. Birkes, J. Colding, and C. Folke (eds), *Navigating Social-Ecological Systems: Building Resilience for Complexity and Change.* [Online]. Cambridge University Press, pp. 352–87.

Forgrave, A. (2019) *Farming anger forces Rewilding Britain to pull out of Summit To Sea project in Mid Wales.* [Online] Available from: www.dailypost.co.uk/news/local-news/rewilding-britain-summit-sea-wales-17119541 (accessed 1 August 2022).

Garforth, C., Angell, B., Archer, J., and Green, K. (2003) Fragmentation or creative diversity? Options in the provision of land management advisory services, *Land Use Policy*, 20(4): 323–33.

Glover, J. (2019) *Landscapes Review.*

Goldsmith, B. (2021) Families who have worked the same land for generations are best placed to breathe life back into our landscapes. *Country Life,* 31 August 2021, available at: www.countrylife.co.uk/nature/ben-goldsmith-families-who-have-worked-the-same-land-for-generations-are-best-placed-to-breathe-life-back-into-our-landscapes-231950.

Gov.uk (2022a) *Environment Secretary Shares Further Information on Local Nature Recovery and Landscape Recovery Schemes.* [Online]. Available from: www.gov.uk/government/speeches/environment-secretary-shares-further-information-on-local-nature-recovery-and-landscape-recovery-schemes (accessed 3 April 2022).

Gov.uk (2020) *Government Commits to Protect 30% of UK Land by 2030.* [Online]. Available from: www.gov.uk/government/news/pm-commits-to-protect-30-of-uk-land-in-boost-for-biodiversity (accessed 3 April 2022).

Gov.uk (2022b) *New Working Group Launched to Support Tenant Farmers.* [Online]. Available from: www.gov.uk/government/news/new-working-group-launched-to-support-tenant-farmers (accessed 17 April 2022).

Harvey, D. and Scott, C. (2022) *Farm Business Survey 2020/2021 Hill Farming in England.*

HM Revenue & Customs (HMRC) (n.d.) *BIM55051—Farming in tax law: definition of farming.* [Online] Available from: https://www.gov.uk/hmrc-internal-manuals/business-income-manual/bim55051 (accessed 30 April 2022).

Hurley, P., Lyon, J., Hall, J., Little, R., Tsouvalis, J., White, V. and Rose, D.C. (2022) Co-designing the environmental land management scheme in England: The why, who and how of engaging 'harder to reach' stakeholders, *People and Nature*, 4(3), June, 744–57.

Jepson, P., Paul, R., and Blythe, C. (2020) *Rewilding: The Radical New Science of Ecological Recovery.* Icon Books Ltd.

Jepson, P. and Schepers, F. (2016) *Making Space for Rewilding: Creating an Enabling Policy Environment.* Policy brief published by Rewilding Europe.

Jepson, P., Schepers, F. and Helmer, W. (2018) Governing with nature: A European perspective on putting rewilding principles into practice, *Philosophical Transactions of the Royal Society B: Biological Sciences*, 373: 20170434.

Jones, C.H. (2022) Climate change: Cold callers shock farmers with tree-plant plea, BBC News, 27 Jan.

Lawton, J.H. et al. (2010) Making space for nature: A review of England's wildlife sites and ecological network. Report to Defra (September).

Marshall, J., Rutter, J., Kane, J., and Goss, D. (2022) *Agriculture after Brexit Replacing the CAP.* Institute for Government.

Monbiot, G. (2013) *Meet the Greatest Threat to our Countryside: Sheep.* [Online] Available from: www.spectator.co.uk/article/meet-the-greatest-threat-to-our-countryside-sheep (accessed 3 April 2022).

NAO (2021) *The Environmental Land Management Scheme.*

Ostrom, E. (1990) *Governing the Commons.* Cambridge University Press.

Ostrom, E. (2000) Collective action and the evolution of social norms, *Journal of Economic Perspectives*, 14(3): 137–58.

Ostrom, E., Burger, J., Field, C.B., Norgaard, R.B. and Policansky, D. (1999) Revisiting the commons: Local lessons, global challenges. *Science*, 284 (5412).

Pretty, J. (2018) Intensification for redesigned and sustainable agricultural systems, *Science*, 362 (6417). doi/10.1126/science.aav0294.

Public Accounts Committee (2021) *Environmental Land Management Scheme Thirty-First Report of Session 2021-22 Report, together with formal minutes relating to the report.*

Reboah, C. (2019) *Faculty of Natural Resources and Agricultural Sciences Barriers to Rewilding on Sussex Farmland – Socio-psychological Implications of Rewilding on Farmers' Sense of Place.*

Rewilding Britain (2021) *The World We Want to See.* [Online]. Available from: www.rewildingbritain.org.uk/about-us/the-world-we-want-to-see (accessed 3 April 2022).

Rust, N.A., Stankovics, P., Jarvis, R.M., Morris-Trainor, Z., de Vries, J.R., Ingram, J., Mills, J., Glikman, J.A., Parkinson, J., Toth, Z., Hansda, R., McMorran, R., Glass, J., and Reed, M.S. (2022) Have farmers had enough of experts?, *Environmental Management*, 69(1): 31–44.

Sandom, C.J., Dempsey, B., Bullock, D., Ely, A., Jepson, P., Jimenez-Wisler, S., Newton, A., Pettorelli, N., and Senior, R.A. (2019) Rewilding in the English uplands: Policy and practice, *Journal of Applied Ecology*, 56(2): 266–73.

Seppelt, R., Verburg, P.H., Norström, A., Cramer, W., and Václavik, T. (2018) Focus on cross-scale feedbacks in global sustainable land management. *Environmental Research Letters*, 13: 090402.

Starnes, T., Beresford, A.E., Buchanan, G.M., Lewis, M., Hughes, A. and Gregory, R.D. (2021) The extent and effectiveness of protected areas in the UK, *Global Ecology and Conservation*, 30: e01640,

Sutherland, L.A., Mills, J., Ingram, J., Burton, R.J.F., Dwyer, J., and Blackstock, K. (2013) Considering the source: Commercialisation and trust in agri-environmental information and advisory services in England, *Journal of Environmental Management*, 118: 96–105.

Tidman, Z. (2022) Tenant farmers could be hit by new rewilding incentives, Union warns, *The Independent*, 7 January.

Torres Aurora, Fernández Néstor, zu Ermgassen Sophus, Helmer Wouter, Revilla Eloy, Saavedra Deli, Perino Andrea, Mimet Anne, Rey-Benayas José M., Selva Nuria, Schepers Frans, Svenning Jens-Christian, and Pereira Henrique M. (2018) Measuring rewilding progress, *Phil. Trans. R. Soc. B* 3732017043320170433.

Vigani, M., Urquhart, J., Black, J.E., Berry, R., Dwyer, J., and Rose, D.C. (2021) Post-Brexit Policies for a Resilient Arable Farming Sector in England, *EuroChoices*, 20(1): 55–61.

24

UNSEEN CONNECTIONS

The role of fungi in rewilding

David Satori and Matt Wainhouse

Introduction

Fungi are foundational to the function of terrestrial ecosystems, carrying out a remarkable array of ecological processes upon which countless other life forms depend (Dighton, 2018). The focal point of rewilding has largely been on reviving ecosystem functionality through the initiation of top–down trophic cascades, the (re)introduction of ecosystem engineers, and the facilitation of vegetational succession to promote the development of self-regulating, biodiverse ecosystems (Pettorelli et al., 2018; Svenning et al., 2016). Approaches to rewilding often share a common thread of working with highly visible species, predominantly animals, whose effects can be monitored from our aboveground vantage point. Fungi, in contrast, remain hidden from view, growing vegetatively through opaque substrates such as soil and wood, but their influence over global nutrient cycles, biodiversity, and ecosystem productivity underpin ecosystem function. Their role in maintaining trophic complexity in the natural world has been overlooked and understudied (Contos et al., 2021). As with other taxa, fungal assemblages are vulnerable to anthropogenic disturbance, particularly habitat loss and nitrogen deposition (Dahlberg et al., 2010), so a dual framework for fungus conservation is needed, one that recognises their utility in ecosystem recovery, and that protects fungal biodiversity in its own right. The full ecological impact of rewilding cannot be assessed with a focus on plants and animals alone, nor are we able to restore many degraded landscapes without considering how fungi facilitate the emerging complexity of recovering ecosystems (Koziol et al., 2018; Policelli et al., 2020).

Flora, Fauna, and Funga—the three Fs terminology proposed by Kuhar et al. (2018) was created to explicitly recognise the fungal kingdom as a fundamental component of biodiversity. Delimiting fungal taxa within an ecosystem and giving them prominence as *funga* is a crucial first step towards their inclusion in global conservation frameworks, signalling a major shift in the policy landscape towards a more holistic approach to nature conservation and ecosystem restoration (Gonçalves et al., 2021). This naturally extends to rewilding frameworks and so here we discuss how mycology can align with its guiding principles (Carver et al., 2021). Fungi exist as decentralised, complex networks, foraging for and connecting nutritional resources, with each additional connection improving their robustness to damage. In many ways, fungal networks serve as an apt metaphor for rewilding that can inspire the discipline at the theoretical

DOI: 10.4324/9781003097822-27

level, reflecting the complexity, interconnectedness, and resilience that rewilding hopes to achieve.

Fungi and ecosystem function

The structure and physiology of a fungus is unlike that of a plant or an animal, though they share the commonality of being eukaryotic organisms that form complex multicellular structures. Fungi may be filamentous (e.g., moulds and macrofungi), or single-celled (e.g., yeasts and chytrids) (Naranjo-Ortiz & Gabaldón, 2019). Like animals, fungi are heterotrophic, deriving their nutrition from the biomass of living or non-living organisms. The 'body' of a filamentous fungus is composed of tubular cells called *hyphae*, each c. 5 μm in diameter, that grow through their substrates, secreting extracellular enzymes and organic acids to break down carbohydrate and mineral sources into smaller molecules that are then absorbed through their chitinous cell walls. Collectively, hyphae form extensively-branched structures, *mycelia*, that allow fungi to explore substrates multidirectionally whilst maintaining intercellular coordination so that hyphal growth can be withdrawn and redirected towards a nutritional resource when encountered on any part of the mycelial front (Boddy, 2021). Fungi may reproduce sexually or asexually through the production of spores, often borne on conspicuous fruiting structures which disperse them into the environment via wind or grazing animals (Figure 24.1).

Terrestrial land was first colonised 1000–700 Mya by fungi adapted to the unique stresses of the Proterozoic (Gross, 2019). It would be some 400–500 Mya before plants emerged onto dry land, facilitated by fungal symbiosis. Mycorrhizal fungi—those that form symbiotic

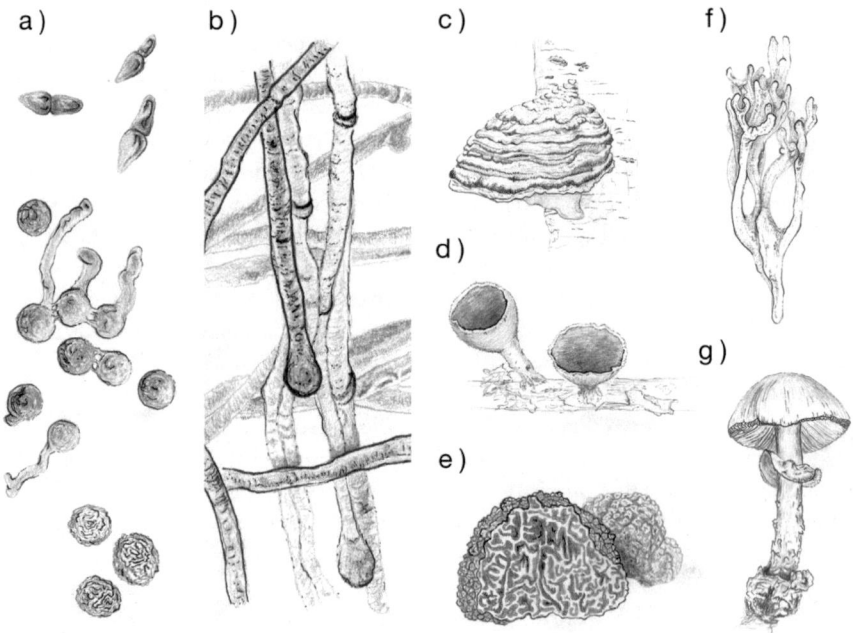

Figure 24.1 Examples of fungal vegetative and reproductive structures. (a) spores; (b) hyphae; (c) bracket (*Fomes* sp.); (d) cups (*Sarcoscypha* sp.); (e) truffles (*Tuber* sp.); (f) coral (*Ramaria* sp.); mushroom (*Amanita* sp.). © Sally Somerville-Woodiwis.

associations with c. 85% of vascular plant species (Brundrett & Tedersoo, 2018)—co-evolved with terrestrial plants. The coupled processes of plant photosynthesis and fungal mineral weathering of rock assist in the formation of soils that form the basis of today's terrestrial ecosystems (Taylor et al., 2009). Mycorrhizal fungi and their plant hosts transform the chemical and physical properties of soils through the continuous deposition of organic matter, channelling of water, and the liberation of essential elements from mineral sources, influencing the landscape's distribution of soil types, vegetational communities, and biomes (Leake & Read, 2017). Fungi secrete enzymes that break down organic matter into soluble inorganic ions, and through the imperfect recapture of those nutrients, enhance soil fertility (Dighton, 2018). Unlike bacteria and soil fauna that also play a role in decomposition, saprotrophic fungi (primarily basidiomycetes) are capable of breaking down some of nature's most recalcitrant lignocellulose complexes that form woody tissues. Thus, fungi are key drivers of the Earth's carbon cycle, regulating fluxes of organic nutrients within soil, the largest terrestrial carbon pool on the planet.

Fungi also have an important, but often neglected, role in food webs and primary productivity. Fungal fruiting bodies and mycelia are a valuable source of food for much of the biota, from mammals (see section: **Safeguarding ectomycorrhizal fungi and forest food web**) to soil-dwelling and saproxylic (deadwood-dependent) invertebrates and cavity-nesting birds (see section: **Tree cavities and wood decay in rewilded landscapes**). Belowground, arthropods, oligochaetes, molluscs, and nematodes consume spores and hyphae, assisting in fungal dispersal and nutrient redistribution in the soil (Crowther et al., 2012). Mycorrhizal fungi directly influence primary productivity by mediating the uptake of mineral nutrients to their plant symbionts, enhancing their photosynthetic capacity and nutritional content. Strikingly, the fingerprint of mycorrhizal activity can be found in the megafauna that roam the grasslands of the Serengeti. Half of the biomass of its iconic migratory herbivores and one-third of the biomass of carnivores is derived from arbuscular mycorrhizal fungi (see section: **Arbuscular mycorrhizal fungi and grassland recovery**) through the enhanced growth and nutritional quality of plants (Stevens et al., 2018). Without the biochemical inputs of mycorrhizal fungi, the carrying capacity of mammals in this ecosystem would certainly be diminished. The return of large mammals to depauperate ecosystems is often an aspiration for rewilded landscapes so the contribution of mycorrhizal fungi to support the productivity of these systems should not be overlooked.

Arbuscular mycorrhizal fungi and grassland recovery

Grasslands are dynamic ecosystems sculpted by the perennial pressures of herbivory and/or periodic wildfires (Bond, 2005), and, just as crucially, the ceaseless activity of arbuscular mycorrhizal (AM) fungi (Read & Perez-Moreno, 2003). AM fungi are mutualistic symbionts of over 80% of terrestrial plant species, including grasses, forbs, most tropical trees, and some temperate trees, exchanging nitrates, phosphates, and other minerals for photosynthetically derived sugars (Smith & Read, 2008). Hyphae of AM fungi grow into root tissue, penetrating cortical cells where they produce nutrient exchange structures known as *arbuscules*—transient branched clusters of hyphae that resemble little trees. Grassland soils have higher rates of organic matter turnover and nitrogen (N) mobility compared to other ecosystems (e.g., forests and heathlands) which results in phosphorus (P) becoming the main growth-limiting nutrient which AM fungi assist in the uptake of (Read, 1991). Additional host benefits include greater tolerance to drought and to some diseases (Smith & Read, 2008). The continual growth and turnover of AM hyphae leads to the deposition of considerable quantities of chitinous residues that turn the soil into a living carbon sink, with AM fungi contributing as much as 15% to the soil carbon pool in grasslands (Leake et al., 2004).

Agricultural intensification and the conversion of natural grasslands to productive arable use imperils many soilborne organisms (Tsiafouli et al., 2015). The deposition of inorganic fertilisers, frequent tillage, and mass simplification of plant communities with monocultures of annual crops are stresses that select for highly adapted ruderal soil fungi (French et al., 2017; Guzman et al., 2021). In particular, elevated N levels select for a small number of N-tolerant AM fungi which tend to have less extensive hyphal networks that require lower quantities of photosynthates to maintain. This leads to a cascade of changes in which fungal species with less robust hyphal networks and lower carbon demands allocate smaller quantities of P to their hosts, reducing plant fitness, photosynthetic capacity, and soil carbon inputs (Treseder et al., 2018), a trend which rewilding should aim to reverse and stabilise. When agricultural activity ceases and an ecosystem is allowed to recover to a more natural state, the proportion of fungi present and the quality of organic matter improves over time (van der Wal et al., 2006). In many cases, it is the exhausted soils (and lost income) that provide landowners with the motivation to move towards a land use that embraces rewilding and the restoration of natural processes such as soil formation.

Animals can directly impact the composition of AM fungal communities (Heyde et al., 2017). The restoration of locally extinct species is a guiding principle of rewilding, the result of which can directly alter soil fungal communities (Gehring et al., 2002). In an arid ecosystem in Australia, the reintroduction of four native mammals, and the removal of exotic vertebrates increased the diversity of the fungal community (Clarke et al., 2015). The likelihood is that microbial taxa faced coextinction with the loss of native vertebrates. What this means for fungi (and AM fungi in particular) is not clear, though the community shift towards complexity may be indicative of a more natural and, ultimately, resilient system.

Restoration of AM fungi in grasslands relies on local source populations from which they can disperse. Some landscapes, however, are so degraded that passive rewilding may not be sufficient to adequately restore their vegetational and soil microbial communities. It can be beneficial to introduce AM fungi via soil inoculation as a cost-effective strategy for restoring the full range of species interactions by increasing plant biodiversity and product-ivity (Koziol et al., 2018; van der Heijden et al., 1998) and accelerating the establishment of late successional plants (Koziol & Bever, 2016; Middleton & Bever, 2012). This may be just as important (though less visible) as the reintroduction of charismatic plants and animals, but landscape-scale inoculations with locally-adapted species are yet to be realised. Evidence from prairie restoration in North America without AM fungi or soil microbial reintroductions often result in lower plant species diversity and richness compared to pris-tine examples, and in some cases this diversity and richness can decline over time (Koziol et al., 2018). Locally-adapted AM fungal inocula typically use soil as a carrier material for spores, root fragments, or mycelia, which are co-introduced with their plant hosts, pref-erentially already in symbiotic association or in direct contact with their root systems and can be sourced from soil samples of reference ecosystems and/or grown as pure cultures (Koziol et al., 2018). Different species of AM fungi differ in their successional stage and capacities to affect host nutrient uptake, making some species more important to utilise in restoration than others (Koziol & Bever, 2018). Inoculation with AM fungi is particularly relevant in severely degraded areas such as mine sites (Vahter et al., 2020), soils with high salinity (Porcel et al., 2012), or drought-prone sites (Wu et al., 2017) where a more passive approach to rewilding may struggle to restore the complexity, resilience, and species inter-action dynamics of intact ecosystems.

Tree cavities and wood decay in rewilded landscapes

Old, mature trees, termed veteran, or ancient, are iconic features of wood pasture and natural forest. However, they are often poorly represented in anthropogenic landscapes and recruitment to this age cohort is declining globally (Lindenmayer et al., 2012). Old trees are a 'keystone structure' since they make a disproportionate contribution to the functioning of ecosystems relative to their area with regards to carbon dynamics, nutrient cycling, and wildlife habitat (Lindenmayer, 2017) (Figure 24.2a). The value of a tree for biodiversity increases with age as its physical structure complexifies and micro-habitats within it accrue and develop (Kõrkjas et al., 2021), the most critical of these being the decaying wood and hollows created by saprotrophic fungi. Decaying wood is a highly versatile substrate. The physical, chemical, and even biotic properties change as decomposition progresses, supporting a diverse community of invertebrates at each stage of wood-decay (Boddy, 2021).

The importance of tree hollows as a habitat resource for terrestrial vertebrates cannot be understated. Globally, 20% of all bird species and more than 1000 mammals are dependent on tree cavities (Stokland et al., 2012; van der Hoek et al., 2017). The legacy of tree loss due to forestry, farming, and fire has led to decline in both taxonomic and functional diversity (Ibarra et al., 2017; Lindenmayer & Sato, 2018). Even in restored landscapes, populations of cavity-nesting birds can continue to decline while other groups increase due to the natural loss of older trees and historically low tree recruitment, no heart rot development, and the natural loss of any remaining older trees (Poessel et al., 2020).

With the exception of long-established pollards, land management has not valued veteran trees or deadwood as a biological resource. Compared to pristine or near-natural habitats, much of today's managed forests are depleted of old trees and large-diameter deadwood. In British forests, for example, the number of standing dead trees over 40cm diameter is <1/ha compared to 14/ha in the strict reserve of Białowieża National Park in Poland (Drozdowski et al., 2017; Kirby et al., 1998). Yet, evidence from the fossil record shows that Britain in the early-mid Holocene supported far greater quantities deadwood resources than the current era (Svenning, 2002; Whitehouse & Smith, 2010). The palaeoecological evidence shows saproxylic Coleoptera disappearing from the fossil record at the same time as the agricultural clearances of the Neolithic, ca. 5000–3000 cal BP (Whitehouse, 2006). Populations of these extinct species persist on the continent where they are associated with heart-rot in mature trees or large-diameter decomposing wood, though never recovered in Britain.

Decomposition is a slow process, so a major challenge for rewilding treescapes is how to restore biodiversity associated with large old trees relative to a pre-agricultural reference point. Restoration of natural heart-rot dynamics allows for the fungus to establish and colonise a tree, but the time lag before cavities develop is considerable. In western Europe, pedunculate oak (*Quercus robur*) is a long-lived tree with more associated species than any other in the temperate zone. However, it can take 200 years for a heart-rot fungus to establish and 400 years before cavities are ubiquitous (Ranius et al., 2009). One of the most widely used interventions in con-servation, the bird box, is an artificial response to this loss, though they tend to support a narrow range of cavity-using fauna. Different species require different, often disparate characteristics in their cavities. The structural diversity of hollows, and the chemical and biological properties of decaying wood are not easily replicated in rewilded landscapes. However, interventions that accelerate natural decomposition processes can be implemented and fit within the rewilding framework where conventional approaches like the bird box are unsuitable.

Deliberately damaging younger trees, termed 'veteranisation', can facilitate fungal colon-isation and accelerate heart-rot formation (Bengtsson et al., 2012; Figure 24.2b). This might

Figure 24.2 (a) Veteran trees in wood pasture are critical for structural and resource complexity of the system. Photo credit: Ted Green. (b) Veteranising trees to create cavity and wood–decay habitat missing in rewilded landscapes. A nest-box cavity cut with a chain saw after eight years (left) and a bored woodpecker cavity after two years (right). Photo Credit: Vikki Bengtsson, Pro Natura. (c) Conservation inoculations of fungi to accelerate heart-rot development and to reintroduce threatened species. Holes are cut into the tree and inoculum inserted (left). The threatened fungus, *Hericium coralloides*, fruits from a conservation inoculation wound after two years (right). Photo Credit: Matt Wainhouse.

include girdling stems, cutting holes in trunks or removing bark and sapwood from the base to simulate damage by large herbivores like horse or moose (Bengtsson et al., 2015). Fungal colonisation following veteranisation is a stochastic process. The natural extension to this is to inoculate trees with heart-rot fungi to further accelerate decomposition and cavity formation (Wainhouse & Boddy, 2022). Additionally, fungal inoculation can be used as a conservation tool to reintroduce threatened saprotrophic species associated with near-natural old growth forests that have become isolated in nature reserves (Nordén et al., 2020; Figure 24.2c). Inoculated fungi can be selected for specific characteristics such as decomposition rate, rot-type, or association with another species. Despite a long experimental pedigree in forestry studies, inoculation with wood-decaying fungi as a tool for habitat restoration is in its infancy, and more long-term research and monitoring is needed to fully appreciate the value of the intervention. On the continuum of rewilding (Carver et al., 2021), recreating some of the structural complexity of ancient trees may shorten the progression from anthropogenic to wild landscapes by restoring some of the missing ecological functions of newly rewilded ecosystems, and in turn, mitigate the loss of cavity using and saproxylic taxa and heart-rot fungi.

Safeguarding ectomycorrhizal fungi and forest food webs

Temperate and boreal forests are major terrestrial biomes. In many instances, the predominant tree species in these, and other wooded ecosystems, form obligate associations with ectomycorrhizal (ECM) fungi (Crowther et al., 2019). ECM fungi form hyphal mantles around roots tips, weaving around epidermal and/or cortical cells to form bidirectional nutrient exchange sites known as Hartig nets (Smith & Read, 2008), and their exploratory hyphae may comprise a third or more of the total microbial biomass of forest soils (Högberg & Högberg, 2002). These fungi associate with a broad range of geographically widespread trees and woody shrubs including the Betulaceae, Fagaceae, and Pinaceae in the Northern Hemisphere, and the Nothofagaceae and *Eucalyptus* in the Southern Hemisphere.

The discovery of common mycorrhizal networks that link ECM fungi to multiple trees has reshaped our understanding of woodland regeneration, revealing the fluxes of carbon, mineral nutrients, and water that govern plant productivity. In primary successional environments with poorly-developed soil spore banks, seedlings connected by pre-existing ECM networks achieve greater biomass than those that are isolated, likely a result of the increased nutrient translocation capacities of extensive mycelial networks (Nara, 2015). In non-forested ecosystems such as heathlands, emerging tree seedlings have limited access to ECM fungi, as the predominant vegetation associates with other mycorrhizal groups (e.g., ericoid mycorrhizal fungi in heathlands). This has been implicated as a mechanism that can restrict woodland establishment, but a small number of spore-dispersed ECM fungi can facilitate woodland succession by improving seedling establishment, replacing resident mycorrhizal communities, and in some cases facilitating tree invasions (Collier & Bidartondo, 2009). From a rewilding perspective, understanding mycorrhizal dynamics can help shed light on woodland successional patterns and soil carbon sequestration, and help assess ECM dispersal limitation to inform woodland creation initiatives.

For several decades, there have been indications that ECM fungi are declining in the extent of root colonisation (Cairney & Meharg, 1999) and in the abundance of fruiting bodies (Arnolds, 1991). Throughout evolutionary history, ECM symbiosis persisted under N-limited conditions wherein the nutritional benefit of enhanced N uptake to host plants outweighed the energetic costs of maintaining mycorrhizal associations. In the present-day, excessive atmospheric N pollution is adversely impacting ECM functional diversity and abundance (Lilleskov et al., 2011). Additionally, climate change is predicted to shift environmental conditions beyond the physiological tolerance of some ECM fungi (Kipfer et al., 2010), and in temperate and boreal Pinaceae forests alone, this is expected to result in a 21% loss in ECM species richness in the next 50 years (Steidinger et al., 2020). The ecological consequences of this ongoing loss of fungal diversity could be significant, impacting seedling recruitment, host nutrient status, drought tolerance, and plant pathogen resistance. Although external threats such as N pollution and climate change cannot be eliminated, their impact could be buffered by rewilding (Corlett, 2016). Many ECM fungi are restricted to old-growth forests and protecting these 'cores' coupled with appropriate efforts to create connectivity via corridors of natural, unmanaged forests is an important step in facilitating ECM persistence in the landscape.

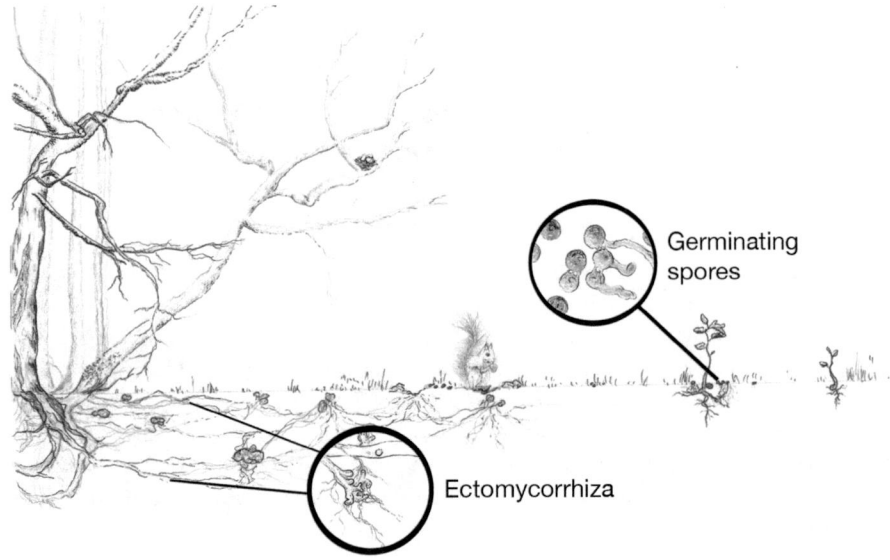

Figure 24.3 Forest regeneration depends on the dispersal of ectomycorrhizal fungi by small mammals which facilitate the development of symbiotic mycorrhizal associations amongst tree seedlings. © Sally Somerville-Woodiwis.

One way in which rewilding may support ECM function is through the reintroduction of spore-dispersing animals. The consumption of ECM fungi is a vital but often overlooked mechanism of forest ecosystem dynamics. Many small animals such as rodents and marsupials collect, unearth, cache, and consume hypogeous (underground) ECM fungal fruiting bodies and inadvertently serve as spore dispersal vectors, encouraging forest regeneration, ecosystem stability, and ECM diversity (Maser et al., 2008) (Figure 24.3). In Australia, the restoration of indigenous mammals is restoring ECM assemblages and as a direct result, contributing to increased productivity and ecosystem recovery (Fleming et al., 2014). Reintroductions of functionally similar species can be useful where species have gone extinct (e.g., cattle in place of aurochs in the rewilding of European landscapes), but fungus-eating animals need careful consideration. Threatened species of *Cortinarius* fungi in New Zealand lost their dispersal vector when the moa went extinct (Boast et al., 2018). However, the consumption of these fungi by exotic mammals such as red deer (*Cervus elaphus*) and brushtail possum (*Trichosurus vulpecula*) do not provide an alternative dispersal mechanism as their spores do not remain viable after excretion (Wood et al., 2015). New Zealand's non-native mammals also consume exotic ECM fungi, which may facilitate the spread of invasive conifers, and the removal of these introduced mammals may help to reverse the 'invasion meltdown' brought about by historic introductions (Wood et al., 2015). Much attention has been paid to the role of large herbivores and apex predators in shaping the biotic communities of their habitats, but the strong interactions between animals, ECM fungi, and trees are important factors to consider when reintroductions or population reinforcements are necessary to enhance forest biodiversity and function.

Conclusion

No view of natural systems is complete without recognising that fungi play important roles in increasing trophic complexity and should fit within pre-existing rewilding frameworks, but there are challenges that need to be overcome. Rewilding highly degraded and unproductive agricultural landscapes feels like an unquestionably sensible intervention. However, some of the most ideal sites for this type of landscape-scale restoration can support important communities of fungi, and the preservation of these anthropogenic landscapes should be weighed against rewilding goals. In post-Brexit Britain, the upland fringe is a candidate for change in land-use and management as farmers look to exchange sheep farming for carbon offset tree-planting. The overgrazed hill pastures—such as those of the Dark Peaks and the Pennines—are internationally important for their assemblages of waxcap fungi (*Hygrocybe* spp.) and their allies. Both rewilding and tree planting schemes must avoid these globally important and threatened habitats (Baird & Pope, 2022). Similarly, beaver reintroductions can have immense value, but can be devastating for epiphytic lichens and deadwood fungi (Scottish Government, 2017).

As with any land-use change, consideration needs to be given to the social context. Collecting fungi is a culturally-significant activity for many peoples across the world and requires the intergenerational transmission of an extensive body of taxonomic, biological, and ecological knowledge (Cunningham & Yang, 2016). Local expertise gives valuable data on species habitat preferences and population trends that may or may not be apparent in scientific literature (Kotowski et al., 2021). Cooperation with local communities can lead to mutual exchanges of ecological information that should inform habitat management. Rewilding offers an optimistic conservation approach to counteract fungal decline that embraces the exchange of local knowledge, mutual learning, and the inclusion of traditions that foster nature connection and eco-reciprocity.

Fungi are unequivocally fundamental to ecosystem processes—from nutrient cycling and carbon sequestration to plant productivity and succession—but fungi are rarely sufficiently monitored following restoration efforts (Contos et al., 2021). Research into the response of fungal communities to rewilding on above- and belowground food webs is underdeveloped (Andriuzzi & Wall, 2018). Fortunately, metabarcoding approaches to studying fungal diversity coupled with classic methods of fruit body surveys is rapidly enriching our understanding of the relationship between fungi and their environment. Rewilding, with its focus on rebuilding natural processes and complete food webs at all trophic levels, offers a new template for exploring fungal relationships. Considerable knowledge gaps remain of our understanding of fungal autecology, diversity, distribution, and population dynamics that lead to a high degree of uncertainty in both their conservation and their roles in ecosystem function. Nevertheless, this should not be a basis for inaction (Molina et al., 2011), but rather an opportunity to open pathways of collaboration between rewilding researchers, practitioners, and mycologists to share insights across disciplinary lines.

References

Andriuzzi, W.S. and Wall, D.H. (2018). Soil biological responses to, and feedbacks on, trophic rewilding. *Philosophical Transactions of the Royal Society B: Biological Sciences*, *373*(1761). https://doi.org/10.1098/RSTB.2017.0448

Arnolds, E. (1991). Decline of ectomycorrhizal fungi in Europe. *Agriculture, Ecosystems & Environment*, *35*(2–3): 209–44. https://doi.org/10.1016/0167-8809(91)90052-Y

Baird, A. and Pope, F. (2022). 'Can't see the forest for the trees': The importance of fungi in the context of UK tree planting. *Food and Energy Security*, *00*: e371. https://doi.org/10.1002/FES3.371

Bengtsson, V., Hedin, J., and Niklasson, M. (2012). Veteranisation of oak—managing trees to speed up habitat production. *Trees beyond the Wood: An Exploration of Concepts of Woods, Forests and Trees*, 61–8. Wildtrack Publishing.

Bengtsson, V., Niklasson, M., and Hedin, J. (2015). Tree veteranisation—using tools instead of time. *Conserv Land Manag*, 14–17.

Boast, A.P., Weyrich, L.S., Wood, J.R., Metcalf, J.L., Knight, R., and Cooper, A. (2018). Coprolites reveal ecological interactions lost with the extinction of New Zealand birds. *Proceedings of the National Academy of Sciences of the United States of America*, *115*(7): 1546–51. https://doi.org/10.1073/PNAS.171 2337115/-/DCSUPPLEMENTAL

Boddy, L. (2021). *Fungi and Trees: Their Complex Relationships*. 306. www.summerfieldbooks.com/prod uct/fungi-and-trees-their-complex-relationships/

Bond, W.J. (2005). Large parts of the world are brown or black: A different view on the 'Green World' hypothesis. *Journal of Vegetation Science*, *16*(3): 261–6. https://doi.org/10.1111/J.1654-1103.2005. TB02364.X

Brundrett, M.C. and Tedersoo, L. (2018). Evolutionary history of mycorrhizal symbioses and global host plant diversity. *New Phytologist*, *220*(4): 1108–15. https://doi.org/10.1111/NPH.14976

Cairney, J.W.G. and Meharg, A.A. (1999). Influences of anthropogenic pollution on mycorrhizal fungal communities. *Environmental Pollution*, *106*(2): 169–82. https://doi.org/10.1016/S0269-7491(99)00081-0

Carver, S., Convery, I., Hawkins, S., Beyers, R., Eagle, A., Kun, Z., van Maanen, E., Cao, Y., Fisher, M., Edwards, S. R., Nelson, C., Gann, G. D., Shurter, S., Aguilar, K., Andrade, A., Ripple, W. J., Davis, J., Sinclair, A., Bekoff, M., …Soulé, M. (2021). Guiding principles for rewilding. *Conservation Biology*, *35*(6): 1882–93. https://doi.org/10.1111/COBI.13730

Clarke, L.J., Weyrich, L.S., and Cooper, A. (2015). Reintroduction of locally extinct vertebrates impacts arid soil fungal communities. *Molecular Ecology*, *24*(12): 3194–3205. https://doi.org/10.1111/ MEC.13229

Collier, F.A. and Bidartondo, M.I. (2009). Waiting for fungi: the ectomycorrhizal invasion of lowland heathlands. *Journal of Ecology*, *97*(5): 950–63. https://doi.org/10.1111/J.1365-2745.2009.01544.X

Contos, P., Wood, J.L., Murphy, N.P., and Gibb, H. (2021). Rewilding with invertebrates and microbes to restore ecosystems: Present trends and future directions. *Ecology and Evolution*, *11*(12): 7187–7200. https://doi.org/10.1002/ECE3.7597

Corlett, R.T. (2016). The role of rewilding in landscape design for conservation. *Current Landscape Ecology Reports 2016*, *1*(3): 127–33. https://doi.org/10.1007/S40823-016-0014-9

Crowther, T.W., Boddy, L., and Hefin Jones, T. (2012). Functional and ecological consequences of saprotrophic fungus–grazer interactions. *The ISME Journal 2012*, *6*(11): 1992–2001. https://doi.org/ 10.1038/ismej.2012.53

Crowther, T.W., van den Hoogen, J., Wan, J., Mayes, M. A., Keiser, A.D., Mo, L., Averill, C., and Maynard, D.S. (2019). The global soil community and its influence on biogeochemistry. *Science*, *365*(6455). https://doi.org/10.1126/SCIENCE.AAV0550/ASSET/B9626FCA-8F3C-471F-9728- 413841D2D44D/ASSETS/GRAPHIC/365_AAV0550_FA.JPEG

Cunningham, A.B. and Yang, X. (eds). (2016). *Mushrooms in Forests and Woodlands: Resource Management, Values and Local Livelihoods*. Routledge.

Dahlberg, A., Genney, D.R., and Heilmann-Clausen, J. (2010). Developing a comprehensive strategy for fungal conservation in Europe: current status and future needs. *Fungal Ecology*, *3*(2): 50–64. https:// doi.org/10.1016/J.FUNECO.2009.10.004

Dighton, J. (2018). *Fungi in Ecosystem Processes*. https://doi.org/10.1201/9781315371528

Drozdowski, S., Brzezeicki, B., and Keczynski, A. (2017). The Forests of the Strict Reserve of Białowieża National Park. *Białowieski Park Narodowy*.

Fleming, P.A., Anderson, H., Prendergast, A.S., Bretz, M.R., Valentine, L.E., and Hardy, G.E.S. (2014). Is the loss of Australian digging mammals contributing to a deterioration in ecosystem function? *Mammal Review*, *44*(2): 94–108. https://doi.org/10.1111/MAM.12014

French, K.E., Tkacz, A., and Turnbull, L.A. (2017). Conversion of grassland to arable decreases microbial diversity and alters community composition. *Applied Soil Ecology*, *110*: 43–52. https://doi.org/ 10.1016/J.APSOIL.2016.10.015

Gehring, C.A., Wolf, J.E., and Theimer, T.C. (2002). Terrestrial vertebrates promote arbuscular mycorrhizal fungal diversity and inoculum potential in a rain forest soil. *Ecology Letters*, *5*(4): 540–8. https:// doi.org/10.1046/J.1461-0248.2002.00353.X

Gonçalves, S.C., Haelewaters, D., Furci, G., and Mueller, G.M. (2021). Include all fungi in biodiversity goals. *Science*, *373*(6553), 403. https://doi.org/10.1126/SCIENCE.ABK1312/ASSET/74416285- C1B3-4BA2-AF95-615A15E859F8/ASSETS/GRAPHIC/373_403A_F1.JPEG

Gross, M. (2019). The success story of plants and fungi. *Current Biology*, *29*(6): R183–R185. https://doi. org/10.1016/J.CUB.2019.02.058

Guzman, A., Montes, M., Hutchins, L., DeLaCerda, G., Yang, P., Kakouridis, A., Dahlquist-Willard, R.M., Firestone, M.K., Bowles, T., and Kremen, C. (2021). Crop diversity enriches arbuscular mycorrhizal fungal communities in an intensive agricultural landscape. *New Phytologist*, *231*(1): 447–59. https://doi.org/10.1111/NPH.17306

Heyde, M. van der, Bennett, J.A., Pither, J., and Hart, M. (2017). Longterm effects of grazing on arbuscular mycorrhizal fungi. *Agriculture, Ecosystems & Environment*, *243*, 27–33. https://doi.org/10.1016/J.AGEE.2017.04.003

Högberg, M.N. and Högberg, P. (2002). Extramatrical ectomycorrhizal mycelium contributes one-third of microbial biomass and produces, together with associated roots, half the dissolved organic carbon in a forest soil. *New Phytologist*, *154*(3): 791–5. https://doi.org/10.1046/J.1469-8137.2002.00417.X

Ibarra, J.T., Martin, M., Cockle, K.L., and Martin, K. (2017). Maintaining ecosystem resilience: functional responses of tree cavity nesters to logging in temperate forests of the Americas. *Scientific Reports*, *7*(4467): 1–9. https://doi.org/10.1038/s41598-017-04733-2

Kipfer, T., Egli, S., Ghazoul, J., Moser, B., and Wohlgemuth, T. (2010). Susceptibility of ectomycorrhizal fungi to soil heating. *Fungal Biology*, *114*(5–6): 467–72. https://doi.org/10.1016/J.FUNBIO.2010.03.008

Kirby, K.J., Reid, C.M., Thomas, R.C., and Goldsmith, F.B. (1998). Preliminary estimates of fallen dead wood and standing dead trees in managed and unmanaged forests in Britain. *Journal of Applied Ecology*, *35*(1), 148–55. https://doi.org/10.1046/J.1365-2664.1998.00276.X

Kõrkjas, M., Remm, L., and Lõhmus, A. (2021). Development rates and persistence of the microhabitats initiated by disease and injuries in live trees: A review. *Forest Ecology and Management*, *482*: 118833. https://doi.org/10.1016/J.FORECO.2020.118833

Kotowski, M.A., Molnár, Z., and Łuczaj, Ł. (2021). Fungal ethnoecology: observed habitat preferences and the perception of changes in fungal abundance by mushroom collectors in Poland. *Journal of Ethnobiology and Ethnomedicine*, *17*(1): 1–23. https://doi.org/10.1186/S13002-021-00456-X/TABLES/4

Koziol, L. and Bever, J.D. (2016). AMF, phylogeny, and succession: specificity of response to mycorrhizal fungi increases for late-successional plants. *Ecosphere*, *7*(11): e01555. https://doi.org/10.1002/ECS2.1555

Koziol, L. and Bever, J.D. (2018). Mycorrhizal feedbacks generate positive frequency dependence accelerating grassland succession. *Journal of Ecology*, *107*(2), 622–32. https://doi.org/10.1111/1365-2745.13063

Koziol, L., Schultz, P.A., House, G.L., Bauer, J.T., Middleton, E.L., and Bever, J.D. (2018). The plant microbiome and native plant restoration: The example of native mycorrhizal fungi. *BioScience*, *68*(12): 996–1006. https://doi.org/10.1093/BIOSCI/BIY125

Kuhar, F., Furci, G., Drechsler-Santos, E.R., and Pfister, D.H. (2018). Delimitation of funga as a valid term for the diversity of fungal communities: The Fauna, Flora & Funga proposal (FF & F). *IMA Fungus*, *9*(2): 71–4. https://doi.org/10.1007/BF03449441/FIGURES/1

Leake, J., Johnson, D., Donnelly, D., Muckle, G., Boddy, L., and Read, D. (2004). Networks of power and influence: the role of mycorrhizal mycelium in controlling plant communities and agroecosystem functioning. Https://Doi.Org/10.1139/B04-060, *82*(8): 1016–45. https://doi.org/10.1139/B04-060

Leake, J. and Read, D.J. (2017). Mycorrhizal symbioses and pedogenesis throughout Earth's history. *Mycorrhizal Mediation of Soil: Fertility, Structure, and Carbon Storage*, 9–33. https://doi.org/10.1016/B978-0-12-804312-7.00002-4

Lilleskov, E.A., Hobbie, E.A., and Horton, T.R. (2011). Conservation of ectomycorrhizal fungi: exploring the linkages between functional and taxonomic responses to anthropogenic N deposition. *Fungal Ecology*, *4*(2): 174–83. https://doi.org/10.1016/J.FUNECO.2010.09.008

Lindenmayer, D.B. (2017). Conserving large old trees as small natural features. *Biological Conservation*, *211*: 51–9. https://doi.org/10.1016/J.BIOCON.2016.11.012

Lindenmayer, D.B., Laurance, W.F., and Franklin, J.F. (2012). Global decline in large old trees. *Science*, *338*(6112): 1305–6. https://doi.org/10.1126/SCIENCE.1231070

Lindenmayer, D.B. and Sato, C. (2018). Hidden collapse is driven by fire and logging in a socioecological forest ecosystem. *Proceedings of the National Academy of Sciences of the United States of America*, *115*(20): 5181–6. https://doi.org/10.1073/PNAS.1721738115/-/DCSUPPLEMENTAL

Maser, C., Claridge, A.W., and Trappe, J.M. (2008). *Trees, Truffles, and Beasts*. https://doi.org/10.36019/9780813544656/HTML

Middleton, E.L., and Bever, J.D. (2012). Inoculation with a native soil community advances succession in a grassland restoration. *Restoration Ecology*, 20(2): 218–26. https://doi.org/10.1111/J.1526-100X.2010.00752.X

Molina, R., Horton, T.R., Trappe, J.M., and Marcot, B.G. (2011). Addressing uncertainty: How to conserve and manage rare or little-known fungi. *Fungal Ecology*, 4(2), 134–46. https://doi.org/10.1016/J.FUNECO.2010.06.003

Nara, K. (2015). *The Role of Ectomycorrhizal Networks in Seedling Establishment and Primary Succession*. 177–201. https://doi.org/10.1007/978-94-017-7395-9_6

Naranjo-Ortiz, M.A. and Gabaldón, T. (2019). Fungal evolution: diversity, taxonomy and phylogeny of the Fungi. *Biological Reviews*, 94(6): 2101–37. https://doi.org/10.1111/BRV.12550

Nordén, J., Abrego, N., Boddy, L., Bässler, C., Dahlberg, A., Halme, P., Hällfors, M., Maurice, S., Menkis, A., Miettinen, O., Mäkipää, R., Ovaskainen, O., Penttilä, R., Saine, S., Snäll, T., and Junninen, K. (2020). Ten principles for conservation translocations of threatened wood-inhabiting fungi. *Fungal Ecology*, 44, 100919. https://doi.org/10.1016/J.FUNECO.2020.100919

Pettorelli, N., Barlow, J., Stephens, P. A., Durant, S.M., Connor, B., Schulte to Bühne, H., Sandom, C.J., Wentworth, J., and du Toit, J.T. (2018). Making rewilding fit for policy. *Journal of Applied Ecology*, 55(3): 1114–25. https://doi.org/10.1111/1365-2664.13082

Poessel, S.A., Hagar, J.C., Haggerty, P.K., and Katzner, T.E. (2020). Removal of cattle grazing correlates with increases in vegetation productivity and in abundance of imperiled breeding birds. *Biological Conservation*, 241: 108378. https://doi.org/10.1016/J.BIOCON.2019.108378

Policelli, N., Horton, T.R., Hudon, A.T., Patterson, T.R., and Bhatnagar, J.M. (2020). Back to roots: the role of ectomycorrhizal fungi in boreal and temperate forest restoration. *Frontiers in Forests and Global Change*, 3: 97. https://doi.org/10.3389/FFGC.2020.00097/BIBTEX

Porcel, R., Aroca, R., and Ruiz-Lozano, J.M. (2012). Salinity stress alleviation using arbuscular mycorrhizal fungi. A review. *Agronomy for Sustainable Development*, 32(1): 181–200. https://doi.org/10.1007/S13593-011-0029-X/TABLES/4

Ranius, T., Niklasson, M., and Berg, N. (2009). Development of tree hollows in pedunculate oak (Quercus robur). *Forest Ecology and Management*, 257(1): 303–10. https://doi.org/10.1016/J.FORECO.2008.09.007

Read, D.J. (1991). Mycorrhizas in ecosystems. *Experientia 1991*, 47(4): 376–91. https://doi.org/10.1007/BF01972080

Read, D.J. and Perez-Moreno, J. (2003). Mycorrhizas and nutrient cycling in ecosystems—a journey towards relevance? *New Phytologist*, 157(3): 475–92. https://doi.org/10.1046/J.1469-8137.2003.00704.X

Scottish Government. (2017). *Beavers in Scotland: Consultation on the Strategic Environmental Assessment*.

Smith, S.E. and Read, D.J. (2008). *Mycorrhizal Symbiosis* (3rd ed.). Academic Press.

Steidinger, B.S., Bhatnagar, J.M., Vilgalys, R., Taylor, J.W., Qin, C., Zhu, K., Bruns, T.D., and Peay, K.G. (2020). Ectomycorrhizal fungal diversity predicted to substantially decline due to climate changes in North American Pinaceae forests. *Journal of Biogeography*, 47(3): 772–82. https://doi.org/10.1111/JBI.13802

Stevens, B.M., Propster, J., Wilson, G.W.T., Abraham, A., Ridenour, C., Doughty, C., and Johnson, N.C. (2018). Mycorrhizal symbioses influence the trophic structure of the Serengeti. *Journal of Ecology*, 106(2): 536–46. https://doi.org/10.1111/1365-2745.12916

Stokland, J.N., Siitonen, J., and Jonsson, B.G. (2012). Other associations with dead woody material. In *Biodiversity in Dead Wood* (pp. 58–81). Cambridge University Press.

Svenning, J.C. (2002). A review of natural vegetation openness in north-western Europe. *Biological Conservation*, 104(2): 133–48. https://doi.org/10.1016/S0006-3207(01)00162-8

Svenning, J.C., Pedersen, P.B.M., Donlan, C.J., Ejrnæs, R., Faurby, S., Galetti, M., Hansen, D.M., Sandel, B., Sandom, C.J., Terborgh, J.W., and Vera, F.W.M. (2016). Science for a wilder Anthropocene: Synthesis and future directions for trophic rewilding research. *Proceedings of the National Academy of Sciences of the United States of America*, 113(4): 898–906. https://doi.org/10.1073/PNAS.1502556112/-/DCSUPPLEMENTAL

Taylor, L.L., Leake, J.R., Quirk, J., Hardy, K., Banwart, S.A., and Beerling, D.J. (2009). Biological weathering and the long-term carbon cycle: integrating mycorrhizal evolution and function into the current paradigm. *Geobiology*, 7(2): 171–91. https://doi.org/10.1111/J.1472-4669.2009.00194.X

Treseder, K.K., Allen, E.B., Egerton-Warburton, L.M., Hart, M.M., Klironomos, J.N., Maherali, H., and Tedersoo, L. (2018). Arbuscular mycorrhizal fungi as mediators of ecosystem responses to nitrogen deposition: A trait-based predictive framework. *Journal of Ecology*, 106(2): 480–9. https://doi.org/10.1111/1365-2745.12919

Tsiafouli, M.A., Thébault, E., Sgardelis, S.P., de Ruiter, P.C., van der Putten, W.H., Birkhofer, K., Hemerik, L., de Vries, F.T., Bardgett, R.D., Brady, M.V., Bjornlund, L., Jørgensen, H.B., Christensen, S., Hertefeldt, T.D., Hotes, S., Gera Hol, W.H., Frouz, J., Liiri, M., Mortimer, S. R., ...Hedlund, K. (2015). Intensive agriculture reduces soil biodiversity across Europe. *Global Change Biology*, 21(2): 973–85. https://doi.org/10.1111/GCB.12752

Vahter, T., Bueno, C.G., Davison, J., Herodes, K., Hiiesalu, I., Kasari-Toussaint, L., Oja, J., Olsson, P.A., Sepp, S.K., Zobel, M., Vasar, M., and Öpik, M. (2020). Co-introduction of native mycorrhizal fungi and plant seeds accelerates restoration of post-mining landscapes. *Journal of Applied Ecology*, 57(9): 1741–51. https://doi.org/10.1111/1365-2664.13663

van der Heijden, M.G.A., Klironomos, J.N., Ursic, M., Moutoglis, P., Streitwolf-Engel, R., Boller, T., Wiemken, A., and Sanders, I.R. (1998). Mycorrhizal fungal diversity determines plant biodiversity, ecosystem variability and productivity. *Nature 1998*, 396(6706): 69–72. https://doi.org/10.1038/23932

van der Hoek, Y., Gaona, G. v., and Martin, K. (2017). The diversity, distribution and conservation status of the tree-cavity-nesting birds of the world. *Diversity and Distributions*, 23(10): 1120–31. https://doi.org/10.1111/DDI.12601

van der Wal, A., van Veen, J.A., Smant, W., Boschker, H.T.S., Bloem, J., Kardol, P., van der Putten, W.H., and de Boer, W. (2006). Fungal biomass development in a chronosequence of land abandonment. *Soil Biology and Biochemistry*, 38(1): 51–60. https://doi.org/10.1016/J.SOILBIO.2005.04.017

Wainhouse, M. and Boddy, L. (2022). Making hollow trees: Inoculating living trees with wood-decay fungi for the conservation of threatened taxa—A guide for conservationists. *Global Ecology and Conservation*, 33: e01967.https://doi.org/10.1016/J.GECCO.2021.E01967

Whitehouse, N.J. (2006). The Holocene British and Irish ancient forest fossil beetle fauna: implications for forest history, biodiversity and faunal colonisation. *Quaternary Science Reviews*, 25(15–16): 1755–89. https://doi.org/10.1016/J.QUASCIREV.2006.01.010

Whitehouse, N.J. and Smith, D. (2010). How fragmented was the British Holocene wildwood? Perspectives on the 'Vera' grazing debate from the fossil beetle record. *Quaternary Science Reviews*, 29(3–4): 539–53. https://doi.org/10.1016/J.QUASCIREV.2009.10.010

Wood, J.R., Dickie, I.A., Moeller, H. v., Peltzer, D.A., Bonner, K.I., Rattray, G., and Wilmshurst, J.M. (2015). Novel interactions between non-native mammals and fungi facilitate establishment of invasive pines. *Journal of Ecology*, 103(1): 121–9. https://doi.org/10.1111/1365-2745.12345

Wu, Q.-S., Zou, Y.-N., Wu, Q., and Zou, Y. (2017). Arbuscular mycorrhizal fungi and tolerance of drought stress in plants. *Arbuscular Mycorrhizas and Stress Tolerance of Plants*, 25–41. https://doi.org/10.1007/978-981-10-4115-0_2

25

REWILDING AND HUMAN HEALTH

Heather VanVolkenburg, Rene Beyers, Cara Nelson, Liette Vasseur, Angela Andrade, Ian Convery, and Steve Carver

Introduction

The World Health Organisation (WHO) defined human health in its constitution in 1948 as 'A state of complete physical, mental and social well-being and not merely the absence of disease or infirmity.'[1] Health outcomes are determined by the social, economic and physical environment and a person's individual characteristics. Nature has often been neglected as a determinant of health but has been coming to the foreground with a growing awareness of the detrimental effects of climate change and the loss of biodiversity on human health, well-being, and prosperity. It is clear that humans are intricately connected to nature through the provision of ecosystem services such as clean air and water, healthy soils, fuel, food, and other renewable natural resources, climate, and disease regulation. Although these services may feel remote or even abstract to many people living in urban environments, which is more than half of the global population (and increasing), they are essential for the survival of the planet. Biodiversity and nature have also been shown to have direct positive impacts on mental health (see below).

The SARS-CoV-2 (COVID-19) pandemic that began in 2019 has been an urgent reminder of the complex interactions between humans and the environment. COVID-19 is a zoonotic disease—meaning a disease that is transmitted from animals to people. These types of diseases are a leading cause of death and illness globally. Gebreyes et al. (2014) estimate that zoonoses are responsible for 2.5 billion cases of human illness and 2.7 million human deaths worldwide each year. Zoonoses are not new (see below) and account for 60% of all infectious diseases in humans (United Nations Environment Programme and International Livestock Research Institute, 2020). The majority of zoonotic transmissions to humans come from domestic animals but spillovers of pathogens from wild reservoirs to humans occur sporadically as may have been the case with Covid-19 (see below). Most new zoonotic Emerging Infectious Diseases (EID) originate in wildlife and they are expected to increase in the future (Titcomb et al., 2019). The COVID-19 pandemic has brought to attention the role of ecosystems in disease regulation and the risk of EID's associated with ecosystem degradation and increased contact between humans and wildlife (Gebreyes et al., 2014).

Domestication of wild animals, habitat encroachment, land conversion, loss of biodiversity, urbanisation, and other forms of degradation may alter human-animal interactions, thereby influencing patterns of EID's (Mackenzie & Jeggo, 2019). Climate change adds another level

DOI: 10.4324/9781003097822-28

of complexity in the emergence of zoonotic diseases, through changes in species distribution and species interactions. This may lead to the transfer of zoonoses to new host combinations in new areas, thus forming novel host communities (Carlson et al., 2022). Legal and illegal wildlife trade and consumption also play a role at increasing risk of zoonotic disease emergence. Salyer et al. (2017) highlight the critical importance of a 'One Health' approach to understanding the impacts of emerging and endemic zoonoses in both humans and animals. Global investments in research and development are improving our understanding about transmission of and prevention and treatment modalities for zoonotic diseases, especially COVID-19 (WHO, 2022). A similar, more recent call to action with respect to understanding or managing the linkages between zoonotic disease emergence and ecosystem degradation and management now exists. This understanding could prove critical for reducing the risk of future pandemics, if it leads to identifying priorities areas and actions for surveillance, conservation, ecological restoration, and rewilding.

This leads us to the question: Can we effectively improve the management of our ecosystems and of human-wildlife interactions to reduce the risk of a spillover event? Indeed, can we afford to not to? In this chapter, we first summarise the current knowledge regarding the relationship between environmental degradation and zoonoses and then offer perspectives on what research and action need to focus on to reduce the possibility of other EID pandemics. Furthermore, we consider the role of rewilding in relation to zoonotic disease/EID mitigation.

Ecosystem degradation and zoonoses

Zoonotic diseases have infected and evolved with humans since the earliest beginnings of the human species (Reperant et al., 2013). Close contacts with newly domesticated species and exposure to pathogens for which humans had no effective immune response caused epidemics of new diseases.

As Jared Diamond notes in his influential book *Guns, Germs and Steel,* (2005, pp. 87–8): 'Of equal importance in wars of conquest were the germs that evolved in human societies with domestic animals. Infectious diseases like smallpox, measles, and flu arose as specialized germs of humans, derived by mutations of very similar ancestral germs that had infected animals. The humans who domesticated animals were the first to fall victim to the newly evolved germs, but those humans then evolved substantial resistance to the new diseases. When such partly immune people came into contact with others who had no previous exposure to the germs, epidemics resulted in which up to 99 percent of the previously unexposed population was killed. Germs thus acquired ultimately from domestic animals played decisive roles in the Europeans conquests of Native Americans, Australians, South Africans, and Pacific Islanders.' These new germs may have killed most humans previously unexposed to them and in some cases, this has led to the extinction of local communities. Therefore, widespread pandemics are not a new phenomenon!

Major increases in food production, including livestock, enabled the human population to grow and disperse rapidly and expanded the human–animal interface, providing fertile ground for new infectious diseases to spread and establish themselves in humans. With an increasing population density and greater movement of people, goods, and animals, zoonotic diseases quickly spread among countries and continents.

The successful emergence of a zoonotic disease is a process that is still poorly understood. It is also a rare event as several conditions have to occur simultaneously in a limited space and time frame. An emerging pathogen (virus, bacteria, or parasite) has to overcome multiple barriers to establish itself in humans (Plowright et al., 2017). These include factors related to the reservoir host such as distribution and abundance of the host species, prevalence and intensity of infection in the host, the host's behaviour and ecology and its physiological and immunological state. Pathogens can be released from the host through excretion (shedding), direct contact through handling and consumption of an infected animal or through the bite of an intermediary vector (an arthropod such as a mosquito or flea). When a human is exposed, the pathogen has to overcome a new set of physical and immunological barriers to survive and replicate. The spreading and establishment in human populations depend on the characteristics of the pathogen, such as its virulence, mode of transmission, transmissibility, and incubation time, the morbidity, mortality, movement, and behaviour of the new human host, and the density of its population.

Land use changes and other ecosystem disturbances can upset the balance in biological communities and remove some of these initial barriers. They affect the dynamics of host and/or vector species populations, food web interactions, biodiversity, and relations between pathogens, hosts, and humans. Increased access through new roads and other infrastructure and influx of people expand the wildlife-human interface. Deforestation in the tropics is particularly problematic as new natural areas are being opened up that are very rich in species, including microbes that humans have never been exposed to. Large-scale land use changes have been linked to zoonotic diseases such as Malaria, Lyme disease, Chagas disease, Yellow Fever, and Leishmaniasis (Karesh, 2012).

The population ecology and community dynamics of the host and/or vector species in emerging diseases are typically understudied (Preston et al., 2013). However, one well-researched example is the spillover of Hendra virus from fruit bats to humans through horses as an intermediary host in Australia (Plowright et al., 2015). Fruit bats are the natural reservoir of Hendra virus. The destruction of their habitat has been linked to periodic food shortages in winter and increased dispersal of bats to semi-urban areas, such as horse farms, in search of food. Nutritional stress of bats feeding on lower quality foods in those areas, is thought to have led to increased virus shedding causing spillover to horses in close proximity to bats. Humans have subsequently been infected through close contact with infected horses. This cascade of events is directly linked to habitat destruction and the resulting food web disruption.

Understanding food webs is essential for rewilding and thus important in a rewilding approach to risk management of zoonoses. However, there is still much to be learned about the role of food webs in regulating disease. Well-connected food webs can control the abundance of reservoir hosts. Disturbance of a food web, for example, due to habitat fragmentation or over-hunting, may lead to an increased risk of spillover disease. For example, the loss of a predator (essentially a decrease in biodiversity) may cause an explosion of the population of a host species that it preys on, for example a disease-carrying rodent.

A loss of biodiversity has been shown to enhance the risk for vector-born zoonotic disease emergence (Bernstein & Chivian, 2008; Dobson, 2004). Often many species are exposed to a pathogen but only a few, or sometimes one, can support its survival and replication. We call these competent hosts as opposed to incompetent species that may be exposed to the pathogen but are otherwise not affected (Martin et al., 2016). Vectors that transmit zoonoses are often generalists and bite both competent and incompetent species. The pathogen is 'diluted' by a range of species and vectors that are less likely to become infected when species diversity is high. With a loss of species diversity, competent, often generalist, species, become more abundant and the level of pathogen transmission in the community goes up thereby increasing the risk for infection in humans. This dilution effect does not always hold and in some cases a greater diversity of host species may actually amplify an epidemic outbreak (Dobson, 2004).

The transmission of Lyme disease in the Eastern United States has been shown to respond to both this dilution effect and the predator-prey dynamics that influence its host. Lyme disease is caused by a bacterium (*Borrelia burgdorferi* and others) and transmitted by a tick (*Ixodes* spp.) to its main host, the white-footed mouse (*Peromyscus leucopus*). Forest fragmentation and degradation has led to a concentration of mice in high densities in small remaining fragments, probably because of a lack of both predators and competitors that could not survive in these forest islands. Because of the loss of other species, including those incompetent for Lyme that would have contributed to a dilution effect, white-footed mice are now the main or only prey for ticks and this results in high infection levels among the mouse and tick population. Combined with the close proximity of humans to these degraded habitats, the risk for contracting Lyme is much higher than in larger undisturbed forests (Bernstein & Chivian, 2008). Highlighting the importance of food webs, there are also suggestions that the outbreaks of Lyme disease in the eastern USA during the late 20th century might have been a delayed consequence of the extinction of the passenger pigeon a century earlier (Blockstein, 1998).

In contrast to their wild counterparts, domestic animals have also been an important source of EID's (Salyer et al., 2017). The increase in high density livestock factory farming in the last decades constitutes a considerable risk for zoonotic disease emergence. Livestock that are in close proximity to wild pathogenic hosts, could be infected and amplify the disease and allow it to evolve and become more infectious to humans. Thus, these farms may act as a catalyzer for a spillover event. For example, in the case of Nipah virus in Malaysia, loss of habitat through deforestation and replacement with palm oil plantations likely forced bats that host the virus to settle near alternative (tree) food sources near pig farms. Pigs were infected through close contact with the bats and their high density provided a living petri-dish for the virus which replicated and spread rapidly through the pig population. Eventually farm workers in close contact with the pigs were infected leading to outbreaks in humans (Preston et al., 2013).

The consumption of bushmeat and other animal products is also a major risk for new zoonoses. Bushmeat consumption has been linked to the emergence of HIV/AIDS, Ebola, and SARS (Karesh, 2012). Increased access to previously remote tropical forests, for example in Central Africa have dramatically increased hunting and trapping of wildlife (see Chapter 26). Because of this and the lack of knowledge and surveillance of the high diversity of wild viruses and other microbes in these places, there is a risk for more EID's. International trade and transport of live animals and animal products make the problem worse and can turn a localised EID into a global pandemic.

Putting wild back into health

'There is an urgent need that physicians and health workers become more intimately familiar with the problems inherent in the zoonoses.'

The above quote could very easily have been written during the COVID-19 pandemic. It is, however, taken from a 1954 World Health Organisation (WHO) report (Meyer, 1954). It demonstrates that whilst human medicine has made significant progress over the last 70 years or so, when it comes to zoonotic diseases, we still have much to learn. The WHO report notes that 'animals expose rural people to certain infections that drain their vitality and make their labours less profitable'. There was relatively little interest in understanding the role of natural ecosystems within zoonotic disease management. The main focus was on how to improve rural human health, welfare and economy. We may argue that very little has changed since 1954. If the COVID-19 pandemic has shown us anything, then it is that we know less about the natural world than we thought.

While no definitive answer has been given as to where COVID-19 came from, evidence suggests that it emerged from the wildlife trade with the 'wet market' of Huanan (where fresh meat and live animals were being sold) as the epicentre of the pandemic (Worobey et al., 2021; Maxmen, 2022). Sources of this latest pandemic remain unclear but possible wildlife hosts include horseshoe bats as primary hosts and raccoon dogs or other susceptible species as intermediary hosts. Raccoon dogs have been implicated in the emergence of SARS, are susceptible to infection with COVID-19 and were being sold at the Huanan market at the time of the outbreak (Opriessnig & Huang, 2021, Worobey et al., 2021). With increasing interactions between humans and wildlife either through legal or illegal trade, and greater encroachment of natural habitat due to agricultural expansion and livestock farming and an ever-increasing population, the risks will continue to grow.

We must, therefore, expand on what we know about the connections between health and the wild. They are, as we have demonstrated, intimately connected. Halting ecological degradation including deforestation, managing the human–wildlife interface and reducing the wildlife trade, increasing surveillance and early disease detection, and improving preparedness to respond to an outbreak will be necessary to reduce the risk of future zoonotic disease emergence (Dobson et al., 2020).

However, the challenge remains that most citizens are not fully aware, nor do they understand the cause and effect of these pressures that humans put on natural ecosystems. Dolman (1954) writes that the huge slaughter-houses and meat packing plants of present-day North America are a far cry from the buffalo hunts of the Great Plains Indians, and notes that 'despite the growing strength of animal welfare societies and wildlife conservation agencies, the average citizen's interest in these questions seems very limited'. Once again, has much changed since 1954? The COVID-19 pandemic has exposed much ignorance about the ecological connection of pandemics. While there was more reporting on this than probably ever before, the political discourse often outweighed the scientific discussion about the most likely cause of the emergence of this disease. It is important that the conservation/rewilding community increases efforts to communicate the connections between health and the remaining wild spaces.

Rewilding and zoonoses

Rewilding has generated widespread interest and debate within the conservation sector and beyond (as discussed extensively elsewhere in this volume). Here we refer to the Carver et al. (2021) definition of the process of rebuilding natural ecosystems...to create a complete food web where all trophic levels function in a sustainable and resilient ecosystem without human interference. Rewilding supports biota that would have been present had the disturbance not occurred.

The idea that we can encourage the reversal of biodiversity decline and create natural habitats and wild landscapes simply by allowing wildlife and natural processes to reclaim areas of land no longer under human management has captured both scientific and public imagination. Rewilding, as part of the 'ecological restoration[2] family' is an important strategy to conserve biodiversity and reduce land degradation. The UN Decade on Ecosystem Restoration runs from 2021 to 2030, and ambitious global targets have been set to restore degraded ecosystems—e.g. 350 million ha of forest restoration by 2030 (Dave et al., 2017). Governments, NGOs, communities, and individuals are actively considering rewilding options, and many rewilding projects have been initiated around the world (see Chapters 14–20). Rewilding calls for a paradigm shift in our relationship with nature, the need for tolerance and coexistence, and a recognition that there may also be associated human health and wellbeing benefits. Might rewilding also reduce the risk for the emergence of new diseases and pandemics?

We know that ecosystem degradation can lead to the emergence of new zoonotic diseases but can reverting degradation prevent it? There is no simple answer to this question. One reason for this is that a recovering ecosystem may follow a different trajectory than its degradation trajectory and may reach different endpoints (see Chapter 12; Speldewinde et al., 2015). Little research has been completed on the mitigation of zoonotic risk through rewilding or restoration. Only well planned and long-term experiments will provide definitive answers. However, it is probably safe to assume that halting or reversing degradation has the potential to reduce zoonotic spillovers based on current knowledge and understanding. The aforementioned transmission of Hendra and Nipah virus is a good working example. By addressing the root causes of this dynamic and restoring original habitat and food sources of the bats involved, it is reasonable to expect that both bat dispersal to human landscapes and nutritional stresses that increased virus shedding can be reduced, thereby decreasing the risk of viral transmission to humans.

Likewise, restoring trophic cascades may reduce disease emergence by restoring the balance of disrupted food webs. The abundance of both competent and non-competent host species is controlled by bottom-up and/or top-down regulations (see other chapters). So, knowing how these species are regulated and how that regulation can be restored may, at least in principle, reduce the risk for pathogens to escape (Preston et al., 2013). For example, the re-introduction of both mammalian and avian predators of rodents, which are reservoirs for many diseases (Hanta virus, Plague, Lassa fever, Leptospirosis etc.) can control or even reduce their abundance and prevent disease emergence (see Lyme example). Predators have been used in the biological control of zoonotic disease, such as malaria and other mosquito-borne pathogens (Bernstein and Chivian, 2008).

One of the principles of rewilding is to shift from an anthropocentric to a more ecocentric lens enabling a paradigm shift in our thinking about our relationship with nature (Carver et al., 2021, Chapter 1 and 5). This puts us on a path to a better understanding of how ecosystems work and how we can live in co-existence with wildlife and wild nature. However, co-existence with wild animals does bring its challenges, including exposure to potentially harmful encounters and disease. The challenge is to find a balance between this coexistence and protecting ourselves by taking individual and public measures to minimise potentially harmful exposures. We need strategies that minimise the risk of contact and improve ecosystem and human health at the same time. As Destoumieux-Garzón et al. (2018) note, there is increased awareness of 'the importance of the human–animal–ecosystem interface in the evolution and emergence of pathogens… a better knowledge of causes and consequences of certain human activities, lifestyles, and behaviours in ecosystems is crucial for a rigorous interpretation of disease dynamics and to drive public policies'. Yet it is clear, given the lessons being learned from the current pandemic, that more needs to be done. It is also clear that coexistence comes at a cost.

We cannot continue with the 'business as usual' model, it is simply not possible to have the coexistence cake and also eat it. Many remnants of wild nature are rapidly being degraded by human activity, stemming at least in part from increasing human populations and interventions near or in these remnants (across a range of contexts, from the involvement of impoverished communities in illegal logging in the Global South to the overuse of national parks by rich tourists in the Global North). If we are going to rewild humanity alongside ecosystems, then we need to recognise that human health is closely connected to the health of other animals (both livestock and wildlife) and our shared environment, and we must consider human carrying capacity, equitable resource use[3] and population growth as part of a 'One Health' approach to planetary survival.

As Mackenzie and Jeggo (2019) note, the concept of One Health is not new and can be traced back for at least two hundred years. Whilst it is defined in a variety of different ways, perhaps the most commonly used definition (One Health Commission[4]) is: 'One Health is defined as a collaborative, multisectoral, and transdisciplinary approach—working at the local, regional, national, and global levels—with the goal of achieving optimal health outcomes recognising the interconnection between people, animals, plants, and their shared environment.' We see strong parallels with the rewilding movement and 'win–wins' in that rewilding can also be beneficial to human well-being and health (physical, mental, and spiritual). Alongside the disease mitigation measures already discussed, time spent in informal green/blue space has been shown to provide significant mental health benefits (Aerts et al., 2018) with 'urban rewilding' (restoring urban greenspace) also identified as a possible means by which to improve resilience and sustainability in urban landscape settings (Lehmann, 2021). There is also evidence that 'dewilding' has contributed to an increase in human immune disorders. Mills et al. (2017) propose a 'Microbiome Rewilding Hypothesis', whereby restoring biodiverse habitats in urban green spaces can also 'rewild' the environmental microbiome and improve outcomes of human disease.

Alongside the One Health concept, Nature-based Solutions (NbS) are also increasingly thought of as potential mechanisms by which rewilding can provide pathways to benefits for human health and wellbeing by using nature to create sustainable socio-ecological systems that address multiple societal challenges (Dick et al., 2019). NbS are not only limited to urban green spaces. Indeed, NbS are increasingly being implemented in agricultural systems as a means to increase biodiversity and system resilience to environmental stressors (climate change, soil degradation, etc.).

Nonetheless, there are two facets of the rewilding story linked to human health and wellbeing that are seemingly at odds. On the one hand, it sees rewilding as an approach to provide nature with spaces separate to and from humans, while on the other, it actively encourages close human participation in a wilder experience. The answer to this apparent paradox lies in its reference to the geographical space/place setting and what rewilding is for. Rewilding for nature at large scales creates a buffer between human spaces and wild spaces and gives wild animals room to thrive with minimal contact with humans and livestock, thus minimising the risks of disease transmission in both directions. Rewilding for people at smaller scales in urban and periurban environments provides people with indirect health benefits from relaxing spaces to escape the stresses of urban lifestyles where nature is experienced in a more controlled and ordered setting. These two scales of rewilding sit along the wilderness continuum, which depends on scale, local, and cultural contexts. They all provide opportunities for improving human health and wellbeing while simultaneously providing space for wild nature. When viewed as part of a holistic, 'joined-up' approach to land-use and land-use planning, the paradox disappears.

How to solve these 'wicked problems'?

There is good evidence to suggest that pandemics (and disasters generally) disproportionately affect the poorest and most disadvantaged and vulnerable members of society (Perry et al., 2021; Convery et al., 2008). When individual or community capabilities to withstand vulnerabilities, disturbances, and shocks are overwhelmed, as is the case with emerging epidemics, it becomes much more difficult to maintain a sustainable livelihood system. The processes that drive disease emergence and transmission are therefore closely related to social justice issues, as well as ecological integrity and socio-economic activities such as agriculture, and trade of animals and animal products and overpopulation.

Understanding these complex interactions presents a problem that moves beyond approaches such as simple counting, requiring knowledge about structure, connection, and dynamic behaviour. These systems are usually associated with high levels of uncertainty, imprecision, partial knowledge, and hard-to-define concepts, which are not uniformly distributed across both time and space at different scales, and difficult to observe directly. It requires adaptive, holistic, and interdisciplinary approaches such as the One Health model or other models integrating human and animal health with socio-ecological issues.

Need for synthetic research

Although multiple aspects of ecosystems and their management driving disease emergence and transmission have been examined, no systematic review of what is and is not known about the relationships between these various aspects exists. Links between some ecosystem components, such as biodiversity, have been explored. Other aspects have received much less attention. Little information on the extent to which ecosystem management, including restoration and rewilding may reduce risk of zoonotic disease emergence can be found. Investments in scientific investigations on the ecological drivers of zoonotic disease emergence are urgently needed.

The Commission on Ecosystem Management of the International Union for Conservation of Nature has embarked on a project that addresses this need to better understand such drivers. The Ignite group has begun to identify and elucidate the relationships that exist in our current knowledge pool between ecosystem degradation, ecological integrity, biodiversity, climate change, and environmental stressors on zoonotic disease emergence and transmission. The ultimate goal of the project is to help inform a strategic research agenda through the identification of current gaps in knowledge. Furthermore, the project will better inform surveillance and long-term EID monitoring programmes and assist in field experimental design that allows for much needed testing of the underlying mechanisms of zoonotic disease emergence and transmission to human populations. A critical component of the project will initiate dialogue on how rewilding can play an important role in preventing disease emergence and transmission.

Conclusions

We come once again to our original questions…can we effectively improve the management of our ecosystems and of human–wildlife interactions to reduce the risk of a spillover event, and does rewilding have a role to play?

Over the last 20–30 years, there has been increasing interest in this question, and the current pandemic has highlighted once again the complex relationships that connect human health and well-being to ecosystem degradation, biodiversity loss, and climate change. Indeed, the COVID-19 pandemic has increased the urgency of understanding these relationships, as well as understanding the extent to which ecosystem management, including ecological restoration and rewilding, could improve human health.

As human populations continue to grow and expand into new geographic areas, our current path of ecosystem degradation caused by land use changes, pollution, climate change, poor land management, and associated stressors will undoubtedly lead to more habitat and biodiversity loss which in turn increases the likelihood of zoonotic disease emergence. Is there a better time than during the UN Decade on Ecosystem Restoration to go wild and embrace the role that self-willed ecosystems play in disease regulation? Carefully planned, scale dependent rewilding strategies can help to bolster biodiversity and pave the way for nature to self-regulate. Understanding better what we already know will help us plan for a future where rewilding becomes a critical component of how we re-envision our global landscapes and our relationships with nature. We also see important connections and cross-fertilisation with growing concepts such as One Health and NbS. The wicked problems we currently face require collaborative, adaptive, multi-scalar problem solving. Rewilding can offer this, and whilst there are many challenges ahead, these are also exciting times for those of use working at the rewilding/health interface.

Notes

1 See (www.who.int/about/governance/constitution).
2 'The process of assisting the recovery of an ecosystem that has been damaged, degraded or destroyed' as defined by the Society for Ecological Restoration International Science & Policy Working Group 2004; available at: https://cdn.ymaws.com/www.ser.org/resource/resmgr/custompages/publications/ser_publications/ser_primer.pdf
3 As the UN Global Resources Outlook indicates, there are significant inequalities in the material footprint of countries; high-income countries have levels of per capita material footprint consumption 60% higher than upper-middle income countries, and more than 13 times the level of low-income countries (IRP, 2019).
4 See www.cdc.gov/onehealth/index.html and https://ohi.vetmed.ucdavis.edu/about/one-health for overviews of the One Health concept.

References

Aerts, R., Honnay, O., and Van Nieuwenhuyse, A. 2018. Biodiversity and human health: mechanisms and evidence of the positive health effects of diversity in nature and green spaces. *British Medical Bulletin*, 127(1): 5–22. https://doi.org/10.1093/bmb/ldy021

Bernstein, Aaron and Chivian, Eric. 2008. *Sustaining Life: How Human Health Depends on Biodiversity*. Oxford University Press.

Blockstein, D.E. 1998. Lyme disease and the passenger pigeon? *Science*, 279(5358): 1831–3. https://doi.org/10.1126/science.279.5358.1831c

Carlson, C.J., Albery, G.F., Merow, C., Trisos, C.H., Zipfel, C.M., Eskew, E.A., Olival, K.J., Ross, N., and Bansal, S. 2022. Climate change increases cross-species viral transmission risk. *Nature (London)* 2022: n.p. https://doi.org/10.1038/s41586-022-04788-w

Carver, S., Convery, I., Hawkins, S., Beyers, R., Eagel, A., Kun, Z., Van Maanen, E., Cao, Y., Fisher, M., Edwards, S., Nelson, C., Gann, G., Shurter, S., Aguilar, K., Adrade, A., Ripple, B., Davis, J., Sinclair, T., Bekoff, M., Noss, R., Foreman, D., Pettersson, H., Root-Bernstein, M., Svenning, J., Taylor, P., Wynne-Jones, S., Featherstone, Watson,A., Fløjgaard, C., Stanley-Price, M., Navarro, L., Aykroyd, T., Parfitt, A., and Soulé, M. 2021. Guiding principles for rewilding. *Conservation Biology*, 35(6): 1882–93. https://doi.org/10.1111/cobi.13730

Convery, I., Mort, M., Baxter, J., Bailey, and C. 2008. Exploring the lifescape. In: I. Convery, M. Mort, J. Baxter, and C. Bailey (eds), *Animal Disease and Human Trauma* (pp. 132–50). Palgrave Macmillan.

Dave, R., Saint-Laurent, C., Moraes, M., Simonit, S., Raes, L., and Karangwa, C. (2017). Bonn Challenge Barometer of Progress: Spotlight Report 2017. IUCN

Diamond, Jared M. 2005. *Guns, Germs, and Steel: The Fates of Human Societies*. Norton, pp. 87–8.

Destoumieux-Garzón, D., Mavingui, P., Boetsch, G., Boissier, J., Darriet, F., Duboz, P., Fritsch, C., Giraudoux, P., Le Roux, F., Morand, S., and Paillard, C. 2018. *The One Health Concept: 10 Years Old and a Long Road Ahead. Vol. 5*. Frontiers Media. https://doi.org/10.3389/fvets.2018.00014

Dick, J., Miller, J.D., Carruthers-Jones, J., Dobel, A.J., Carver, S., Garbutt, A., Hester, A., Hails, R., Magreehan, V., and Quinn, M. 2019. How are nature based solutions contributing to priority societal challenges surrounding human well-being in the United Kingdom: a systematic map protocol. *Environmental Evidence*, 8(1): 1–11. https://doi.org/10.1186/s13750-020-00208-6

Dobson, A.P., Pimm, S.L., Hannah, L., Kaufman, L., Ahumada, J.A., Ando, A.W., Bernstein, A., Busch, J., Daszak, P., Engelmann, J., Kinnaird, M.F., Li, B.V., Loch-Temzelides, T., Lovejoy, T., Nowak, K., Roehrdanz, P.R., and Vale, M.M. 2020. Ecology and economics for pandemic prevention. *Science*, 369(6502): 379–81. https://doi.org/10.1126/science.abc3189

Dobson, A. 2004. Population dynamics of pathogens with multiple host species. *American Naturalist*, 164(5): S64–S78. https://doi.org/101086/424681

Dolman, C.E. 1954. Some ways in which animal health affects human health. *Canadian Journal of Comparative Medicine and Veterinary Science*. 18(2): 35–50.

Gebreyes, W.A., Dupouy-Camet, J., Newport, M.J., Oliveira, C.J., Schlesinger, L.S., Saif, Y.M., et al. 2014. The global One Health paradigm: challenges and opportunities for tackling infectious diseases at the human, animal, and environment interface in low-resource settings. *PLoS Neglected Tropical Diseases,* 8: e3257. https://doi.org/10.1371/journal.pntd.0003257

IRP. 2019. Global Resources Outlook 2019: Natural Resources for the Future We Want. Oberle, B., Bringezu, S., Hatfield-Dodds, S., Hellweg, S., Schandl, H., Clement, J., Cabernard, L., Che, N., Chen, D., Droz-Georget, H., Ekins, P., Fischer-Kowalski, M., Flörke, M., Frank, S., Froemelt, A., Geschke, A., Haupt, M., Havlik, P., Hüfner, R., Lenzen, M., Lieber, M., Liu, B., Lu, Y., Lutter, S., Mehr, J., Miatto, A., Newth, D., Oberschelp, C., Obersteiner, M., Pfister, S., Piccoli, E., Schaldach, R., Schüngel, J., Sonderegger, T., Sudheshwar, A., Tanikawa, H., van der Voet, E., Walker, C., West, J., Wang, Z., and Zhu,B. *A Report of the International Resource Panel. United Nations Environment Programme.* Nairobi, Kenya.

Karesh, W.B. 2012. Ecology of zoonoses: natural and unnatural histories. *The Lancet*, 380(9857): 1936–45. https://doi.org/10.1016/S0140-6736(12)61678-X

Lehmann, S., 2021. Growing biodiverse urban futures: renaturalization and rewilding as strategies to strengthen urban resilience. *Sustainability*, 13(5): 2932. https://doi.org/10.3390/su13052932
https://doi.org/10.1093/icb/icw064

Mackenzie, J.S. and Jeggo, M. 2019. The One Health approach—why is it so important? *Tropical Medicine and Infectious Disease*, 4(2): 88. https://doi.org/10.3390/tropicalmed4020088

Martin, L.B., Burgan, S.C., Adelman, J.S., and Gervasi, S.S. 2016. Host competence: an organismal trait to integrate immunology and epidemiology. *Integrative and Comparative Biology*, 56:6, 1225–37.

Maxmen, Amy. 2022. Wuhan market was epicentre of pandemic's start, studies suggest. *Nature (London)*, 603(7899): 15–16.

Meyer, K.F. 1954. Seventh World Health Assembly. Accessed: 04/05/22, Available from: https://apps.who.int/iris/bitstream/handle/10665/103756/WHA7_TD-3_eng.pdf

Mills, J.G, Weinstein, P., Gellie, N.J.C., Weyrich, L.S., Lowe, A.J., and Breed, M.F. 2017. Urban habitat restoration provides a human health benefit through microbiome rewilding: the Microbiome Rewilding Hypothesis. *Restoration Ecology*, 25, 866–72. https://doi.org/10.1111/rec.12610

Opriessnig, T. and Huang, Y.W. 2021. Third update on possible animal sources for human COVID-19. *Xenotransplantation*, 28(1). https://dx.doi.org/10.1111%2Fxen.12671

Perry, B.L., Aronson, B., and Pescosolido, B.A. 2021. Pandemic precarity: COVID-19 is exposing and exacerbating inequalities in the American heartland. *PNAS*, 118(8). https://doi.org/10.1073/pnas.202 0685118

Plowright, R.K., Parrish, C.R., McCallum, H., Hudson, P.J., Ko, A.I., Graham, A.L., and Lloyd-Smith, J.O. 2017. Pathways to zoonotic spillover. *Nature Reviews Microbiology*, 15: 502–10. https://doi.org/10.1038/nrmicro.2017.45

Plowright, R.K, Eby, P., Hudson, P.J., Smith, I.L., Westcott, D., Bryden, W.L., Middleton, D., Reid, P.A., McFarlane, R.A., Martin, G., Tabor, G.M., Skerratt, L.F., Anderson, D.L., Crameri,G., Quammen, D., Jordan, D., Freeman, P., Wang, L.F., Epstein, J.H., Marsh, G.A., Kung, N.Y., and McCallum, H. 2015. Ecological dynamics of emerging bat virus spillover. *Proceedings of the Royal Society B: Biological Sciences*, 282(1798): 2014–24. https://doi.org/10.1098/rspb.2014.2124

Preston, N.D., Daszak, P., and Colwell, R.R. 2013. The Human Environment Interface: Applying Ecosystem Concepts to Health. In Mackenzie, J.S., Jeggo, M., Daszak, P., and Richt, J.A. (eds), *One Health: The Human-Animal-Environment Interfaces in Emerging Infectious Diseases. The Concept and Examples of a One Health Approach. Current Topics in Microbiology and Immunology* Springer.

Reperant, L.A., Cornaglia, G., and Osterhaus, A.D.M.E. 2013. The importance of understanding the human–animal interface. from early hominins to global citizens. In Mackenzie, J.S., Jeggo, M., Daszak, P., and Richt, J.A. (eds), *One Health: The Human-Animal-Environment Interfaces in Emerging Infectious Diseases. The Concept and Examples of a One Health Approach. Current Topics in Microbiology and Immunology* . Springer.

Salyer, S. J., Silver, R., Simone, K., and Barton Behravesh, C. 2017. Prioritizing zoonoses for global health capacity building-themes from One Health Zoonotic Disease Workshops in 7 countries, 2014–16. *Emerging infectious diseases*, 23(13): S55–S64. https://doi.org/10.3201/eid2313.170418

Speldewinde, P.C., Slaney, D., and Weinstein, P. 2015. Is restoring an ecosystem good for your health?. *The Science of the Total Environment*, 502: 276–9. https://doi.org/10.1016/j.scitotenv.2014.09.028

Titcomb, G., Young, H., and Jerde, C.L. 2019. High-throughput sequencing for understanding the ecology of emerging infectious diseases at the wildlife–human interface. *Frontiers in Ecology and Evolution*, 7(126). https://doi.org/10.3389/fevo.2019.00126

United Nations Environment Programme and International Livestock Research Institute (UN). 2020. Preventing the next pandemic—Zoonotic diseases and how to break the chain of transmission. Nairobi, Kenya.

World Health Organization. COVID-19 Research and Innovation, Powering the world's pandemic response – now and in the future. February 2022. Available at https://cdn.who.int/media/docs/default-source/blue-print/achievement-report-_grif_web_finalversion15.pdf?sfvrsn=39052c73_9&download=true

Worobey, M. 2021. Dissecting the early COVID-19 cases in Wuhan. *Science (American Association for the Advancement of Science)*, 374(6572): 1202–4.

26

REWILDING, THE WILDLIFE TRADE AND HUMAN CONFLICT

Rene Beyers and Sally Hawkins

Introduction

Exploitation of wildlife by humans has become increasingly unsustainable, with examples of extirpations and extinctions increasing through history. Excessive or unsustainable hunting, poaching, or harvesting of wildlife for subsistence or trade is referred to as overexploitation (WWF, 2016). This includes flora and fauna and affects populations of target species, but can also indirectly affect populations of non-target species, for example through bycatch in the fisheries industry or through loss of habitat.

Humans hunt, poach or harvest other species for a number of reasons, including for food consumption, medicine, ornaments/trophies, religious reasons, fuel or pets, or if they are perceived to pose a threat or nuisance to people's lives. Due to a number of complex factors, including increases in human population, human movement, commodification of natural resources, improved access to wildlife through development of infrastructure, increased wealth and globalisation, exploitation of wildlife has become increasingly unsustainable, and subsistence hunting has been replaced by large-scale commercial trade.

Rewilding aims to restore degraded ecosystems to the point that they are self-sustaining, persistent and resilient. In doing so, rewilding recognises the importance of restoring interactions between all non-human species, from soil microbiota to large apex predators. Humans also rely on other species and ecological processes, so in order to sustain our lives and these fundamental processes, we need to ensure that humans and other species can coexist in landscapes. Overexploitation is one of the main factors threatening species at this time (Maxwell et al., 2016). This chapter focuses on overexploitation of wildlife, by which we mean undomesticated non-human species of animals and plants living in natural conditions, although extraction and overexploitation of other natural resources, such as water or minerals, can also dramatically affect ecological systems.

Two key drivers of overexploitation

There are a number of reasons that exploitation of natural resources can become unsustainable. Two key drivers are explored in this chapter—international wildlife trade and human–human

DOI: 10.4324/9781003097822-29

conflict. These are discussed separately but are linked, as the commodification of natural resources through trade can cause, exacerbate and/or perpetuate human conflict.

Wildlife trade

There is a huge and growing international market for wildlife species, which are extracted to fulfil increasing demand, often with vast distances across supply chains. The reasons for the exponential growth in trade are complex and need to be assessed both at the point of demand and point of extraction. Factors which influence demand include an increasing human population, increased purchasing power, emigration and globalisation. Rates of extraction are affected by the market price of target species, poverty, and a lack of economic opportunities in less developed areas, a lack of alternative sources of protein, lack of law enforcement or governance, and loss of traditional hunting or fishing controls (Wilkie et al., 2016; Milner-Gulland & Bennett, 2003). Development of transport networks in rural areas, forest fragmentation and access to firearms and other modern techniques enable and exacerbate commercial-scale extraction (Abernethy et al., 2013).

As a result, wildlife trade—both legal and illegal—has become big business that is difficult to regulate, requiring national and international laws and policy. The United Nations' Convention on International Trade in Endangered Species of Wild Fauna and Flora (CITES) is an intergovernmental agreement regulating international trade of endangered species. The agreement also addresses illegal trade, as does INTERPOL and the United Nations Office on Drugs and Crime. While legalising some wildlife trade has benefits, including regulation and improved monitoring, species may continue to be threatened by legal trade because decisions about international trade within CITES are often based on political compromise as countries with strong economic and/or cultural interests in certain species push to continue to harvest and trade them, even if they become rare or endangered. Political tensions with international conventions such as CITES also exist when, as is the case with elephants, some countries (Southern Africa as with elephants) have healthy animal populations and effective law enforcement while others (Central Africa and some countries in Eastern Africa) face decline or even extinction of the species while not having the means to protect them. Furthermore, CITES only covers international trade and protected species may continue to be traded domestically. Legal and illegal operations are not totally distinct either, with some species being harvested illegally, but sold in other countries legally (UNODC, 2020).

The illegal wildlife trade has been estimated by the UN to be worth between US$7 and 23 billion per year (Nellemann et al., 2016; Morton et al., 2021). At the same time, wildlife tourism generated as much as 5.2 times more revenue than trade. This shows that in many cases it could be more profitable to keep wildlife alive than to harvest it (WTCC, 2019). A large portion of international wildlife trade is illegal and therefore uncontrolled. Criminal networks provide the links between source countries (for example in Asia and Africa) and markets. It is thought that the greatest volume of non-edible commodities such as for medicinal and ornamental purposes are marketed to specialists rather than in local markets, and are increasingly traded online. Criminals tend to operate in countries with weak government institutions and judicial systems, with the result that wildlife crime is displaced to these countries. Goods are transported hidden in containers on ships and airplanes. Corruption is rife and bribes are paid to customs officials and operators working for airline or shipping companies (Nelleman et al., 2016; Bennett, 2011). Traffickers don't get caught easily, often get away with low penalties or sentences or are not persecuted to the full extent of the law. They are highly specialised and adapt easily to change in demand or new restrictions, regulations and enforcement measures (UNODC, 2020).

To counter criminal syndicates that supply wealthy markets in Asia and elsewhere, governments have to devote adequate resources and staff. Law enforcement officials need appropriate training in the identification of animal species, applicable laws and regulations and use effective technologies such as DNA sequencing, X-ray machines, sniffer dogs etc. to detect illegal merchandise. The judiciary needs to be educated about the mechanics and scale of wildlife crime and prosecute and sentence criminals at an appropriate level. National and international experts and specialised NGOs can play an important role in providing technical assistance.

International coordination and support are crucial as criminal networks do not respect borders. Several agreements have been reached in the last decade. For example, an International Consortium on Combating wildlife crime was created in 2010 between Interpol, CITES, the World Customs Organisation, the UN Office on Drugs and Crime, and the World Bank. Networks and trafficking routes have been exposed through intelligence gathering and joint operations have led to multiple seizures of wildlife products and arrests.

On a national level, because of the COVID-19 crisis in 2020, China, for example, banned the consumption of almost all terrestrial wild animals, except for non-edible animal products for medicinal purposes that continue to be regulated under the Traditional Chinese Medicine Law.

About one in five vertebrate species are traded for food, medicine, as pets or luxury items (WTTC, 2019) and demand along the commodity chain is biased towards larger and rarer species (Abernethy et al., 2013). Wild meat or 'bushmeat' is a popular commodity. It is consumed as an essential local source of protein (subsistence hunting), especially in poorer countries, but more recently it has become an important source of revenue through commercial trade (Petrozzi et al., 2016). In Central Africa, for example, wild meat is a desirable commodity in towns and cities for cultural reasons. Fa et al. (2009) estimate that 4 million tons of bushmeat are extracted from the Congo Basin every year for consumption within the region but some is shipped overseas as well.

Human–human conflict

Globally between 2000 and 2014 there were on average 35 active armed conflicts (between states or between a state and (a) rebel group(s)) (Hanson, 2018), showing an upward trend (Strand et al., 2021). The presence of valuable natural resources can cause, exacerbate or perpetuate conflict. Eliason (2019), for example, reports that poaching can occur as a form of resistance or rebellion against the government in response to conflict over resources and that private ownership affecting access to land or resources can be a driver of conflict. The bankrolling of military operations through wildlife trade can perpetuate conflict. For example, the Renamo in Mozambique partly funded their operations against the government in the 1980s with the sale of ivory. There have also been cases where overexploitation of valued species has been used as a tool to enable colonisation or to drive out extant human populations. For example, the systematic killing of bison in North America by European colonialists in the 19th century was due, in part, to intentions to disrupt Native American culture and livelihoods (Whyte, 2017). The connections between conflict, both during and after, and overexploitation are complex and good overviews can be found in Hanson (2018), with conflict and poaching specifically addressed in a review by Eliason (2019).

Inversely, overexploitation and wildlife trade can be exacerbated by conflict, for example when conflict affects national and local governance; emigration/immigration of people, including into protected areas; livelihoods, including sustainable and ecotourism businesses; suspension of conservation projects; and habitat destruction through illegal logging or land conversion.

Areas or countries with conflict often overlap with biodiversity hotspots. The majority of armed conflicts occur in Africa and Asia where most of the remaining large mammal species are found (Pettersson & Wallensteen, 2015). Conflict was a highly significant predictor for the decline in large herbivores in protected areas in Africa between 1946 and 2010 (Daskin & Pringle, 2018). Where there is intense conflict, the toll on wildlife can be very high, as for example in the DRC where the recent wars caused significant declines in elephants, forest antelopes, okapi and other large mammals. One well documented case was the Okapi reserve in the Ituri region where the elephant population was halved and forest antelopes declined by up to 84% between 1995 and 2006 (Beyers et al., 2011).

The impacts of overexploitation of wildlife on social-ecological systems

Ecological impacts

The loss of species through overexploitation has significant consequences for the ecosystems that they live in and decreases ecological integrity. Other chapters of this book describe how species interactions in the food–web maintain and regulate ecosystems. The decline or disappearance of keystone species leads to instability and degradation. Apex predators and many large herbivores, pollinators, seed and fruit dispersers are keystones that are all affected by hunting, trade, and conflict.

Where wildlife is overexploited, the large mammals are the most sensitive to hunting because of their low reproduction rates and their appeal. In Central Africa, roughly 178 wildlife species are being hunted, threatening half of them (Abernethy et al., 2013). Not only big mammals are affected—smaller ungulates and rodents can make up to 60% of the village hunt in this region. Eventually not much of the original diversity and abundance remains resulting in an impoverished ecosystem. The effects of this depletion, could take years or even decades to manifest. We can observe this in vast tracts of tropical forest through a phenomenon known as the empty forest syndrome. The fact that the vegetation still looks intact creates the illusion that the forest is healthy, while in fact it is distorted and degraded.

The consequences of the decline of keystone species, especially in tropical ecosystems, are poorly known. However, the elimination of apex predators is expected to cause an unravelling of ecosystems which is often slow at the beginning but accelerates over time. For example, extinctions of large mammal species in protected areas in Ghana after 30 years were much greater than what was predicted by size of the reserves. Extinctions were initially slow but increased as the effects of the disappearance of carnivores cascaded through the system. The elimination of lions and leopards led to an increase of olive baboons, which are mesopredators. The resulting rise in predation by baboons caused a decline of five primate and nine antelope species and also affected several bird and other taxa, including plants (Brashares et al., 2010).

In the oceans, sharks are being rapidly depleted by overfishing. Up to 100 million sharks are killed each year for the shark fin trade alone. The bigger shark species are top predators and their decline releases mesopredators such as smaller sharks and other predatory fish. This is causing fundamental changes in marine food webs with possible far-reaching consequences for the marine environment and those (including humans) that depend on it (Pacoureau et al., 2021).

The disappearance of megaherbivores affects everything from carnivores and scavengers to ecosystem processes like fire regimes, nutrient cycling, vegetation, and hydrology (Ripple et al., 2015). Most current losses of large herbivores, due to the trade in meat and body parts, occur in developing countries and 59% are threatened with extinction (Ripple et al., 2019). A sharp increase in demand for rhino horn used in traditional Chinese medicine is pushing many rhino populations to the brink. South Africa, which has the largest populations of rhinos left in the world, saw poaching climbing from 13 in 2007 to more than 1000 annually in 2013. White rhinos consume a large amount of plant material and keep grasses short, creating grazing lawns for smaller herbivores. The interaction with these smaller grazers reduces the severity and frequency of fire. Because they also selectively feed on certain species, they leave room for others to grow thereby promoting a higher diversity of species (Cromsigt & te Beest, 2014).

In Central Africa healthy populations of forest elephants constitute between 33 and 89% of the total mammal biomass. As ecosystem engineers, elephants have a profound influence on the vegetation structure and composition (Campos-Arceiz & Blake, 2011). They are the most important dispersers of large fruits and seeds of African tropical forest trees. The recent reduction of forest elephants of more than 86% is expected to lead to a decrease in larger tree species to the benefit of smaller species with lighter seeds which are dispersed by smaller animals or wind. Without elephants there is less trampling and browsing of the vegetation creating a denser understory of many small trees, which gives larger tree species fewer chances to grow tall. Both the lack of dispersal and the change in vegetation structure result in fewer big trees and less carbon storage.

Many species from primates to birds and rodents are involved in the seed dispersal of tropical trees. But many of them are also heavily hunted and their decline favours trees and shrubs that depend on wind or that are dispersed by animals not affected by hunting, such as smaller birds and bats. For example, in the Atlantic forests in Brazil the extirpation of the bigger fauna benefits light-wooded species that sequester less carbon (Bello et al., 2015).

Socio-cultural impacts

People establish relationships with the plants, animals, physical entities, and ecosystems of the landscapes in which they live. They can derive benefits from or may be interdependent with these species and processes. Overexploitation undermines the ecological conditions required for these communities to exercise their cultures, economies, and political self-determination (Whyte, 2017). The most prevalent impact discussed in the literature is related to impacts on livelihoods. It is estimated that more than one billion people depend on wild meat for their survival (Ripple et al., 2015). Because overhunting is exhausting this resource, millions of people are faced with food insecurity. Overexploitation can also erode cultures or affect cultural attachments to nature in the long term. This is demonstrated by the shifting baseline syndrome, where cultures change their perceptions of biological systems due to a loss of experience of past conditions which are more biodiverse or ecologically complex (Papworth et al., 2009).

Overexploitation causes wealth inequality where those who overexploit these resources experience disproportional benefits compared to their peers or future generations (Hauser et al., 2014). The loss of wildlife is usually felt most heavily by the rural poor, directly reducing the amount of animal protein available to them and affecting access to resources that they can sell (Milner-Gulland & Bennett, 2003).

Scientists have predicted the increased risk of zoonotic diseases due to overexploitation and the wildlife trade for many years (Milner-Gulland & Bennett, 2003) and the recent global Covid-19 crisis is considered to have been the result of trade in wildlife in China. As mentioned before, this led to more strict controls on wildlife trade, though there are concerns over the longevity and impact of these restrictions as well as the impact on those who have no alternative sources of income, increasing the risk of illegal activities (Koh et al., 2021).

Indirectly, losses to livelihoods and attachment to place can increase movement of people, potentially leading to conflict with extant human populations elsewhere.

Rewilding and mitigating the impacts of overexploitation caused by wildlife trade and conflict

Rewilding is interpreted in different ways based on the ecological and social conditions of the system in question, and there are a number of rewilding interventions that could be suitable for mitigating overexploitation. Changing conditions will create new opportunities for, or barriers to, rewilding. For example, the recent covid pandemic created opportunities to prevent overexploitation through stricter policies in China and provided some opportunities for spontaneous rewilding where there has been a reduction in recreation or human access. The following section refers to practical interventions as well as identified aims of rewilding (Hawkins, Chapter 5) and the rewilding principles (Carver et al., 2021). We present these in the context of the foundational three C's (Cores, Corridors, and Carnivores (or keystone species)) of rewilding and Co-existence between humans and wildlife through cultural change.

Core areas

Core natural areas should be large enough to accommodate the natural food web and ecological processes, such as natural disturbance regimes. Large areas that are either protected or remote and sparsely or un-inhabited also provide a natural barrier against overhunting. Many studies have found a relationship between distance from roads, navigable rivers, and other access points and the abundance of wildlife. Hunters in Central Africa for example usually travel less than 10km from a village although they may go as far as 40km from the nearest road (Abernethy et al., 2013). Thus, regulating access and leaving natural areas roadless, is often the best way to protect them. However, the livelihoods of people inhabiting the area should be maintained or improved while keeping the exploitation of wildlife and other resources at sustainable levels. If people are forced to move, either intentionally or as a consequence of increased poverty, they may increase hunting pressure elsewhere.

Roads inside protected areas should be strictly controlled. Although extractive industries such as logging, mining and oil exploitation cause significant disturbance to natural landscapes, the roads built for these operations may have a greater impact on biological communities. Hunting should be regulated and monitored in these concessions, poverty of local populations should be alleviated by programmes to provide alternative livelihoods and this should be part of an agreement between governments and extractive companies.

Peace-parks provide a tool for protecting and rewilding large core areas. These transboundary conservation parks serve to protect ecosystems, to build trust and encourage peaceful cooperation between countries, and foster socio-economic development in accordance with conservation objectives (Biosafety Unit, 2017). Donadio et al. (Chapter 16) and Pringle and Goncalves (Chapter 17) provide two examples of the use of protected areas in rewilding.

Connectivity

Connecting core areas allows for movement of species and recolonisation of places where species were depleted by overhunting. However, wildlife corridors are often narrow and prone to edge effects, including illegal hunting, so it is important to take this vulnerability into account when designing and managing them. Carruthers-Jones et al. (Chapter 8) explore this topic in more detail.

Keystone species

Where keystone species were extirpated or have declined because of overexploitation, they can be re-introduced or their populations can be augmented. This could be an active or a passive process through immigration from other areas. However, re-introductions will only be successful when ecological and socio-economic conditions are suitable. The drivers of the decline of the species should have been properly addressed and/or sufficient protective measures should be in place. For example, re-introducing rhinos is either unsuccessful or very costly as long as demand for rhino horn remains high.

Areas depopulated due to conflict, may provide opportunities for conservation and rewilding. For example, with the Demilitarised Zone (DMZ) between North and South Korea, a largely uninhabited buffer of approximately 250km by 4km was created between the two countries in 1953. Nature thrives in this unpopulated space and adjacent protected areas, benefiting the natural recovery of habitat and vulnerable and endangered species such as Siberian musk deer, white-naped crane, Asiatic black bear and long-tailed goral (Harbage, 2019).

Cultural change/coexistence

While ecological aims and impacts of rewilding are well discussed in the literature (Carver et al., 2021; Pettorelli et al., 2019), the socio-cultural aims are less so. However, rewilding acknowledges that, in order for the ecological aims to be realised, rewilding requires and promotes transformational cultural change, both within society at large and in how we practise conservation and rewilding (see Hawkins, Chapter 5, and Carver et al., 2021).

Many of the issues discussed in this chapter stem from the pervasive neoliberal discourse with its commitment to growth in GDP, promoting unconstrained exploitation of people, other species and natural resources (Longo & Clausen, 2011; Riedy, 2020). Rewilding promotes ecocentric culture rather than ego- or anthropocentric culture, which is promoted and perpetuated by neoliberalism. Rewilding encourages humans to acknowledge other species' intrinsic value but also to appreciate what we gain from it, in order that we can reciprocate and understand our responsibility to give back (see Hawkins, Chapter 5 and Maffey and Arts, Chapter 55).

Ecocentrism promotes solutions based on common good rather than self-interest and rewilding therefore aligns with community-based, grassroots approaches including adaptive co-management and co-governance. These methods can be used to find effective solutions to diversifying economic opportunities, reduce poverty and find alternative sources of protein. This includes improvement in local agricultural practices, domestic livestock management, and disease control and easier access to markets to reduce dependence on hunting and trade of wild animals (Milner-Gulland & Bennett, 2003). Involving local communities in the protection of wildlife usually yields better outcomes. For example, during the war in DRC (Democratic Republic of Congo), informal community organisational structures such as the authority of village chiefs, played an important role in reducing poaching and illegal bushmeat sales as they controlled urban wild meat markets (De Merode & Cowlishaw, 2006). Incentivising communities to protect wildlife from hunting and trade and involving the participation of communities in conservation planning improves compliance with wildlife protection and hunting policies (Andrade & Rhodes, 2012).

Wildlife exploitation should not only be sustainable at a population level (to ensure that a species persists) but also at an ecosystem level, taking into account the interdependencies between the target species and others in the system. Various methods and indices to determine sustainable hunting have been designed but much work remains to be done in the research and development of good hunting practices (Abernethy et al., 2013). Community-based approaches combined with a social-ecological systems approach will allow people to understand the impact that their actions have on natural resources, their own livelihoods and other people. The concept of social-ecological traps, which describes processes that drive a system towards an unsustainable or undesirable state through feedback loops between social and ecological systems, can be used to identify activities that are unsustainable and to implement change. Governance through community-based natural resource management can slow down or reverse social-ecological traps (Baker et al., 2018).

Cultural and traditional beliefs in the health benefits of animal products like rhino horn, tiger bones, pangolin scales, and many others, are a strong driver of the international wildlife trade. Entrenched habits and ignorance about the impacts of this consumption make it challenging to bring about the large-scale behavioural changes necessary to prevent the decline or extinction of these species.

Ultimately, to reduce overexploitation we must reduce consumption and increase our ability to work together to identify and prevent these unsustainable activities. This requires cultural transformation, which in turn requires collaboration, cross-cultural and interdisciplinary support and a blending of traditional scientific and other forms of knowledge (as discussed in Chapter 12 by Fenton and Playdon and promoted by many, including Robin Wall Kimmerer (2013)). Insight and knowledge from existing ecocentric cultures (whether indigenous or emerging) is especially valuable and the cultural practices and ways of knowing that could contribute to coexistence and reducing overexploitation must be respected, protected, and shared. Consumers must become aware of the dire consequences both for the species and ecosystems concerned and for ourselves, who are at risk if we continue to degrade and lose the systems we depend on. The Covid-19 crisis catalyzed the discourse about links between wildlife trade and epidemic spillover disease in humans and provides an opportunity to rethink our relationships with other animals and nature. Rewilding provides a framework for a paradigm shift from seeing ourselves as separate from nature to a new relationship of interdependence, reciprocity and co-existence with nature and wildlife.

References

Abernethy, K.A. et al. (2013) Extent and ecological consequences of hunting in Central African rainforests in the twenty-first century, *Philosophical Transactions of the Royal Society B: Biological Sciences*, 368(1625): 20120303. doi: 10.1098/rstb.2012.0303.

Andrade, G.S.M. and Rhodes, J.R. (2012). Protected areas and local communities: an inevitable partnership toward successful conservation strategies? *Ecology and Society*, 17(4): 14.

Baker, D., Murray, G., and Agyare, A. K. (2018) Governance and the making and breaking of social-ecological traps, *Ecology and Society*, 23(1). doi: 10.5751/ES-09992-230138.

Bello, C., Galetti, M., Pizo, M.A., Magnago, L.F.S., Rocha, M.F., Lima, R.A F., Peres, C.A., Ovaskainen, O., and Jordano, P. (2015). Defaunation affects carbon storage in tropical forests. *Science Advances*, 1(11): e1501105.

Bennett, E.L. (2011). Another inconvenient truth: the failure of enforcement systems to save charismatic species. *Oryx*, 45: 476–9.

Beyers, Rene L., Hart, John A., Sinclair, Anthony R. E., Grossmann, Falk, Klinkenberg, Brian, and Dino, Simeon. (2011) Resource wars and conflict ivory: The impact of civil conflict on elephants in the Democratic Republic of Congo: The case of the Okapi Reserve. *PLoS ONE*, 6(11): e27129. doi:10.1371/journal.pone.0027129.

Biosafety Unit (2017) *Peace Parks*. Secretariat of the Convention on Biological Diversity. Available at: www.cbd.int/peace/about/peace-parks/ (accessed: 19 August 2021).

Brashares, J.S., Prugh, L.R., Stoner, C.J., and Epps, C.W. (2010). Ecological and conservation implications of mesopredator release. In Terborgh, J. and Estes, James A. (eds), *Trophic Cascades. Predators, Prey and the Changing Dynamics of Nature*. Island Press, pp. 221–40.

Campos-Arceiz, Ahimsa, and Blake, Steve (2011). Megagardeners of the forest: the role of elephants in seed dispersal. *Acta Oecologia*, 37: 542–53.

Carver, Steve, Convery, Ian, Hawkins, Sally, Beyers, Rene, Eagle, Adam, Kun, Zoltan, Van Maanen, Erwin, Cao, Yue, Fisher, Mark, Edwards, Stephen R., Nelson, Cara, Gann, George D., Shurter, Steve, Aguilar, Karina, Andrade, Angela, Ripple, William J., Davis, John, Sinclair, Anthony, Bekoff, Marc, Noss, Reed, Foreman, Dave, Pettersson, Hanna, Root-Bernstein, Meredith, Svenning, Jens-Christian, Taylor, Peter, Wynne-Jones, Sophie, Featherstone, Alan Watson, Fløjgaard, Camilla, Stanley-Price, Mark, Navarro, Laetitia M., Aykroyd, Toby, Parfitt, Alison, and Soulé, Michael. (2021). Guiding principles for rewilding. *Conservation Biology*: 1–12.

Cromsigt, J.P.G.M. and te Beest, M. (2014). Restoration of a megaherbivore: Landscape-level impacts of white rhinoceros in Kruger National Park, South Africa. *Journal of Ecology*, 102(3): 566–75.

Daskin, J.H. and Pringle, R.M. (2018). Warfare and wildlife declines in Africa's protected areas. *Nature*, 553(7688): 328–32.

De Merode, E. and Cowlishaw, G. (2006). Species protection, The changing informal economy, and the politics of access to the bushmeat trade in the Democratic Republic of Congo. *Conservation Biology*, 20(4): 1262–71.

Eliason, S.L. (2019) Poaching, social conflict, and the public trust: Some critical observations on wildlife crime. *Capitalism Nature Socialism*, doi: 10.1080/10455752.2019.1617325

Fa, J.E. and Brown, D. (2009). Impacts of hunting on mammals in African tropical moist forests: A review and synthesis. *Mammal Review*, 39(4): 231–64.

Hanson, T. (2018). Biodiversity conservation and armed conflict: A warfare ecology perspective. *Annals of the New York Academy of Sciences*, 1429(1): 50–65.

Harbage, C. (2019) In Korean DMZ Wildlife Thrives. Some Conservationists Worry Peace Could Disrupt it. Available at: www.npr.org/2019/04/20/710054899/in-korean-dmz-wildlife-thrives-some-conservationists-worry-peace-could-disrupt-i. Accessed 5 May 2022.

Hauser, O.P. et al. (2014) Cooperating with the future, *Nature*, 511(7508): 220–3. doi: 10.1038/nature13530.

Koh, L.P., Li, Y., and Lee, J.S.H. (2021) The value of China's ban on wildlife trade and consumption, *Nature Sustainability*, 4(1): 2–4. doi: 10.1038/s41893-020-00677-0

Longo, S.B. and Clausen, R. (2011) The tragedy of the commodity: The overexploitation of the Mediterranean bluefin tuna fishery, *Organization & Environment*, 24(3): 312–28. doi: 10.1177/1086026611419860

Maxwell, S., Fuller, R.A., Brooks, T.M., and Watson, J.E.M. (2016) The ravages of guns, nets and bulldozers. *Nature*, 536(7615): 143–5.

Milner-Gulland, E.J. and Bennett, E.L. (2003) Wild meat: the bigger picture: Trends *Ecology & Evolution*, 18(7): 351–7. doi: 10.1016/S0169-5347(03)00123-X

Morton, Oscar, Scheffers, Brett R., Haugaasen, Torbjørn, and Edwards, David P. (2021) Impacts of wildlife trade on terrestrial biodiversity. *Nature Ecology & Evolution*, 5: 540–8.

Nellemann, C. (Editor in Chief). Henriksen, R., Kreilhuber, A., Stewart, D., Kotsovou, M., Raxter, P., Mrema, E., and Barrat, S. (eds). (2016). *The Rise of Environmental Crime—A Growing Threat to Natural Resources Peace, Development And Security*. A UNEP- INTERPOL Rapid Response Assessment. United Nations Environment Programme and RHIPTO Rapid Response–Norwegian Center for Global Analyses, www.rhipto.org

Pacoureau, Nathan, Rigby, Cassandra L., Kyne, Peter M., Sherley, Richard B., Winker, Henning, Carlson, John K., Fordham, Sonja V., Barreto, Rodrigo, Fernando, Daniel, Francis, Malcolm P., Jabado, Rima W., Herman, Katelyn B., Liu, Kwang-Ming, Marshall, Andrea D., Pollom, Riley A., Romanov, Evgeny V., Simpfendorfer, Colin A., Yin, Jamie S., Kindsvater, Holly K., and Dulvy, Nicholas K. (2021). Half a century of global decline in oceanic sharks and rays. *Nature*, 589(7843): 567–71.

Papworth, S.K. et al. (2009) Evidence for shifting baseline syndrome in conservation, *Conservation Letters*, 2(2): 93–100. doi: 10.1111/j.1755-263X.2009.00049.x

Petrozzi, F., Amori, G., Franco, D., Gaubert, P., Pacini, N., Eniang, E. A., Akani, G.C., Politano, E., and Luiselli, L. (2016). Ecology of the bushmeat trade in West and Central Africa. *Tropical Ecology*, 57(3): 545–57.

Pettorelli, N, Durant, S, and Du Toit, J (eds) (2019). *Rewilding (Ecological Reviews)*. Cambridge University Press.

Pettersson, T. and Wallensteen, P. (2015) Armed conflicts, 1946–2014. *J. Peace Res.*, 52: 536–50.

Riedy, C. (2020) Discourse coalitions for sustainability transformations: common ground and conflict beyond neoliberalism', *Current Opinion in Environmental Sustainability*, 45: 100–12. doi: 10.1016/j.cosust.2020.09.014.

Ripple, W.J., Newsome, T.M., Wolf, C., Dirzo, R., Everatt, K.T., Galetti, M., Hayward, M.W., Kerley, G.I.H., Levi, T., Lindsey, P.A., Macdonald, D.W., Malhi, Y., Painter, L.E., Sandom, C.J., Terborgh, J., and Van Valkenburgh, B. (2015). Collapse of the world's largest herbivores. *Science Advances*, 1(4): e1400103.

Ripple, W.J., Wolf, C., Newsome, T.M., Betts, M.G., Ceballos, G., Courchamp, F., Hayward, M.W., Van Valkenburgh, B., Wallach, A.D., and Worm, B. (2019). Are we eating the world's megafauna to extinction? *Conservation Letters*, e12627.

Strand, H. and Hegre, H. (2021) Trends in armed conflict, 1946–2020, *Conflict Trends*, 3. Oslo: PRIO.

UNODC (2020) *World Wildlife Crime Report 2020: Trafficking in Protected Species*. Available at: www.unodc.org/unodc/en/data-and-analysis/wildlife.html (accessed: 28 June 2021).

Wall Kimmerer, R. (2013) *Braiding Sweetgrass: Indigenous Wisdom, Scientific Knowledge and the Teachings of Plants*. Penguin Books.

Whyte, K. (2017) *The Dakota Access Pipeline, Environmental Injustice, and U.S. Colonialism*. SSRN Scholarly Paper ID 2925513. Rochester, NY: Social Science Research Network. Available at: https://papers.ssrn.com/abstract=2925513 (accessed: 28 June 2021).

Wilkie, D.S., Wieland, M., Boulet, H., Le Bel, S., van Vliet, N., Cornelis, D., BriacWarnon, V., Nasi, R., Fa, J.E. (2016). Eating and conserving bushmeat in Africa. *African Journal of Ecology*, 54(4): 402–14.

World Travel & Tourism Council (2019) The Economic Impact of Global Wildlife Tourism. Available at: https://wttc.org/Portals/0/Documents/Reports/2019/Sustainable%20Growth-Economic%20Impact%20of%20Global%20Wildlife%20Tourism-Aug%202019.pdf?ver=2021-02-25-182802-167. Accessed April 2022.

WWF (2016) *Living Planet Report 2016*. Risk and Resilience in a New Era. Gland, Switzerland: WWF International.

27

REWILDING CHILDREN AND YOUNG PEOPLE

The role of education and schools

Heather Prince

Rewilding children and young people: The role of education and schools

In the context of education and schools, rewilding has a different connotation to the traditional understanding of the term in an ecological sense. It does however share debate over its implementation, and its framing as an educational practice is contested.

If rewilding is the process of rebuilding natural ecosystems following a history of major human disturbance for sustainability and resilience, then in educational terms, rewilding education might be seen almost as a metaphoric interpretation of this. It provides opportunities for pedagogy and learning in 'nature', following a period where that relationship or 'connectedness' has been diminished particularly for children and young people. Rewilding is a response to key environmental challenges at global and at landscape scales, for example, the biodiversity crisis, conventional conservation and the wider issues of environmental inequality. Education is critical to the manifestation of a harmonised juxtaposition of people and nature for a sustainable and resilient future.

If a narrative of rewilding is 'failure of conventional conservation' then rewilding education could be viewed as compensatory to conventional education. However, 'wild' is a 'state of nature and a state of mind' (Harrold, 2020) and subject to different cultural, moral, political and ontological considerations.

This chapter explores rewilding in the context of enabling children and young people to engage with nature in educational settings. It debates the provision of outdoor learning in curricula as a pedagogic approach to realising this ambition and the facilitation and enactment of these experiences for best practice. It situates outdoor learning in wider pedagogical frameworks and debates the positioning of rewilding education as radical pedagogy, with particular regard to inclusion and enabling all children and young people to benefit from outdoor experiences. Their voices are included to balance adult dispositions and to present an egalitarian approach.

Relationships with nature in formal, non-formal, and informal education

There is no doubt that the outdoor and natural is being displaced in childhood by the indoor and virtual (Macfarlane, 2015). Recent research has shown that we are losing the language of

DOI: 10.4324/9781003097822-30

nature. Words such as 'tweet', 'cloud', 'web', and 'stream' now have a greater technological meaning for children and young people (National Trust, 2019).

The term 'rewild the child' was probably first coined by George Monbiot (2013). Capitalising undoubtedly on a rhyming catchphrase, Monbiot extolled the virtues of outdoor learning with research evidence for improved outcomes for children and young people as well as through direct experience outdoors in mid-Wales over two days with a group of disadvantaged ten year olds from London. This is in contrast to the 're-wild the child' movement, which is a social endeavour embedded in a particular ideology of childhood, nature and modern life that is middle class and says very little about entitlement, access to nature and inclusion across a range of demographics (Bates, 2020). Thus, outdoor learning and 'rewilding education' are not synonymous although have some relationship to each other that will be explored later.

The importance of spending time in nature for children goes beyond nature literacy and cognitive knowledge, with wide-ranging outcomes to support, for example, physical and mental health and wellbeing, care and concern for the environment, and personal, social and emotional development. However, there are many challenges to enabling children and young people to re-engage with nature (if we acknowledge its diminution) from concerns about risk including 'stranger danger', limited suitable areas and green spaces, to young people's accessibility to these. Even when more 'natural' areas are available, it is important that people feel secure and safe in using them. Familiarity with place may result in repeated visits, whereas unfamiliar locations may feel more intimidating. It is not just about availability and access to natural settings; it is about the knowledge and confidence in how to use them through education and learning, including spontaneous play.

Government policy in England supporting outdoor learning has been episodic at best and with a neo-liberalist approach and focus on performativity in schools, has witnessed a downward trend in provision in England (Prince, 2019; Waite, 2010). There are glimmers of positive intent albeit with the same underlying agendas, with the Manifesto for Learning Outside the Classroom (DfES, 2006) recognising that if all children and young people were given these opportunities, this would make a significant contribution to raising achievement. The 25-year Environment Plan (DEFRA, 2018) seeks to achieve frequent, progressive outdoor experiences for all, particularly children and young people. The independent review of national parks and areas of outstanding natural beauty (Glover, 2019) recommended giving every child the opportunity 'to spend a night "under the stars" in a national landscape' (p. 17) to help more children to connect with nature.

Scotland's 'Curriculum for Excellence' (CfE) includes policy and practice documents to support outdoor learning including *CfE through Outdoor Learning* (Learning and Teaching Scotland, 2010), which 'could be considered as offering some of the strongest outdoor learning policy language, arguably anywhere in the world' (Christie, Higgins, & Nicol, 2016: 116). The CfE also includes Learning for Sustainability integrating global citizenship and outdoor learning as an entitlement for all learners and is also embedded in teacher professional standards. At the core of Scotland's National Performance Framework are the UN Sustainable Development Goals. The 'New' Curriculum for Wales introduced from 2022 has excited educators with the possibilities to embed outdoor learning and environmental responsibility (Wales Council for Outdoor Learning, 2022). Building on the current progressive curriculum in the area of health and wellbeing, it identifies the importance of outdoor learning in the environment or experience to develop learners' motivation, resilience, and empathy and decision-making abilities to produce healthy, confident individuals (the third 'purpose' of the curriculum).

Teachers report an enabling curriculum for the Early Years Foundation Stage (0–5 years) in England (Prince, 2019) with the necessity to provide an area for outdoor play, or access to outdoor activities on a daily basis. The recognition of the importance of outdoor play in the early years is mirrored across the curricula of the four nations of the UK. However, globally, countries vary in their philosophies and policies on early childhood education and care (ECEC), from a western-centric valuing of that phase for education and learning to limited or perhaps non-existent entitlement (Ärlemalm-Hagsér & Elliot, 2018). If ECEC is seen as a social pedagogic necessity, then outdoor learning and play will be mandated or encouraged; if readiness for formal schooling is the major priority, then such activities may be seen as detracting from valuable preparation time (Moss, 2013). In a survey of school-based outdoor learning in 19 countries, early years outdoor activities were one of the three most reported forms of practice (with field studies, and outdoor and adventure education) (Waite, 2020).

Implementing outdoor learning in schools

Following a nature-led, human-enabled approach, teachers can facilitate innovative and imaginative approaches to learning in natural environments to foster children's curiosity and enjoyment. However, attempts at rewilding children and young people through outdoor learning in schools are subject to other systemic challenges. In England, sustained teacher commitment to an enabling curriculum (Prince, 2020), the use of safe local spaces and places, key initiatives and a risk benefit approach were shown to be key ingredients of successful outdoor learning in primary schools (Prince, 2019).

Teachers' professional expertise in outdoor learning has diminished in recent years as few initial teacher education programmes offer a specialism in this area. However, many teachers have strong personal interests and motivation for being and working outdoors and within the teaching profession, the inhibitors to enabling practice relate more to teacher confidence in working outdoors with students, and making curriculum links (Ager, 2019; Waite, 2020). Although time and cost are major barriers throughout the western world, many teachers make use of school grounds and local spaces to mitigate these factors.

There are many examples of schools transforming their grounds to facilitate children's engagement with nature and free play, and to provide alternative teaching spaces particularly in deprived areas where there are limited opportunities to access other such areas, (Bates, 2020; Dyment, 2005). Several studies comparing green play spaces with schoolyards have shown that children play differently in green spaces—they play more creatively, in more egalitarian ways, and with a sense of wonder, inventing games that they continue from day to day (Louv, 2010). Prince & MacGregor (2021) describe possible features of a varied outdoor classroom including a pond, wildlife, woodland, and growing areas, which can be valuable for primary school children, linking directly to Science in the National Curriculum with different levels and textures for physical movement and sensory experiences.

Bates (2020) worked with a landscape architect who re-designed the school grounds of a primary school with pupils as co-designers. Interestingly there were adult impositions in respect of perceived safety and ease of management and because of a funding application, although the resultant design challenged legacy attitudes to surveillance and risk. Through multimodal methods, she researched the effect that such a transformation had on children's imagination as they explored places. She found that the children saw their new space as a place of possibility, identifying their own 'special places' in the area and that it was interpreted differently by different ages of children. She suggests that such initiatives might allow children to shape their own childhoods and relationships with nature.

Safe practice in outdoor learning involves recognising risks associated with the experiences and activities, minimising those risks and balancing these with an identification of the benefits. A risk benefit analysis of outdoor activities not only helps with risk management but provides a basis for evaluating programmes and activities. A new culture of resilience has emerged in the last ten years, certainly in the UK, which embraces risk and danger as essential ingredients of a rounded childhood (Gill, 2011). Children will test their own capabilities through experimentation through physical activity and imaginary play and develop their own 'risk thermostat' (Moss, 2012: 14).

Schools in England reported initiatives such as 'Welly Wednesdays', 'Forest Fridays' and 'Grandparents Gardening weeks' to ensure that outdoor learning had a space and place in the curriculum of primary schools (Prince, 2019). They felt supported by shared resources and expertise across school consortia and groupings, as well as from outside partners such as Forestry England (a government agency), in respect of access to woodland areas. The key to increasing confidence and supporting progressive learning outdoors is through regular opportunities within formal education and the school day, rather than on school visits that might be a one-off and less inclusive for all pupils. The challenge is to sustain time in nature into adolescence, with performativity priorities and within complex social and economic situations.

Forest School had entered the lexicon of teachers' practice in schools in England by 2017 (Prince, 2019). Partly, this might have been because there has been funding support for training for teachers from local authorities for recognised leader qualifications, but it has certainly engendered confidence for those teachers and practitioners who have not had previous experience in working outdoors with their pupils. Forest School as a pedagogy was introduced in the UK in the early part of the 21st century from Scandinavia and in 2012 the professional body, the Forest School Association, was launched. Forest School engages children (mainly preschool, early years and primary aged) in regular and repeated learning in natural environments to stimulate their curiosity and interest in learning using child-centred approaches (Knight, 2013) It has gained a reputation as educational practice that can mitigate concerns about the effects of children's diminishing relationship with the natural world (O'Brien, 2009). Many researchers have examined the outcomes and benefits of Forest School, for example, in health and wellbeing (Harris, 2012), personal and social development (Ilea, Mos-Butean, & Holmec, 2016) and self-determination (Barrable & Arvanitis, 2019). Increasingly, however, it is being justified through alignment with mainstream priorities with specific learning outcomes and educational for sustainable development (Waite, Bølling, & Bentsen, 2016).

The Danish concept of 'udeskole' (literally 'outdoor school') is for children 7–16 years and is characterised by compulsory educational activities outdoors on a regular basis. It is practised by teachers as an additional and alternative learning space and follows their pedagogical practice as facilitators across a range of curriculum areas in an educational system dominated by teacher autonomy. Beyond cognitive outcomes, it has been shown to have similar benefits to Forest School with an emphasis on enhanced sensory awareness and an understanding between green environments and community. Udeskole can take place in both natural and cultural settings but as in forest school, there have been schools that experience difficulties and additional costs in accessing 'natural' areas particularly in urban areas.

Rewilding children and wider pedagogical frameworks

Although the ethos and philosophy of facilitating learning outdoors with children and young people is similar across different western cultures, there are variations in the interpretation and enactment of practice in schools and other educational settings. Cultural-historical perspectives are influential too to the ways in which educational systems have developed across the world. Different political and educational systems influence the manifestation of outdoor learning practices across the globe. Research also highlights the benefits for the environment itself (e.g. Chawla, 1988) but there is a counter argument that spending time in the natural environment does not automatically result in pro-environmental attitudes or behaviour (Davis, 1998) and that the relationship between outdoor learning experiences and pro-environmental behaviour is complex (Prince, 2017). This is particularly important if we consider rewilding through education and schooling. What exactly are we wanting the outcomes to be for children and young people, and for the environment?

One of the dangers of juxtaposing the contemporary spatial control of children with the mission to 'rewild the child' is that it risks romanticising childhoods past, in which children were free to roam, and childhood was a time of innocence in nature (Bates, 2020: 366). Sitka-Sage et al. (2017) offer 'rewilding' as a means to think about education that moves beyond the romantic vestiges of 'nature' without lapsing into delusions of human exceptionalism. It is important for children and young people to recognise that human influences on 'natural' areas are not all positive and have led to major environmental dilemmas: biodiversity loss, climate change, and negative management strategies. Educators, as conservationists, arrive at spaces and places with certain individual, professional, or culturally framed perspectives and ontologies. There is a need to acknowledge the agency and activeness of the 'more-than-human' world and the co-existence of all beings, human, and non-human (Bekoff, 2014). Sitka-Sage et al. (2017) suggest that, for environmental educators, rewilding should be a shared project for mutual flourishing rather than unidirectional human endeavour. Educators need to demonstrate reflective practice and modify any actions and language that are antithetical to an ecological orientation. They suggest that environmental educators should be challenged to move towards the 'right' kind of 'post-nature' world; one where we work to move beyond tendencies or rhetoric towards a new natural contract with a more-than-human world. Bekoff (2014) sees rewilding as a mindset and its realisation requires a change in education. 'Rewilding our hearts is about becoming re-enchanted with nature. It is about nurturing our sense of wonder' (p. 5).

Monbiot's experience reflects the lack of a mandatory curriculum for outdoor learning in England and a focus on other priorities:

> One boy stood out: he had remarkable powers of observation and intuition. When I mentioned this to his teacher, her reply astonished me: 'I must tell him. It's not something he will have heard before.' When a child as bright and engaged as this is struggling at school, the problem lies not with the child but with the education system. We foster and reward a narrow set of skills.
>
> *(Monbiot, 2013)*

Outdoor learning sits comfortably within Dewey's seminal work (1963) on experiential learning. The emphasis of the importance of natural environments for children's health and wellbeing and learning has been influenced by theorists who were educationalists (e.g. Pestalozzi in Switzerland; Froebel in Germany) and social theorists and reformists (Rousseau). The resulting educational practices are accommodated within mainstream education. Some educationalists have established alternative pedagogies (e.g. Montesorri in Italy; Steiner in Austria) or radical education movements (e.g. Freire in Brazil; Neill in the UK) out with mainstream education that focus on the emotional, social, intellectual, and spiritual links between the natural environment and children's naturalistic/natural, holistic development (Garrick, 2009).

Rewilding as radical education

In the UK, radical education emerges from political systems in which educational change is a key aspect of radical social change. Its expression comprises a range of principles and practices from 'progressives' who provide a 'more humane' version of mainstream education, to 'emancipators' who explore the limits of educational freedom (Fenton, Playdon, & Prince, 2020). 'Child-centred' progressive approaches are the focus, as Dewey emphasised, 'the potentialities of education when it is treated as intelligently directed development of the possibilities inherent in ordinary experience', which is 'always the actual life-experience of some individual' (Dewey 1963, 13.61).

As Fenton, Playdon, and Prince (2022) have argued for bushcraft education might rewilding education be theorised as radical education with central foci suggesting a 'conscientisation' (Friere) that develops a critical awareness, transformative of society's relationship with ecosystems and providing autonomous, individual learning? The relationality, described by Neill as positioning individual development within the community an individual has chosen to occupy, together with 'conscientisation', is extended to the natural world. In relation to responding to key environmental issues, a new way of thinking is required, a new relationship with our ecology, and a new educational approach to support that. If this vision is necessarily emancipatory, then it focusses on 'the 'positive' freedom of all individuals to determine those matters which affect their own lives' (Carr & Hartnett, 1996: 28).

The 'Rewilding Education' organisation suggests that there is a need to radically change our education and create healthier, fairer, and wilder ways of educating children, young people, and adults. As radical educators, they argue for a greater freedom of education for children to allow them to be self-directed and self-willed (one definition of 'wild') with wild experiences and immersion in 'deep nature'. They are 'convinced we need to incorporate some of the wisdom from wild educators into the ways that we educate all children, young people and adults' (Hope, 2020). However, this ideology of childhood questions the entitlement and equitability of access to learning for all. Rewilding might be part of outdoor learning or outdoor learning can be positioned as the beginning of that deep engagement with nature. There are also questions about the feasibility of rewilding education and the challenges that young people may face in participating. True co-creation of learning opportunities involves listening to the voices of those who should have that entitlement.

Children and young people's voices

Outdoor learning in schools is facilitated by teachers who place value on regular or intensive experiences and work in educational systems that allow them to do so. However, as they grow older children and young people might rewild themselves and engage in spontaneous activities outdoors. Hayball et al's (2018) research showed that Scottish children 10–12 years spent more time outdoors if they could engage in autonomous and risky play with friends. If friends were not available, they were more likely to spend time indoors. However, there are issues of accessibility to 'natural' spaces and places. Pound, Larkins, and Pound (2019) in a study with young people aged 11–18 in an area of deprivation in the UK close to the coast sought to identify the factors affecting whether or not they would travel there. Of the young people sampled only 3% indicated that it was 'not their thing'; the majority were motivated to access the coast but were influenced by factors that were needed to 'make it easier to go': their own attitudes and confidence, accessibility, having someone to go with, feeling safe and belonging and the time of year and weather. The research indicated that young people need to be able to 'get there' and be able 'to afford to go'. This was reflected in another study of a town in the UK where the car (in the absence of a safe bike network or public transport) was critical in delivering access to the experiences that young people value and enjoy, travelling with family or with friends (Ford et al., 2020). It is also important to use the language of engagement in the sense to which adults intend—nature 'connection' means some sort of relationship to their phone to many young people.

And this is the view of Generation 'Z':

Wild is a child

Wild is a child who stays out until dark
Wild is the child that lights fire with bark
Wild is a child with mud on their knees
Wild is the child who climbs up in the trees
Wild is a child a long way from home
Wild is the child with no need for a comb
Wild is a child who wipes their bum with a leaf
Wild is the child who uses a stick to brush their teeth.
Wild is a child who sleeps under the stars
Wild is the child who keeps tadpoles in jars
Wild is a child who fell out of a tree
Wild is the child with their own parking space at A&E[1]
Wild is a child that I would like to be.

(Rowan Ashworth (9), Winner of the Wordsworth Poetry Prize (2017))

Conclusion

Outdoor learning in schools and other educational settings is an inclusive starting point for children and young people's engagement with nature. Western educational systems and pedagogic frameworks enable forms of outdoor learning reflective of different political, historical, and cultural perspectives. Rewilding children and young people is a challenge for many schools if it involves a deeper and more sustained time in, and connectedness with nature. It is for this reason that rewilding the child initiatives take place in informal education or as radical pedagogy. However, if rewilding in the context of education is to have the same transformative purpose as rewilding as an ecological concept to address major environmental issues, then perhaps this is what is needed. Young people, if they can be enabled to access these spaces, places, and pedagogies with others, appear motivated to participate.

Note

1 A & E is the 'Accident and Emergency' department in UK hospitals.

References

Ager, J. (2019) Can I do it outside? How to introduce a CIDIO approach in a primary school. *Horizons*, 84: 33–5.
Ärlemalm-Hagsér and Elliott, S. (2018) Transcultural explorations in nature-based early childhood education. In P. Becker, B. Humberstone, C. Loynes, and J. Schirp (eds), *The Changing World of Outdoor Learning in Europe*, pp. 89–103. Routledge.
Barrable, A. and Arvanitis, A. (2019) Flourishing in a forest through a self-determination lens. *Journal of Outdoor and Environmental Education, 22*(1): 39–55.
Bates, C. (2020) Rewilding education? Exploring an imagined and experienced outdoor learning space. *Children's Geographies, 18*(3): 364–74.

Bekoff, M. (2014) *Rewilding Our Hearts: Building Pathways of Compassion and Coexistence*. New World Books.

Carr, W. and Hartnett, A. (1996) *Education and the Struggle for Democracy*. Buckingham: Open University Press.

Chawla, L. (1988) Children's concern for the natural environment. *Children's Environments Quarterly*, *5*(3): 13–20.

Christie, B., Higgins, P., and Nicol, R. (2016) Curricular outdoor learning in Scotland. In, B. Humberstone, H. Prince, and K.A. Henderson (eds), *International Handbook of Outdoor Studies* (pp. 113–20). Routledge.

Davis, J. (1998) Young children, environmental education, and the future. *Early Childhood Education Journal, 26*(2): 117–23.

Department for Education and Skills (DfES) (2006) *Learning Outside the Classroom. Manifesto*. Retrieved from:: www.lotc.org.uk/wp-content/uploads/2011/03/G1.-LOtC-Manifesto.pdf

Dewey, J. (1963) *Experience and Education*. Collier-Macmillan.

Dyment, J. (2005) Green school grounds as sites for outdoor learning: Barriers and opportunities. *International Research in Geographical and Environmental Education, 14*(1): 28–45.

Fenton, L., Playdon, Z., and Prince, H.E. (2022) Bushcraft education as radical pedagogy. *Pedagogy, Culture & Society*, 30(5), 715–729.

Ford, C., Convery, I., Harvey, D., Hallam, S., Loynes, C., and Prince, H. (2020) *Sustainable rural town performance: Place value, attachment and migration*. A research and impact project, unpublished.

Garrick, R. (2009) *Playing Outdoors in the Early Years*. Continuum.

Gill, T. (2011) *Nothing ventured… Balancing risks and benefits in the outdoors*. Retrieved from: www.englishoutdoorcouncil.org/wp-content/uploads/Nothing-Ventured.pdf

Glover, J. (2019*) Landscapes review: Final report*. Retrieved from: www.gov.uk/government/publications/designated-landscapes-national-parks-and-aonbs-2018-review

Harrold, J. (2020) *Rethinking Rewilding*. Wales: Snowdonia Society. Retrieved from: www.rgs.org/geography/online-lectures/revisiting-rewilding-john-harold/

Harris, F. (2012) The nature of Forest School: Practitioners' perspective'. *Education, 3-13, 45*(2): 272–91.

Hayball, F., McCrorie, A., Kirk, A., Gibson, A., and Ellaway, A. (2018) Exploring children's perceptions of their local environment in relation to time spent outside. *Children and Society*, 32(1): 14–26.

Hope, M.A. (2020) *Rewilding Education*. Retrieved from: https://maxhope.co.uk/current-projects

Ilea, M., Mos-Butean, A., and Holmec, L. (2016) Forest School. A modern method in educational process. *ProEnvironment, 9*(27): 197–202.

Knight, S. (2013) *Forest School and Outdoor Learning in the Early Years*. 2nd ed. SAGE.

Learning and Teaching Scotland (2010) *Curriculum for Excellence through Outdoor Learning*. Retrieved from: https://education.gov.scot/documents/cfe-through-outdoor-learning.pdf

Louv, R. (2010) *Last Child in the Woods*. Algonquin Books.

Macfarlane, R. (2015) *Landmarks*. Hamish Hamilton.

Monbiot, G (2013) Rewild the child. *The Guardian*, 7 October 2013. Retrieved from: www.monbiot.com/2013/10/07/rewild-the-child/

Moss, P. (2013) *Early Childhood and Compulsory Education: Reconceptualising the Relationship*. Routledge.

Moss, S. (2012) *Natural childhood*. Swindon: National Trust. Retrieved from: https://nt.global.ssl.fastly.net/documents/read-our-natural-childhood-report.pdf

National Trust (2019) Are we losing nature language? National Trust/Engine Mischief. Retrieved from: www.youtube.com/watch?v=vbCCR4kCllc

O'Brien, L. (2009) Learning outdoors: The Forest School approach. *Education 3-13,37*(1): 45–60.

Pound, D., Larkins, C., and Pound, J. (2019) *Living Coast—Youth Voice*. Natural England Commissioned Reports, No 2264.

Prince, H.E. (2019) Changes in outdoor learning in primary schools in England, 1995 and 2017: Lessons for good practice. *Journal of Adventure Education and Outdoor Learning, 19*(4): 329–42.

Prince, H.E. (2020) The sustained value teachers place on outdoor learning. *Education 3-13, 48*(5): 597–615.

Prince, H.E. (2017) Outdoor experiences and sustainability. *Journal of Adventure Education and Outdoor Learning, 17(*2): 161–71.

Prince, H. and MacGregor, L. (2021) *Outdoor Learning*. In Elton-Chalcraft, S. and Cooper, H. (eds), *Professional Studies in Primary Education* (pp. 357–78). 4th Edn. SAGE.

Sitka-Sage, M.D., Kopnina, H., Blenkinsop, S., and Piersol, L. (2017) Rewilding education in troubled times; or, getting back to the wrong post-nature. *Visions for Sustainability*, 8: 20–37.

Waite, S. (2010) Losing our way? The downward path for outdoor learning for children aged 2–11 years, *Journal of Adventure Education and Outdoor Learning, 10*(2): 111–26.

Waite, S. (2020) Where are we going? International views on purposes, practices and barriers in school-based outdoor learning. *Educ. Sci.*, 10: 311.

Waite, S., Bølling, M., and Bentsen, P. (2016) Comparing apples and pears? A conceptual framework for understanding forms of outdoor learning through comparison of English Forest Schools and Danish *udeskole*. *Environmental Education Research*, *22*(6): 868–92.

Wales Council for Outdoor Learning (2020) *Shaping the Curriculum for Wales*. Retrieved from: www.walescouncilforoutdoorlearning.org

28

WILD ADVENTURE

A restorying

Chris Loynes

Introduction

'Wild' is appearing increasingly as an adjective to describe children, play, pedagogies and communities as well as places, natural processes, and environmental interventions. It is appearing in the brands and descriptions of outdoor recreation and learning interventions, whether providing experiences in remote wilderness settings or urban playgrounds. During the 20th century outdoor organisations across the western world branded their experiences as 'adventure'. They drew on the legacy of historical adventurers of the British empire, stories of the exploration and conquest of nature popular with aspirant populations of the industrial revolution. These scientists, entrepreneurs, soldiers, merchants, and, later, adventurers, noticed the transformative effect of these experiences. Valuing the perceived impacts on the character of themselves and those around them some, such a Baden-Powell, Murray Levick, and John Hunt, became determined to reproduce the experiences of and benefits arising from adventure for the young citizens of a fast-changing society (Loynes, 2008).

However, 'adventure' typically constructs nature as of instrumental value, providing experiences that build identity and character, and opportunities for socialisation, maturation, escape, and renewal. 'Wild', on the other hand, typically describes an approach that constructs nature as of intrinsic value and with which people have a two-way relationship and an ethical responsibility (Molsher & Townsend, 2016). Perhaps 'wild' experiences are adventures with nature in mind; adventures that look outwards at the world around as well as inwards at the inner landscape of the person. This change is widespread in western cultures and may indicate a significant shift in human nature relations, one that could help change cultural narratives about nature and effect human responses to and engagement with the environment including the promotion of pro-environmental values and behaviours.

This chapter sets out to consider the potential of places of environmental and adventurous opportunities for people that are newly narrated as 'wild'. The narratives of adventure and nature may be being superseded. Embedded in their old stories are ideas about ourselves and about nature that may be helpful but are also possibly harmful to an emerging 'wild' pedagogy of experience more in tune with the emergence of a post-humanist culture. New practices and the lessons from minor narratives hidden from view by the dominant stories of adventure and nature-based tourism are reviewed for what they can reveal about the new possibilities of a 'wild' adventure.

DOI: 10.4324/9781003097822-31

In this chapter the term 'nature' will be applied with various meanings, what Castree (2014) calls epistemic communities, each using nature in a particular epistemological way. For lack of alternatives, 'nature' will apply variously to 'human nature', 'nature' as 'other than human' and 'nature' in which humans are immersed and a part. I will attempt to make it as transparent as possible which meaning is being conveyed at any one time.

Nature bites back

In 2015, like many in the British Isles, I was affected by Storm Desmond. My partner was stuck when the rail lines went under water in a city where flooding had put out all the lights. I could not reach her as the roads were just as badly affected. Others had it much worse losing their homes and businesses. This storm is one of many recent catastrophic events worldwide, floods, fires, droughts, pandemics, that have increased in intensity and frequency due to climate change. They have become the events of a new collective and globalising narrative of change, even disaster, precipitated by climate change. The wilder acts of nature caused by human pollution are no longer small scale, incremental, or elsewhere. The chains of cause and effect have come full circle (Stammers, 2020). Natural forces are again noticeable to all however urban, and these forces cannot be interpreted as benign, as nature has often been portrayed in the modern world's media, but wild, self-willed, agentic, disruptive, even hostile. Lives are being lost. That we are a part of these natural processes and are directly implicated in these catastrophic events is undeniable.

For others, nature is being noticed for its absence. McCarthy (2015), in his eulogy to British nature, laments the loss of the 'moth snowstorm' that would cover the windscreen of any car driving at night in the countryside during his, and my, youth. Such declines used to be gradual, hardly noticed so that the moving baseline of the human sense of a healthy environment incrementally crept unnoticed towards catastrophe. McCarthy's book is a recent contribution to this increasingly insistent call begun by Rachel Carson and her book *Silent Spring* (1962), which was my own wake up call to environmental losses so substantial that they could be noticed in a lifetime or much less. Over 50% of wildlife has been lost on the watch of the current generation (Almond et al., 2020). The UK, frequently cited as a nation of nature lovers, is one of the most nature depleted countries in the world (Hayhow et al., 2019). UK nature lovers are unintentionally loving their nature to death through the lifestyles they choose to lead. The impacts of the pollutants and resource extraction modern lifestyles demand know no boundaries. Along with the rest of the world's people living high consumption lifestyles, UK citizens are disproportionally responsible for the harm being done to nature globally (Berners-Lee, 2019). There is no place unaffected by humans. Pollutants, and people, are everywhere.

Wild as other than home

Perhaps it is the use of fire or the domestication of animals and plants by Mesolithic peoples that led to the development of settled communities that some describe as domesticated, along with the animals and plants that were brought along with the human story (Brody, 2002). It is around these events that human development is first considered as 'separating' from nature and that the wild nature of previous lifestyles became domestic, tame. Perhaps the term 'separation' is not the best description of all that had taken place, or what human societies had hoped had taken place. 'Subjugation', in many cases, captures the shift in human nature relations that occurred. Bees became pollinators, snakes became pest control, dogs became hunters and guards, fire a source of heat, water a source of power, all implicitly instrumentalised as useful in the minds of humans. Those elements of nature that were not also tamed were eradicated or receded to the edge of human habitation. Forests were cleared with fire and axes, barriers erected to keep nature at bay, especially those elements of nature considered dangerous because humans no longer knew how to live alongside them. Bears and wolves entered the stories told to children to keep them away from 'the wild'. Traditional ecological knowledge (TEK) was lost to all but a few, forgotten or even pilloried as primitive. Where TEK had instrumental value for the wider community a few, hunters, gatherers, herbalists, foresters, guides, and traders, retained TEK but as 'civilisation' 'progressed', more and more of their knowledge was transcended by scientific knowledge and abandoned.

The modern industrial revolution, Toffler's (1980) 'third wave', initiated in the UK and spreading rapidly worldwide, further dispossessing people from subsistence farming and turning them into labour for the factories, releasing the land to be taken into private ownership for industrial farmscapes. Remnants of the wild were further reduced by clearance or techno-logical alternatives to organic soils and pest controls. The resultant growth economies did, to a degree, trickle down their wealth, providing the time and disposable income to launch consumer lifestyles with recreation as a significant opportunity and product. As a consequence of this raising of living standards a return to nature, as outdoor recreation, eco and adven-ture tourism, began. In America, the railroad companies developed wilderness areas, soon to become national parks, as destinations to encourage passengers onto their trains in order to support their business model by funding the push through all the way to the west coast. In many cases native people were cleared from these areas in order to create a 'wilderness', genocides caused by national park development. In the UK, the railways facilitated a surge of urban people travelling to the few wilder open spaces they could still explore, the seaside, the moors and hills. Battles for the 'right to roam' were fought and, partially, won, though, in England, not enshrined in law until 2000.

Cheap air travel is currently combining with the breaking of Toffler's third wave in other countries as the 'world becomes the oyster' of the new adventurers and travellers. The same phe-nomenon has also fuelled a neo-colonial rush of travellers seeking wild places and experiences, paradoxically but, until recently, unintentionally contributing considerable emissions to the atmosphere driving climate change and the resulting biodiversity loss. The very places that have become destinations for the remnants of the wild they hold are threatened even further as a result. This has even led to the phenomenon of 'last chance to see' tourism as people rush on yet more flights to see a glacier carving or a polar bear in the wild before they are gone. Wild places have increasingly become destinations that represent what has been lost. The wild has become a gym for the adventurer and a zoo for the wildlife lover. Cheung et al. (2019), commenting on tourists visiting Antarctica, found that seasoned western tourists were moved to adopt pro-environmental behaviours post-voyage whilst the new tourists from Asia were, much like the first explorers, moved by the sublime experience of a terra incognita. Even now, for some, the climate change narrative has not penetrated their narratives of travel.

The nature that has not become deeply othered in remote settings behind the boundaries of parks has been emasculated in back gardens. In local parks and gardens the remnants of nature have largely become benign. In this context, nature is paradoxically being appreciated for the wellbeing it provides for people seeking release from a busy life. This rediscovery has been captured by research as the phenomenon of 'nature connection', a psychological orienta-tion to a largely green and benign nature. Lumber et al's (2017) findings indicate that contact, emotion, meaning, compassion, and beauty are five pathways for improving nature connected-ness. The potential of 'green solutions' to the growing demands for physical and mental well-being have not gone unheard by governments keen on low-cost preventative interventions for public health. Nature is instrumentalised as good for the wellbeing of people. Less thought has been given to the wellbeing of the remnants of the wild or the barely functional habitats that sustain them and, because humans are also part of nature, sustain us. Little wonder that there is a counterculture, a call to rewild the landscape (Tree, 2019) as well as a call for people to become feral in a re-connection with the land and sea (Monbiot, 2013). The five pathways identified by Lumber et al. are close to the claims made by Colin Morlock (1998) for adventure education of experiences that promote love and respect for self, others, and the environment, and may act as a guide to the construction of nature-friendly and holistic adventures.

Adventure as a return to the wild

Adventure may have something to offer this project, a new ecological knowledge (NEK) and a new knowing and valuing of the wild. However, the English-speaking world's version of out-door adventure is, at first glance, problematic. It is founded on a European and especially British imperial tradition of adventure in foreign lands. The desire for wealth led to global explor-ation followed by trading and then their entrainment as resources, workers, slaves and markets. Growing wealth fed a desire to return to these lands for new purposes—an adventurous phase of exploration. The stories of the exploits of these adventurers helped inspire a generation to be aspirational, to progress. People sought betterment and, amongst the trappings of an emerging middle class, freedom, self-willed time, often in wild places, even wildernesses. Nature was 'other' but had its uses. Adventure became an acceptable release—a pressure valve in an increas-ingly hothouse society (Beames & Brown, 2016).

The spirit of adventure established its own self-disciplines and morality (Beames & Brown, 2016). This moral structure was seen as efficacious for emerging democratic societies. Whether the moral equivalent to war epitomised by the Scout and Guide movements or the progressive, self-reliant adventures of Ransome's *Swallows and Amazons*, adventure became valued for its potential for personal development and character education, for transitions to adulthood and citizenship in the UK and, increasingly, in the western and rapidly democratising world beyond. After the Second World War, in its Outward Bound form, adventure education spread to over 50 countries as one of the institutions essential for developing the leadership of an emerging democracy. In the second half of the 20th century this 'education' was hijacked by neo-liberal ideas of self-determinism and individualism. Trends in adventure education and recreation sim-ultaneously supported a growing market economy and became a market in its own right. In these commodified forms in which risk, technologies, and activities are extracted from the richer adventure experience, it has been labelled 'adventure in a bun' (Loynes, 1996) epitomised by the adventurous stop overs of the gap year traveller.

Adventure re-imagined as place

The first time that I remember being afraid of a natural phenomenon was on Lundy, a tiny out-crop of granite in the Bristol Channel, home to birds, seals and a flock of visitors each week. A thunderstorm headed up the channel straight for the island. It was ferocious and all that could be done was stay put. The lightning was constant and the thunderclaps deafening. It felt entirely random whether we would be struck by lightning or not. A yacht in the bay was struck though no one was on board. It was awesome, both wonderful and terrifying at the same time, perhaps truly a sublime experience.

Spretnak (1997) argues for a 'resurgence of the real', by which she means a realisation of our full immersion in nature and the full range of emotional responses this elicits. She understands this authentic 'romantic' engagement, full of drama and consequence, as distinct from the emasculated ideas of the appreciation of 'sublime' views and situations adopted as the heritage of the Romantic Movement and the tourist gaze. She argues for a consequential engagement in nature, through hands on, embodied work or recreation, that transcends the view as the backdrop to an aesthetic appreciation.

Like Spretnak's re-interpretation of the sublime, there are other narratives emerging within the adventure world that shift the story from a heroic fantasy in a wild space to a realist encounter with a wild place. Maynard (2021) recently commented that 'little bandwidth now remains for explorers who come back from the wilderness with a story about themselves, rather than our planet'. An expedition has been defined as a journey with a purpose. Setting aside the potential negative aspects of the imperial foundations of the overseas expedition, many ventures were inspired by a curiosity to find out about the world, for explorers to discover for themselves and report on for their cultures the world they were living in. The economic, scientific or religious implications of these 'discoveries' were, in some cases, ethically dubious or downright exploitative. However, they laid the foundations for the idea of adventures in unfamiliar settings and with uncertainty of outcome. The concept of adventure with a purpose has spawned a new wave of travellers who care about their carbon footprint and return from their travels with stories of the impacts our modern way of life is having on far flung and especially wild places. They have the skills to visit these places, a deep compassion for the impacts of humans on each other and on the world and compelling narratives to share with their audiences. They have become our canaries in the mine. Their journeys, and their carbon footprints, justified as they become advocates and defenders of remote and wild places and peoples, speaking up for their value and their need for protection.

Extreme sports, once understood as performed for the sense of aliveness that comes from the risk taking, have been re-interpreted by Brymer and Schweitzer (2013). For his interviewees, the 'buzz' comes from the sense of complete immersion and the loss of self that total concentration in a thoroughly dynamic act requires when engaging in this way with the elements. For them, they become actors immersed in nature by choice in contrast to my passive and unintended thunderstorm experience, something they can only achieve through rigorous skill development so that risk is largely controlled, moved to one side.

The microadventures advocated by Alistair Humphreys (2014) capture some of this philosophy encouraging short, local, simple overnight experiences to savour, for example, the night, the stars, the tides or the dawn chorus. More embedded in nature and celebrating the wild, they have the added advantage of a low carbon footprint.

Adventure narratives with other cultural roots are also emerging. Bush walking from Australia has long held that hikes in the wild are for the purpose of learning the natural and cultural history of a landscape. 'Bush' is sometimes equated with 'wilderness' but, as an aboriginal ranger remarked to me 'how can I be lost in my own home'. At its best the bush is becoming a site of transcultural learning between two cultural views of a landscape.

The Finnish concepts that celebrate the two aspects of rural and wild in the Finnish landscape are elo (adventure at home) and erä (adventure in the wild) (Kujala, 2018). Like 'bush', 'erä' is often translated as 'wilderness'. However, it is better explained as land largely under the influence of natural processes and with which people engage as hunters and gatherers; and rely on for clean water and firewood. In other words, like 'bush' it is not an unfamiliar or hostile space but a familiar place. Both elo and erä capture an outdoor life that is simple, safe, and fulfilling and that works sustainably with the affordances of nature. The Finn is at home in both settings. 'Risk' is decoupled from 'adventure', and 'home' is no longer opposed to 'away'.

The idea of living in and with nature peacefully and sustainably also underpins the growing global interest in Bushcraft. This approach restores the wild as a home for people learning from the TEK of indigenous peoples, and either recreating them or transforming them into a new ecological knowledge with relevance to our own modern cultures. The knowledge of an ecology that comes from hunting finds a new expression in tracking and photographing wildlife. Fenton et al. (2020) comment:

> Bushcraft emerged from indigenous knowledge with a skill-base used for military, commercial and recreational purposes. We identify it as embodied contextual learning, for and with the environment, arising from a deep inter-subjective relationship with the natural world. This focus suggests a 'conscientisation' developing a critical awareness, transformative of society's relationship with ecosystems and providing autonomous, individual learning.
>
> *(p. 1)*

If Bushcraft can avoid the pitfalls of hyper-masculinity, escapism and mis-appropriation, then it has much to contribute to developing an embodied knowing that can be integrated with modern concepts of wild systems.

In the emerging time of environmental emergencies place based approaches have arisen to challenge the space-based ideas of adventure in which human centred stories of achievement and discovery were constructed. The wild could again become a place in which other actors have agency and write their stories, sometimes together with humans. The 'other', previously constructed by adventure stories as a fantasy space in which to enact narratives of personal transformation, could become home again, occupied and storied by non-human actors in whose stories humans are enmeshed. In this new entanglement, people could become feral restoring a wilder, interactive and more equitable relationship with the other. These alternatives represent a post-humanist landscape for adventures in nature that value ecological as well as personal knowledge and set other human life on a more equal footing. The adventure, and the joy, is to be found in the relationship with rather than a dominion over nature. They exhibit vulnerability rather than invulnerability, seeking interdependence with the environments and cultures visited rather than independence from them. They also have the potential, through their simplicity and localism, to have a lower carbon footprint.

Wilder landscapes

Traditionally, the adventure field has focussed on the experiences and the activities treating the land or seascape as a passive setting exploited for personal development or business advantage. However, as indicated above, trends are activating wild places as dynamic, fragile, and in need of care and restoration. In the UK recent calls have been made to develop wilder, publicly accessible land closer to the centres of population as a way to overcome barriers of inequity that exclude whole sections of society from experiences in nature. The term urban rewilding has entered the language (Owens & Wolch, 2019).

Calls for 50% of the planet to be put aside for nature are being heard (Wilson, 2017). Whilst on the face of it, this is problematic, maintaining a separation of nature from culture, it does capture imaginations. In the UK, the Government has indicated 30% of land for nature as a target. Rewilding Britain is calling for 10% to be rewilded whilst making it clear that this land would not exclude people but rather embrace them as participants in the naturalising habitats. In Scotland, the John Muir Trust has mapped Scottish wild lands and successfully lobbied for stricter planning laws to apply to these places. They also own or partner with owners of wild land in order to create exemplars of restored wild landscapes and learn how this can best be done. They are increasingly not alone as landowners with such aspirations. The Avalon Marshes in England are undertaking landscape scale restoration of nature on a former peat extraction site by a partnership of voluntary sector and private landowners. In winter the place is visited by sometimes over 1000 people daily to observe the murmurations of starlings at dusk flying in to roost in the developing reed beds. Such spectacles are the basis for an emerging nature-based economy founded on the restoration of habitats and facilitation of visits for the increasing number of people keen to see iconic species, especially reintroductions—the great white stork at Avalon, and habitats. Beavers and sea eagles, lauded as icons of the wild, are joining ospreys as the sights/sites to see. The same aspiration of the revitalisation of rural communities and landscapes is held by the Rewilding Europe initiative, where lynx, wolf and bison have joined the list of 'must sees'. Visionary publications (e.g. Foreman, 2004) and organisations across the world, many highlighted in this book, are amplifying the rewilding baton creating a movement that suggests people are learning to revalue ecosystemic as much as egocentric experiences in the wild.

Visitors are not only interested in seeing and learning about wild places. Increasingly, there is a demand for opportunities to be a part of the movement supporting crowdfunding to buy land for rewilding as happened recently at Langholm in Scotland or volunteering to contribute to the work of restoration. Community buy-outs and community forests, land taken into the collective ownership by local communities, are reversing the trend to privatise land. These projects are consistently rooted in the ethics of sustainability and conservation and encourage nature based and adventure recreation to spread the word as much as to make a living. Leaving a legacy is catching up with and may well overtake the vicarious experience as a reason to be in wild places. And these wild places may increasingly be closer to home and allow for repeated encounters that allow people to experience the return of the moth snowstorm and know that their efforts are contributing to a much needed reversal of biodiversity and carbon trends. Perhaps the term 'recreation' can be applied to both the wild lands as much as the increasingly feral people who benefit from the wild.

The value for wild places

An international movement developing and encouraging wild pedagogies is attempting to articulate the meaning, both to people and nature, of these shifts in both practices and the meanings that are given to practices (Jickling et al., 2018). For Jickling et al. there are three central concepts. First, in relation to place, 'wild' means self-willed land, not pristine from human touch, but where land and the nature are the dominant processes influencing the evolution of the landscape and the habitats within it. Second, the term 'wild' reflects the agency more-than-human factors have in making dynamic and spontaneous connections with the human world. Third, in the sense of 'wilding' existing pedagogies, the term reflects a desire to disrupt the domestication of the current education system, and to start to think about how best to educate in an era of new uncertainty. It is not a long stretch to add 'recreation' to 'education' in this manifesto and to requote Fenton et al. (2020) that '(t)his focus suggests a "conscientisation" developing a critical awareness, transformative of society's relationship with ecosystems and providing autonomous, individual learning'. For the wild, it can be argued that diverse interactions with people influence, like the wolves in Yellowstone, the evolution of habitats adding to the diversity of opportunities for species, engaging with the dynamics of ecosystems and enhancing rather than detracting from biodiversity.

Conclusion

Nature is both increasingly present in modern urban human lives as storms, floods, and unseasonal weather; and, at the same time, increasingly absent as species, wild biomass, and habitats disappear. It is increasingly fragile and yet increasingly powerful. It has been understood as a 'nice to have' and yet is now being re-storied as a necessity for human life. The 'other than humans' are increasingly being afforded moral standing. These changes are reflected in the recreational and tourist activities, destinations, and stories of those of us who visit, adventure in, travel through, and explore 'wild nature'. And, for those who enjoy wild land and seascapes or who help facilitate these experiences, there is the opportunity to develop experiences and offers that 'leave no trace', inform participants of the harm being done, 'consider your trace', and even to contribute to a restoration, a reversal of that harm, to 'leave more trace' (Loynes, 2018) of the right kind as people take responsibility for reversing the damaging actions we have inflicted on our planet.

Acknowledgements

I would like to thank Richard Ensoll, Dr Lisa Fenton, Dr Jamie Mcphie, and Amy Smallwood for significant formative conversations in the preparations for writing this chapter.

References

Almond, R.E.A., Grooten, M., and Petersen, T. (eds). WWF (2020) *Living Planet Report 2020—Bending the Curve of Biodiversity Loss*. WWF, Gland, Switzerland.

Beames, S. and Brown, M. (2016). *Adventurous Learning: A Pedagogy for a Changing World*. Routledge.

Berners-Lee, M. (2019). *There Is No Planet B: A Handbook for the Make or Break Years*. Cambridge University Press.

Brody, H. (2002). *The Other Side of Eden*. Faber and Faber.

Brymer, E. and Schweitzer, R. (2013) The search for freedom in extreme sports: A phenomenological exploration. *Psychology of Sport and Exercise, 14*(6): 865–73. DOI: 10.1016/j.psychsport.2013.07.004

Carson, R. (1962). *Silent Spring*. Houghton Mifflin.

Castree, N. (2014). *Making Sense of Nature*. Routledge.

Cheung W.Y., Thomas Bauer, T., and Deng, J. (2019) The growth of Chinese tourism to Antarctica: a profile of their connectedness to nature, motivations, and perceptions, *The Polar Journal, 9* (1): 197–213, DOI: 10.1080/2154896X.2019.1618552

Fenton, L., Playdon, Z. and Prince, H. (2020). Bushcraft education as radical pedagogy. *Pedagogy, Culture & Society,* DOI: 10.1080/14681366.2020.1864659

Foreman, D. (2004). *Rewilding North America*. Island Press.

Hayhow, D.B., Eaton, M.A., Stanbury, A.J., Burns, F., Kirby, W.B., Bailey, N., Beckmann, B., Bedford, J., Boersch-Supan, P.H., Coomber, F., Dennis, E.B., Dolman, S.J., Dunn, E., Hall, J., Harrower, C., Hatfield, J.H., Hawley, J., Haysom, K., Hughes, J., Johns, D.G., Mathews, F., McQuatters-Gollop, A., Noble, D.G., Outhwaite, C.L., Pearce-Higgins, J.W., Pescott, O.L., Powney, G.D., and Symes, N. (2019). *The State of Nature 2019*. The State of Nature partnership.

Humphreys, A. (2014). *Microadventures*. Glasgow, Collins.

Jickling, R., Blenkinsop, S., Timmerman, N., and de Danann Sitka-Sage, M. (2018). *Wild Pedagogies*. Palgrave Macmillan.

Kujala, J. (2018). From Erä to Elo, loss or gain? A brief history of Finnish outdoor education. In P. Becker, B.Humberstone, C. Loynes, and J. Schirp (eds), *The Changing World of Outdoor Learning in Europe*. Routledge.

Loynes, C. (1996). Adventure in a bun. *Journal of Experiential Education*, 21: 35–9. DOI: 10.1177/105382599802100108

Loynes, C. (2008). Social reform, militarism and other historical influences on the practice of outdoor education in youth work. In Becker, P. and Schirp, J. *Other Ways of Learning,* Marburg, Germany: BSJ, pp. 75–102.

Loynes, C. (2018). Leave more trace. *Journal of Outdoor Recreation, Education and Leadership.* *10*(3): 179–86. DOI: 10.18666/JOREL-2018-V10-I3-8444

Lumber, R., Richardson, M., and Sheffield, D. (2017) Beyond knowing nature: Contact, emotion, compassion, meaning, and beauty are pathways to nature connection. *PLoS ONE*, 12(5). DOI: 10.1371/journal.pone.0177186

Maynard, M. (2021). Explore: the purpose of adventure. *The Geographical Magazine*, *93*(6): 41–6.

McCarthy, M. (2015). *The Moth Snowstorm: Nature and Joy*. John Murray Press.

Molsher, R. and Townsend, M. (2016). Improving wellbeing and environmental stewardship through volunteering in nature. *EcoHealth*, 13: 151–5. DOI: 10.1007/s10393-015-1089-1

Monbiot, G. (2013). *Feral*. Penguin Press.

Mortlock, C. (1998). *The Adventure Alternative*. Cicerone Press.

Owens, M. and Wolch, J. (2019). Rewilding cities. In N. Pettorelli, S.Durant, and Johandu Toit, *Rewilding*, pp. 280–302. Cambridge University Press.

Spretnak, C. (1997). *The Resurgence of the Real: Body, Nature, and Place in a Hypermodern World*. Routledge

Stammers, T. (2020). Nature bites back, *The New Bioethics, 26*(2): 81. DOI: 10.1080/20502877.2020.1787689

Toffler, A. (1980). *The Third Wave*. William Morrow.

Tree, I. (2019). *Wilding*. Picador.

Wilson, E.O. (2017). *Half-Earth*. Liveright.

PART IV

Wilder values

The ethics and philosophy of rewilding

29

WILDER VALUES

The ethics and philosophy of rewilding

Kate Rawles (section editor)

What would be the equivalent of fiddling while Rome burns in the context of the ecological and climate emergencies? Trying to freeze a habitat in time to protect a single species in a limited area, perhaps, or figuring out whether the carbon footprint of a hand drier or a paper towel is lower, while in the toilets of an international airport departure lounge.[1] Or possibly, the equivalent of approaching the urgent question of whether and how to put the flames out, with the mindset and values of an arsonist.

Two questions energise and underpin the chapters in this section. The first is whether the spectrum of approaches to nature conservation known as 'rewilding' has—deliberately or by default—taken on the task Robert MacFarlane identifies when he writes that, '[t]here is no more urgent intellectual task facing the human species ... . than thoroughly to reimagine its relationship with nature' (MacFarlane, 2006). Does it advance, or is it underpinned by, a philosophical position about the relationship between humans and the rest of nature that is significantly different from those that dominate mainstream thinking in industrialised 'Western' societies? Why this might be considered of supreme importance is sketched out below.

The second question—or set of questions—is about how to think about how we should treat the other-than-human world. All approaches to the conservation of nature are values-based,[2] and they all take a position on a range of ethical and (small 'p') political issues. This begins to emerge as soon as it is asked what conservation is trying conserve, why, and for who. How? Who gets to decide? And so on. Traditionally, the value judgements embedded in nature conservation, and the ethical issues that inevitably accompany them, have not always been noticed, let alone critically debated.

This matters for two reasons. First, decisions about how to act on behalf of other species and natural systems will arguably be made better if the values and ethics they are based on are considered.

Second, the Western philosophical tradition of thought about ethics and values has been developed to grapple almost exclusively with questions about how individual humans should treat other individual humans, or about how human-created organisations and societies should treat individual humans. The ethical systems of thought that we have as a result are, not surprisingly, human-centric—in their design as well as their application.

DOI: 10.4324/9781003097822-33

Until relatively recently (in terms of the history of Western philosophy[3]) ethical questions about what counts as acceptable/unacceptable treatment of other-than-humans were simply not asked. When they are asked, it fast becomes clear that the ethical frameworks we routinely use for our inter-human questions do not work all that well when applied to the other-than-human world. To take just one example, it's difficult and ungainly to use the notion of 'rights', designed to delineate obligations to individual people, to think about an entity that is not best thought of as an individual at all, such as a species or an ecosystem. Leopold's 'Land Ethic' provides one possible alternative, requiring us to think instead of ecosystems, or the land, as communities that we belong to on much the same terms as the other-than-human members or citizens.

The discipline of environmental ethics has been at the forefront in challenging the assumption that, ethically, only humans register or that only humans have intrinsic as well as instrumental value; and in tackling questions about how we might best think about what constitutes decent treatment of other-than-humans, how they should be valued and how best to articulate the relationships between us. A host of further questions are raised in turn. Who counts, ethically? Are individual organisms that (arguably) lack the capacity for consciousness an appropriate locus of ethical concern? What about ecological systems and processes? Does the answer depend on a characteristic—like sentience or the ability to self-regulate—or on the relationship between the entity and the ethical agent? How do we deal with conflicts between those we deem ethically significant? Who or what should we value intrinsically as well as instrumentally? What are the values that underpin conservation more widely, and what should they be? Whose views about this should be considered? Whose interests are privileged, and whose are discounted?

It's a highly contested area, with significant conceptual work still be done about how we even ask the questions—and an urgent need for the sort of progressive thought about both questions and answers that an approach like rewilding inevitably provokes and contributes to.

The current, planetary scale ecological emergency is the context in which debates about rewilding are playing out. Multiple ecological boundaries, critical to the stability of earth's major systems, have been transgressed. The trajectories in relation to extinctions; wildlife populations in our oceans, land and freshwater; greenhouse gas emissions; ocean acidification; plastic pollution and nitrogen use amongst others are all still going in the wrong direction. The latest Living Planet Report details a 68% loss of wild animal populations since 1970 (WWF, 2020), while the UN tells us that one million organisms are now headed for extinction. Plastic has been found everywhere, from arctic ice to riverbeds to zooplankton to human blood (Leslie et al., 2022). We are on track for a catastrophic three degree rise in the average global temperature of the only planet we know to be habitable. Overall, the negative impacts of humanity—or parts of humanity—on the earth are now so great that geologists propose that the current era be renamed the Anthropocene. The human era. It is not normally meant as a complement.

That we're not going to tweak our way out of this is no longer an 'extreme' view—organisations as diverse as the United Nations and Extinction Rebellion are now calling for urgent, systemic change across all major sectors of society including transport, energy, agriculture, and land use generally. The response needs to be commensurate in scale. It also needs to be deep-rooted. What are the drivers of the planetary emergency? Or, to put it another way, why is the (allegedly) most intelligent species on the planet systematically and knowingly undermining its own life support system? The answers to this question must inform the ways we seek to tackle the Anthropocene.

Arguably, one answer is that many of us have been trapped. The trap is (at least) two-fold, involving our economic systems, and our worldviews or philosophies—the collection of beliefs, values, and assumptions that underpin and shape the way we perceive the world, and the way we think and act within it; including in relation to other species and natural systems.

Growth-dependent economic systems, the context in which many of us live and work, require businesses to pursue economic profit as the primary value or goal, at almost any cost. The corresponding culture of consumerism tells us that the good life is primarily constituted by possessions and money. The ever-increasing extraction of resources demanded to sustain this, and the by-products this linear 'take, make, waste' system creates, drive habitat degradation, extinctions, climate change, and pollution alike. This is unsustainable in a very literal sense: as WWF put it some years ago, if everyone lived the life of an average Western European … . by 2050, we would need three planet earths (WWF, 2012). Yet taking a critical approach to growth economics, or to capitalism and consumerism, is often still dismissed as 'unrealistic', despite the evident lack of realism in an economic system that requires additional earths.

Another trap is the dominant worldview of 'western' industrialised societies, the intellectual ocean that we spend our entire lives swimming in, but that is often as invisible and unnoticed as water to a fish. Amongst the topics it shapes our thoughts and actions on, is nature. The dominant, western worldview tells us a multi-faceted story about our relationship with other species and with ecological systems and processes. Particularly troubling themes within this story include that we are not just separate to the rest of nature, but superior to it. Apart from, not a part of. That nature, while it may have value as a resource in various ways to us, only has value in this instrumental, often commodified, way; for example, as nature that provides raw materials for growth. Whether we are thinking of nature in terms of resources or ecosystem services, it's basically there for us and we can and should manage it solely for our own ends. The entire living world, from oak trees to ants, from starfish to blue whales, from zooplankton to the many million microscopic organisms in a teaspoon of soil and the relationships between them all—all reduced to a set of resources and services for a single species. Humans, the centre of value, the centre of the world.

If this, often unnoticed, philosophical mindset is indeed one of the root causes of our multiple environmental crises, or, to put it another way, if Einstein was right when he declared that you can't tackle a problem with the same thinking that caused it, (to paraphrase), then Macfarlane has to be correct in his assertion. A thorough rethinking of the human relationship with nature is indeed amongst the most urgent intellectual tasks we face. Without it, we are left seeking to understand and tackle a fire emergency with the mindset of an arsonist. This, for example, is what trying to motivate nature conservation by referring to 'Natural Capital' arguably does—articulating the value of other living beings, species, and natural systems in a way that serves to reinforce an instrumentalised view of nature as a set of commodities for us.

This takes us to the heart of the first question that animates the chapters in this section—the question of whether rewilding can be understood as rising to the challenge Macfarlane issues. Against the backdrop outlined above, rewilding could be considered, like a keystone species, to have the potential for influence far beyond what might be expected given its abundance. If rewilding has burst onto the scene, not just with a different set of practices in relation to what has conventionally been called nature conservation, but a different philosophy, a paradigm shift in our underlying mindset, then its potential to contribute to urgently needed, radical change could far outreach more conventional approaches to the protection of nature and environment. And in unexpected ways—as much to do with its underpinning philosophy as its actions on the ground.

Within this book, let alone the wider world of rewilding literature and practice, there are reasons to think that this might be the case; that rewilding brings with it mindsets as well as practices that could contribute significantly to the 'great transition' needed if we are to tackle our ecological crises, rather than tweak them. Some of these are summarised below.

- Rewilding advocates 'bold' and 'audacious' action, that is large-scale and long term, in response to those challenges—i.e. it offers a response that is commensurate. And it presents a positive and inspirational vision, in the context of doom and gloom, while still being honest about the urgency and reach of the multiple challenges we face
- Rewilding calls for widespread re-engagement between people and nature, at a time when millions of people in industrialised societies experience disconnection from nature in a way that has, however misleading, become a cliché. (We are all, of course, still in fact connected with nature so long as we eat drink and breathe.) And it calls openly for 're-examining or renegotiating [the] human relationship with ecology' (Carver et al., 2021)
- With regard to the latter, it eschews solely anthropocentric ways of valuing nature and endorses ecocentric ones. Ecocentric approaches acknowledge that other-than-humans, including species and ecosystems, have intrinsic value alongside their undeniable instrumental value. This shift from seeing 'nature' primarily as a set of instrumentally valued resources to something more akin to a living community that we are part of and utterly depend on, then changes how we think of these entities and beings from an ethical perspective
- Within this, there is an emphasis on figuring out what positive co-existence between humans and the rest of nature entails, without denying the inevitable dilemmas—several of which are explored in the chapters to follow
- Rewilding offers at least glimpses of what a recrafted, less consumerist notion of what it could mean for humans to life well on earth—with ideas about rewilding ourselves—and gaining quality of life as much if not more from our relationships and experiences, including with the other-than-human, than our possessions and material wealth
- Rewilding emphasises the need for humans to relinquish at least some of their power over nature, backing off and allowing or promoting 'self-sustaining, self-willed, or self-regulated ecosystems' instead …'A wildness with outcomes determined by the self-assembly of wild nature' (Carver et al., 2021)
- Rewilding acknowledges the need to liberate ourselves from damaging political and economic systems, as well as intellectual ones—rather than to work from within their confines
- Rewilding is inclusive in relation to knowledge as well as engagement, 'informed by science, traditional ecological knowledge (TEK), and other local knowledge' (Carver et al., 2021, principle 7). This is important not least (though not only) in light of the evidence that indigenous peoples[4] are the best guardians of e.g. rainforest ecosystems—or at least, when their rights are respected (Food and Agricultural Organisation of the United Nations, 2021)

The chapters that follow offer a selection of perspectives and debates rather than a systematic and comprehensive exposition and discussion. Many explore a variety of the ethical questions, questions about values and inevitable conflicts that arise within rewilding, as well as its potential to contribute to meaningful, positive change—including philosophical—and what this means.

To what extent is human intervention a legitimate means to rewilding ends? How paradoxical—or otherwise—is this, given the value placed on self-willed, autonomous ecological systems? How should we act when rewilding aims at an ecological level come into conflict with compassion for individual sentient animals? Can inflicting suffering or death on individual animals in the interests of ecological integrity overall be justified, for example, in relation to attempts to eradicate invasive species? Can we make sense of the notion of eco-democracy and how should members and constituents of the other than human world best be represented within it? What ethical and evaluative assumptions is rewilding itself based on and are concerns about environmentalism as a new form of colonialism in any way justified?

How do we approach, with justice and fairness, resistance to rewilding, and conflicts between advocates of rewilding and those who identify with the cultural landscape and its history, or those who currently own land and manage it in accordance to different values, aims, and outlooks? What metaphor should we use in trying to understand our own place in nature? Are we best understood as conquerors or community members? Stewards, managers or animals to be located somewhere around level two on a trophic pyramid?[5] What does rewilding offer in relation to an arguably necessary shift in human consciousness, and what would rewilding ourselves really mean?

Above all, these chapters explore some of the all-important questions about how we take ethical responsibility for our devastating, global impacts. We hope they offer valuable insights into what it means to act with ambition, audacity, courage, respect, and compassion in relation to human and other-than-human individuals; to species; and to ecosystems and processes disrupted by the Anthropocene.

Notes

1 With thanks to Mike Berners-Lee for that great example.
2 Using 'values' here in a wide and descriptive sense to mean goals or things that are considered to be of value. In the context of nature conservation, these often range widely, and may include rarity, particular species, biodiversity, nature, wildness, ecological health, ecological integrity, and so on. Many of these concepts, e.g. ecological integrity, embody further value-judgements in turn.
3 This is not in any way meant to imply that the western philosophical tradition is the only one of any significance or value—but it is, arguably, the one whose legacies have facilitated the environmental crises, and the one most of the authors of this book, including the author of this introduction, are primarily shaped by.
4 Fully recognising that these peoples are numerous and distinct and should not be treated as a homogenous group.
5 Yirka (2013) reports that a team of researchers in France worked out average eating habits worldwide and, from this, calculated the trophic level of humans to be around 2.21—about the same as anchovies or pigs.

References

Carver, S., Convery, I., Hawkins, S., Beyers, R. et al. (2021) Guiding principles for rewilding. *Conservation Biology*, 35(6): 1882–93.

Food and Agriculture Organisation of the United Nations (2021) Forest governance by indigenous and tribal peoples. An opportunity for climate action in Latin America and the Caribbean. Available at: www.fao.org/policy-support/tools-and-publications/resources-details/en/c/1472651/

Leslie, H.A. (2022) Discovery and quantification of plastic particle pollution in human blood. *Environment International*, 163: 107199.

Macfarlane, R. (2006) Turning points. In David Buckland et al. (eds), *Burning Ice; Art and Climate Change*. Cape Farewell

WWF (2020) Living Planet Report. Available at: www.worldwildlife.org/publications/living-planet-report-2020

WWF (2012) Living Planet Report. Available at: https://wwf.panda.org/discover/knowledge_hub/all_publications/living_planet_report_timeline/lpr_2012/?msclkid=168aad5cc6d411ec8f81ca43e76474db

Yirka, B. (2013) Researchers calculate human trophic level for first time. *Phys.Org*, 3 December. Available at: https://phys.org/news/2013-12-human-trophic.html#:~:text=(Phys.org)%20%E2%80%94A,relates%20to%20the%20food%20chain.

30

REWILDING FROM THE INSIDE OUT

A personal commitment to other animals and their homes during the anthropause and afterwards

Marc Bekoff

It's common knowledge that we are losing species and habitats at an unprecedented rate in a geological epoch known as the 'anthropocene'—the age of humanity. However, the so-called age of humanity is anything but humane. In fact, it's extremely violent and I prefer to call it 'the rage of inhumanity'. One hope is that because of the COVID-19 pandemic and a decrease in human mobility during a period called the 'anthropause' by Christian Rutz and his colleagues, the violence with which many humans interact with nonhuman animals (animals) and their homes will decrease and that as wild animals enter into urban areas, places that were once their homes, things will change for the better as people meet these animals as they come back to town. It is hoped these individuals will help people bridge the empathy gap and that they will display the same caring and compassion they direct towards companion animals to their new neighbours, who rightfully deserve to be there.

Concerning the term 'anthropause' science writer Graham Lawton aptly states, 'I like the term anthropause. It captures the current hiatus in human domination of the planet, but also reminds us that the worst aspects of the Anthropocene could simply come roaring back... The pandemic presents a unique opportunity to put it on a more secure scientific footing.' Of course, as various nonhuman animals move into cities and towns and reclaim what was once theirs, there are positive and negative consequences.

In their essay, 'COVID-19 lockdown allows researchers to quantify the effects of human activity on wildlife' Dr Rutz and his colleagues write about the COVID-19 Bio-Logging Initiative. They note, 'Reduced human mobility during the pandemic will reveal critical aspects of our impact on animals, providing important guidance on how best to share space on this crowded planet.' Also, 'what is clear is that humans and wildlife have become more interdependent than ever before, and that now is the time to study this complex relationship. A quantitative scientific investigation is urgently needed.' I fully agree.

DOI: 10.4324/9781003097822-34

Rewilding our hearts

Let's first consider what 'rewilding' has traditionally meant and how I want to personalise it. The word 'rewilding' became an essential part of talk among conservationists in the latewhen two well known conservation biologists, the late Michael Soulé and Reed Noss, wrote a now classic paper called 'Rewilding and Biodiversity: Complementary Goals for Continental Conservation'. In her book *Rewilding the World*, conservationist Caroline Fraser noted that rewilding basically could be boiled down to three words: cores, corridors, and carnivores. According to Dave Foreman, the director of the Rewilding Institute in Albuquerque, New Mexico and a true and some might say radical and occasionally irreverent visionary, rewilding is a conservation strategy based on three premises: '(1) healthy ecosystems need large carnivores, (2) large carnivores need big, wild, roadless areas, and (3) most roadless areas are small and thus need to be linked'. Conservation biologists see rewilding as a large-scale process involving multiple projects of different sizes that may focus on carnivores but ultimately include the panoply of wildlife.

In my book *Rewilding Our Hearts* (Bekoff, 2014) and elsewhere, I've asked people to become re-enchanted with the natural world, to act from the inside out—from their hearts—and to allow their hearts to guide them and to dissolve false boundaries so they could truly connect with both nature and themselves. I argued that by personally rewilding—by undoing the unwilding that can easily happen in the hustle-bustle of daily life and by reconnecting and becoming re-enchanted with nature including other animals—we can overcome negativity and see the world in more positive ways.

Rewilding basically means to make wild again, to rehabilitate our hearts and tap into our biophilic instincts. Personal rewilding can lead to some sort of emotional affinity for, and is one way to reconnect with, other nature. We, humans, are all over the place and there aren't many, if any, places in which our destructive footprints aren't playing significant roles in affecting the lives of countless other animals, representing a dizzying array of species.

Individual rewilding personalises what conservation projects try to accomplish in the world by building wildlife bridges and underpasses so that animals can move freely between fragmented areas. I see rewilding our hearts as a dynamic, intimate, and personal process that fosters corridors of coexistence and compassion for animals and their homes at the same time that it facilitates corridors in ourselves that connect our heart and brain, our caring, and awareness. In turn, these connections, or reconnections, can result in making wiser choices, and help us to pursue heartfelt actions that make the lives of all beings better.

Rewilding our hearts and rewilding the human dimension also mean redefining the borders in our interactions with other animals and overcoming the cognitive dissonance that abounds globally. I prefer the word 'borders' to 'boundaries' or 'barriers' because the latter words imply a less permeable interface between 'them' and 'us'. Redefining and softening these borders and distinctions is what rewilding is all about. Because of this, rewilding demands that we employ humility in our interactions with other animals and their homes. We really are the dominant and dominating species, and to achieve more equality in our interactions with nature, we need to control ourselves. We need to 'be humble in the face of nature's awesomeness'. We should regard and approach nature as a friend, one whose welfare matters for its own sake—and because it matters for our sake, too.

All in all, rewilding our hearts is a positive and inspirational social movement about what we can and must do, as individuals within a global community, working in harmony for common goals, to deal with the rampant and wanton destruction of our planet and its innumerable and awe-inspiring residents and their homes. It is a personally transformative process. It is about nurturing our sense of wonder. Rewilding is about being nice, kind, compassionate, empathic, and harnessing our inborn goodness and optimism. We really do need wild(er) minds and wild(er) hearts to make the changes that need to be made right now, so that we can work towards having a wild(er) planet.

Rewilding our hearts calls for a social revolution based on a personal commitment to change how we interact with other animals, with other humans, and the places where they live. It mandates a global paradigm shift on a deeply personal level. Rewilding is about melting the ice in our hearts so that we might all work together to solve the dilemmas posed by the melting ice of climate change.

The Earth is tired and broken and is not infinitely resilient. Like a fatigued person who is teetering on burning out, our wondrous and magnificent planet needs all the help it can get. Every second of every day we decide who lives and who dies; we are *that* powerful. Of course, we also do many wonderful things for our magnificent planet and its fascinating inhabitants, but right now, rather than patting ourselves on the back for all the good things we do, we need to take action to right the many wrongs before it is too late for other animals and ourselves. We need to keep the wild wild. Protected areas and wilderness are the foundation for conservation, and to make sure we preserve whatever wilderness remains we need to respect not only the land on which we and other animals live, but also the animals themselves. Our interests should not and must not always trump theirs.

To sum up, 'rewilding' is a mindset. It reflects the desire to (re)connect intimately with animals and landscapes in ways that dissolve borders. Rewilding means appreciating, respecting, and accepting other beings and landscapes for who or what they are, not for who or what we want them to be. It means rejoicing in the personal connections we establish and need so badly. It is inarguable that if we are going to make the world a better place now and for future generations, personal rewilding is central to the process. Laws and public policy won't do it anytime soon, so while many people are working very hard in this area, each of us must also undergo a major personal paradigm shift in how we view and live in the world and how we behave.

Personal rewilding also is a guide for action. As a social movement, it needs to be proactive, positive, persistent, patient, peaceful, practical, powerful, passionate, playful, present, principled, and proud.

Rewilding our hearts by minding other animals

When I mind animals in this way, I practice what I consider 'deep ethology'. That is, as the 'see-er', I try to become the 'seen'. When I watch coyotes, I become coyote. When I watch penguins, I become penguin. I will also try to become tree and even rock. I name my animal friends and try to step into their worlds to discover what it might be like to be a given individual—how they sense their surroundings, how they move about, and how they behave in myriad situations. This isn't just a flight of fancy. These intuitions and moral imaginations can sometimes be the fodder for further scientific research and lead to verifiable information, to knowledge. As a scientist, I know that it's never enough to simply imagine another animal's perspective. But as a person, I know that it's never enough to accept supposed unclarity or uncertainty about animal minds as a reason not to care for them, or as an excuse for inaction or willful harm.

A paradigm shift in our approach to other animals is vital because of what we now know and have really known for many decades about the cognitive and emotional capacities of other animals and their ability to suffer. We need to use what we know on their behalf, not only when it benefits us.

Compassionate conservation: The lives of *individual* animals matter

Compassionate conservation is a rapidly growing global and transdisciplinary field. The four guiding principles are: First do no harm, individuals matter, value all wildlife, and strive for peaceful coexistence. Compassionate conservation is based on the ethical position that actions taken to protect biodiversity should be guided by compassion for all beings. It stipulates that we need a conservation ethic that prioritises the protection of other animals as *individuals*: not just as collective members of populations or species. Individuals aren't merely objects or metrics who can be traded off for the good of populations, species, or biodiversity. Researchers in various disciplines are working closely together to put this into practice and there are numerous success stories. I fully realise that some gnarly dilemmas can arise when we try to 'fix' nature especially where we have previously intruded and made messes that won't easily go away without our interfering once again—or at least some people believe that—but using these guiding principles puts non-killing front and centre, as a viable option that demands serious consideration. My own personal experiences and those of others is that many people working on the ground think of killing as being 'business as usual' and do not seriously consider non-killing options that might work.

With a guiding principle of 'First do no harm', compassionate conservation offers a bold, virtuous, inclusive, and forward-looking framework that provides a meeting place for different perspectives and agendas to discuss and solve issues of human–animal conflict. Peaceful coexistence with other animals and their homes is needed in an increasingly human-dominated world if we are to preserve and conserve nature the best we can. Resisting human domination has not really worked.

Compassionate conservation also can play a leading role in fostering peaceful coexistence between urban animals and humans. It isn't only about our interactions with wild animals, and more and more people are learning that many of the nonhumans who wind up in their backyards or on the streets of their cities and towns aren't harmful as long as they're given the space they need. In and around my hometown of Boulder, Colorado, black bears, cougars, and red foxes have become more prevalent and wide-ranging and a large number of people have been tolerant of their presence. These sorts of connections can also be extremely important for humans' well-being.

Similar to compassionate conservation, personal rewilding calls for an important paradigm shift, one that values compassion above all. My hope is if we as a species can proceed from compassion, such that the life of every *individual* animal matters whenever we make choices, societal or personal, we might begin to undo the alienation and fragmentation that currently defines our damaged relationship to the natural world. Compassion will also help us heal our damaged, alienated, and fragmented relationships with each other. I believe that only then, when we honour coexistence for all beings, human and nonhuman, will we be able to find the ecological solutions that we so desperately need. And compassion begets compassion.

Where to from here?

We are the re-generation. Over many years I have come to see that we are always 're-ing' one thing or another. Rewilding in the real world requires us to try to restore and re-create ecosystems, for example, by reintroducing or repatriating animals into areas where they once lived. But since we really cannot re-create or restore ecosystems 'to what they were', it has been suggested that we rebuild rather than rewind as we move into the future. Rebuilding surely is part of the process of rewilding. The main reason I say we can't re-create of restore ecosystems 'to what they were' is because ecosystems are dynamic entities and develop and evolve in relationship to who is living there and ecological conditions that also vary over time.

We also talk about the need to rekindle, rebalance, refine, reconnect, reenvision, reintegrate, reimmerse, reeducate, rehabilitate, rethink, and reshape our relationships with other nature. Many of these efforts are reactions to environmental and ecological problems we can no longer ignore, but being reactive—the 'putting out the fire' mentality—does not work. As we rewild our hearts, there is an urgent need to be proactive. Instead of looking to the past as a guide, we have to envision the positive future we want and actively work towards it.

Ultimately, what we need is not only more information about animals, but also a social movement and revolution in how we interact with animals and nature, a movement based on peace, compassion, empathy, and social justice. As you will see, my vision of this movement is not that it represents a single idea or a specific programme. There is no 'membership.' Instead, we are all already members, as living, breathing human beings who move in circles of coexistence. Peace, compassion, empathy, and social justice are all part of a much-needed revolution in thinking and acting with kindness for all.

Regardless of scale, ranging from huge areas encompassing a wide variety of habitats that need to be reconnected or that need to be protected to personal interactions with animals and habitats, the need to rewild and reconnect centres on the fact that there has been extensive isolation and fragmentation in nature, between ourselves and (M)other nature, and *within* ourselves. Many, perhaps most, human animals, are isolated and fragmented internally concerning their relationships with nonhuman animals, so much that we're alienated from them. We do not connect with other animals, including other humans, because we cannot or do not empathise with them. The same goes for our lack of connection with various landscapes. We don't understand they are alive, vibrant, dynamic, magical, and magnificent. Alienation often results in different forms of domination and destruction, but domination is not what it means 'to be human'. Power does not mean license to do whatever we want to do because we can.

When all is said and done, and more is usually said than done, we need a heartfelt revolution in how we think, what we do with what we know, and how we act. We can no longer act as *Homo denialius*, as if nothing is really happening. Rewilding can be a very good guide. The revolution has to come from deep within us and begin at home, in our heart and wherever we live. I want to make the process of rewilding a more personal journey and exploration that centres on bringing other animals and their homes, ecosystems of many different types, back into our heart. For some they're already there or nearly so, whereas for others it will take some work to have this happen. Nonetheless, it is inarguable that if we're going to make the world a better place now and for future generations, personal rewilding is central to the process, and it is worth repeating that this will entail a major paradigm shift in how we view and live in the world, and how we behave. It's not that hard to expand our compassion footprint and if each of us does something the movement will grow rapidly.

The time is right, the time is now, for an inspirational, revolutionary, and personal social movement that can save us from doom and keep us positive while we pursue our hopes and dreams.

So, what do we need to do? We must rewild now and continue into the future. We need to take the leap. It will feel good to rewild because compassion and empathy are very contagious. *Ecocide is suicide.* When 'they' (other animals) lose, we all lose. We suffer the indignities to which we subject other animals. We can feel their pains and suffering if we allow ourselves to do so.

Compassion begets compassion and violence begets violence. There really is hope if we change our ways. We owe it to ourselves and to future generations who will inherit the world we leave them long after we're gone.

We live in a magnificent yet wounded world. Despite all of the rampant destruction and abuse, it remains a magnificent world filled with awe and wonder. If you are not in awe, you're not paying attention. So let's get on with it. Open your heart to nature and rewild as you go through your daily routines and rituals. The beginning is now. We can always do more as we rewild. Rewilding is a work in progress from which we must not get deflected. How lucky we are that we are able to partake in this process, gratefully and generously blurring borders between 'them' and 'us' and their homes and ours.

One of my favourite bumper sitckers is 'Nature Bats Last'. We can try to outrun and out-smart nature, but in the end she always wins. We know full well that species come and go, yet nature survives. She evolves and moves on. Will we allow ourselves to become one of the species who didn't make it? Or worse, will we continue to be the one species who threatens all others and who allows uncounted species and individuals to perish so we can live where and how we please? I hope not.

Let's make personal rewilding all the rage. We are all intimately interconnected, we are all one, and we all can and must work together as a united community to reconnect with nature and to rewild our hearts.

Some of this essay was excerpted from *Rewilding Our Hearts*, with permission of New World Library.

I thank Kate Rawles and an anonymous reviewer for extremely helpful comments on a previous draft of this essay. Each raised valuable questions that couldn't possibly be dealt with in this short essay. However, I have taken note of them all and surely will incorporate them into future forays into personal rewilding and compassionate conservation.

References

Bekoff, Marc. (2014). *Rewilding Our Hearts: Building Pathways of Compassion and Coexistence.* New World Library.

Fraser, Caroline. (2010). *Rewilding the World.* Picador.

Lawton, Graham. (2020). Lockdown is a unique chance to see how human activity affects wildlife. *New Scientist*, 25 July.

Ramp, Daniel and Bekoff, Marc. (2015). Compassion as a practical and evolved ethic for conservation. *BioScience*, 65, 323–7.

Rutz, C., Loretto, M., Bates, A.E. et al. (2020). COVID-19 lockdown allows researchers to quantify the effects of human activity on wildlife. *Nature Ecology & Evolution*, 202.

Selhub, Eva and AlanLogan. (2012). *Your Brain on Nature.* Collins.

Soulé, Michael and Noss, Reed. (1998). Rewilding and biodiversity: Complimentary goals for continental conservation. <u>*Wild Earth*</u>. Fall, 18–28.

Wallach, Arian, Marc Bekoff, Michael Nelson, and Daniel Ramp. (2015). Promoting predators and compassionate conservation. *Conservation Biology*, 29, 1481–4.

Wallach, Arian, Marc Bekoff, Chelsea Batavia, Michael Nelson, and Daniel Ramp. (2018). Summoning compassion to address the challenges of conservation. *Conservation Biology*, 32.

Wallach, Arian, et al. (2020). Recognizing animal personhood in compassionate conservation. *Conservation Biology*. https://conbio.onlinelibrary.wiley.com/doi/epdf/10.1111/cobi.13494

31

REWILDING AND CULTURAL TRANSFORMATION

Healing nature and reweaving humans back into the web of life

Peter Taylor, Alan Watson Featherstone, Simon Ayres, Adam Griffin, and Eric Maddern

Introduction

At the 10th World Wilderness Congress in Salamanca in 2013, the organisers had invited indigenous elders, leaders and conservation workers to contribute and showcase how they were working successfully to conserve the wild places in which they lived. It was striking how many had temporarily left their tribal communities to study, gain scientific qualifications and then return to their ancestral lands. However, one Iranian herdsman who had gained a doctorate, addressed the panel from the floor during discussions of the continuing crisis, and said 'what is needed is a change in consciousness' and he meant—of the 'western' mind. The intervention did not register and almost went unrecorded. The panel did not really know what was meant (Taylor, 2013).In this contribution, we attempt to address what that 'change of consciousness' might mean.

We suggest that an element of human consciousness has been lost—one that has existed within a range of different indigenous communities, all of which have experienced themselves as an integral part of nature. This state of consciousness is not easy to recover. The UN Permanent Forum on Indigenous Issues tells us indigenous peoples are culturally varied and not readily definable—other than that they are a people living today on land they have lived upon for a long, largely pre-industrial or pre-colonial time, and in the same way as for thousands of years. Such original cultures are few—with forest gardens or a hunter-gatherers' existence, for example, whereas considerably more were militarily and culturally colonised and controlled, and although they remain on their land and self-identify as indigenous people, they live largely 'western' lifestyles. In this case, the UN seeks not to define and accepts self-identification as the key criteria with respect to human rights (UN Permanent Forum on Indigenous Issues, 2007).

However, neither of these descriptions really conveys what we mean. More relevant to our own dilemma is the mentality of the indigenous mind in contrast to the 'western' and the scientific. The indigenous mind does not see itself as *separate* from nature, and the consequence

DOI: 10.4324/9781003097822-35

of that cannot be readily understood by the scientific mind—the mind to which modern conservation has allied itself.

'Science' in its linguistic root means 'to separate'. If the mind does *not* separate itself from nature, then it automatically expands and begins to *extend* the human quality of sentience such that it *experiences* the Earth itself as sentient—and it is this sentience, we would argue, that gets symbolised as Mother, the greenery almost as Lover, the Sun as Grand Father or Grand Mother, though the latter term is more likely to be reserved for the dark spiral of power at the Galactic Centre. Western psychology, in obeyance to science, would assume with some arrogance that this is merely an imaginative psychological projection, but in the actual *experience* of oneness is generated gratitude in the heart, humility in the face of Nature's ways and an acceptance of her cycles of abundance and scarcity.

The indigenous mind cannot therefore *dominate* Nature, nor exploit, not tear up her body, or cut down her beauty. The indigenous mind celebrates Nature with ritual, it *feels* the reality as much as it sees, and has no barriers to a flow of life-energy hardly known to science, that we might call *ki* after the ancient teachings of the Vedas, that has influenced yoga, healing and martial arts, and most crucially, is the source and creator of sexual union and the great mysteries that are there generated.

To this we would add that beneath the surface of a European or modern American veneer lies an indigenous Celt or Slav, and that many initiatives exist in Europe to recover ancient knowledge and practices.

During the birth of Natural History as a scientific discipline, its pursuit and some of the deepest revelations were set within a strong spiritual context—evinced by Agassiz among many others, including, for example, the early years of Charles Darwin's discoveries. However, modern science was born into a patriarchal and sexually repressed culture ingrained with the drive of colonial domination, hierarchy, and control. Nature came to be seen as mechanism, perhaps as a reflection of the way the human body and mind were also de-contextualised by science. With no 'ghost in the machine' nature and human nature were despiritualised both externally and internally with 'spirit' reduced to some private and subjective but largely irrelevant experience. Human consciousness had thus separated itself not just from 'spirit' but from the very nature that had given it birth. And conservation, with all its policies and practices, has tended to follow that scientific path. Peoples who have lived for millennia in a wild balance with nature do not suffer such a separation in consciousness—and this we believe, is not only the root of the current crisis of ecology, but also its salvation.

We here outline the social, political and spiritual nature of the projects we have been involved in, with a plea that the nascent movement towards changing consciousness that they represent can influence the trajectory of rewilding as it now becomes drawn into the mainstream of conservation practice and policy.

Rewilding—the contribution from Britain

Rewilding as a movement has taken off in the past decade—a word that in England has been on the lips of cabinet ministers and a concept that has inspired some major landowners to give over large tracts to let nature take over in both England, for example at the Knepp Estate, and in Scotland as at Glen Feshie in the Highlands.[1] However, Britain also has a unique history of much smaller grass-roots initiatives which are many and varied, and we raise here some concerns that this heritage may be overlooked in the focus upon large scale re-designations and changes of managerial practice.

The success of rewilding—both in terms of areas of land under more natural processes and lighter management, as well as in the public eye and in new and inspiring literature, for example, Isabella Tree's *Wilding* (2018), the story of Knepp—comes at a time when 'conservation' has been examining its failings in the face of a continuing loss of natural abundance, the fragmentation of landscapes and an ongoing Sixth Mass Extinction event. Where conservationists were at first sceptical, they have in recent years begun to listen and seek to incorporate the lessons of rewilding.

However, conservation has arguably become dominated by science, with success measured in terms of species or habitats conserved and accounted as part of 'functional systems'. In contrast, it is our experience that science, and in particular academia, has been of little relevance to the major successes of rewilding. Our concern is that the reductionist scientific approach will dominate rewilding. Science relies upon separation and analysis, and is not without its value, but we would argue that it does not foster closer connections to nature and to building that intangible quality we call *wildness*. Scientific conservation needs to be complemented by workable methods of healing such separation, processes to which we are, in our different ways, developing in our projects.[2]

Perhaps there is a way forward for the conservation sector, despite its corporate mentality and scientific orthodoxy. In order to explore this potential, we ask conservationists, academics, and journalists who have addressed the issues of rewilding to think of it as a social movement and part of a paradigm shift in values. In this respect, goal-oriented corporate plans largely bypass the process of reconnecting to the natural world—preferring car parks and interpretation centres, rather than tree-nurseries and programmes that involve children and camping out; or, as some of us have done, to invite indigenous elders into our community to share their wisdom and ways of relating to the natural world.

Fifteen years ago, we called on government agencies and conservation bodies to work together and create a landscape-scale wildland zone in each of the nations of England, Scotland, and Wales, to be designated a Natural Sanctuary Area where natural processes predominate, economic exploitation ceases and human infrastructure is removed. We asked that this be done as a conscious act of reverence for nature.[3]

We argue that these flagship projects should be coupled to a major shift in education that extends understanding and experience of nature beyond the sciences to include its healing powers and the indigenous forms of knowledge that expand human awareness. This initiative could be showcased internationally by Britain as a step towards a new relationship with the rest of nature.

Why we need a shift in consciousness

In our experience the desire to be engaged in rewilding is a growing expression of a social movement concerned with the loss of *wildlife* and the lack of direct personal experience of wildness, particularly as communicated almost daily in the mainstream media. However, this concern gets translated into indices of 'biodiversity', a professional term subject to much interpretation. The *number* of species gives no indication of the 'wild' quality of the habitat, nor the appeal of scale, lack of human dominance, or the charisma of some species.

For example, one of the greatest concerns over loss of diversity and abundance of species relates to increased agricultural intensity in farmed landscapes, where original assemblages had adapted to traditional and extensive farming methods providing an apparently rich assemblage of grassland, wet meadow, wood copse, hedgerow, and arable land species, particular of birds, flowering herbs, and insects. In the face of agricultural intensification there have been steep declines in these familiar species, losses that cannot be addressed by a system of isolated 'nature' reserves.

However, we would argue that the species richness of traditional farmed landscape, wood pastures, and grazed heather moorland should be compared to an 'original' mid-Holocene landscape with wild grazers and predators, now thought to be a more open and species-rich landscape due to the abundance of grazers and browsers. Thus, even traditional landscapes still resonant in the public memory, are severely denuded, depending upon how the index of species is weighted, for example, towards mammals and birds, or lichens and herbs.

Similar changes with attendant losses have occurred in extensive grazing regimes in the uplands as livestock numbers increased, and in Scottish moorlands devoted to sporting interests where deer numbers have multiplied and suppressed natural succession. There is a growing awareness that the old paradigm of stemming this loss of 'biodiversity' is failing, whereas rewilding initiatives without specific biodiversity goals are growing and engaging the public imagination.

The majority of these initiatives have grown organically from the grassroots of local communities (for example, Moor Trees on Dartmoor (Figure 31.1), Coetirian in Wales, and Carrifran in Scotland), and although we have no doubt that conservation bodies with large memberships and corporate governance can learn from a bottom-up approach, there is a risk of developing a sanitised version of rewilding, based on the old paradigm of scientific ecology with no relation to the social and spiritual dimensions of many local projects with which we are familiar.

We feel there is a need for a change of consciousness that includes 'rewilding the human', where we shift from head to heart, from mental construct to our feeling nature, from analytics to intuition and from taking *from* to giving *back*. Above all, this change must foster a reconnection

Figure 31.1   Teenagers from the city planting native tree seeds at Moor Trees, Dartmoor, a charity engaged in youth work. © Moor Trees.

to heal the separation that has occurred between the modern civilised human and the nature that supports us all. 'Nature' in its Latin root, actually means 'mother'—or 'she who has given birth'.

British rewilding has its own origin

There is a current drift towards defining rewilding as large scale ecosystem restoration. This is the American model of a very worthy but largely top-down approach schooled in trophic dynamics, keystone species, and predator–prey relations. It is a valid way forward in North America where old-style wilderness still exists (though without the Native American forms of range management) and where it is essentially ecological restoration by a more catchy name.

For us on the ground in Britain, rewilding is a broad church and we have sought to network all manner of initiatives from the smallest local to the largest landowner-based, top-down, professionally led, scientific, and economic of projects—on the principle that sharing experience would further a more progressive paradigm.[4]

We have one first principle of rewilding—it is to sit down and humbly *listen* to and re-establish a deep connection with the land and its nature and to the people who live and have lived in our locality. We shared our visions and collective ways forward emerged. At the outset we were engaged in a social movement that is particular to Britain. A shift in consciousness was taking place and we sought to communicate and network through education and the media.

For example, at Cae Mabon, situated in the wildwoods of Snowdonia National Park, courses have been run for over 30 years introducing indigenous forms of relatedness through art, poetry, time alone on the mountain, ancestral stories, and practical skills of survival in wild terrain. In addition, there has been teaching of 'plant spirit medicine' and relating to 'power animals' as in the Native American tradition, and in our own ancient Celtic heritage.

Eric Maddern's unique centre has been described as the 'number one natural building project in the UK' and called 'a Welsh Shangri-La'. It plays a vital role—in Eric's words—'in healing the crippling disconnection within Western culture between body, soul, spirit, and place'. The Chief Druid of the Order of Bards, Ovates, and Druids has said that Cae Mabon is 'the most druid-like place he knows anywhere'. One teenage lad said: 'Being here is like being high on nothing. Like being wild and free.'

Around Dartmoor, and in particular the town of Totnes, Adam Griffin's pioneering work with Moor Trees has grown to include a wilder landscape consultancy and teaching in the 'deep dive' and 'soul motion' dance community, as a way of deepening a heart and body connection to nature. Three of us—Adam, Simon, and Peter, have worked closely with indigenous elders in bringing a different perspective on animal and plant species through the medium of ritual and dance.

Areas of natural sanctuary

We decided that as educated ecologists and denizens of one of the richest countries in the world, one that has pioneered science and conservation and which, for many, still stands as a world leader in many fields of human endeavour, we could make a difference if we took the lead with flagship projects that heralded the new paradigm.

As already indicated, we have advocated that the three nations of England, Wales, and Scotland finance the rewilding of a significant national 'core' area, in England and Wales, 400 square kilometres would necessarily be the largest; in Scotland nearer 1500 (see Taylor 2005, *Beyond Conservation*). The mix of species is not so important—wolf and lynx in Scotland, lynx in Northern England, beaver and wild boar, eagles, crane, marten, wild cattle, wood bison, moose, and forest horses. The mix doesn't matter. None of it will be fully functional. There are boundaries and buffer zones that will require management, barren areas that will benefit from tree-planting. And there will be rural employment—in education, health, and recreation.

In the world of today, we constantly hear of decline and conflict—yet there are many positives: new land under wildland management, with thousands of hectares in Scotland regenerating and extending forests; new marshlands and colonising egrets and introduced crane; the return of great bustard on Salisbury Plain; the return of the sea eagle to Scotland, the Isle of White, and Ireland by well organised re-introduction; the augmentation of red kite populations; the unofficial release of beaver in Devon and Scotland, along with government approved computer-modelled and very slow trials; and the escape of wild boar, goshawk, and eagle owl. And although not officially recognised and much dismissed as a public delusion, the regular and widespread sightings of black leopard, puma, and lynx.[5]

We are at a point of re-evaluation. The old rules need to be examined in the light of modern reality—of climate change, invasive species, an ever greater sterility of agricultural lands and pressure on space, with food chains in Britain, at least, largely built upon non-native species (e.g. brown hare, brown rat, pheasant, rabbit, and red-legged partridge). There exist huge problems in a youth culture of drug dependency, mental health, and lack of contact with nature. Rewilding can address these problems and is already engaged—for example, at Cae Mabon, in the trance-dance culture and the festival scene and in working with inner city children and adults creating small inroads of nature, rewilding river corridors, and creating wild space. None of this has been quantified, but we believe it is an *essential* dimension of rewilding.

How the old paradigm seeks to possess the new

A new paradigm can change the world, but first it must survive the tendency to be absorbed into the old paradigm. We have been party to the creation of rewilding as a movement that is gaining momentum and we urge that careful reflection must now be given by the organisations seeking to incorporate its lessons. In particular, much can be learned from the grass-roots initiatives.

Britain is entirely different from North America in respect of scale and wildness—most of our wild areas have been denuded of native vegetation, with both herbivore and carnivore guilds absent, and no option in most cases of restoration or repopulation by migration.[6] These circumstances led to a British rewilding that had its own origins and philosophy—one quite distinct from the USA and Holland, the other main centres that have garnered much attention.

In England, Scotland, and Wales, even the 'wild' landscape is heavily exploited for livestock, game shooting, recreational activities, water supplies, and forestry. There is usually a patchwork of ownership and regulation. All projects that would effect change in this landscape of necessity become social projects and, as we have documented in the book *Rewilding* (2011) with articles from 60 authors drawn from the BANC journal ECOS, they are of great variety with regard to partnerships and ownership, as well as design and ultimate goal.

As a recommendation for how large conservation bodies and government agency can interface with rewilding initiatives, we suggest a model based on our experience in Wales. Coetir Anian (Cambrian Wildwood) has partnered with the Woodland Trust, a large national conservation body. The Woodland Trust purchased the property freehold, following a significant contribution from a major donor for the project, and then leased the property to Cambrian Wildwood, giving the local charity a free rein to deliver the project on the site. The local group keeps its autonomy and roots in the community whilst having tenure of land and access to expertise from the larger charity.

This unique contribution that Britain has made, where rewilding is a social and sometimes spiritual phenomenon stemming from the grassroots activity of local people, is often neglected in academic writing and broad-brush journalism.[7]

This is an important matter. As rewilding gains momentum and fires the public imagination, conservationists are keen to incorporate its precepts into the conservation portfolio as a management option. Journalists proselytise with an eye to book sales and establishing themselves in a niche market, and academics must also have an eye to funding, reputations, and their own niche in a competitive system. This is not to gainsay the importance of conservationists, academics, and journalists, all of whom we work with closely—rather, it is to point out potential limitations, constraints and the pitfalls that ensue. In that respect, conservationists can become mired in a scientific managerial philosophy that demands definitions, objectives, and targets (see Taylor, 2008).

Furthermore, many conservation organisations work now with a model of corporate governance, management targets, and models of corporate growth. As professionals within and dependent upon a scientific and political system, they have to keep a wary eye on the boundary of what is normative and socially acceptable. Through this lens, conservationists have come to see nature as a 'system'—as a functional ecology, and lately, as a provider of 'services' or 'natural capital', and thus perhaps unconsciously as a corporate entity within a system of corporate entities. In this environment, concepts of the Earth as 'mother' or 'teacher' are foreign and even cause embarrassment. An integrated conception of nature has to have a balance between the analytical mind, and the receptive—only then does nature 'speak' and the awareness dawn that a loving sentient being reaches out to further our own evolution as part of her own evolving 'self'. This receptive part of the mind has to be cultivated from an early age (see later: programmes at Cae Mabon, in Wales, have been developed to further this aim).

Journalists bring a different challenge—all write for a political audience of one kind or another, often seeking influence upon policy and, as with academics, may have very little connection to specific places and people in and of the land. Given the power of journalism to influence the public mind, this disconnection can prove detrimental.[8]

The personal connection

The beginning of rewilding in Britain started for all of us as a personal dream for a landscape with which we were familiar and to which we felt a belonging. It was only by establishing a deep personal connection with the land itself that we could begin the process of rewilding from the stance of being an integral and conscious part of the 'ecosystem' rather than acting as external and distant 'managers'.

Our foremost connection was to trees—they carried for us a special feeling to landscape, a healing quality that spanned generations, and provided the habitat and home for many species including those that have been extirpated from the UK. Yet old trees were few and far between and thus rewilding started in our mind's eye as regeneration of ancient forests.

In our modern world, it was as if trees had become enemies—they had long been eradicated over large stretches of Scottish, English, and Welsh mountains. Bare hillsides felt haunted by the ghosts of oak and pine, and empty of animal spirits, especially the large grazing animals and their predators. We did not see 'dysfunctional ecosystems' or restoring a particular past assemblage—in any case, what baseline would we use? In Britain, the mid-Holocene had a rich fauna that included wild cattle, forest horses, wolf, lynx, and bear, but it had lost its mega-herbivores to the first wave of European hunters at 30–40 kyr Before Present.

The intact forests of Britain evolved over several million years, twenty ice-ages, to create what ecologists would call functional or original, and it had forest elephants at its heart, with a supporting cast of forest rhino, wood bison, giant elk, and very big cats. The large herbivore guild had, in addition to elk and bison: herds of horse, wild cattle, red and roe deer, boar, and beaver—the vegetation adapted to the powerful grazing pressure of these guilds, creating a mosaic of habitats. The loss of 'megafauna' began with the expansion of human hunters out of Africa about 60,000 years ago. Not a single forest in North or South America, nor the savannahs and the plains, no Arctic tundra, no Carpathian wilderness, Boreal forest, or steppe, nor anywhere in Australasia, is now even close to its 'natural' cycle of grazing and fertilisation, migrations, and predations. As humans, we preside over a *continuing* Sixth Mass Extinction—where 80% of the large mammal fauna, whole genera of species on these continents, has been extinguished.

As the first rewilders, we already knew these ecological histories. We all had that grounding and it led to a dream not just of trees growing old and a returning forest symphony of large herbivores and predators but processes that could often only be fully expressed in a large-scale rewilded landscape.

It is well appreciated by all of us that some of the most fundamental scene-shifters could never return and the British forest elephant, rhino, and giant elk are gone for ever—as are the sabre-toothed cats that kept them in check. The real lesson of the past is that it *cannot* be recreated, nor should it be. What is required, though, is to help the Earth's own inherent process of recovery from disturbance and to create, where feasible, large areas subject only to natural processes with minimal human intervention. There, the evolutionary journey of species and their interactions can continue on nature's terms, rather than being controlled, and almost invariably limited, by human agendas.

We were starting from a deep personal connection with the landscapes we were seeking to rewild and we realised that to manifest our dream a key element involved education, as the vast majority of the public are still unaware of how much has been lost in Britain. In Scotland, Alan Watson Featherstone founded *Trees for Life*—a programme at the Findhorn Foundation, a spiritual community based on the recognition that nature has intelligence, consciousness, and a higher purpose in terms of evolutionary potential.[9]

Trees for Life started with a dual strategy: Alan and his team took parties of often young people out into Glen Affric, to engage in hands-on practical work to help the recovery of the forest there, at the same time as helping them understand the depletion and loss that had taken place. He explained to them why there were so few old pines, no younger generation of native trees at all but many non-native conifers grown in regimented plantations as a crop; alongside this, he found common cause in his dream of restoring the ancient Caledonian pine forest with some like-minded foresters and managers in the Forestry Commission, which was responsible for management of the best remaining fragments of old forest (Watson Featherston, 1997, 2004). The project attracted volunteers from all over the world, who took part in weeklong work sessions involving seed collection, native tree planting, removal of invasive non-native species, biodiversity surveys, and monitoring. They also provided time and space for each participant to connect deeply with the land and the life there.

There followed planting programmes and fencing to exclude over-browsing by red deer (*Cervus elaphus*). This happened every year from 1989 onwards and involved thousands of people. In 2008, through outreach and fundraising, Trees for Life purchased the 10,000 acre

Dundreggan Estate in Glenmoriston, immediately south of Glen Affric, which is strategically located to connect up with the Forestry Commission's holdings and those of the National Trust for Scotland and the Royal Society for the Protection of Birds, with the potential to create a large landscape scale extent of restored forest and other natural habitats that could also provide a home for some of the country's extirpated wildlife species.

In a parallel initiative, Adam Griffin and others set up native tree nurseries and began to work with the network of landowners on Dartmoor in Southwest England. Many youngsters came as volunteers from deprived urban backgrounds. Adam also moved into the realms of shamanic dance and 'movement medicine', setting up a local teaching centre as a rewilding of a wider dance culture that has thousands of followers reconnecting to nature on an *inner* level.[10]

In North Wales, initiatives in Snowdonia (see Coed Eryri in *Beyond Conservation*) moved very slowly, through discussions with the National Trust, a major landowner, who began purchasing and rewilding degraded mountain tops (under tenant farmers) through removal of grazing or the use of hardy cattle breeds rather than sheep. The focus eventually moved to the Cambrian Mountains. Simon Ayres had been inspired by Trees forLife and Coed Eryri and set up Coetir Anian/ Cambrian Wildwoodwhich in cooperation with the UK national charity The Woodland Trustnow manages over 350 acres of restored peatland and regenerating native woodlands. Wild Konik horses have been introduced to graze the mountain, pine marten has been translocated to the area, and there are plans for the return of water vole, red squirrel, and beaver.

In Wales, there was not the luxury of a large forest, secluded glens, and large estates, as in Glen Affric. Welsh mountains had hardly any native woodland. The dominant habitat is bare grassland for sheep and cattle (feral ponies and goats occur in Snowdonia) and fenced-off plantations of spruce. Even the red deer has gone. And there are no eagles to be seen from Snowdon in a land called *Eryri*—place of eagles.

By 2010, we had 25 years of experience, pilot projects in Scotland, Wales and England, cooperative strategies with big landowners such as the National Trust and Forestry Commission, and an ongoing dialogue with farmers to reduce grazing on the high tops, replace sheep with hardy cattle breeds and ponies, safeguard and subsidise the old meadows, regenerate native woodlands, and talk about re-introductions—for example, of the sea eagle in Wales and the lynx in Scotland. And nature itself was helping—ospreys, a symbol in Britain for regeneration of the wild, were arriving along the wilder estuaries of West Wales and occupying nest sites established by a local man, Steve Watson, who also pioneered effective no-fence tree planting in the presence of domestic livestock.

Indigenous wisdom

What we have lost is not to be found through science. The scientific approach to conservation is useful for monitoring and has had some success at informing management decisions, but science has often been inadequate at predicting and steering nature. Letting nature look after itself can produce more favourable and more interesting outcomes. For example, at Knepp the returning abundance of turtle dove and nightingale had not been predicted: the estate now hosts the greatest concentration of these birds in Britain. Scientists advising the project recommended growing specific arable crops for the turtle doves, but Knepp decided to allow wild nature to provide. The outcome is that the estate now has the only increasing population of turtle doves in Britain. It also has tree-nesting peregrine falcon, and the first white storks to nest in Britain since medieval times.

The lesson here is not to put ourselves above nature. What we have lost is the attitude of the indigenous soul.[11] It is a humility and a joy with respect to an Earth that is alive and conscious and loving in its presence. In the desperate words of an Amazonian indigenous leader:

> You forced your civilisation upon us and now look where we are: global pandemic, climate crisis, species extinction and, driving it all, widespread spiritual poverty. In all these years of taking, taking, taking from our lands, you have not had the courage, or the curiosity, or the respect to get to know us. To understand how we see, and think, and feel, and what we know about life on this Earth. As Indigenous peoples, we are fighting to protect what we love—our way of life, our rivers, the animals, our forests, life on Earth—and it's time that you listened to us. The Earth does not expect you to save her, she expects you to respect her. And we, as Indigenous peoples, expect the same.
>
> *(Nemonte Nenquimo (2020))*

We have begun to listen—but the listening has to be accompanied by an inner *experience,* followed by a shift in outer practical actions that reflect and embody the inner change.In that Amazonian heart-fullness, the kind of raping of the world around us that all conservationists have witnessed, is simply not possible. We have explored techniques to recover that wild heart. We get to know the presence of animals and trees as *teachers* or *guides* and in that transformational process, we become wilder.

As rewilders, we are not trying to save the natural world. In whatever form, it will persist for millions of years. We want to save our humanity, and to transform our relationship with the rest of nature, so that we become conscious participants in the web of all life, instead of its destroyers. We can change in our own land—show deep respect to Nature, be willing to make economic sacrifices, let some cultural and iconic landscapes go back to wilder time—we can help re-start the wondrous journey of evolution here, even if the whole thing is not fully functional, and some species lose abundance on the way. It is for the wild heart that we work.

We want Britain to be the leader in that. We need no definitions. No authorities. Anyone can engage in rewilding—it is always a two-way process. Plant trees, carry water… .and listen around the campfire to the old stories. And we have learned from some indigenous peoples, for example from North, Central, and South America. Their teachers have come to our land and taught us *how* to get closer to Nature—to go alone on the mountain with no food or shelter, and to ask for personal strength and growth with help from plant and animal powers. Wild nature helps us discover who we really are—our purpose and our creativity.

Dedication

We dedicate this article to the loving memory of Alison Parfitt of the Wildland Research Institute, who worked tirelessly to balance the scientific endeavour with these elements we have called a social movement; together with heartfelt thanks to the co-ordinator of would-be revolutionaries—Rick Minter, editor of *Ecos.*

Notes

1 Glen Feshie is a large private forested estate in the Cairngorms of Scotland and recently changed ownership (bought by a Danish billionaire) with the new owner declaring it a 'rewilding project'—however, new trackways have been bulldozed up the glen, and thus it is not clear how rewilding is thus

defined. Glen Feshie: www.glenfeshie.scot/Glenfeshie/Glenfeshie_Estate_Welcome.html and also at www.rewildingbritain.org.uk/rewilding-network.

2 For example—Coetir Anian (Cambrian Wildwood) employs an educational officer and work in local schools is complemented by programmes for refugees, asylum seekers, drug recovery, and people from disadvantaged city environments; Cae Mabon hosts similar programmes including 'rites of passage' for young people, sweat lodge, vision quest, and the development of poetry, songs, and music connected to the Earth. Healing the separation of mind and body, thinking and intuition, the human spirit and spirit of nature, is a lifelong process of rebalancing. Exposure to wild nature is a beginning, but for many, the process requires a physiological shift—brought by fasting and vision quest, sweat lodge, yogic meditation, trance dance, and in advanced training, the ingestion of psycho-active 'plant medicines' (see Taylor, 2017).

3 The British Association of Nature Conservationists funded research for *Beyond Conservation, A Wildland Strategy*, which advocated one large wildland zone as a Natural Sanctuary for each of Scotland, Wales, and England (Taylor, 2005).

4 Following publication of *Beyond Conservation,* the *Wildland Network* was set up to organise regional seminars and site-visits, with the aim of connecting a large range of different initiatives. This culminated in the setting up of the *Wildland Research Institute* at Leeds University. These initiatives are documented in Taylor (2011). The new group *Rewilding Britain* has continued the networking role.

5 Though dismissed by some commentators as a mass delusion, the occurrences have been painstakingly documented, and openly now admitted for example, by the Forestry Commission in the Forest of Dean. See Minter (2011).

6 In a series of major conferences—with the National Trust at its centennial celebration in 1995, for example, and at Newcastle University in 2000 (Design for the Wild) we sought to move away from 'wilderness' as a useful term for Britain towards 'wildland'—a term that would focus upon *process*—whether small-scale, as for example the reconstitution of meanders over a section of river, or larger, such as in Glen Affric and Ennerdale in the Lake District, where major landowners came together to coordinate a wilder approach to landscape management.

7 For example, in the otherwise engaging story told by the journalist George Monbiot (2013), there is little mention of grassroots projects or the efforts to network with Welsh farming communities with regard to the potential for rewilding the uplands—which Monbiot labelled provocatively as 'sheep-wrecked'.

8 Following Monbiot's popular and inspirational book, a national organisation was set up—Rewilding Britain, with representation of most initiatives on the ground. However, its major initiative in Wales, 'Summit to Sea' in the North Cambrians, encountered difficulties. It has been seen as 'outsiders' imposing rewilding upon a traditional farming community. In West Wales, for example, farmers have taken to putting up placards saying 'conservation yes, rewilding no'. Time is now being invested in more effective community participation and Rewilding Britain has left the project.

9 Trees for Life hosted a regional meeting of the Wildland Network at Findhorn's Universal Hall in September, 2008, on the theme of returning key species to Scotland. The gathering drew land managers and conservationists together for wide-ranging discussions and field visits to Glen Affric and other rewilding projects in Alladale and Carrifran.

10 See: www.shamanictrancedance.global/ and the work of Zelia Pye; and also www.schoolofmovementmedicine.com/ and the work of Susannah and Ya'Acov Darling Khan.

11 Such work began perhaps with Robert Graves' *The White Goddess: An Historical Grammar of Poetic Myth* (Faber and Faber, 1999) in a recovery of ancient runic knowledge and tree-lore. The neo-Druid *Oak Dragon Project* and camps of the *Order of Bards, Ovates and Druids* brought many people into nature and the study of indigenous lore—such as herbalism, vision quest, and power animal dreaming, thus creating a new relationship to wildlife and 'species'. See also: Taylor, P. (1995) Coed Eryri *Reforesting Scotland* 13, Winter 1995; and Taylor P. (1996) Return of the animal spirits, *Reforesting Scotland* 15, Autumn, 1996; and Taylor P. (2017) *The Spirit of Rewilding: Steps toward a Shamanic Ecology.* Ethos.

References

Minter, R. (2011) *Facing Britain's Wild Predators.* Whittles Publishing.
Monbiot, G. (2013) *Feral: Rewilding the Land, Sea and Human Life.* Penguin, UK.

Nenquimo, N. (2020) This is my message to the western world—your civilization is killing life on earth. *The Guardian*, 12 October. Available at: www.theguardian.com/commentisfree/2020/oct/12/west ern-worldyour-civilisation-killing-life-on-earth-indigenous-amazon-planet (accessed April 2022).

Taylor, P. (2005) *Beyond Conservation: A Wildland Strategy*. Earthscan.

Taylor, P. (2008) Rewilding the grazers—obstacles to the 'wild' in wildlife management. *British Wildlife*, 20(5), Special Issue: Naturalistic Grazing and Rewilding in Britain.

Taylor, P. ed. (2011) *Rewilding: ECOS Writing on Wildland and Conservation Values*. Ethos.

Taylor, P. (2013) The road to Salamanca, *Ecos*, 34 (3/4): 21–7.

Taylor, P. (2017) *The Spirit of Rewilding*. Ethos.

Tree, I. (2018) *Wilding: The Return of Nature to a British Farm*. Picador.

United Nations Permanent Forum on Indigenous Issues (2007) *Indigenous Peoples, Indigenous Voices*. United Nations Department of Public Information—DPI-2454—07-27514.

Watson Featherstone A. (1997) The wild heart of the Highlands. *ECOS*, 18(2): 48–61.

Watson Featherstone A. (2004) Rewilding in the North Central Highlands—an update, ECOS, 25(3/4): 4–10.

Links

Cae Mabon: www.caemabon.co.uk

Coetir Anian / Cambrian/Wildwood: www.cambrianwildwood.org and www.coetiranian.org

Knepp Estate: https://knepp.co.uk/home

Moor Trees: www.moortrees.org/ and see also: www.facebook.com/Moor-Barton-Rewilding-Project-1826448290947661/

Rewilding Britain: www.rewildingbritain.org.uk

School of Movement Medicine and the work of Susannah and Ya'Acov Darling Khan. www.schoolofm ovementmedicine.com/

Shamanic Trance Dance: www.shamanictrancedance.global/ and the work of Zelia Pye.

Trees for Life: www.treesforlife.org and https://en.wikipedia.org/wiki/Trees_for_Life_%28Scotland%29

Reading

ECOS. *Journal of the British Association of Nature Conservationists*. www.banc.org.uk

32

WILD DEMOCRACY

Ecodemocracy in rewilding

Helen Kopnina, Simon Leadbeater, and Anja Heister

Introduction: Linking ecodemocracy and rewilding

The concept of 'environmental justice' is conventionally used concerning the distribution of natural resources and burdens (such as pollution) among human groups, and rarely includes justice between species (Schlosberg, 2007; Celermajer et al., 2020). This has raised questions concerning the ability of current anthropocentric (human-centred) democratic systems to respond adequately to pressing environmental challenges, from climate change to biodiversity loss (Lidskog & Elander, 2010; Gray et al., 2020). In particular, there is often little recognition of the intrinsic value of individual nonhuman animals and nature as a collective entity (Kopnina, 2019a, 2019b; Meijer, 2019). The interests of other species are seldom given any, let alone equal, consideration (Dobson, 2010; Ares, 2022). In our view, conservation that does not recognise the intrinsic value of nature is not only inadequate from the perspective of ethics and justice but will also fail to protect vulnerable ecosystems (Washington et al., 2017; Piccolo et al., 2018). Both elements of this claim are, however, contested. For example, Tony Juniper (2013) asserts that decades of arguing for the intrinsic value of nature have clearly failed and that the way forward is to reveal that nature has a much higher instrumental value than previously recognised. We have argued elsewhere that neither the failings in conservation nor in ethics and justice will be remedied by the further instrumentalisation of nature (e.g. Kopnina et al., 2018, 2019a; Taylor et al., 2020).

Anthropocentrism extends to biodiversity conservation (Shoreman-Ouimet and Kopnina, 2015; Taylor et al., 2020) and rewilding policy-making (Merckx, 2016). When human welfare and the instrumental utility of nature solely determines value, rewilding is then perceived as less valuable than cultural landscapes or agricultural lands and, consequently, the justification for ecosystem protection or restoration weakens significantly (Creasy, 2019); extinctions of marine and terrestrial species continue unabated (Wilson, 2003; Kopnina, 2016; Ceballos et al., 2017). The urgent need to replace the anthropocentric with an Earth-centred paradigm necessitates broadening the definition of 'stakeholders'. We posit that any party holding an interest in a habitat or ecosystem, with the potential to affect or be affected by changes in this given environment, comprise stakeholders. They thus include animals collectively (species) and individually, and more broadly nature (Cullinan, 2003). Earth-centred policies in contrast with those underpinned by an inherent anthropocentric predilection would seek to balance

DOI: 10.4324/9781003097822-36

these various stakeholder interests explicitly rather than implicitly favour outcomes privileging human interests.

Rewilding aims are broad and can fall under ecological restoration, recreation of historical landscapes and reintegrating humans and nature (Bekoff, 2015; Merckx, 2016; Gann et al., 2019; Hawkins, this book). To promote rewilding application, human interests and potential benefits to society are integrated into decision-making in projects (for examples see case studies in this book). The critical question we ask in this chapter is: if ecocentrism and non-instrumental animal ethics were integrated into decision-making, what would rewilding look like? To address this question, we discuss some existing forms of eco-representation, or suggestions for possible models, and explore the practical implications of ecodemocracy using ecological integrity and animal welfare perspectives (Shoreman-Ouimet & Kopnina, 2016; Kopnina et al., 2019a, 2019b) building on previous research on the case of the failed rewilding experiment at the Dutch Oostvaardesplassen (OVP; e.g., Kopnina et al., 2019a, 2019b). In this chapter we explore whether aspects of eco-justice, ecodemocracy or earth-jurisprudence, could have averted OVP's ecological collapse and the violation of animal rights and/or welfare.

Definitions

First, however, it is important to clarify terminology. 'Ecodemocracy' is sometimes equated with 'eco-representation', and the terms are often used interchangeably (Eckersley & Gagnon, 2014). It can also be closely associated with 'ecological justice' (Baxter, 2005), and 'Earth jurisprudence', the latter of which can be included within countries' constitutions, statutes, local laws, and other legal provisions (Cullinan, 2003; Gray & Curry, 2020). Generally, ecodemocracy, which is our preferred 'umbrella term' refers to the political processes that can be reached by establishing legal and political levers to protect other species. The terms 'nonhumans' and 'multispecies' are both used to refer to other than human animals, including all vertebrates and invertebrates (Chan, 2011). The term 'nonhumans', however, aggregates 'millions of species by an absence; as if they were missing something' (de Waal, 2016, pp. 27–8). For this reason, we will use the term 'multispecies'.

'Rights' refers to legal and/or the political recognition of interests, often based on the intrinsic value of (either) individuals within species, entire species, or habitats (Garner, 2015). The notion of rights is different from other ways in which the interests of multispecies might be recognised ethically and politically, e.g. those based on a utilitarian approach to ethics. Ecojustice and ecodemocratic concepts do not specifically entail that individual animal interests are taken into consideration. 'Nature rights' refer to 'ecocentric' or ecology-centred (biotic and abiotic) perspectives (Chapron et al., 2019). A centuries-old idea, the concept of natural rights holds that nature and its components are not 'things' or human property but rather, living beings with intrinsic value and an inherent right to exist and flourish. 'Animal ethics' and 'compassionate conservation' (Bekoff, 2015; Wallach et al., 2020) are broadly based on the recognition of individual animal sentience (Griffin, 1976; Safina, 2015) and linked to 'animal rights' or animal welfare (Nibert, 1992). We note that animal rights and animal welfare have very different starting points and implications. The 'animal turn' in political theory reaches beyond some select welfarist notions that treat animals as resources and instead argue vigorously against any type of exploitation of nonhuman animals. 'Interspecies justice' (which is also at times used interchangeably with ecodemocracy) refers to prioritising 'the interests of individuals over their relational position', affirming how the interests of animals (as sentient beings, individuals with moral worth and intrinsic value) 'ought to shape the aims and structure of politics' (Cochrane, 2018: 3–5).

'Ecocide' refers to the phenomenon of anthropogenic-driven ecosystem decline, at a certain scale and degree of seriousness tracked in terms of laws or conventions. Here we use the term ecocide to emphasise the imperative to change anthropocentric decision-making processes. These have created an unholy alliance between climate change, the destruction of natural habitats, and the systemic violence against animals exploited in food systems, combining to result in widespread species declines and extinctions.

'Rewilding' was originally understood as the restoration of natural processes and wilderness areas, based on the three Cs—Cores, Carnivores, Corridors, involving keystone species, including predators, and connectivity between habitats (Soulé & Noss, 1998). Originally envisaged on a continental scale, its subsequent applications to much smaller parcels of land have resulted in ecological and ethical anomalies that bring into question either the use of the term or the validity of the concept. Rewilding had a transitory association with reintroducing 'prehistoric' (e.g. Pleistocene) species (Donlan et al., 2005; Nogués-Bravo et al., 2016). It can also refer to active restoration and ecological self-regulation through the replacement of missing ecological processes, involving active management (Seddon et al., 2014). Alternatively, rewilding can be passive (the 'letting nature be' approach), allowing the natural recovery of ecosystem processes through reduced human management (Gillson et al., 2011). Rewilding has been criticised as being stretched to the point where it risks becoming an all-encompassing blanket solution to environmental problems (Jørgensen, 2015). Carver et al. (2020) have now helpfully melded the ideas from previous debates and experience of practical examples into a series of 'Guiding principles.'

The following discussion will address different forms of eco-representation. These include existing ones—the application of the precautionary principle; the Council of All Beings; Parties for Animals; grassroots organisations—and possible ones, such as proxy representation.

Oostvaardersplassen: The case of fenced rewilding

OVP covers about 56 square kilometers (6,000ha). The grassland area of OVP also includes the adjacent marshland, which is the primary reason OVP was declared a protected area as an important bird habitat. OVP was partially financed by institutions and programmes supporting rewilding and nature conservation (https://rewilding.org/european-experiments-in-rewilding-oostvaardersplassen/) and is managed by the Dutch Forest Service (Staatsbosbeheer). OVP was created on a polder (constructed in 1968), intended to restore prehistoric landscapes principally inspired by the work of Dutch ecologist Frans Vera (2000). This was based on the belief that historically large herbivores 'managed' landscapes. It was also assumed that limited human management would be required, letting nature take its course (Vera, 2009; Lorimer & Driessen, 2014; Barkham, 2018). Konik horses (*Equus ferus*) and Heck cattle (*Bos taurus*) were introduced from the 1980s onwards. The horses were brought from breeding farms in Poland as they resembled extinct tarpans, and cattle from Germany, as analogues for extinct aurochs (Svenning et al., 2016; Kopnina et al., 2019a).

In the 1980s Vera's 'Dutch Serengeti' (The Economist, 2018) appeared to be a success (Barkham, 2018; Carey, 2016), with pioneering plants, the return of bird species such as kingfishers, common egrets, herons, and spoonbills, as well as insects, fish, amphibians, and reptilians, all settling in this 'new wilderness' (Barkham, 2018; Yin, 2019). Soon, however, this perceived wildlife oasis was transformed into a nature-denuded wasteland. The large herbivores exhausted the newly emerging vegetation and began to starve (Shoreman-Ouimet & Kopnina, 2016; Henley, 2018). As grasslands became bare due to overgrazing, flowers, insects, and birds disappeared (Rubenstein & Rubenstein, 2016; Svenning et al., 2016).

This raises the following question. If the project decision-makers had included animals and plants as stakeholders, what would have happened? In raising this question we refer to ecojustice (Baxter, 2005; Washington et al., 2018), ecodemocracy (e.g., Mathews, 1996; Gray et al., 2020), and interspecies justice (Cochrane, 2018, Meijer, 2019). Interspecies justice is related to 'the animal turn' in political theory (Meijer 2019; Garner & O'Sullivan, 2016). As two of the leading animal turn proponents have noted, this reaches beyond animal ethics, our obligations to animals, and what we owe them, and requires statutory 'involvement in the protection of animal interests because the focus of the political theorist is directed not at voluntary personal lifestyles but the state's coercive power' (Garner & O'Sullivan, 2016: 2).

The failure of OVP is not due to one single issue, but rather owing to multiple related criteria. Thousands of animals died of starvation, resulting in considerable public disquiet, because grassland biodiversity was depleted. The animals could not migrate to search for alternative food sources due to the fence around OVP and an inherent lack of connectivity to the outside world. This failure was compounded by mass killing, and the numbers of animals within OVP are now being decided and reduced by human management with a focus on hunting rather than on 'natural processes'.

Paleoecological evidence casts doubt on the idea that wild herbivores alone could control landscape processes; it is much more likely that open grasslands started with land cultivation (for the ongoing debate about open savannahs versus closed-canopy forests see Merckx, 2016; Hodder et al., 2009; Vera, 2009). While the evidence of the 'baseline' landscape remains contested, it is unlikely that thousands of confined grazers could have ever lived 'naturally'.

In the early 2000s, the management of OVP attempted an ethical assessment, based primarily on the work of Dutch authors, Klaver, Keulartz, and van den Belt (2002) and Swart and Keulartz (2011), concerning how to deal with de-domesticating cattle and horses. There was, however, no meaningful engagement with animal ethics or animal rights issues, rather a focus on the animal welfare literature.

OVP herbivore management was also the focus of advice from two international expert committees with the first—the International Commission on Management of the Oostvaardersplassen (ICMO1)—established in 2005. The 2010 report by the second International Commission on Management of the Oostvaardersplassen (ICMO2) concluded that 'long periods of food restriction resulting in large scale, unnecessary suffering and subsequent starvation of animals as a result of living conditions partially created by man is morally not acceptable, and has to be prevented'.[1] To implement this finding, OVP's management tried to emulate the activity of predators—the so-called *predator model*—later morphing into shooting animals on an *early reactive culling* basis as soon as it became apparent that they would not survive the winter. Consequently, between 2017 and 2018, over 89% of OVP herbivores were shot (Barkham, 2018; Kopnina et al., 2019a, 2019b). The resulting public outcry led to calls for closer management of OVP (Kopnina et al., 2019a; Yin, 2019), and in 2019 the Staatsbosbeheer set an upper population limit of 1100 for large grazers, leading to more than 1800 deer being shot, with much of their flesh sold as meat (Kopnina et al., 2019a; Yin, 2019).

We have argued (Kopnina et al., 2019a) that OVP failed because the project ignored the 'golden rules' of rewilding, the three Cs, and the fourth C of Compassion, or consideration for individual animals' interests (Bekoff, 2015, Wallach et al., 2020). Essentially, animals starved as they overgrazed and could not then migrate in search of food. The competition for land for economic development had prevented the Dutch government from creating wildlife corridors. Could eco-representation help prevent these types of problems in the future?

Existing forms of eco-representation

Existing forms of eco-representation, the representation of individual animals and/or collectives (e.g. species, taxa, ecological systems, and processes) include the precautionary principle, Parties for Animals, and grassroots organisations—refer to Table 32.1 for wider explanations.

Aside from the mechanisms discussed above, a few other initiatives have recently emerged, including The Global Alliance for the Rights of Nature (GARN https://therightsofnature.org/); and The Global Ecocentric Network for Implementing Ecodemocracy (GENIE www.ecodemocracy.net). Both organisations tackle legal and political mechanisms of eco-representation.

In 2009, the United Nations General Assembly adopted its first Resolution on Harmony with Nature (www.harmonywithnatureun.org/rightsOfNature/), which supported Earth Jurisprudence (Cullinan, 2003). It is not yet clear whether this Resolution had any practical outcomes, other than itself being seen as progress and offering some possibilities for the conceiving of proxy representation or granting rights (Strang, 2020). The recent case in Ecuador, where mining plans were found to violate the rights of nature, is also widely publicised (Surma, 2021; Greenfield, 2021). We would argue that applying eco-representation to rewilding could lead to very different outcomes for nonhuman animals and their ecosystems than the ones witnessed in OVP.

Discussion: How can ecodemocracy inform rewilding projects?

The UN Harmony with Nature Resolution also discussed Indigenous land rights. If the Indigenous status of other species is to be considered on par with Indigenous (human) rights, OVP's paleoecological history could then become important in determining which species have a 'right' to remain. The flaw here of course is that natural systems are not static. Also favouring certain species based on their (debated) 'native' status, raises several difficult questions, not least how long must a species inhabit a place before being considered 'indigenous', and how to account for historical migrations and introductions. Instead, we would argue that all species, native or otherwise, are deserving of compassion (Wallach et al., 2020). The question is not whether modern analogs of extinct species have a greater right to remain in OVP, but whether any animal, plant, or natural entity is accorded a 'good life' (Pepper, 2019).

It should be acknowledged that challenging questions concerning the application of both the notion of 'compassion' and 'a good life' to entities such as species and ecological systems and also to individual non-sentient life forms still need to be addressed. For example, should the quality of a life be judged by its ending alone? Death by starvation is doubtless unpleasant, but before this point, an animal may have 'lived a good life'. Perhaps the key issue here is the degree of avoidable suffering. Further research and deliberation can address how the fourth C of Compassion relates to the inevitability of death; each life will eventually come to an end, and it is a simple biological fact that in a stable ecosystem, the animal populations' birth rate is balanced against the death rate. A comprehensive discussion of 'avoidable suffering' is outside the scope of this chapter. However, whilst anecdotal, videos have been posted on social media appearing to show young healthy deer at OVP being harassed by marksmen in 4 x 4 vehicles, and wounded animals becoming stressed as the vehicles move in to finish them off, and in some cases driving over dead or dying animals in the process. This does not appear to be the actions of a compassionate organisation intent on trying to avoid animal suffering.

In the case of involving predators in a small territory, Frans Vera has suggested introducing wolves to OVP (The Economist, 2018). However, as 'eco-representatives' we would question whether a 56km² area is sufficiently large to give prey animals any chance of finding refuge from

Table 32.1 The possibilities of democratic representation via multispecies proxies

Form of eco-representation	Definition
Precautionary principle	The precautionary principle typically applies where an activity threatens human interests, such as public health through pollution, or economic prosperity through climate change (Myers, 1993; O'Riordan, 2013). At present, the precautionary principle is not yet extended to animal wellbeing (Cameron & Abouchar, 1991). If extended to nonhuman beings, the precautionary principle might mean that even when certain characteristics purportedly required for representation, such as sentience and consciousness, are not proven, and we are uncertain whether they possess these qualities, nonhumans should be represented nonetheless to avoid unjustified exclusions (Chan, 2011:323).
Parties for Animals	Parties for Animals, present in over a dozen countries, focus on animal protection issues (Morini, 2018; Kopnina, 2019a, 2019b). Similar to Green parties, they have extended their campaigning to broader and interconnected issues of sustainability and environmental justice (Kopnina, 2019a, 2019b). The Dutch Party for Animals, for instance, has supported the establishment of migration corridors rather than lethal population control in OVP, stating that they 'will continue to oppose policies whereby recreation, tourism, and hunting interests take precedence over those of the animals' (PvdD, 2018).
Grassroots organisations and indigenous/local councils	The Council of All Beings (Macy, 1983; Seed et al., 1988) employs rituals and workshops in which participants learn to step aside from their human identity and speak on behalf of nonhuman community members (Macy, 1983; Seed et al., 1988). The Council seeks to further develop an emotionally richer response to environmental problems (Gray and Curry, 2020). Abiotic community members' interests, such as those of rivers and mountains, could be represented within Councils of All Beings. Indigenous perspectives are valuable as the root causes of the planetary emergencies we face lie in the dominant 'western' view of humans as separate and superior to the rest of nature, from whence the instrumentalised understanding of nature's value is derived.
Proxies	Nonhuman stakeholders represented through proxies—local or not—are of paramount importance to the prospect of ecodemocracy (Lundmark, 1998; Baxter, 2005; Dobson, 2010) and involve an advocacy mandate, with 'representatives being appointed for the specific task' (Lidskog & Elander, 2010: 37). Proxies could operate similarly to Parties for Animals, expanding their focus from conditions of domestic animals to wild animals, collectives, and ecosystems. Nonhuman proxies might need to create and maintain continuous affirmative action programmes as a correction for the fact that nonhumans cannot represent themselves in anthropocentric institutions. Some existing organisations and initiatives, such as The Nonhuman Rights Project (www.nonhumanrights.org/who-we-are/) already attempt 'to change the common law status of great apes, elephants, dolphins, and whales from mere "things", which cannot possess any legal right, to "legal persons", who possess such fundamental rights as bodily liberty and bodily integrity'.

predator attack, could they possibly live a 'good life' in such a scenario? Would the resulting spectacle of the hunt attract voyeuristic 'adventure safari' spectators, and more fundamentally should safari tourists be allowed to visit OVP?

Should humans be viewed as natural predators and thus permitted to hunt within OVP? In terms of animal quality of life, how might this compare to modern (intensive) farming practices where animals are confined and effectively tortured within a large-scale factory system? What would happen to the wolves within OVP, for example, how might they disperse and find new territory? Is there sufficient (human) tolerance for peaceful co-existence? Eco-representatives could argue that humans, unlike wolves (Bullock, 2019), are over-represented in The Netherlands (above 17 million), and have a choice not to eat meat. Eco-representatives could note that meat-eating is highly contested—e.g., it is linked to human-induced climate change, which in turn is destroying natural habitats and contributing to species extinction, and that animals in the food chain are exposed to systematic violence.

Vera's proposal to introduce wolves to OVP is highly questionable, particularly given that OVP was originally designated as a wildlife sanctuary rather than some weird Pleistocene fantasy park tourist spectacle (Kopnina et al., 2019a). Eco-representatives could then consider options for megafauna, such as contraception and migration, or finding a true sanctuary for the larger herbivores while leaving OVP to the smaller herbivores, birds, and mesopredators such as foxes.

The implications of OVP's failure for animal stakeholders raise some additional questions, for example: would translocating larger herbivores to more appropriate and sustainable areas cause less suffering than simply shooting them? How far should the political representation of interests and the right to a good life extend? What is the longer-term moral responsibility of the OVP human founders towards the animals they translocated, without any apparent consideration of their quality of life? How would ecodemocracy inform rewilding practices and prevent an OVP-type failure from reoccurring? How can the interests of different stakeholders be balanced—can wolves be introduced or killed in the name of 'multispecies stock management' or to preserve 'healthy ecosystems', particularly in compromised and inadequate areas such as the OVP?

A further challenge is that eco-representation through proxies to represent billions of nonhumans would require significant numbers of human representatives. Representation via proxies (ecologists) may need to be proportionate to the number of individuals within each species currently present in OVP, including mice and insects.

In response to these questions, we suggest a focus on interspecies justice (Cochrane, 2018), for example, by examining whether proxies or 'dedicated animal representatives' should represent animals as a single species, related taxa, or by viewing animals as residents of a specific geographical location. Deliberative democracy can expose hidden conflicts so that 'silenced voices are heard' (O'Neill, 2006, p. 276). While the precautionary principle has been used in OVP (Broekmeyer et al., 2017), it appears that 'precaution' was mostly deployed to avert scandals concerning how Dutch and European funds for nature rewilding were used. OVP rewilding exemplifies the dearth of any non-anthropocentric decision-making. Tragically, even in a progressive and democratic country like The Netherlands, it is due to the parochial interests of just one species that other sentient and intelligent species continue to be treated as 'things' and 'objects' and relegated to targets for hunting and resources for food, touristic attractions, and entertainment to name a few.

Future research needs to explore, whether ecodemocracy, despite some conflicts in implementation, could help challenge and shape attitudes and values to shift the societal and institutional debate away from the preoccupation with human concerns to ways in which both human and multispecies interests can be balanced. The natural requirements for respective species, that is, the amount of territory and resources needed, would be of critical importance in such deliberations (Mathews, 2016).

Eco-representatives could help to navigate tensions and conflicts within 'eco' approaches on the one hand, and the animal rights approach (interspecies justice), on the other. In some future ecocentric democracy proxy representation, either species-specific or focused on individuals or ecosystems, could create a voting majority whose constituents might demand that human interests (e.g., concerning tourism or agriculture) should not dominate decision-making processes. Instead, the magnitude of problems caused by large-scale agricultural-industrial farming-packaging, food waste, catastrophic animal suffering and death—and the high risks for slaughterhouse workers (at present, being exposed to a coronavirus causing Covid-19); pollution, and waste of land (Crist et al., 2017), all need to be fully considered.

An exploration of the relationship between sustainability and the protection of non-human community members provides insights into the role that ethics can play in politics and vice versa (Garner, 2015). Scientists, from ecologists to paleo-botanists, as well as ecocentric and animal ethicists and political theorists, can play a role in considering the best options for areas like OVP. At present, while the Dutch Party for Animals (PvdD) takes an active stand in rewilding questions, its influence is limited in comparison with other (human-focused) parties and interests. Multispecies should not be voiceless in decision-making processes, and our continued disregard for their interests is likely to put us on the wrong side of history. A more progressive step towards affirmative action to represent animals and nature is needed.

Reflection: Limitations and ways forward

Jørgensen's (2015) concern that rewilding is becoming conceptually vague might at least in part be addressed by the use of ecodemocratic approaches to balance conflicting anthropocentric interests. Viewed from an ecodemocracy perspective, it is highly unlikely that OVP would have been approved. Considering the rewilding perspective of Cores, Carnivores, Corridors, and Compassion (Kopnina et al., 2019a, 2019b), and foregrounding animal autonomy that Prior and Ward (2016) believe should be at the core of all rewilding, it is also highly unlikely that OVP would have been approved.

How realistic is ecodemocracy for rewilding and beyond? Progress achieved in the past by civil rights movements, coupled with gradual institutional change, provides hope that a transition towards recognising multispecies rights and justice is a natural step in the moral development of democracies, considering the interconnectedness of human and animal lives in sustainable futures. This would require a collective understanding that we are part of multispecies communities. The well-being and health of animal life and human life are inextricably linked (planetary health, animal health, human health are all interconnected) and, given that we are all already part of multispecies communities (local and global), this ought to be reflected in decision-making processes.

In the case of OVP, political will and the economic means to allow nature corridors and to expand protected areas, as well as human intervention and support for the animals when needed, would be required. Such actions are likely to be seen as political and economic concessions rather than just and fair resolutions both in pragmatic and ethical terms in multispecies representation.

Conclusion

This chapter has delved into the existing mechanisms of eco-representation, such as the precautionary principle, grassroots organisations, Parties for Animals, and other possibilities, including proxy representation. In addressing the failures of OVP, we have established that a multispecies stakeholder mechanism was not considered in the decision-making process. Examination of eco-representation in OVP also exposed a range of complex questions and the need to consider how to balance multiple interests and identify convergences, in multispecies decision-making processes. Ultimately, any project potentially impacting multispecies should first undergo a rigorous ethical review, with the leading tenet of compassionate conservation, 'First, do no harm', being foremost in decision-maker's minds.

Note

1 http://docplayer.net/8001575-Natural-processes-animal-welfare-moral-aspects-and-management-of-the-oostvaardersplassen.html

References

Ares, E. 2022. Animal Welfare (Sentience) bill. House of Commons library. https://researchbriefings.files.parliament.uk/documents/CBP-9423/CBP-9423.pdf

Barkham, P. 2018. Dutch rewilding experiment sparks backlash as thousands of animals starve. *The Guardian*. www.theguardian.com/environment/2018/apr/27/dutch-rewilding-experiment-backfires-as-thousands-of-animals-starve

Baxter, B. 2005. *A Theory of Ecological Justice*. Routledge.

Bekoff, M. 2015. A Rewilding Manifesto: Compassion, Biophilia, and Hope. www.psychologytoday.com/us/blog/animal-emotions/201512/rewilding-manifesto-compassion-biophilia-and-hope

Broekmeyer, M.E.A., Bastmeijer, C.J., and Kamphorst, D.A. 2017. 'Towards an improved implementation of the Birds-and Habitats Directive': An inventory of experiences...(No. 2833). Wageningen Environmental Research.

Bullock, A.M. 2019. Wolves return to the Netherlands after 140 years. BBC. www.bbc.com/news/science-environment-47838162

Cameron, J. and Abouchar, J. 1991. The precautionary principle: a fundamental principle of lawand policy for the protection of the global environment, *Boston College International andComparative Law Review*, 14(1): 1–27.

Carey, J. 2016. Core concept: rewilding. *Proceedings of the National Academy of Sciences*, 113(4): 806–8.

Carver, S., Convery, I., Hawkins, S., Beyers, R., Eagle, A., Kun, Z., Van Maanen, E., Cao, Y., Fisher, M., Edwards, S.R., Nelson, C., Gann, G.D., Shurter, S., Aguilar, K., Andrade, A., Ripple, W.J., Davis, J., Sinclair, A., Bekoff, M., Noss, R., Foreman, D., Pettersson, H., Root-Bernstein, M., Svenning, J-C., Taylor, P., Wynn-Jones, S., Watson Featherstone, A., Fløjgaard, C., Stanley-Price, M., Navarro, L.M., Aykroyd., T., Parfitt, A., and Soulé., M., (2020), Guiding principles for rewilding, *Conservation Biology*, DOI: 10.1111/cobi.13730, 26 February 2021.

Ceballos, G., Ehrlich, P. R., and Dirzo, R. 2017. Biological annihilation via the ongoing sixth mass extinction signaled by vertebrate population losses and declines. *Proceedings of the National Academy of Sciences*, 114(30): E6089–96.

Celermajer, D., Schlosberg, D., Rickards, L., Stewart-Harawira, M., Thaler, M., Tschakert, P., Verlie, B., and Winter, C. 2020. Multispecies justice: theories, challenges, and a research agenda for environmental politics. *Environmental Politics*, 1–22.

Chan, K. 2011. Ethical extensionism under uncertainty of sentience: duties to non-human organisms without drawing a line. *Environmental Values*, 20(3): 323–46.

Chapron, G., Epstein, Y., and López-Bao, J.V. 2019. A rights revolution for nature: Introduction of legal rights for nature could protect natural systems from destruction. *Science,* 363(6434): 1392–3.

Cochrane, A. 2018. *Sentientist Politics: A Theory of Global Interspecies Justice*. Oxford University Press.

Creasy, C. 2019. Contending with New Conservationism. In Kopnina, H. and Washington, H. (eds), *Conservation—Integrating Social and Ecological Justice*. Springer, pp. 33–42.

Crist, E., Mora, C., and Engelman, R. 2017. The interaction of the human population, food production, and biodiversity protection. *Science*, 356(6335): 260–4.

Cullinan, C. 2003. *Wild Law: A Manifesto for Earth Justice*. Green Books.

DeWaal, F. 2016. *The Summer Isles: A Voyage of the Imagination*. Granta Books.

Dobson, A. 2010. Democracy and nature: Speaking and listening. *Political Studies*, 58(4): 752–68.

Donlan, J., Greene, H.W., Berger, J., Bock, C.E., Bock, J.H., Burney, D.A., et al. 2005. Re-wilding North America. *Nature*, 436: 913–14.

Eckersley, R. and Gagnon, J.P. 2014. Representing nature and contemporary democracy. *Democratic Theory*, 1(1): 94.

The Economist. 2018. Starving the beasts: A Dutch park that mimics nature angers, animal–rights activists, www.economist.com/europe/2018/06/07/a-dutch-park-that-mimics-nature-angers-animal-rights-activists

Gann, G.D., McDonald, T., Walder, B., Aronson, J., Nelson, C.R., Jonson, J., Hallett, J.G., Eisenberg, C., Guariguata, M.R., Liu, J. and Hua, F., 2019. International principles and standards for the practice of ecological restoration. *Restoration Ecology*, 27(S1): S1–S46.

Garner, R. 2015. Environmental politics, animal rights, and ecological justice. In H. Kopnina and E. Shoreman-Ouimet (eds), *Sustainability: Key issues*. Routledge Earthscan, pp. 331–46.

Garner, R. and O'Sullivan, S. (eds). 2016. *The Political Turn in Animal Ethics*. Rowman & Littlefield.

Gillson, L., Ladle, R.J., and Araújo, M.B. 2011. Baselines, patterns, and process. In R.J. Ladle and R.J. Whittaker (eds), *Conservation Biogeography*, Oxford: Wiley-Blackwell, pp. 31–44.

Gray, J. and Curry, P. 2020. 'Representation for nature': Ecodemocratic decision-making as a practical means of integrating ecological and social justice. In H. Kopnina and H. Washington (eds), *Conservation: Integrating Social and Ecological Justice*. Springer, pp. 155–66.

Gray, J., Wienhues, A., Kopnina, H., and DeMoss, J. 2020. Ecodemocracy: Operationalizing ecocentrism through political representation for non-humans. *Ecological Citizen*, 3(20): 166–77.

Greenfield, P. 2021. Plans to mine Ecuador forest violate rights of nature, court rules, *The Guardian*. www.theguardian.com/environment/2021/dec/02/plan-to-mine-in-ecuador-forest-violate-rights-of-nature-court-rules-aoe

Griffin, D.R. 1976. *The Question of Animal Awareness: Evolutionary Continuity of Mental Experience*. Rockefeller University Press.

Henley, J. 2018. About 1,000 deer to be culled at controversial Dutch rewilding park. *The Guardian*. www.theguardian.com/environment/2018/sep/20/about-1000-deer-to-be-culled-at-controversial-dutch-rewilding-park

Hodder, K.H., Buckland, P.C., Kirby, K.K., and Bullock, J.M., 2009. Can the mid-Holocene provide suitable models for rewilding the landscape in Britain? *British Wildlife*, 20(5): 4–15.

Jørgensen, D. 2015. Rethinking rewilding. *Geoforum*, 65: 482–8.

Juniper, T., 2013. *What Has Nature Ever Done for Us?: How Money Really Does Grow on Trees*. Profile Books.

Kopnina, H. 2016. Half the earth for people (or more)? Addressing ethical questions in conservation. *Biological Conservation*, 203: 176–85.

Kopnina, H., Taylor, B., Washington, H., and Piccolo, J. 2018. Anthropocentrism: more than just a misunderstood problem. *Journal of Agricultural and Environmental Ethics,* 31(1): 109–27.

Kopnina, H. 2019a. Party for animals: Introducing students to the democratic representation of non-humans. *Society and Animals*: 1–21.

Kopnina, H. 2019b. If animals could talk: Ecological citizenship and the Dutch Party for Animals. *Animal Studies Journal*, 8(1): 158–89.

Kopnina, H., Leadbeater, S., and Cryer, P. 2019a. The golden rules of rewilding: Examining the case of Oostvaardersplassen. *ECOS,* 40(6). www.ecos.org.uk/ecos-406-the-golden-rules-of-rewilding-examining-the-case-of-oostvaardersplassen/

Kopnina, H., Leadbeater, S., and Cryer, P. 2019b. Learning to rewild: Examining the failed case of the Dutch 'New Wilderness' Oostvaardersplassen. *The International Journal of Wilderness*, 25(3). https://ijw.org/learning-to-rewild/

Klaver, I., Keulartz, J., and van den Belt, H. 2002. Born to be wild. *Environmental Ethics*, 24(1): 3–21.

Lidskog, R. and Elander, I. 2010. Addressing climate change democratically. Multi-level governance, transnational networks, and governmental structures. *Sustainable Development*, 18(1): 32–41.

Lorimer, J. and Driessen, C. 2014. Wild experiments at the Oostvaardersplassen: Rethinking environ-mentalism in the Anthropocene. *Transactions of the Institute of British Geographers, 39*(2): 169–81.

Lundmark, C. 1998. *Eco-democracy: A Green Challenge to Democratic Theory and Practice.* Umeå Universitet, Doctoral dissertation.

Macy, J. 1983. *Despair and Personal Power in the Nuclear Age.* New Society Publishers.

Mathews, F. (ed.). 1996. *Ecology and Democracy.* Frank Cass.

Mathews, F. 2016. From biodiversity-based conservation to an ethic of bio-proportionality. *Biological Conservation,* 200: 140–8.

Meijer, E. 2019. *When Animals Speak: Toward an Interspecies Democracy.* New York University Press.

Merckx, T. 2016. Rewilding: an enrichment to Flemish nature conservation. *Natuur focus,* 15(1): 28–33.

Morini, M. 2018. 'Animals first!' The rise of animal advocacy parties in the EU: a new party family. *Contemporary Politics, 24*(4): 418–35.

Myers, N. 1993. Biodiversity and the precautionary principle. *Ambio,* 74–9.

Nibert, D. 1992. *Animal Rights/Human Rights: Entanglements of Oppression and Liberation,* Rowman and Littlefield.

Nogués-Bravo, D., Simberloff, D., Rahbek, C., and Saunders, N.J. 2016. Rewilding is the new Pandora's box in conservation. *Current Biology,* 26: R83–R101.

O'Neill, J. 2006. Who speaks for nature? In Y. Haila and C. Dyke (eds), *How Nature Speaks: The Dynamics of the Human Ecological Condition.* Duke University Press Books, pp. 261–78.

O'Riordan, T. 2013. *Interpreting the Precautionary Principle.* Routledge.

Pepper, A. (2019) adapting to climate change: what we owe to other animals. *Journal of Applied Philosophy,* 36(4): 592–607.

Piccolo, J., Washington, H., Kopnina, H., and Taylor, B. 2018. Why conservation scientists should re-embrace their ecocentric roots. *Conservation Biology, 32*(4): 959–61.

Prior, J. and Ward, K.J., 2016. Rethinking rewilding: A response to Jørgensen. *Geoforum, 69*: 132–5.

PvdD [Partij voor de Dieren]2018. Oostvaardersplassen: dieren bescherm je niet door ze af te schieten. www.partijvoordedieren.nl/nieuws/oostvaardersplassen-dieren-bescherm-je-niet-door-ze-af-te-schieten

Rubenstein, D.R. and Rubenstein, D.I. 2016. From Pleistocene to trophic rewilding: A wolf in sheep's clothing. *Proceedings of the National Academy of Sciences, 113*(1): E1–E1.

Safina, C. 2015. *Beyond Words: What Animals Think and Feel.* London: Profile Books Ltd.

Schlosberg, D. 2007. *Defining Environmental Justice: Theories, Movements, and Nature.* Oxford: Oxford University Press.

Seddon, P.J., Griffiths, C.J., Soorae, P.S., and Armstrong, D.P. 2014. Reversing defaunation: restoring species in a changing world. *Science,* 345: 406–12.

Seed, J., Macy, J., Fleming, P., and Naess, A. 1988. *Thinking Like a Mountain: Towards a Council of All Beings.* Heretic Books.

Shoreman-Ouimet, E., and Kopnina, H. 2015. Reconciling ecological and social justice to promote bio-diversity conservation. *Biological Conservation,* 184: 320–6.

Shoreman-Ouimet, E., and Kopnina, H. 2016. *Culture and Conservation: Beyond Anthropocentrism.* Routledge Earthscan.

Soule, M.E. and Noss, R.F. (1998) 'Rewilding and biodiversity: complementary goals for continental conservation', *Wild Earth,* 8(3): 18–28.

Strang, V. 2020. The rights of the river: Water, culture, and ecological justice. In H. Kopnina and H. Washington (eds), *Conservation: Integrating Social and Ecological Justice.* Springer, pp. 105–21.

Surma, K. 2021. Ecuador's high court affirms constitutional protections for the rights of nature in a land-mark decision. https://insideclimatenews.org/news/03122021/ecuador-rights-of-nature/?fbclid=IwAR10DypxG0VUnqsJlSfTIHmkUYw0L-ZRPTy-d0ShALmBa811LAHJOsiZ8_8

Svenning, J.C., Pedersen, P.B., Donlan, et al., 2016. Science for a wilder Anthropocene: Synthesis and future directions for trophic rewilding research. *Proceedings of the National Academy of Sciences,* 113(4): 898–906.

Swart, J.A. and Keulartz, J. 2011. Wild animals in our backyard. A contextual approach to the intrinsic value of animals. *Acta Biotheoretica,* 59: 185–200.

Taylor, B., Chapron, G., Kopnina, H., Orlikowska, E., et al., 2020. The need for ecocentrism in biodiver-sity conservation. *Conservation Biology,* 34(5): 1089–96. https://conbio.onlinelibrary.wiley.com/doi/epdf/10.1111/cobi.13541

Vera, F. (2000) *Grazing Ecology and Forest History.* Oxford: CABI.

Vera, F.W. 2009. Large-scale nature development – The Oostvaardersplassen. *British Wildlife*, 20(5): 28.

Wallach, A., Batavia, C., Bekoff, M. et al. 2020. Compassionate conservation raises a debate on personhood. *Conservation Biology*, 34(5): 1097–1106. https://conbio.onlinelibrary.wiley.com/doi/abs/10.1111/cobi.13494

Washington, H., Taylor, B., Kopnina, H. Cryer, P., and Piccolo, J. 2017. Why ecocentrism is the key pathway to sustainability. *Ecological Citizen*, 1(1): 32–41.

Washington, H., Chapron, G. Kopnina, H., Curry, P., Gray, J., and Piccolo, J. 2018. Foregrounding ecojustice in conservation. *Biological Conservation*, 228: 367–74.

Wilson, E.O. 2003. *The Future of Life*. Vintage Books.

Yin, S. 2019. The Netherlands' grand rewilding experiment, gone haywire. *PBS*. https://whyy.org/segments/the-netherlands-grand-rewilding-experiment-gone-haywire/

33

REWILDING AND THE ETHICS OF PLACE

Martin Drenthen

What is rewilding?

Rewilding is an increasingly popular strategy in landscape management and nature conserva-tion. In 1989, Michael Soulé and Reed Noss introduced the concept of 'rewilding' as a term for the scientific argument for the restoration of great wilderness based on the regulatory role of large predators. According to Soulé and Noss, contemporary rewilding is characterised by three independent characteristics: in short: Cores, Corridors, and Carnivores (Soulé & Noss, 1998: 19). Other definitions have since emphasised other aspects, for example that rewilding is a forward-looking approach (Hughes et al., 2011; Carver, 2012), that rewilding focuses on restoring natural processes and is characterised by an 'experimental' approach (Lorimer & Driessen, 2014), or that rewilding involves a hands-off approach to ecological management. Although Soulé and Noss's scientific use is still dominant, over time, the term rewilding itself has gained more and more meanings, that stretch the breath of the concept even further, leading environmental historian Dolly Jørgensen to the conclusion that the term rewilding has become a fundamentally asocial and ahistorical 'plastic' word without specific content (Jørgensen, 2015). For example, certain anarcho-primitivist groups see rewilding not just as a strategy in landscape management, but also as an antidote against an excess of civilisation (Urban Scout, 2008).

The growing popularity of rewilding is part of a broader cultural trend to defend the value of wild nature against an overly humanised world. However, even though rewilding is popular among many people, concrete rewilding projects often give rise to controversies, especially when they are situated in cultural landscapes. Those opposing rewilding often stress that rewilding in essence is the attempt to rid the landscape of humans. According to Dolly Jørgensen, 'taken as a whole, rewilding discourse seeks to erase human history and involvement in the land and flora and fauna' (Jørgensen, 2015: 482). And, indeed, it seems that some rewilders think of rewilding as the effort of freeing nature from human interference, and create 'new wildernesses'.

An element common to all traditional definitions of wilderness is the absence of human traces on the land. For that reason, it does not come as a surprise that for those who identify with the landscapes that have resulted from centuries old human influences, rewilding is seen as nothing less than a threat, not just to the landscapes they cherish, but also to their identity that is based upon these places.

DOI: 10.4324/9781003097822-37

Rewilding in cultural heritage landscapes gives rise to intense conflicts between different views of landscapes and self that cannot be easily reconciled. Focusing on the strictly scientific meanings of rewilding does not help in understanding or addressing these conflicts. Andrea Gammon (2018) argues that in order to understand 'the wider interest in rewilding as an emerging environmental phenomenon', we should acknowledge the breath of meanings of rewilding and treat the term as a cluster concept.

In this chapter I argue that we should take the cultural and moral dimension of rewilding more seriously. By making explicit the cultural and moral dimension of rewilding as a human endeavour, rewilding advocates can even strengthen the moral base of the nature conservation movement that sets out to correct an overly anthropocentric perspective on landscape. But in order to do that, it is helpful to pay attention to the ethical dimension of place. The way people feel, think, and interact with specific places cannot be reduced to purely geometric relationships. Therefore, it is important to distinguish between abstract 'space' and personally experienced 'place'. Unlike 'space', 'place' is much more than just a physical location and can be described as a location filled with meanings and human experiences. Acknowledgment of the significance of place will not necessarily help solve the conflicts about rewilding, but at least it can help understand what is at stake in rewilding.

Wilderness philosophy

Wilderness is commonly defined as 'a tract or region uncultivated and uninhabited by human beings' and 'an area essentially undisturbed by human activity' (Merriam Webster dictionary). The famous 1964 Wilderness Act, that created the legal definition of wilderness in the United States, and protected 9.1 million acres (37,000km²) of federal land, defines wilderness: 'in contrast with those areas where man and his works dominate the landscape, […] as an area where the earth and its community of life are untrammelled by man, where man himself is a visitor who does not remain'. The idea of wilderness has played a key role in American environmental ethics.[1]

Early conservationists such as Henry-David Thoreau valued wilderness not merely as a source of enjoyment, but also because it can teach people something profound—either through its astonishing beauty or by putting human lives in perspective. For Thoreau wild nature also had a moral and a spiritual meaning: good and wise nature stood against the contrived and hypocritical human society. The study of nature could make you a better person; it gave insight into what mattered and what was trivial. Thoreau's famous quote from his essay *Walking*, 'In wildness is the preservation of the world' proclaims that wildness is the highest ethical ideal. In this view, nature represents a moral order to which people and society should conform.

However, today this 'transcendental' aspect of early wilderness philosophy has since all but faded from view. Most nature conservationists today no longer claim that 'nature' refers to a transcendental meaningful moral order with which people should attune their lives, or that nature as such necessitates a fundamental reflection on the meaning of one's own existence (Van de Gronden, 2015). Instead, they are much more hands-on and focus on more pragmatic reasons to protect plant and animal species and natural areas against the detrimental effects of urbanisation, infrastructure, industrialisation, and agricultural intensification. Today, the protection of wild nature is typically justified by referring to the need to protect biodiversity and safeguard the many ecosystem services that nature provides. In recent years, there is also increasing attention to the beneficial influence of nature on personal well-being. Besides these utilitarian pleas, conservationists also call upon the intrinsic value of nature, that is: the value of nature in itself, independent of any value nature has for humans. Both of these forms of argument tend to abstract from the profound *significance* wild nature can have (James, 2016).

In contrast, in today's popular culture one can still see the deep fascination with the profound meaning of the wild that was so typical for early wilderness philosophers. For example, many popular books and films about wilderness, books like Robert Macfarlane's *The Wild Places* (2007), Cheryl Strayed's *Wild* (2012), and non-fiction books such as Jon Krakauer's book *Into the Wild* (1996), or Richard Louv's *The Last Child in the Woods,* many nature documentaries, but also movies like *Touching the Void* (2003), *Grizzly Man* (2005), *Into the Wild* (2007), *Wild* (2014), *A Walk in the Woods* (2015), and *The Revenant* (2015) have shared themes, exploring ways in which confrontations with wild nature can become significant events that place one's life in perspective (Drenthen, 2009a).

Overall, the idea that untouched wilderness is the kind of nature that is most valuable and most worthwhile protecting has remained influential, even though it is usually not explicitly acknowledged in the debate among conservationists. In his influential paper on the notion of wilderness in environmental thought, William Cronon (1996) shows that wilderness thinking is actually a relic of 18th-century European Romanticism. Cronon argues that the suggestion that the most pristine wilderness is also most worth protecting, in fact perpetuates a form of human-nature dualism that is not very helpful in thinking about a sustainable relationship between people and their non-human environment. The idealisation of wilderness may seem appealing, but in practice it makes us less concerned about our impact on those parts of the world in which we humans reside. According to Cronon, in reality we value supposed wildernesses not so much because they do not contain human traces, but rather because they possess a quality he calls 'wildness': a certain independence of nature that has not been tamed, that escapes human control. In other words, Cronon clearly distinguishes between wilderness and wildness. While criticising the concept of wilderness and the idea that nature is most valuable if untrammelled or untouched by humans—he draws attention to the notion of *wildness* as autonomy (or, to use another term, self-willedness) as something that is valued by most humans. That quality of wildness, however, does not only reside in pristine, uninhabited areas but can also be found in our backyards.

> When we visit a wilderness area, we find ourselves surrounded by plants and animals and physical landscapes whose otherness compels our attention. In forcing us to acknowledge that they are not of our making, that they have little or no need of our continued existence, they recall for us a creation far greater than our own. In the wilderness, we need no reminder that a tree has its own reasons for being, quite apart from us. The same is less true in the gardens we plant and tend ourselves: there it is far easier to forget the otherness of the tree. […] The tree in the garden is in reality no less other, no less worthy of our wonder and respect, than the tree in an ancient forest that has never known an ax or a saw—even though the tree in the forest reflects a more intricate web of ecological relationships. […] Both trees stand apart from us; both share our common world. The special power of the tree in the wilderness is to remind us of this fact. It can teach us to recognize the wildness we did not see in the tree we planted in our own backyard.
>
> *(pp. 23–4)*

Moreover, Cronon argues that we should not forget that the so-called wildernesses, just like our backyards, are part of a world that is and has been influenced by people[2] and for which people must bear some responsibility.

Whereas the classic ideal of wilderness presupposes the absence of humans, the concept of wildness lends itself to acknowledging the fact that wildness can exist in cultural landscapes. Conversely, whereas wilderness thinking could lead us to think that rewilding would necessarily involve the destruction of human traces, a focus on wildness opens up a more inclusive idea of rewilding (Ward, 2019).

The critique on the wilderness ideal by Cronon and others has led many rewilders to reframe rewilding as bringing back wildness rather than restoring wilderness. Yet, the classic wilderness idea is still influential in the way rewilding projects are being discussed today, ironically enough especially by those who oppose rewilding. The reason for that is that many of those who oppose rewilding projects start from a radically different perspective on the value of landscapes and on the role of humans therein.

Ethics of place, identity, and the value of historic landscapes

What rewilders often fail to appreciate is that landscapes are not only valued for their ecological significance but also because they contribute to the well-being and identity of their human inhabitants. Some local residents react angrily when rewilders want to 'give farmland back to nature' or reintroduce beavers in places where these animals were wiped out long ago. Underlying the disputes about rewilding are conflicts of interest (economic and political) between farmers, conservationists, water managers, and other stakeholders, and much is invested in trying to resolve these conflicts by seeking smart compromises and win–win situations. However, underlying these conflicts are also differences that have to do with clashing notions of what is worthy of protection. These kinds of differences in meaning are not easily reconciled.

Many opponents of rewilding start from an entirely different vantage point than wilderness philosophy. In environmental philosophy this alternative approach is known as 'ethics of place'. At the core of ethics of place is the idea that human life in an environment can in principle be characterised as 'storied residence'[3] and is characterised by a deeply contextualised 'discourse about places'.

'Place' is a central concept in social geography. In the 1970s, Yi-Fu Tuan, inspired by phenomenology, claimed that social geography must do justice to the way people perceive and react to their environment (Tuan, 1977). According to Tuan, geography as a science must therefore distinguish between abstract 'space' and personally experienced 'place'. Being in place cannot be reduced to purely geometric relationships, because people's spatial behaviour is a reflection of their values, feelings, and desires: 'Space is abstract. It lacks content; it is broad, open, and empty, inviting the imagination to fill it with substance and illusion; it is possibility and beckoning future. Place, by contrast, is the past and the present, stability and achievement.'[4]

O'Neill et al. argue that people make sense of their lives because they can place it within a larger narrative of what came before and what will come after. And one of the reasons environments matter to people is that landscapes provide just such a larger context from which they can make sense of themselves (O'Neill et al., 2008).

According to Jim Cheney (1989) the task of environmental ethicists is to deconstruct mystical images of nature and instead articulate moral approaches that emerge from the lived reality of the place in question. Cheney criticises a simplistic and ahistorical concept of the value of nature and argues for a pluralistic postmodern ethics of place, the goal of which would not be to establish ethical principles in a speculative way, but rather to develop moral narratives that allow for a normative understanding of how specific places provide a context for concrete, physical people who find themselves in those places. Whereas a universalist would emphasise that a specific place (e.g., the forest next to my home) is a particular instance of something more general (e.g., oak-beech woodland), a place-based perspective focuses on the unique and contingent character of each place and on our connection to it. I feel connected to the forest near my home, because it is part of my world and plays a role in my life—as the place that I visit several times a week to encounter fellow beings and experience that my life is part of a larger whole, in which I, my community and other lifeforms have become intertwined in a way that is unique to this specific place. This forest is a unique place with a particular history, that is different from other places that might look similar. I cherish this particular place, not just any forest, because it is the deeply contingent whole that provides my life with a wider context that helps me understand who I am and where I am.

In a similar vein, Mick Smith in his book *An Ethics of Place* (2001) argues that 'place' provides a foundation upon which a person can construct a local identity. Drawing on a complex and multifaceted concept of place, Smith establishes an environmental ethic that prioritises the moral considerations of the individual subject on the ground over abstract top-down management. Ethics of place, then, is primarily a concept for resistance against the abstract ethical reasoning of modern environmental ethics that understands places merely as particular instances of a universal 'nature', and loses sight of people's connectedness to specific places. Based on this idea, Smith explores the phenomenon of NIMBY (Not-In-My-In-Backyard) activism. Politicians often respond negatively to NIMBY activists because the nature of their engagement is perceived as inappropriate for political debate. The dominant political discourse within liberal democracy requires citizens to formulate their political position, either with reference to the 'common good' or in the form of personal preferences that can be weighed against other preferences. Many NIMBY activists, however, take a radically different approach. They challenge the idea of the comparability of places in general. According to the NIMBY activists' worldview, the world consists of different, unique places, each with their own character. From this perspective, it is not possible to make general statements about places—it requires a place-based recognition of the relevant local conditions. Such place-based rationality is difficult to incorporate into current political processes, since in liberal democracies, site-specific arguments appear to be either 'irrational'—because they cannot be generalised—or purely personal and subjective, and thus politically non-binding preferences.

In contrast, an ethics of place explores how the world can appear as a moral home for embodied, world-open beings in the context of a place that is always specific. An ethics of place seeks to understand what it means to live in and feel connected to a particular place.

Place-based ethics emphasises that a morally meaningful relationship to the world presupposes that mere space is constituted as a structure of meaning. The starting point of an ethics of place is thus the observation that people who engage with their environment have *appropriated* the world culturally and materially and have incorporated that environment into a symbolic order. An ethics of place assumes that a moral engagement with one's environment presupposes an understanding of the world as an ethos (ηθος), that is, as a morally meaningful home, a place of living and being, a meaningful, significant place in which we can live as moral beings. People appropriate the world through interpretation, but necessarily always in a specific fashion, that is: reflecting certain assumptions, leaving aside other possibilities. The prime goal of an ethics of pace is to explicate, examine, and reflect upon specific interpretations of place, including possible criticisms of any particular interpretations of place.

An ethics of place emphasises the importance of those types of places that can appear *as a meaningful world* in the first place. With globalisation, the 'inspired landscapes of generations of farmers, monks and landowners' are becoming an 'anonymous by-product of the global economy' with 'local characteristics' increasingly giving way to 'interchangeable stereotypes'

(Pedroli et al., 2007). Out of concern for this development, the *European Landscape Convention* was signed in 2000. According to that convention, a landscape is 'an area, as perceived by people, whose character is the result of the action and interaction of natural and human factors', which is 'a fundamental part of Europe's natural and cultural heritage' and 'contributes to human well-being and the consolidation of European identity' (Council of Europe, 2000) and should therefore be protected. However, recognising that each place has a non-interchangeable meaning and identity, does not necessarily mean that we should leave everything as it is.

An adequate understanding of a place should not simply reinforce existing place-based identities, but stimulate a critical self-examination in light of a critical interpretation of the landscape. Place narratives cannot be rewritten at will, but they should somehow be 'grounded' in an understanding of the 'objectivity' of a place: its history, soil composition, hydrology, habitat, food, climate, etc. Some rewilders provide the basis for a critical re-interpretation of a landscape by focussing the attention on certain essential ecological processes that have been suppressed in recent times and that any genuine interpretation of a place should acknowledge. Rewilding implies a radical non-anthropocentric normative reinterpretation of place and human history that calls for a critical re-examination of the cultural identities that are based on that history (Drenthen, 2018).

The loss of the old cultural landscapes is not primarily due to the rise of rewilding. More common causes are the intensification of agriculture, expansion of infrastructure, and urbanisation. Yet many critics of rewilding seem to blame rewilders in particular for the perceived loss of traditional landscapes, perhaps because rewilding is seen as a deliberate attempt to change existing landscapes, whereas other forms of landscape change are seen as mere inevitable side-effects of modernisation. With the loss of old cultural landscapes, the assemblages of species that are typical of traditional cultural landscapes tend to diminish, often causing a deep sense of loss among its inhabitants. In the Dutch language, this feeling has been dubbed 'landscape ache'.[5] Moreover, besides specific local species, something else gets lost: an—often centuries-old—way of living with and in the landscape. Several writers have given voice to the sense of loss when an old cultural landscape in which a story could be told about every place disappears, for example James Rebanks' *A Shepherd's Life* (2015). Rapid changes in old cultural landscapes can lead to feelings of loss and disorientation amongst residents (Buijs, 2009).

Rewilders and traditional landscape defenders do not simply emphasise different values. Whereas defenders of cultural landscapes see the landscape as a meaningful reflection of human history and value landscape features that reflect sense of place, rewilders value non-anthropocentric values such as biodiversity, ecological fidelity, and wildness (Higgs, 2003). However, the conflict about the landscape cannot be reduced to a disagreement about which functions and landscape elements are valuable or not (wetlands or farmers' fields, wolves or sheep), but ultimately also entails a clash between different narratives about the landscape, and about humanity's place within nature. It is important to not only look at the various 'values' in the landscape, but also pay attention to the role of history and stories in people's relationships with their environment (Deliège, 2011; Deliège & Drenthen, 2014; Drenthen, 2018).

The narrative significance of wildness

During protests against large-scale rewilding (or 'nature development'[6]) along the rivers in the Netherlands in the 1980s, writer and activist Willem van Toorn stressed that old landscapes remind us 'that there is a past [in which] people lived who had to deal with the world just like us, who had to protect themselves from nature and at the same time use its resources' (Van Toorn, 1988; also see Drenthen, 2009b). Staying in touch with this past was important 'because we owe our existence, our identity, our representation of the world to the past, and because we can think about the future only by drawing on past experiences'. By getting rid of all 'legible' human traces from the landscape, rewilding would fail to appreciate the role these human traces can have for the meaning of landscapes.

As already indicated, O'Neill, Holland, and Light (2008) argue that environments matter to people because they embody a larger context within which people can make sense of their lives. This is clearest in the cultural landscapes that 'specifically embody the lives of individuals and communities' (p. 198). But natural landscapes also have a narrative dimension: they not only help put into perspective human history by contextualising it as part of an older history 'that extends to before humans appeared and will continue after the human species has disappeared again' (p. 164), but they also provide a context that helps us understand ourselves and our historic role in the landscape.

> Unintentional natural processes provide part of the context in which intentional human activities take place and through which we understand their value.
>
> *(p. 198)*

This latter narrative dimension shows that rewilding projects, if done right, might actually help to provide a deeper ground for a sense of place. By bringing back the wildness of the natural processes that humans have suppressed for so long, rewilding can restore a sense of historical and narrative continuity, not so much by restoring an original, primaeval, untrammelled version of nature, but rather by giving a glimpse of what it must have meant for the first humans to inhabit (and cultivate) a landscape. Thus, bringing back wildness, rather than wilderness, can help enrich the sense of place that is so important for people to feel attached to a particular place. In this sense, rewilding can be a form of place-making (Gammon, 2019).

Some rewilders seem to be aware of this. For example, Wouter Helmer, founder and former director of Ark Nature Foundation, the most prominent Dutch rewilding organisation (and also co-founder of Rewilding Europe), argues that rewilding should be based on the 'genius of place', respecting the existing geomorphology, local characteristics *and* local history (Helmer et al., 1995). In this approach, rewilding in historically saturated landscapes does not have to mean that the cultural-historical elements are all erased. The Ark Nature Foundation acknowledges on its website that rewilding ('nature development') also has a cultural context.

> It is a new phase in the development of a landscape. And just like in a book, a new chapter is easier to read after the previous ones have been understood. That is why ARK also tries to keep the cultural history of an area visible as much as possible. Historical elements are given meaning again in the present wherever possible.

Examples of this interweaving of nature and culture in nature development are plentiful. In the Gelderse Poort, a rewilding area in Netherlands along the banks of the river Rhine, some historical elements such as cold war bunkers or the remains of old brick factories are expressly preserved and protected from decay, in order to keep the cultural history of a landscape visible. In the same region, in the Millingerwaard, half fossilised 8,500-year-old hardwood trees that were dredged up from the river, were used to make a Stonehenge-like work of art that quite literally recalls a past that preceded human habitation. Another fascinating example is to be found nearby, in the Geitenwaard, where the decision was taken to reinforce old Hawthorn and Blackthorn hedges (whose history dates back to the Iron Age and Roman times), not in order to preserve them forever, but to allow them to slowly fade with the ravages of time. Where the choice seemed at first to be between letting history disappear unnoticed or stubbornly preserving and freezing it, a third option was found. On the one hand, it was recognised that the historic wood banks had lost their function in the contemporary landscape and preserving it would not fit in with the new vision on rewilding the floodplain, on the other hand it was recognised that a meaningful historical relic could contribute to the story of the place. And for that story, the literal preservation of that planting structure is not a necessary condition.

Sometimes landscape artists can help to keep various layers of meaning visible within the context of rewilding. For example, by 'translating' meaningful cultural-historical structures into a different, more natural material so that they continue to tell their story as part of the new nature: a row of trees can mark a vanished wall, a simple wooden swing the spot where a school building once stood. Landscape art can also provide a problematic past history—of a former toxic dump, battlefield or other disaster area—with critical commentary and thus save it from oblivion (Drenthen, 2015). In all these cases, the line between nature and culture blurs and it becomes clear that the so-called 'new wilderness' is not outside our culture, but rather a new phase in our cultural history. Through a broad, people-inclusive interpretation of rewilding, the place attachment of the local community can even be strengthened. In this way, many contradictions in the landscape debate can be softened.

New wild places as cultural landscapes

There is an element in the current fascination with wildness that emphatically wants to leave behind human history, because human 'civilisation' would necessarily entail a suppression of nature. Some radical rewilding groups, adherents of so-called 'primitivism', seek not just the rewilding of nature, but also a radical social upheaval and a rewilding of humans.[7] Numerous studies have shown that it is mainly city dwellers who are attracted to 'wilderness'; in contrast, rural dwellers usually prefer more orderly, Arcadian landscapes (Van den Berg et al., 2006). There is a clear connection between urban culture and human rewilding, which seems to stem from a desire to break with history, and to be able to break free from modern society with all its conventions and limitations. Some critics point out that such a romantic flirtation with wild nature can only exist because others are farming the food for these rewilders. I think it is important nevertheless to recognise that the desire for wildness also contains an aspect that could give a sharper cultural and social profile to the nature conservation movement, as already argued.

The cultural landscape is like a layered text, a palimpsest, in which the recent human traces are written over older ones that precede human activity. Rewilding is primarily about setting free the natural processes that have formed the existing landscape, not with the aim of return to primeval or untouched nature, but rather to show the wildness lying dormant underneath even the most cultivated landscapes. And if our traditional sense of place does not feel at home with this, then perhaps it is time to critically examine that sense of place and to deepen it (ecologically and culturally-historically) into a 'Sense of place 2.0' (Drenthen, 2009b, 2018).

For many people, the current fascination with wildness is related to a critical understanding of the limitations of the all-too-human. Wild places are not so much places without people, but rather places where humans are not centre stage. Wild places can therefore serve as a mirror that allows us to look at ourselves and society from the perspective of non-human nature. Such a look in the mirror can teach people to put into perspective modern civilisation:

> Rewilding expresses a new appreciation of wild nature. It represents a growing
> movement in Europe of people seeking a counterweight to our increasingly regulated

lives, society and landscapes. It signifies a desire to rediscover the values of freedom, spontaneity, resilience and wonder embodied in Europe's natural heritage and to revitalise conservation as a positive, future-oriented force.

(Jepson & Schepers, 2016)

Rewilding can create places where we are confronted once again with the vital power of natural processes and thus learn to put the human, all-too-human world into perspective. Helmer considers these new wild places as 'insane oases' (Helmer, 1996): places where we can recover from the pervasive rationality of modern society.

But let's not be too naive either. Ultimately, the new wild places will also be part of our everyday landscape and will not totally evade its practical, administrative and financial context. One may well dream of 'new wild places' as a counter space, but the new wild will inevitably remain a part of the modern landscape with its economic pressures and human activities such as tourism and biodiversity policies. We will probably never get rid of a certain uneasy feeling that rewilding is never free enough.

Rewilding fundamentally questions the current human–nature relationship by reviving that age-old idea that a better understanding of nature will lead to a fundamental reflection on one's own existence and to a critical perspective on humans and society. Rewilding and the contemporary desire for wildness thus offer the nature movement the opportunity to reconnect with its moral roots and develop into a broader social movement.

If we understand rewilding as part of a social and moral movement and practice in which people learn again to take care of their habitat, while also making room for wildness—the autonomy of non-human beings—then rewilding is not about the creation of wilderness. Rather rewilding is about the sincere attempt of a political and moral community to develop a new relationship with the non-human environment. In that process, biologists and ecological experts can play a role because they have knowledge of the ecosystems with which humans interact. But the new developments must also fit within a meaningful historical narrative of landscape development, and be carried out in such a way that people can re-establish a meaningful relationship with the non-human world around them. The new rewilding zones will not be wilderness in the traditional sense of the word. Even our wildest natural areas are in a sense cultural landscapes, because they are the expression of human choices and value judgments. The new wild places are our newest cultural landscape. But that does not make them less wild.

Notes

1 Nash, 2014; Oelschlaeger, 1981.
2 And of course many 'wildernesses' as classically defined, including North American ones, have long been influenced by indigenous peoples.
3 Cf. Rolston 1988, p. 341.
4 Tuan, 1975: 164–165.
5 Landschapspijn (De Boer, 2017).
6 Rewilding is a relatively recent term, but projects that today are labelled as rewilding have a longer history. In the Netherlands in particular, rewilding has a history of several decades, although the most commonly used term for projects that today would be labelled 'rewilding' is 'nature development' ('*natuurontwikkeling*') or 'new nature' (Bulkens et al., 2016). In this chapter I use the terms interchangeably.
7 Urban Scout, 2008, see also: rewild.com, rewildu.com, rewildportland.com, and fireflygathering.org/manifesto/.

References

Buijs, A.E. (2009). Public support for river restoration. A mixed-method study into local residents' support for and framing of river management and ecological restoration in the Dutch floodplains. *Journal of Environmental Management*, 90(8): 2680–9.

Bulkens, M., Muzaini, H., and Minca, C. (2016). Dutch new nature: (re)landscaping the Millingerwaard. *Journal of Environmental Planning and Management*, 59(5): 808–25.

Carver, S. (2012). (Re)creating wilderness: Rewilding and habitat restoration. In P. Howard, I. Thompson, and E. Waterton (eds), *The Routledge Companion to Landscape Studies* (pp. 383–94). London: Routledge.

Cheney, J. (1989). Postmodern Environmental Ethics. Ethics of Bioregional Narrative. *Environmental Ethics*, 11(2): 117–34.

Council of Europa (2000). *European Landscape Convention*. Florence.

Cronon, W. (1996). The trouble with wilderness; or: getting back to the wrong nature. *Environmental History*, 1(1): 7–28.

DeBoer, J. (2017). *Landschapspijn: over de toekomst van ons platteland*. Atlas Contact.

Deliège, G. (2011). Over natuurvervalsing in de Doelse polders. Robert Elliots antirestauratiethesis. *Tijdschrift voor Filosofie*, 73(3): 421–44.

Deliège, G. and Drenthen, M. (2014). Nature restoration: Avoiding technological fixes, dealing with moral conflicts. *Ethical Perspectives*, 21(1): 101–32.

Drenthen, M. (2009a). Fatal Attraction; Wildness in Contemporary Film. *Environmental Ethics*, 31(3): 297–315.

Drenthen, M. (2009b). Ecological restoration and place attachment; emplacing nonplace?, *Environmental Values*, 18(3): 285–312.

Drenthen, M. (2015). Layered landscapes, conflicting narratives and environmental art. In D. Havlick and M. Hourdequin (eds), *Restoring Layered Landscapes* (pp. 239–62). Oxford University Press.

Drenthen, M. (2018). Rewilding in layered landscapes as a challenge to place identity. *Environmental Values*, 27(4): 405–25.

Gammon, A. (2018). The many meanings of rewilding: An introduction and the case for a broad conceptualisation. *Environmental Values*, 27(4): 331–50.

Gammon, A. (2019). The unsettled places of rewilding. In S. Pinto et al. (eds), *Interdisciplinary Unsettlings of Place and Space* (pp. 251–4). Springer.

Helmer, W. (1996). Waanzinnige oases. In: A. Barendregt, M. Amesz, and J. van Middelaar (eds), *Natuurontwikkeling: zin of waanzin?*(pp. 81–90). Universiteit Utrecht.

Helmer, W., Litjens, G., and Overmars, W. (1995). Levende natuur in een cultuurlandschap. *De levende natuur*, 96(5): 182–7.

Higgs, E. (2003). *Nature by Design: People, Natural Process, and Ecological Restoration*. MIT Press.

Hughes, F., Stroh, P., Adams, W., Kirby, K., Mountford, J., and Warrington, S. (2011). Monitoring and evaluating large-scale, 'open-ended' habitat creation projects: A journey rather than a destination. *Journal for Nature Conservation*, 19(4): 245–53.

James, S.P. (2016). The trouble with environmental values. *Environmental Values*, **25**(2):131–44.

Jepson, P.R. and Schepers, F. (2016). Making space for rewilding: Creating an enabling policy environment. *Rewilding Europe*. DOI: 10.13140/RG.2.1.1783.1287

Jørgensen, D. (2015). Rethinking rewilding. *Geoforum*, 65: 482–8.

Lorimer, J. and Driessen, C. (2014). Wild experiments at the Oostvaardersplassen. Rethinking environmentalism in the Anthropocene. *Transactions of the Institute of British Geographers*, 39(2): 169–81.

Nash, R.F. (2014). *Wilderness and the American Mind*. 5th ed. Yale University Press.

O'Neill, J., Holland, A. and Light, A. (eds). (2008). *Environmental Values*. Routledge.

Oelschlaeger, M. (1981). *The Idea of Wilderness*. Yale.

Pedroli, G.B.M., Elsen T. van, and Mansvelt, J.D. van. (2007). Values of rural landscapes in Europe: inspiration or by-product? *NJAS Wageningen Journal of Life Sciences*, 54(4): 431–47.

Rebanks, J. (2015). *The Shepherd's Life*. Allen Lane.

Rolston, H. (1988). *Environmental Ethics. Duties to and Values in the Natural World*. Temple University Press.

Soulé, M. and Noss, R. (1998). Rewilding and biodiversity: complementary goals for continental conservation. *Wild Earth*, 8: 19–28.

Tuan, Y.F. (1975). Place: An experiental perspective. *The Geographical Review*, 65 (April): 151–65.

Tuan, Y.F. (1977). *Space and Place: The Perspective of Experience*. University of Minnesota Press.

Urban Scout (2008). *Rewild or Die: Revolution and Renaissance at the End of Civilization*. Myth Media.

Van de Gronden, J. (2015). *Wijsgeer in het wild*. Polak & Van Gennep.

Van den Berg, A.E. and Koole, S.L. (2006). New wilderness in the Netherlands: An investigation of visual preferences for nature development landscapes. *Landscape and Urban Planning*, 78: 362–72.

Van Toorn, W. (1988). *Leesbaar landschap*. Querido.

Ward, K. (2019). For wilderness or wildness? Decolonising rewilding. In: N. Pettorelli, S.M. Durant, and J.T. du Toit (eds), *Rewilding* (pp. 34–54). Cambridge University Press.

34

KNEPP WILDLAND

The ethos and efficacy of Britain's first private rewilding project

Simon Leadbeater, Helen Kopnina, and Paul Cryer

We dedicate this chapter to the memory of Father Thomas Berry, whom Paul Cryer knew personally. We hope our writing honours his legacy. His few words here provide the recurring theme to our discussion about *Knepp Wildland*:

> ... this immense effort that has been made over these past ten thousand years to bring the natural world under human control. Such an effort would even tame the inner wildness of the human itself.
>
> *(Berry, 1999: 50)*

Introduction: Rewilding our relationship with Nature

Rewilding may be the natural world's last hope set against seemingly irreversible downward spiraling trajectories of wild fauna and flora extinctions caused by a coalescence of drivers, headed by habitat losses (IPBES, 2019). For 200 years UK conservation efforts have only slowed the pace of species' reductions (Kirby, 2020a: 349). *Knepp Wildland*, however, a 1,400ha privately owned estate in southern England (see Figure 34.1), successfully transformed a traditional agrarian landscape and thereby achieved a step change in conservation outcomes. Guided by methodology derived from the 6,000ha Dutch state-owned park, Oostvaardersplassen (OVP), these changes were implemented under the guise of *rewilding*.

While recognising Knepp's significant gains in both diversity and species' abundance we have previously argued that its forerunner—OVP—never met any rewilding criteria (Kopnina et al., 2019). Applying Carver et al.'s 'Guiding principles' (2020) we accept that Knepp strives to fit within the 'continuum of scale, connectivity... level of human influence [which] aims to restore ecosystem structure and functions to achieve a self-sustaining autonomous nature' (Carver et al., 2020: 2). Knepp was one of the first private rewilding initiatives, and as such the estate experienced a steep learning curve. That all said, whilst the land-use at Knepp represents a significant improvement on previous agri-industrial practices, the estate's management ethos is firmly embedded within the destructive human ideology that is contributing causally to the environmental crisis. Lauding Knepp as a rewilding exemplum, therefore, endorses a flawed formula.

DOI: 10.4324/9781003097822-38

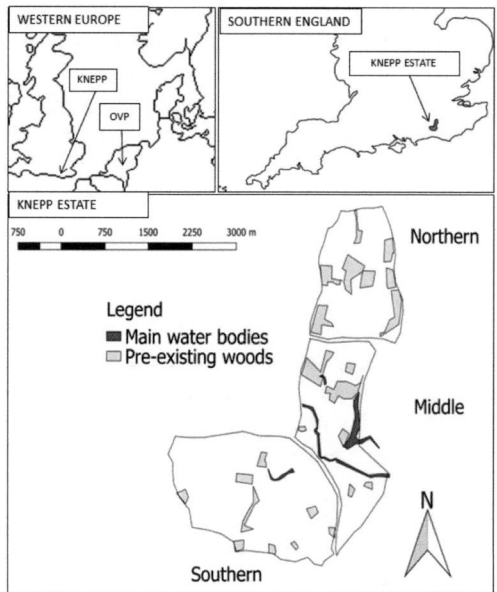

Figure 34.1 Knepp Estate drawn by Keith Kirby; map positioning Knepp in relation to OVP by Paul Cryer.

Agreeing with Hambler that rewilding's 'central, rational reason' must be to reduce extinction rates (Hambler, 2015: 23), this chapter unpacks the ecological assumptions behind Knepp and OVP's approaches to rewilding. An examination of OVP and the Knepp estate's ethical underpinnings exposes two interconnected consequences: firstly, ethical introspection points to rewilding's destiny and greatest potential being the rewilding of ourselves—in changing our relationship with nature (as alluded to in Carver et al's *Principle 10* (Carver et al., 2020: 9)). Secondly, this shift in thinking not only salvages the core ethos of rewilding but also offers a bridge between the original emphasis on large scale ecosystem protection and subsequent interpretations involving the restoration of farmland.

Knepp as OVP's progeny

Rewilding's roots are most closely associated with three people. Dave Foreman coined the term in 1992 to address continental-scale ecosystem recovery, and worked particularly closely with the late Michael Soulé and Reed Noss (Soulé & Noss, 1998: 1–11). In the UK, George Monbiot's *Feral* (2013) influenced the founding of NGO *Rewilding Britain*, which sought to apply similar principles at a smaller scale, while Knepp provided a practical model for other private landowners to emulate. Knepp is a country estate owned by baronet Sir Charles Burrell and his wife, Isabella Tree, the author of *Wilding: The Return of Nature to a British Farm* (2018). Since 2002 (Tree, 2018: xiv) three separate estate areas, the northern, middle, and the largest southern block, divided by roads, have introduced Long Horn cattle, Exmoor ponies, deer, and other mammals including rare breed pigs, with varying densities (Kirby, 2020b: 230–1). These produce some 75 tonnes of meat annually (Tree, *How Rewilding Can Save the Environment*).

Knepp has a role model (Tree, 2018: 56). In the 1980s and 90s herbivores were introduced to the OVP reclaimed land reserve (Theunissen, 2019: 342). The project initially appeared successful, but, after some time, the introduced grazing animals began to starve. A number of commentators (see Kopnina et al., 2919) have since argued this was because OVP failed to include what Soulé and Noss argued were the essential preconditions to successful rewilding: the 3 Cs of cores, carnivores and migration corridors (Kopnina et al., 2019: 74). By 2018 (Theunissen, 2019: 342) OVP's management authority had abandoned the term 'rewilding' (Kopnina et al., 2019: 83).

OVP, and Frans Vera, the ecologist behind this project, were the inspiration for Knepp's wilding programme, with the critical difference that the estate leap-frogged OVP's starvation phase and jumped straight into meat production. Harvesting animals is common practice within *Rewilding Britain*'s project network, of which Burrell is a trustee (Rewilding Britain, 2022). The requirement to control animal numbers within limited areas is acknowledged, and the intensity of such management magnified with diminished scale and connectivity. We argue, however, that when culling animals morphs from an undesirable and thoroughly scrutinised necessity to a primary management goal, the validity of terms like 'protected' or 'rewilded' must be questioned. This applies to enormous 'game' reserves in Africa (where the same practice thrives) as well as smaller concerns like OVP and Knepp.

Wilding (2018) details how the Burrells incrementally rewilded their estate, beginning and ending with the return of the turtle dove: 'that unmistakable purring: soothing, inviting, softly melancholic (Tree, 2018: 1) … that gentle *turr-turr-ing* tugging at the heart strings is also a signal of repair, recovery and rebirth, the re-braiding of unravellings' (Tree, 2018: 308). Such evocative language captures the *raison d'être* of *Knepp Wildland*, the return of rare species thriving in unexpected habitat combinations (Environmental Funders Network, 2020), and at every level, from soil, water and rare species' comebacks, significant improvements have been recorded. It is nevertheless important to critically consider the three interlocked premises upon which Knepp is founded. The first is that a Serengeti-type landscape is natural to lowland Britain and forest is not; second, that the latter is species-poor (Tree, 2018: 64); third, that it is acceptable to utilise farm animals as analogues of their extinct ancestors to create open habitats for the benefit of a wide range of species, including rare ones.

From being trailblazers nearly two decades ago, Knepp is now being emulated, for example, on Lord Somerleyton's 2,000ha estate in Suffolk (Madden, 2020) where Welsh black cattle 'beef up revenue' (*BBC Countryfile*, 2020). In 2020 HEAL was founded, aiming to purchase land for rewilding, and intending to 'graze cattle, pigs, ponies and deer' (www.healrewilding.org.uk/rewilding).The proliferation of this model has been chartered by Fisher, dubbing them 'safari park rewilding' (Fisher, 2020) enterprises.

We will begin our critical examination by asking whether the case against forests can be substantiated.

The forest as myth

That European closed-canopy forests are as much a component of myth as its European fairy tales (Tree, 2018: 82) is adopted from Vera's *Grazing Ecology and Forest History* (2000). This debunked Sir Arthur Tansley's theory of succession leading to climax vegetation. Vera's *alternative hypothesis* maintains that over a cycle of 500 years scrub species develop in grassland, which shelter light loving oaks and hazel. These grow, and shade out the scrub. Herbivores prevent shade loving species, notably beech, from in turn out shading the oak, resulting in groves of trees unable to regenerate owing to grazing pressures. These trees eventually die, which in time results in the grassland environment ready for scrub recolonisation, and the cycle starts again (Robinson, 2014: 293).

Figure 34.2   Bialowieża, October 2019.

We struggle with the presentation contesting the prevalence of tree cover. 'Thorn as the mother of oak', the cornerstone of Vera's hypothesis, is being challenged by Knepp itself, as dense thickets of thorn and bramble appear to be restricting the growth of oak seedlings (Kirby, 2020b: 234). Tree also enlists Białowieża Forest, Poland, to support her case. She suggests oaks are dying out, presumably as forests are unnatural environments and oaks only thrive in open landscapes. But in *Half-Earth* (2016) E.O. Wilson included Białowieża as 'one of the best places in the biosphere' (Wilson, 2016: 136), as a reserve of the Earth's biodiversity, 'authentic original wilderness' (Wilson, 2016: 135) retaining many natural elements such as large herbivores in the shape of bison and red deer, and their predators. It is difficult to understand why shade intolerant oaks and hazel, in a forest with such a minimum anthropocentric footprint (Jaroszewicz et al., 2019: 849), have not already become locally extinct, but instead persist, as illustrated in Figures 34.2 and 34.3 (Leadbeater, 2020). Some researchers disclaimed Vera's hypothesis (e.g., Mitchell, 2005) while others took a more nuanced view (e.g. Whitehouse & Smith, 2009). Our synopsis is that 6,000 years ago ca. 75% of the UK was covered in forest (Riutta et al., 2014); subfossil evidence shows the early Holocene was largely closed forest (Hambler et al., 2011) with subsequent Neolithic clearances being visible in the pollen record (Robinson, 2014: 291). Keith Kirby, reflecting on the Vera thesis since at least 2003 (Kirby, 2003), wrote in 2020 that 'arguments still come down in favour of the predominately wooded landscape model' (Kirby, 2020a: 239) as the pre-Neolithic baseline.

Ultimately Tree's envisionment of a savannah-like landscape dominated by open grown oaks is to envisage the wrong tree. She neglects to consider the prevalence of small-leaved lime in southern Britain until 5,000 years ago (Greig, 1982). This provides the final fatal objection to her theory as lime trees are shade tolerant and grow well in forest environments, such as in Białowieża today (Faliński, 1986). Vera's own treatment is hardly compelling. He discussed this

Figure 34.3   A forest within a forest; oak seedlings are abundant in parts of Białowieża's Strict Protection Area.

species in terms of his alternative hypothesis in less than one page (Vera, 2000: 329–30) while his final synthesis emphasised oak and hazel (Vera, 2000: 376–78). Interestingly Knepp's soil enhancement—though undoubtedly beneficial and to which Tree dedicates a whole chapter (Tree, 2018: 268–90)—possibly bears little resemblance to the area's former condition as ancient lime preferred loess, a fertile windblown soil prone to dispersal with deforestation (Greig, 1982: 27), and encourages base-rich substrates (Pigott, 1989: 23).

Forests as species poor

Alongside asserting that forests belong to the realm of myth instead of being historical realities, Tree claims that 'Closed-canopy forest is demonstrably species-poor...' (Tree: 2018: 64). Let us examine this proposition.

Europe's untrammelled nature would have been primarily characterised by eternally evolving high forest containing glades (Gray, 2021: 21). Would this naturally developing environment, termed 'uncontrolled succession' by HEAL, 'lead to [a] loss of diversity of habitats and any niche species in them' (www.healrewilding.org.uk/rewilding)? As Schieltz and Rubenstein point out, in their review of studies across the world, grazing can be positive (Schieltz & Rubenstein, 2016). Wooded environments are also special, however, and should not be denigrated. Forest specialists—the epitome of niche species—require closed canopy environments; openings in tree canopies can encourage other species to usurp them (Leadbeater, 2016). Many species already lost were woodland ones requiring deadwood and forest habitats (Hambler, 2015). Vera's version of rewilding would not help them. Knepp succumbs to what Hambler and Speight might describe as the 'traditional' focus on a relatively small number of generalist light loving creatures at the expense of the more vulnerable 28,500 invertebrates, 15,000 fungi and an unknown diversity of micro-organisms which require the damp, shady conditions provided by high forest environments, something highlighted 25 years ago (Hambler & Speight, 1995: 137–47). Knepp's light dependent animals may not continue to benefit from rewilding as natural landscapes mostly return to forests. Knepp is itself gradually evolving, in the southern block at least—where the recovery of turtle doves has been particularly marked—towards an increasingly forested environment (see Kirby, 2020b: 235).

Rewilding as predation

How Tree writes reveals much about the distance she places between herself and the animals living on her estate: 'We would prefer to shoot our cattle and pigs… . And to be able to make wonderful charcuterie from our breeding herd of Exmoors' (Tree, 2018: 292). We need to understand why Tree feels this way.

At Knepp, and its precursor OVP, there is an underlying if unacknowledged rationale supporting open landscapes besides a shared faith in Vera's alternative hypothesis: their owners' relationship with nature. This assumed relationship is characterised by what Eileen Crist has termed 'human supremacy'—synonymous with anthropocentricism—'the pervasive belief that the human life form is superior to all others and entitled to use them and their habitats', which 'entrench[es] violence as a way of life' (Crist, 2019: 45). The delusion of humanity's dominion over nature has contributed causally to broad-scale environmental destruction (Swimme & Berry, 1992: 184, 229–30).

Tellingly, Tree suggests that the open landscapes of country parks represent the 'ghost of the savannah in our heads', replicating what our hunter gatherer ancestors saw when they first arrived in Europe, gazing on 'gigantic herds of grazing animals, just like Africa, the richest resources…' (Tree, 2018: 298–9). In other words, Tree looks upon her herbivores as prey just as she envisages her forebears did. This is a shame. As even Captain Kirk said, 'we've come a long way in five thousand years' (*Who Mourns for* Adonais, 1967), since Neolithic people began to fell the wildwood where Knepp is now situated (Greig, 1982). Tree attempts to return us not just to how she thinks landscapes once were but to how she thinks *we* once were. She is not alone. At the start of his rewilding journey Monbiot picked up a dead muntjac and, hoisting it on his shoulders, he realised 'why we were here'. This, he says, made him want to roar (Monbiot, 2013: 33). Monbiot's roar may have been heralding a newly realised communion with the radically interconnected whole but this wild connection with our predatory or scavenging ontogeny can too easily be corrupted into the opposing perspective of human supremacy, a space where environmental protection and commercial meat production can be conflated without contradiction.

What is the alternative to, as Stevan Harnad (2019, personal comm.) puts it, 'trad[ing] off the lives of domestic animals against those of wildlife?' His view, that 'rewilding land should be devoted to wildlife, and domestic animals should not be consumed as an exchange: they should be sterilized…' (Personal Communication, 2019) is implicitly rejected by Tree with respect to the horses Knepp felt forced to castrate. The herd, she explains, lost its dynamism, 'the spark of natural interaction and acquired wisdom halted in its tracks—a "wild" animal going nowhere' (Tree, 2018: 261). If only Knepp could find a market for British horsemeat—'we might again…' (Tree, 2018: 263) 'have the joy of seeing foals at foot' (Tree, 2018: 261). This raises an important ethical dilemma, and also challenges what practices may be deemed compatible with rewilding. Theoretically Knepp could apply immunocontraception to a range of herbivores with minimal behavioural changes and as implementation improves, using less invasive measures. Cohn and Kirkpatrick maintain the main barriers to birth control are now political not practical (Cohn & Kirkpatrick, 2015: 22–9). Over 15 years ago research suggested that while not effective in reducing abundance, as Fagerstone et al. (2006) report, providing deer numbers are already at a desirable level injections every five years reduced fawning by nearly 90% (Fagerstone et al.,

2006: 45–54). As animals are in any case rounded up (Barkham, 2018), doing so once every few years would certainly be less stressful than capture and transport to the abattoir. Fertility control is not new, with one centre convening its ninth annual conference in May 2022 (*The Botstiber Institute for Wildlife Fertility Control*, 2021). It is regrettable that Knepp appears not to have explored this relatively benign management tool.

There might be a number of impediments to Knepp adopting related approaches. There would be no income from meat sales only cost, chiefly for winter feed, but also for experts to implement a contraception programme. Tree would also argue that such measures somehow lessen what wild animals are, and she prefers to slaughter them. She does so, we believe, in part owing to her relationship with nature, as discussed, but also because of the political economic framework within which Knepp is embedded. The Burrells partly explored rewilding as conventional farming was loss-making; the estate needs to make money. But by not considering means by which herbivores may be prevented from becoming ever more abundant, in the dearth of the 3 Cs (cores, carnivores, corridors), Knepp lacks our introduced fourth C, that of compassion (Kopnina et al., 2019). Carver et al. amended Soulé and Noss's triptych to cores, corridors,

Where Knepp fits within rewilding's continuum

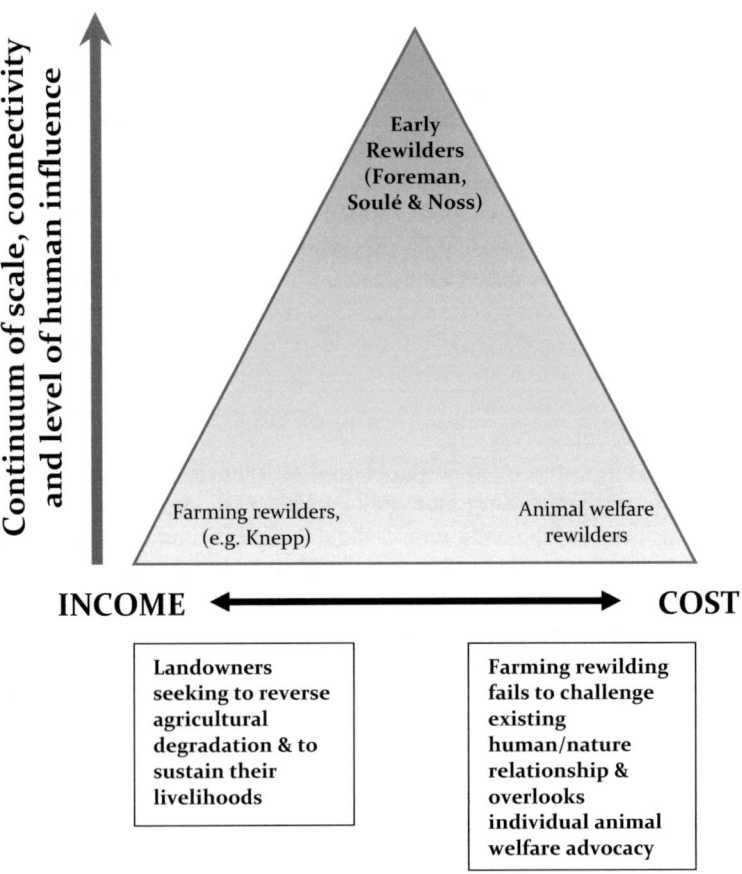

Figure 34.4   Knepp Wildland vs 'animal welfare rewilding'.

coexistence with compassion and 'nonlethal means wherever possible', included within their *Principle 9* (Carver et al., 2020: 9). This leads us to ask whether Knepp is wildland in name only. For Carver et al. rewilding's central aim must be to cede human control, dominance, suppression, and instead allow eco-systems to develop in a self-willed and self-regulated way (Carver et al., 2020: 6). As depicted in Figure 34.4, Knepp achieves this no more by slaughtering their animals than management would by maintaining herbivore numbers through forms of fertility control, a key determinant being landowners' cost-benefit assessment of their estate's financial viability.

Differences of degree not kind...

The attitude Tree adopts towards nonhuman animals is not developed but rather assumed. Knepp courts experts in a range of areas, including Temple Grandin, whose impact on animal welfare has been queried (see Bekoff, 2016). Importantly, so far as we are aware, Tree consults no one to challenge her assumptions about animals. Why, considering all their background research before establishing *Knepp Wildland*, did the Burrells remain so incurious, and fail to explore, as Carl Safina does in his book *Beyond Words* (2015), 'who' their introduced animals might be (Safina, 2015: 1)?

This intimates again that Knepp is rooted in an outmoded and unscientific paradigm, this time concerning animal ontologies. It is over one and a half centuries since Charles Darwin forced us to confront that 'the difference... between man and the higher animals... is one of degree and not of kind' (Darwin, 1871). Human beings are animals (Safina, 2015: 1) and Bekoff and Pierce assert that 'the idea that the differences between species are differences in degree rather than differences in kind...aren't meaningful differences at all' (Bekoff & Pierce, 2009: xi) as 'new information that's accumulating daily is blasting away perceived boundaries...' (Bekoff & Pierce, 2009: x). So far as Knepp's farm animals are concerned, 'it is becoming increasingly clear that [they] have complex thoughts, deep emotions, and social skills and rituals not unlike our own' (Hatkoff, 2009: 16).

Rewilding ourselves: Compassion for our nonhuman kin

If it is our relationship with nature that has placed nature in such a parlous state that we need to restore habitats and reduce extinction rates, it could be argued that rewilding landscapes is of secondary concern. Rewilding ourselves should be the priority. What does this mean? To allow nature to take the driving seat, to relax, let go, and to accept the boom and bust cycles nature entails; this, according to Tree, is rewilding ourselves (Tree, *How Rewilding Can Save the Environment*). We find it strangely incongruent for a rewilding project which purports to dispel many of the misconceptions we hold for conservation not to question our relationship with nature, other than ostensibly urging us to take a backseat.

The 'deepest cause' wrote cultural historian Thomas Berry, 'of the present devastation' is to be found 'in a mode of consciousness that has established a radical discontinuity between the human and other modes of being and the bestowal of all rights on... humans' (Berry, 1999: 4). He went on to explain that the 'other-than-human modes of being' have no rights and are only valued in terms of their utility, that the human order is 'predatory on the very sources from whence we came' and that to initiate our children—'enculturation' (Crist, 2019: 45)—into the economic order which exploits natural resources, we first make them unfeeling towards the natural world (Berry, 1999: 15). To properly change this relationship we need to overcome what Rachel Corby describes as the *great illusion*, which tells us not only that we are separate from nature, but superior as well, and the only ones possessing *livingness,* 'feelings, intelligence...' (Corby, 2019: 51). For Corby, rewilding ourselves entails discarding the conceit of human superiority and separation from nature for parity and kinship; from attributing *otherness* and utility to nonhuman nature and instead acknowledging sentience, cognition and agency of equal value with ourselves. Achieving this *volte-face* transformation inexorably leads us closer to who we really are. To return to Berry, to understand that the 'remarkable continuity' between human and animal 'relatives' within the natural world comprises a 'single sacred community' (Berry, 1999: 22–3).

Berry's 'Great Work' aimed to 'transition ...[to] when humans would be present to the planet in a mutually beneficial manner' (Berry, 1999: 3). *Wilding's* unspoken but clear *leit-motiv*, ourselves as predators, Monbiot's 'why we are here', arguably confers human supremacy over nature, reinforcing the relationship which underpins Berry's 'present devastation'. It is, however, unscientific, as it presupposes differences of kind. Berry and Corby's realignment celebrates differences of degree, whose cognisance ineluctably results in empathy and compassion being at the core, or heart, of rewilding ourselves. This is best expressed by Marc Bekoff who claims the concept of rewilding is 'grounded in the premise that caring... is essential' (Bekoff, 2014: 5). *Rewilding Our Hearts* (2014) argues that rewilding is a 'personal journey and transformative exploration that centres on bringing other animals..., back into our heart' (Bekoff, 2014: 13). *Corridors* for Bekoff are 'between our heart and brain, our caring and awareness' (Bekoff, 2014: 12). In changing our relationship with nature through rewilding our hearts, we heal ourselves as we heal the natural world.

Conclusion: The return of the forest

It is fortuitous for England's large landowners to have fallen upon a new business model which enables them to harvest animals as they have always done, this time with wildlife dividends. The political and economic context of the British country estate is perhaps central to the numerous quandaries owners face in relation to rewilding. They need to be profitable, and the Knepp 'safari park' formula enables them to create several income streams; meat sales and profitable access arrangements wrapped in a purportedly ethical business, in which nature recovers relative to previous agrarian land-uses. But, ultimately, will we continue to perceive this as rewilding if the country estate business case requires the preservation of landscapes little more natural than a Capability Brown parkland? Trapped by the imperative to earn or die, landowners like the Burrells do not have as much room for manoeuvre as we might imagine. Recognition of Knepp's trail-blazing innovation and bravery needs to be balanced with an appreciation of the burdens and dread of failure owning such an estate must entail.

It is perhaps the onerous nature of the Burrells' responsibilities to ensure the estate survives beyond their generation that unravels *Knepp Wildland*, whose version of 're-braiding, repair, recovery, rebirth', does rewilding such a disservice. Middle Ages' romances alluded to enchanted forests (Porteous, 1928). The forests of myth and imagination are vast, potentially dangerous, but also numinous presences (see Berry, 1999: 49). A truly wild landscape—in the sense of not being reigned over by us—is the forest. When rewilders *à la* Vera and Tree tell us forests never existed it is perhaps insufficient to allow they merely overstate their case. Their thesis is not limited to ecology; it thwarts imagination, of woods, lovely dark and deep, impenetrable, above all understood as not being human-dominated landscapes. Are they not actively trying to prevent us from envisioning wildness, and in so doing threaten to cloak a good part of the UK in an anthropocentric prohibition of seral succession masquerading as nature?

Sacrificing herbivores to benefit rare species implies that 'some animals are more equal than others' (Orwell, 1945) and presumes that nonhuman farmed analogues possess an *otherness* in subjective feeling distinct from that of rare wild species and from ourselves. Knepp's unarticulated but underlying premise not only *encultures* children but also glampers, safari-goers, and *Wilding*'s readership, to a profound solipsism with respect to nonhuman animal experiences; rather than reconnect, this reinforces our dominion over nature. *Knepp Wildland* cannot be reconciled with Corby's parity between all beings or Bekoff's 'rewilding our hearts values compassion above all' (Bekoff, 2014: 4). These are not variations of rewilding, but fundamentally divergent views of what it is to be human and our place within the natural world. It was our ancestors who initiated the extinction crisis and felled the forests; recapturing the reconnect with this version of our wilder selves is to embark on a recidivist journey, the reverse direction required for addressing the root cause of extinction, namely our predatory relationship with nature, arising from human supremacy over all species and landscapes and maintained through violence.

Appreciating that the forests of yore once existed, and acknowledging that our sentience and cognitive capabilities are shared with nonhuman relatives, compels us to reject the anthropocentric suppression of vegetation and the utilitarian exploitation of analogue herbivores. Moreover, limited purpose will ultimately be served by rewilding landscapes, notwithstanding how much they may benefit rare species, if rewilding fails to transform our relationship with nature. Rewilding should above all rewild our hearts; replacing predatory violence with compassion; supremacy for integrality with the larger Earth community (see Berry, 1999: 48). Therein also lies the promise of deep emergence in the truly wild; the fathomless verdant forest, of landscape, imagination and myth. That is our vision. We acknowledge difficulties surrounding the intrusion and cost-effectiveness of contraception techniques. There is the important question of who pays to explore their efficacy within a rewilding model, and we accept that our approach will always involve nature being significantly human-willed—if benignly—rather than self-willed. Equally, our vision of wildness is predicated more on the self-autonomously evolving forest rather than on large numbers of grazing herbivores, and that is perhaps the fundamental flaw in Knepp's approach to begin with. But trees alone—particularly wild species like lime with limited timber value—earn relatively little, the payback decades in the making, and productive forestry is in any case not any more compatible with rewilding than Knepp's existing practice. Questions then drill down to how many landowners could afford to encourage the rejuvenation of the wildwood, and who would pay to visit?

Acknowledgements

The authors would like to thank Professor Marc Bekoff and Professor Stevan Harnad for their observations on an earlier draft of this chapter, Kate Rawles and an anonymous reviewer, and particularly Dr Keith Kirby for his more detailed commentary and the inclusion of his map of the Knepp estate.

References

Barkham, P., (2018) 'The magical wilderness farm: raising cows among the weeds at Knepp,' *The Guardian*, 15 June 2018.

BBC *Countryfile*, 23August 2020.

Bekoff, M. (2013), My beef with Temple Grandin: Seemingly humane isn't enough. While I enjoy Dr. Grandin's feistiness I disagree with her views on cows', *Psychology Today*, 14 April 2013: www. psychologytoday.com/gb/blog/animal-emotions/201304/my-beef-temple-grandin-seemingly-humane-isnt-enough

Bekoff, M. (2014), *Rewilding Our Hearts: Building Pathways of Compassion and Co-existence*. New World Library.

Bekoff, M. (2016), 'Stairways to heaven, temples of doom, and humane-washing', *Psychology Today*, 17 November 2016.

Bekoff, M. and Pierce, J. (2009), *Wild Justice: The Moral Lives of Animals*, The University of Chicago Press.

Berry, T. (1999), *The Great Work: Our Way into the Future*, Bell Tower.

Botstiber Institute for Wildlife Fertility Control, Personal Communication to the lead author of 24 September, 2021

Carver, S., Convery, I., Hawkins, S., Beyers, R., Eagle, A., Kun, Z., Van Maanen, E., Cao, Y., Fisher, M., Edwards, S.R., Nelson, C., Gann, G.D., Shurter, S., Aguilar, K., Andrade, A., Ripple, W.J., Davis, J., Sinclair, A., Bekoff, M., Noss, R., Foreman, D., Pettersson, H., Root-Bernstein, M., Svenning, J-C., Taylor, P., Wynn-Jones, S., Watson Featherstone, A., Fløjgaard, C., Stanley-Price, M., Navarro, L.M., Aykroyd., T., Parfitt, A., and Soulé, M. (2020), Guiding principles for rewilding, *Conservation Biology*. DOI: 10.1111/cobi.13730, 26 February 2021.

Cohn, P. and Kirkpatrick, J.F. (2015), History of the science of wildlife fertility control: Reflections of a 25-year international conference series', *Applied Ecology and Environmental Sciences*, 3(1): 22–9.

Corby, R. (2019), *Rewilding and the Art of Plant Whispering*. Amanita Forrest Press.

Crist, E. (2019), *Abundant Earth: Toward and Ecological Civilisation*. The University of Chicago Press.

Darwin, C. (1871), *The Descent of Man, and Selection in Relation to Sex*, John Murray.

Environmental Funders Network (2020), 'Rewilding: inspiring people', with Ben Goldsmith, Isabella Tree, Alistair Driver and Derek Gow. Comment from A. Driver. https://vimeo.com/432584530

Fagerstone, K.A., Miller, L.A., Bynum, K.S., Eisemann, J.D., and Yoder, C. (2006), When, where and for what wildlife species will contraception be a useful management approach? *USDA APHIS Wildlife Services, National Wildlife Research Center, Fort Collins, Colorado*, Proc. 22nd Vertebr. Pest Conf. (R.M. Timm and J.M. O'Brien, eds), Univ. of Calif., Davis.

Faliński, J.B. (1986), *Vegetation Dynamics in Temperate Lowland Primeval Forests. Ecological Studies in Białowieża Forest*, Dr W. Junk Publishers.

Fisher, M. (2020), Faking the wild—safari park rewilding, www.self-willed-land.org.uk/articles/safari. htm, 27, 29 May, 4 June 2020.

Gray, J. (2021), *Thirteen Paces by Four: Backyard Biophilia and the Emerging Earth Ethic*, Dixi Books.

Greig, J. (1982), Past and Present Lime Woods of Europe, in Limbrey, S. and Bell, M (eds), *Archaeological Aspect of Woodland Ecology, British Archaeological Reports*.

Hambler, C. (2015), Evidence-based or evidence-blind? Priorities for revitalising conservation, *ECOS*, 36(3/4): 22–5.

Hambler, C. and Speight, M. (February 1995), 'Biodiversity Conservation in Britain: Science Replacing Tradition', *British Wildlife*, 6: 137–47.

Hambler, C., Henderson, P.A., and Speight, M.R. (2011), 'Extinction rates, extinction-prone habitats, and indicator groups in Britain and at larger scales', *Biological Conservation*, 144(2011): 713–21.

Harnad, Professor Stevan, Personal Communication to the lead author of 29 August, 2019

Hatkoff, A. (2009), *The Inner World of Farm Animals: Their Amazing, Social, Emotional, and Intellectual Capacities*, Stewart, Tabori and Chang, New York *Heal*, the official website: www.healrewilding.org.uk

IPBES (Intergovernmental Science-Policy Platform on Biodiversity and Ecosystem Services) Media release: Nature's dangerous decline 'unprecedented;' species extinction rates 'accelerating.' May 2019. Retrieved from https://ipbes.net/news/Media-Release-Global-Assessment.

Jaroszewicz, B., Cholewińska, O., Gutowski, J.M., Samojlik, T., Zimmy, M., and Latolowa, M. (2019), Białowieża forest—a relic of high naturalness of European forests, *Forests*, 10(849): https://doi.org/ 10.3390/f10100849.

Kirby, K. (2003), What might a British forest-landscape driven by large herbivores look like?, *English Nature*, English Nature Report Number 530.

Kirby, K. (2020a), *Woodland Flowers: Colourful Past, Uncertain Future*, Bloomsbury Wildlife.

Kirby, K. (2020b), Tree and shrub regeneration across the Knepp Estate in Sussex, Southern England, *Quarterly Journal of Forestry*, 114(4): 230–236 (October).

Kopnina, H., Leadbeater, S., and Cryer, P. (2019), Learning to rewild: Examining the failed case of Dutch 'New Wilderness' Oostvaardersplassen, *International Journal of Wilderness'*, 25(3) (December): 72–89.

Leadbeater, S.R.B. (2016), Reaching forward to the past: Rewilding the heart of a PAWS woodland to reverse species decline and extinction, *Quarterly Journal of Forestry*, 107, 110(2) (April): 115–22.

Leadbeater, S.R.B. (2020), Białowieża and back again, *Quarterly Journal of Forestry*, 114(3): 205–211 (July).

Madden, R. (2020), Going wild for Britain, *The Telegraph*, 15 August 2020.

Mitchell, F.J.G. (2005), How open were European primeval forests? Hypothesis testing using palaeoecological data, *Journal of Ecology*, 93: 168–77.

Monbiot, G. (2013), *Feral: Rewilding the Land, Sea and Human Life*, Penguin Books.

Orwell, G. (1945), *Animal Farm: A Fairy Story*. Secker & Warburg.

Pigott, C.D. (1989), The growth of lime *Tilia Cordata* in an experimental plantation and its influence on soil development and vegetation, *Quarterly Journal of Forestry*, 83(1): 14–24.

Porteous, A. (1928), *The Forest in Folklaw and Mythology*, Dover Publications, Inc.

Rewilding Britain, 'Our Trustees', accessed at www.rewildingbritain.org.uk/about-us/who-we-are-and-how-were-run/our-trustees (September 2022).

Riutta, T., Slade, E.M., Morecroft, M.D., Bebber, D.P., and Malhi, Y. (2014), Living on the edge: Quantifying the structure of a fragmented forest landscape in England, *Landscape Ecology*. DOI 10.1007/s10980-014-0025-z, 29 March 2014.

Robinson, M. (2014), The ecodynamics of clearance in the British Neolithic, *Environmental Archaeology*, 19(3): 293.

Safina, C. (2015), *Beyond Words: What Animals Think and Feel*, Henry Holt and Co..

Schieltz, J.M. and Rubenstein, D.I. (2016), Evidence based review: positive versus negative effects of livestock grazing on wildlife. What do we really know?, *Environmental Research Letters*, 11 November 2016, 113003.

Soulé, M. and Noss, R. (1998), Rewilding and biodiversity: Complementary goals for continental conservation, *Wild Earth*, 8(3) (Fall 1998): 19–28.

Star Trek: The Original Series (1967), *Who Mourns for Adonais?* Season 2, Episode 2.

Swimme, B. and Berry, T. (1992), *The Universe Story: From the Primordial Flaring Forth to the Ecozoic Era—A Celebration of the Unfolding of the Cosmos,* Harper Collins.

Theunissen, B. (2019), The Oostvaardersplassen Fiasco, *ISIS*, 110(2) (June): 342.

Tree, I. (2018), *Wilding: The Return of Nature to a British Farm*. Picador.

Tree, I. (2018) *Isabella Tree @5x15 How rewilding can save the environment*: www.youtube.com/watch?v=4cug0KcTnXI.

Vera, F.W.M. (2000), *Grazing Ecology and Forest History*, CABI.

Whitehouse, N.J. and Smith, D. (2009), How fragmented was the British Holocene wildwood? Perspectives on the 'Vera' grazing debate from the fossil beetle record, *Quaternary Science Reviews*, 29(3–4): 539–53.

Wilson, E.O. (2016), *Half-Earth; Our Planet's Fight for Life*, Liveright Publishing Corporation.

35

HUMAN REWILDING

Practical pointers to address a root cause of global environmental crises

Georgina Maffey and Koen Arts

Introduction[1]

Since its inception in the 1980s, the concept of rewilding has been primarily concerned with ecological dimensions, particularly the restoration of self-regulating ecosystems (Soulé & Noss, 1998). While there has always been some reference to human or societal dimensions of rewilding since the idea was conceived (Foreman, 2018; Kopnina et al., 2019), it is only in recent years that these dimensions have been receiving more elaborate attention (Bekoff, 2014; Durant et al., 2019; Kopnina et al., 2019; Carver et al., 2021; Martin et al., 2021). Indeed, it is explicitly recognised that the success of rewilding efforts is highly dependent on both ecological and social contexts (Torres et al., 2018), and that rewilding has 'become a social as much as an ecological phenomenon' (Martin et al., 2021). In rewilding literature, humans largely appear to fall into three more or less distinct categories. First, as stakeholders in rewilding efforts (Martin et al., 2021; Schulte to Bühne et al., 2021); second, as project leaders and mediators (Light & Higgs, 1996; Foreman, 2018; Jepson 2019); and, third, as engaged components of rewilding projects that desire some form of rewilding themselves (Monbiot, 2013; Bekoff, 2014; Clayton, 2019).

In this chapter, we are concerned with the final grouping. We argue that human rewilding is a comprehensive agenda capable of addressing a root cause of current global environmental crises, namely the disconnection between humans and nature. Ecological rewilding, in a strict sense, will restore nature but struggles to move beyond symptomatic relief, as the crucial underlying cause of ecosystem destruction and biodiversity loss remains unaddressed. The call for human rewilding, however, often remains a rather abstract suggestion. What could human rewilding actually, and practically, entail?

Drawing on personal experiences, this chapter offers two very tangible case studies: rewilding daily life and rewilding education. With regard to the first case, we detail a personal experiment in which we aimed to spend a year living outside. In the second case we explore our ongoing work in rewilding educational frameworks in a university setting. These case studies are anchored in the six touchstones of the 'wild pedagogies' framework by Jickling et al. (2018)— (1) agency and the role of nature as co-teacher; (2) wildness and challenging ideas of control; (3) complexity, the unknown, and spontaneity; (4) locating the wild; (5) time and practice;

DOI: 10.4324/9781003097822-39

and (6) cultural change—and where appropriate we indicate parallels between the experiences and the touchstones for reference and further inspiration. For context, we suggest that human rewilding, like ecological rewilding, is not about turning back the clock (Jørgensen 2015), but about recognising that sensitivity to human evolutionary history is pivotal in restoring relationships with nature in a contemporary context. Human rewilding, we assert, focuses on a relational (head, heart, hands) approach to nature experience, and nurtures qualities of technique over technology (as deceleration over acceleration, and immersion over short-lived experiences).

Evolutionary aspects of human rewilding

Through the unprecedented, destructive impact that humans have had on Planet Earth, the current geological epoch has infamously earned the title of 'the Anthropocene'. Although, as Harroway (2016) suggests, the term 'Capitalocene' may be a more appropriate reflection of how negative environmental impacts occur through existing global economic systems. Rapid biodiversity loss indicates that a sixth mass extinction is currently underway (Ceballos et al., 2015), and the planetary boundaries model presents a stark overview of the dire state of the planet; emphasising 'the need to address multiple interacting environmental processes simultaneously' (Steffen et al., 2015). Yet, it should also be acknowledged that in the 250 years since the beginning of the industrial revolution, global living standards have been positively altered, with reductions in extreme poverty and improvements in health and literacy standards (Roser, 2020). This peak of prosperity is a happy consequence of the human evolutionary path that sought to disconnect from, or overcome, the challenging conditions of the natural environment—a path of 'dewilding'.

If we accept that, in a modern-day context, the vast majority of humans no longer live in a 'wild' manner, it seems logical to ask 'when did they?' Similar questions asked in ecological restoration—'when was nature wild?'—led to discussions on temporality (Higgs, 2003; Marris, 2009; Jørgensen, 2015), authenticity (Arts et al., 2016; Drenthen, 2018) and on epistemologies of assessing restoration success (Torres et al., 2018). However, at least one common thread permeates these discussions, that of valuing the complex web of life on Earth and the long evolutionary path that precedes it (Carver et al., 2021).

In this instance, if we adopt the industrial revolution as an indicative point of 'dewilding', then we can trace a potted history through our evolutionary past to find the emergence of contemporary humans. This could begin with the appearance of *Homo sapiens* as a separate species around 200,000 years ago in East Africa, or, much later, with the arrival of language approximately 70,000 years ago, which marked the cognitive revolution and the spread of *sapiens* out of Africa (Harari, 2014; Ellis et al., 2021). Whichever point in time we choose, it follows that humans have spent over 99.99% of their evolutionary time in close interaction with their local natural environment. Thus, a friction emerges between the world in which people have evolved to exist and how many experience modern day life (Louv, 2005), which plays out in every societal sphere, from health (Triguero-Mas et al., 2017; Bratman et al., 2019; Antonelli et al., 2021) to politics (Light and Higgs, 1996; Schulte to Bühne et al., 2021).

Subdisciplines of evolutionary psychology, biology, and anthropology, among others, have been studying human behaviour for many years, and it is important to note that, from an evolutionary perspective, the past does not need to serve as a moral guideline in human rewilding. Instead, it should be a factor in understanding both human nature, and human relationships with nature. Just as in ecological rewilding, human rewilding is about restoring a connection with nature in a contemporary context, bridging the rigid human–nature dichotomy (Birch, 1990), and acknowledging the role that human evolutionary history plays in this relationship.

Alongside evolution, it is also worth considering how language fits in to this idea of bridging the human–nature dichotomy. Viewed through a Western lens, the etymological history of 'wild' is most likely derived from the Old English root of 'will', which refers to something uncontrollable or self-willing. Importantly, before the origin of the word 'wilderness'—that is, wil-dēor-ness, the place of the self-willing animals, from the Old English poem *Bēowulf*— 'will' probably referred to the human sphere (Nash, 2001; Jørgensen, 2015). Consequently, human rewilding has both an evolutionary and etymological stem. Moreover, if we also distinguish between 'wilderness' as biophysical reality and 'wildness' as a quality thereof which must be experienced or interpreted, it falls to the subject, to us, to find or recognise wildness (McMorran et al., 2008; Arts et al., 2012). It is from this perspective of 'finding wildness', that we undertook a personal experiment in human rewilding.

Rewilding daily life

If we try to define nature as 'natural' i.e. independent of humans (McKibben, 2003; Ellis et al., 2021), then we can conclude that there is no 'real' nature any more. Attempts to find 'real' nature—or wilderness—are often paradoxical. The 'last wilderness' can be found in Alaska, Siberia, Antarctica, or the Himalayas. Yet, for most reaching these areas requires an element of preparation, travel, and expense, in exchange for a short-lived experience in nature. In the Netherlands, where we (the authors) are based, this kind of rhetoric is abundant, alongside the idea that there needs to be 'new wilderness' (Kopnina et al., 2019) or 'new nature' (Owens & Wolch, 2019). These ideas feed in to a construct that nature—and by extension, wildness—no longer exists (cf. Touchstone 4). In an effort to take a positive, constructive approach to this

Figure 35.1 Rewilding daily life; Koen Arts and Gina Maffey during the year that they spent living outside. Photo by Otto Kalkhoven.

problem, we formulated a simple research question: In a land without wilderness, is it possible to find wildness in everyday life?

We first addressed the root cause of the problem (Figure 35.1). To find wildness we would need to immerse ourselves more in natural environments than human made ones (cf. Touchstones 1, 3, 5). We opted for the timeframe of a year, as this ensured the experience covered a recognisable natural cycle, and stipulated that during this year we would need to spend at least 50% of our time outside (adopting the broadest definition of natural environments). In addition, we added that the 50% rule had to be applied each season, to avoid time outside being biased to more appealing periods of the year. This immediately created a secondary problem in employing the experiment as part of everyday life. The vast majority of our working hours were spent indoors. To compensate for this a second condition arose, to sleep outside for 365 nights consecutively.

As the year began, on the autumn equinox, we quickly embraced learning techniques over employing technological interventions, as it offered a logical way to engage with the natural world more fully. Primary needs such as shelter and warmth leapt to the fore—sleeping outside is an enriching experience until the cold and the damp set in—and fire quickly embodied a central role in our daily life. The necessity of fire also underpinned our decision to adopt a simple shelter for the winter. Here, we paid heed to European communities that retain a lifestyle embedded in nature and used a modern tipi modelled on the lávuu of the Sami people (Skogvang, 2021). Importantly, the tipi acted just as a shelter, not a barrier to the outdoors, and accommodated a fire in its centre. The environmental and health impact of the trend in using indoor wood-burners is currently questioned (le Page, 2017), but outdoors, during the winter months, fire held invaluable roles beyond comfort for heating, cooking, drying, and socialising.

From a human evolutionary perspective, fire has altered sapiens physiology (e.g., smaller jaws, shorter gut) and behaviour (e.g., hunting, language development; Scott et al., 2014). Pausas and Keeley (2009), hypothesise that 'the world cannot be understood without considering fire' and we certainly saw how this could be true. Fire use has had a historical impact on both ecosystems and natural processes, but also on the human relationship with that environment. During the year, preparation became imperative: locating dead trees, collecting tinder, felling, transporting, and splitting were all physical processes that required time. Yet, the repetition of these tasks could become almost meditative in nature (cf. Touchstone 5). Contrary to adding pressure to a working day it provided time to decelerate, focus, and relax more fully. The mental benefits of nature engagement are well known (Triguero-Mas et al., 2017; Bratman et al., 2019; Antonelli et al., 2021) yet many activities require an individual to 'make time' to engage with nature. During the year we found that enforced interaction through necessity was rapidly normalised and enjoyable.

As the previous examples demonstrate, the experiment gave nature a prominent role in our everyday life, but also a feeling of reciprocity that brought meaning and happiness (cf. Touchstone 1; Buijs and Jacobs, 2021). Foraging for edible plants highlighted variation and diversity in local flora, and stimulated reflection on consumption patterns, food production chains and broader sustainability issues. Our actions had an impact globally and on the species in our immediate surroundings. We observed changes in species behaviour over time and identified local individuals, at different scales, from orb-weaver spiders on balconies, to red deer in nature reserves. We engaged with the natural world at all times of the day, dawn, noon, dusk, and night and we slept better and felt more rested, which, in turn, our senses seemed to function better for; we were quicker to spot things, heard more, smelt more (Touchstones 3, 4, 5).

These responses were all ones that we hoped for and contributed to a feeling of optimism in being able to 'find wildness'. There were however, unanticipated insights during the year

(Touchstones 2, 3). Firstly, our shared experience did not result in a shared perspective of the environment. Our biological sex influenced different physiological responses to air temperature (Kim et al., 1998) and psychological responses to solo excursions (Van den Berg & Ter Heijne, 2005). This variation caused us to reconsider the role and value of social connection in relation to nature connection. Such a spectrum of experiences is also important to consider in order to ensure that discussions on human rewilding are framed in an inclusive and accessible way (cf. Touchstone 6).

In ecological rewilding Carver et al. (2021) recognise a wilderness continuum. During the year, we saw how human experiences can also sit on a wildness continuum. There is a collection of instances where we spent all of our time outside, foraged more and actively avoided digital interaction. However, much of the year we took a pragmatic approach, making use of hot showers, cars and laptops. Yet, it was precisely in the grey areas between nature and culture that we identified more opportunities to find wildness than we originally thought possible (cf. Touchstones 3, 4). To this end, we reflect on the experiment as a success. Nevertheless, it remains an experiment rather than a blueprint for human rewilding. The legacy of the year has been to demonstrate how easy it is to both dilute *and* restore a connection with the natural world (cf. Touchstone 6). It is this understanding that we have used to inform further experiments in higher education settings in the Netherlands.

Rewilding education

Nature-based learning is garnering increasing interest—with growing confirmation in academic literature of the multiple benefits for students, such as 'academic learning, personal

Figure 35.2 Rewilding education; students work together to make fire bows following discussions on the role of fire in human evolutionary history and ongoing conservation efforts. Photo by Georgina Maffey.

development and environmental stewardship' (Kuo et al., 2019). Much research appears to be focused on young learners, which likely reflects societal concerns on increasing digital engagement (Selwyn, 2009) and nature deficit disorder (Louv, 2005). This focus is one that we could relate to. For much of our childhood we were engaged with nature through play, hobbies and studies. Yet, as we began to work in the discipline of nature conservation, we increasingly became 'digital conservationists'—viewing nature as a series of data points, while being inspired to protect the natural world on a meta level. The insight was painful; that we were embodying the strict imperialistic, dichotomy between humans and nature (cf. Touchstone 6; Birch, 1990; Cronon, 1995).

It is striking that practical nature interaction becomes increasingly rare as individuals enter the higher education spheres. Imbued by our own personal learning experiences in university environments we began to question *why* the accepted approach for higher education was a theoretical one. Beyond data collection, passionate students who will go on to become future leaders in conservation and sustainability are rarely encouraged to engage with their own local environments through learning (cf. Touchstone 6). In 2021, we developed a free elective course at Wageningen University under the title of 'Anthropology of elementary natural skills' (Figure 35.2). 36 students participated full-time in a month-long, multidisciplinary course. The course was divided between practicing basic natural skills such as making fire, tracking and natural navigation, and theoretical learning through analytical exercises.

As in our year outside, we adopted a relational approach, connecting cognitive, emotional, and physical elements with didactic interaction. Teaching was conducted outside in green spaces as much as possible and self-study outdoors encouraged (cf. Touchstones 4, 5). There was a degree of criticism in the pursuit, as interested parties questioned what more this course offered than, for example, scouting (cf. Touchstones 6). However, contrary evidence was apparent in the exam results and course evaluations.

Students valued the teaching approach for how they were able to ingest knowledge: 'I don't think I'll forget anything I've learned, unlike normal subjects where you forget half of it after the exam.' With some referring to feelings of being 'very calm', 'more alive', and 'more free'. They reflected on 'how important someone's connection with nature is. At first, I felt like nature was something that was around me, I did not take part of it', and how they were 'part of the larger story', with one student saying 'that, even though people state that they are equal to nature, our behaviour, techniques and mechanisms are still grounded in the old ideas that people are above it'.

As with ecological rewilding, rewilding education is a context dependent pursuit. Students will bring their own perceptions and understanding to teaching. However, nature as a setting for relational learning can be a great enabler, providing a place to develop more grounded, creative, and reflective pupils, who have a holistic understanding of their own place in the (natural) environment (cf. Touchstones 1, 3, 6; Lotz-Sisitka et al., 2015; Loynes, 2018; Fenton et al., 2020).

In this instance, we developed the course from scratch to look specifically at topics that align with human-nature interactions, and purposefully creating opportunities for deeper, transformative connections by adopting qualities of technique over technology, for example (Fenton et al., 2020). However, there is scope to apply a wild pedagogies approach to other domains, such as architecture, urban planning, diet, food production, and economics, in order to challenge the frameworks that fuel environmental crises. In this respect, the label of 'rewilding' may be irrelevant, but it does offer a captivating approach to unite multi-disciplinary efforts.

Conclusion

During the inception of ecological rewilding, Soulé and Noss (1998) foresaw that the 'greatest impediment to rewilding is the unwillingness to imagine it'. There is a need to address this impediment by ensuring that human rewilding and ecological rewilding can embody a reciprocal or cyclic relationship. Consequently, ecological rewilding can occasionally be criticised for treating only the symptom rather than the cause, for not paying enough attention to societal context, which fortifies the necessity for ecological rewilding in the first place. As such, people can only ever adopt roles as initiators, stakeholders, managers, or spectators. While ecological rewilding may inspire people to connect with nature, human rewilding has the potential to *involve* people on physical, emotional, and cognitive dimensions. It can be a form of deep restoration that forges bonds beyond a solely human sphere, thus perhaps helping humans to enter what Haraway (2016) calls the era of the Cthulucene.

Human rewilding resolves a human–nature dichotomy, combines the knowledge of the old with the innovation of the new and acknowledges that this is a point in an ongoing socio-ecological journey. The potential for ecological rewilding to take place in all manner of spaces (Ward, 2019) makes it one of the most optimistic environmental narratives (Jepson, 2019) that captivates people's imagination (Durant et al., 2019). In the same thread, human rewilding can be a motivating endeavour, as no matter where on the spectrum someone sits there is a chance to be 'more' wild. In what is often a bleak planetary perspective, human rewilding offers optimism and hope.

Note

1 This chapter is an adapted version of a Dutch text, Arts & Maffey (2022) 'Menselijke rewilding'. In: *Rewilding in Nederland. Essays over een offensieve natuurstrategie.* Arts, Bakker, Buijs (eds), Zeist, KNNV Uitgeverij, pp. 73–81.

References

Antonelli, M., Donelli, D., Carlone, L., Maggini, V., Firenzuoli, F., and Bedeschi, E., 2021. Effects of forest bathing (shinrin-yoku) on individual well-being: an umbrella review. *International Journal of Environmental Health Research*, 32(8): 1842–67.

Arts, K., Fischer, A., and Van der Wal, R., 2012. The promise of wilderness between paradise and hell: A cultural-historical exploration of a Dutch national park. *Landscape Research*, 37(3): 239–56.

Arts, K., Fischer, A., and Van der Wal, R., 2016. Boundaries of the wolf and the wild: A conceptual examination of the relationship between rewilding and animal reintroduction. *Restoration Ecology*, 24: 27–34.

Bekoff, M., 2014. *Rewilding Our Hearts*. New World Library.

Birch, T.H., 1990. The incarceration of wildness. *Environmental Ethics*, 12(1): 3–26,

Bratman, G.N., Anderson, C.B., Berman, M.G., Cochran, B., de Vries, S., Flanders, J., Folke, C., Frumkin, H., Gross, J.J., Hartig, T., Kahn, P.H., Kuo, M., Lawler, J.J., Levin, P.S., Lindahl, T., Meyer-Lindenberg, A., Mitchell, R., Ouyang, Z., Roe, J., Scarlett, L., Smith, J.R., Van den Bosch, M., Wheeler, B.W., White, M.P., Zheng, H., and Daily, G.C., 2019. Nature and mental health: An ecosystem service perspective. *Science Advances*, 5(7): eaax0903.

Buijs, A. and Jacobs, M., 2021. Avoiding negativity bias: Towards a positive psychology of human–wildlife relationships. *Ambio*, 50(2): 281–8.

Carver, S., Convery, I., Hawkins, S., Beyers, R., Eagle, A., Kun, Z., Van Maanen, E., Cao, Y., Fisher, M., Edwards, S.R., Nelson, C., Gann, G.D., Shurter, S., Aguilar, K., Andrade, A., Ripple, W.J., Davis, J., Sinclair, A., Bekoff, M., Noss, R., Foreman, D., Pettersson, H., Root-Bernstein, M., Svenning, J.C., Taylor, P., Wynne-Jones, S., Featherstone, A.W., Fløjgaard, C., Stanley-Price, M., Navarro, L.M., Aykroyd, T., Parfitt, A., and Soulé, M., 2021. Guiding principles for rewilding. *Conservation Biology*, 35(6): 1882–93.

Ceballos, G., Ehrlich, P.R., Barnosky, A.D., García, A., Pringle, R.M., and Palmer, T.M., 2015. Accelerated modern human-induced species losses: Entering the sixth mass extinction. *Science Advances*, 1(5): e1400253.

Clayton, S., 2019. The psychology of rewilding. In N. Pettorelli, S.M. Durant, and J.T. du Toit (eds), *Rewilding*. Cambridge University Press.

Cronon, W., 1995. The trouble with wilderness; or, getting back to the wrong nature. In *Uncommon Ground: Rethinking the Human Place in Nature*. W. W. Norton & Co, 69–90.

Drenthen, M., 2018. Rewilding in cultural layered landscapes. *Environmental Values*, 27(4): 325–30.

Durant, S.M., Pettorelli, N., and du Toit, J.T., 2019. The future of rewilding: fostering nature and people in a changing world. In: N. Pettorelli, S.M. Durant, and J.T. du Toit (eds), *Rewilding*. Cambridge University Press.

Ellis, E.C., Gauthier, N., Klein Goldewijk, K., Bliege Bird, R., Boivin, N., Díaz, S., Fuller, D.Q., Gill, J.L., Kaplan, J.O., Kingston, N., Locke, H., Mcmichael, C.N.H., Ranco, D., Rick, T.C., Shaw, M.R., Stephens, L., Svenning, J.-C., and Watson, J.E.M., 2021. People have shaped most of terrestrial nature for at least 12,000 years. *PNAS*, 118(17): e2023483118.

Fenton, L., Playdon, Z., and Prince, H.E., 2020. Bushcraft education as radical pedagogy. *Pedagogy, Culture & Society*, DOI: 10.1080/14681366.2020.1864659

Foreman, D., 2018. Dave Foreman on the history and definition of rewilding. *Rewilding Earth Podcast*.

Harari, Y.N., 2014. *Sapiens: A Brief History of Humankind*. Harper.

Haraway, D., 2016. *Staying with the Trouble*. Duke University Press.

Higgs, E., 2003. *Nature by Design*. MIT Press.

Jepson, P., 2019. Recoverable Earth: A twenty-first century environmental narrative. *Ambio*, 48(2): 123–30.

Jickling, B., Blenkinsop, S., Morse, M., and Jensen, A., 2018. Wild pedagogies: Six initial touchstones for early childhood environmental educators. *Australian Journal of Environmental Education*, 34(2): 159–71.

Jørgensen, D., 2015. Rethinking rewilding. *Geoforum*, 65: 482–8.

Kim, H., Richardson, C., Roberts, J., Gren, L., and Lyon, J.L., 1998. Cold hands, warm heart. *The Lancet*, 351(9114): 1492.

Kopnina, H., Kopnina, H.N., Leadbeater, S.R.B., and Cryer, P., 2019. Learning to rewild: Examining the failed case of the Dutch 'new wilderness' Oostvaardersplassen. *International Journal of Wilderness*, 25(3): 72–89.

Kuo, M., Barnes, M., and Jordan, C., 2019. Do experiences with nature promote learning? Converging evidence of a cause-and-effect relationship. *Frontiers in Psychology*, 10 (Feb.): article 305.

Light, A. and Higgs, E.S., 1996. The politics of ecological restoration, *Environmental Ethics*, 18(3): 227–47.

Lotz-Sisitka, H., Wals, A.E.J., Kronlid, D., and McGarry, D., 2015. Transformative, transgressive social learning: Rethinking higher education pedagogy in times of systemic global dysfunction. *Current Opinion in Environmental Sustainability*, 16: 73–80.

Louv, R., 2005. *Last Child in the Woods*. Algonquin Books.

Loynes, C., 2018. Leave more trace. *Journal of Outdoor Recreation, Education, and Leadership*, 10(3): 179–86.

Marris, E., 2009. Conservation biology: Reflecting the past. *Nature*, 462(7269): 30–2.

Martin, A., Fischer, A., McMorran, R., and Smith, M., 2021. Taming rewilding—from the ecological to the social: How rewilding discourse in Scotland has come to include people. *Land Use Policy*, 111: 105677.

McKibben, B., 2003. *The End of Nature*. Rev. Ed. Bloomsbury.

McMorran, R., Price, M., and Warren, C.R., 2008. The call of different wilds: The importance of definition and perception in protecting and managing Scottish wild landscapes. *Journal of Environmental Planning and Management*, 51(2): 177–99.

Monbiot, G., 2013. *Feral: Searching for Enchantment on the Frontiers of Rewilding*. Penguin Books.

Nash, R.F., 2001. *Wilderness and the American Mind*. Yale University Press.

Owens, M. and Wolch, J., 2019. Rewilding cities. In N. Pettorelli, S.M. Durant, and J.T. du Toit (eds), *Rewilding*. Cambridge University Press.

le Page, M., 2017. Where there's smoke. *New Scientist*, 233(3111): 22–3.

Pausas, J.G., and Keeley, J.E., 2009. A burning story: the role of fire in the history of life. *BioScience*, 59(7): 593–601.

Roser, M., 2020. The short history of global living conditions and why it matters that we know it [online]. Our World in Data. Available from: https://ourworldindata.org/a-history-of-global-living-conditions-in-5-charts [accessed 31 October 2021].

Schulte to Bühne, H., Pettorelli, N., and Hoffmann, M., 2021. The policy consequences of defining rewilding. *Ambio*, 51: 93–102.

Scott, A.C., Bowman, D.M.J.S., Bond, W.J., Pyne, S.J., and Alexander, M.E., 2014. *Fire on Earth: An Introduction*. Wiley Blackwell.

Selwyn, N., 2009. The digital native—myth and reality. *Aslib Proceedings*, 61(4): 364–79.

Skogvang, B.O., 2021. Development of Cultural and environmental awareness through Sámi outdoor life at Sámi/Indigenous festivals. *Frontiers in Sports and Active Living*, 3. https://doi.org/10.3389/fspor.2021.662929.

Soulé, M. and Noss, R., 1998. Rewilding and biodiversity: Complementary goals for nature conservation. *Wild Earth*, 8: 19–28.

Steffen, W., Richardson, K., Rockström, J., Cornell, S.E., Fetzer, I., Bennett, E.M., Biggs, R., Carpenter, S.R., de Vries, W., de Wit, C.A., Folke, C., Gerten, D., Heinke, J., Mace, G.M., Persson, L.M., Ramanathan, V., Reyers, B., and Sörlin, S., 2015. Planetary boundaries: Guiding human development on a changing planet. *Science*, 347(6223). https://doi.org/10.1126/science.1259855.

Torres, A., Fernández, N., Ermgassen, S.Z., Helmer, W., Revilla, E., Saavedra, D., Perino, A., Mimet, A., Rey-Benayas, J.M., Selva, N., Schepers, F., Svenning, J.C., and Pereira, H.M., 2018. Measuring rewilding progress. *Philosophical Transactions of the Royal Society B: Biological Sciences*, 373(1761): 20170433.

Triguero-Mas, M., Donaire-Gonzalez, D., Seto, E., Valentín, A., Martínez, D., Smith, G., Hurst, G., Carrasco-Turigas, G., Masterson, D., van den Berg, M., Ambròs, A., Martínez-Íñiguez, T., Dedele, A., Ellis, N., Grazulevicius, T., Voorsmit, M., Cirach, M., Cirac-Claveras, J., Swart, W., Clasquin, E., Ruijsbroek, A., Maas, J., Jerret, M., Gražulevičienė, R., Kruize, H., Gidlow, C.J., and Nieuwenhuijsen, M.J., 2017. Natural outdoor environments and mental health: Stress as a possible mechanism. *Environmental Research*, 159: 629–38.

Van den Berg, A.E. and Ter Heijne, M., 2005. Fear versus fascination: An exploration of emotional responses to natural threats. *Journal of Environmental Psychology*, 25(3): 261–72.

Ward, K., 2019. For wilderness or wildness? Decolonising rewilding. In N. Pettorelli, S.M. Durant, and J.T. du Toit (eds), *Rewilding*. Cambridge University Press, 34–54.

INDEX

Note: Page numbers in **bold** refer to tables and *italics* refer to figures